# SYSTEMS THINKING IN EUROPE

# SYSTEMS THINKING IN EUROPE

**Edited by**

**M.C. Jackson**
*University of Hull*
*Hull, United Kingdom*

**G.J. Mansell**
*Polytechnic of Huddersfield*
*Huddersfield, United Kingdom*

**R.L. Flood**
*University of Hull*
*Hull, United Kingdom*

**R.B. Blackham**
*Polytechnic of Huddersfield*
*Huddersfield, United Kingdom*

**and**

**S.V.E. Probert**
*Polytechnic of Huddersfield*
*Huddersfield, United Kingdom*

SPRINGER SCIENCE+BUSINESS MEDIA, LLC

Library of Congress Cataloging-in-Publication Data

United Kingdom Systems Society Conference on Systems Thinking in
  Europe (1991 : Huddersfield, England)
    Systems thinking in Europe / edited by M.C. Jackson ... [et al.]
      p.   cm.
    "Proceedings of the United Kingdom Systems Society Conference on
  Systems Thinking in Europe, held September 10-13, 1991, in
  Huddersfield, United Kingdom"--Copyright p.
    Includes bibliographical references and index.
    ISBN 978-0-306-44013-7    ISBN 978-1-4615-3748-9 (eBook)
    DOI 10.1007/978-1-4615-3748-9
    1. Management information systems--Congresses.  2. System
  analysis--Congresses.  3. Management--Europe--Congresses.
  I. Jackson, Michael C., 1951-   .  II. Title.
  T58.6.U546  1991
  658.4'032'094--dc20                                    91-20952
                                                             CIP

ISBN 978-0-306-44013-7

Proceedings of the United Kingdom Systems Society Conference on
Systems Thinking in Europe, held September 10-13, 1991,
in Huddersfield, United Kingdom

© 1991 Springer Science+Business Media New York
Originally published by Plenum Press in 1991

# PREFACE

The theme of the conference at which the papers in this book were presented was 'Systems Thinking in Europe'. Members of the United Kingdom Systems Society (UKSS) were conscious that the systems movement flourishes not only in the UK, America and the Antipodes, but also in continental Europe, both East and West, and in the USSR, a nation increasingly being welcomed by the European comity. Membership of the UKSS had not perhaps had the opportunity, however, of hearing important new ideas from continental Europe, and this conference provided an opportunity to do so. Some interesting papers are to be found here from both the West and the East, if the editors may be forgiven for perpetuating what may be an increasingly irrelevant dichotomy. One lesson to be learned from this conference, though, is that systems thinking is truly international.

This is not to say that there is one systems paradigm uniformly applied, however. Perhaps the core of systems thinking is that one is interested in complex 'wholes' with emergent properties, to which cybernetic ideas can be applied. Examples of such systems thinking can be found in these proceedings, for example in the section entitled "Applications of Systems Thinking". Attempts to bring about change with these ideas, however, have given rise to a diversity of approaches, as is evidenced by the papers dealing with the application of methodologies in the 'hard' and 'soft' systems traditions. The essence of the dichotomy between hard and soft approaches lies in the degree of uncertainty or disagreement about objectives. Hard systems thinking assumes unitary objectives, whereas soft systems thinking allows for pluralism and the reaching of accommodations. Given that the systems practitioner has apparent choice of approach depending on circumstances (and the hard/soft dichotomy may be unduly simplistic), there is a need for theory that guides the choice. Papers in the section entitled "Problem Structuring and Critical Systems Thinking" explore this area. Critical Systems Thinking has matured in the years that have passed since the first international conference of the UKSS, and the reflective systems practitioner will welcome articulation of this relatively new area in this book.

One important 'market' for systems ideas is the subject area of Information Systems. Advanced nations have embarked on a project to apply information technology to increase organisational efficiency and productivity, and enhance organisational and individual performance. Theory to assist in this project has been developed by academics and practitioners, and much of it can be seen to be a variation on hard systems thinking. Information Systems Engineering is merely a specialisation of Systems Engineering. As in the wider systems sphere, soft systems ideas have infiltrated information systems thinking, and several papers represent the struggle to bring about a synthesis of hard and soft approaches. The papers on information systems have been divided into those mainly concerned with systems development issues, and those mainly concerned with the nature and purpose of information systems, and the elicitation and structuring of knowledge.

Finally, the distinguished plenary speakers at this conference can be seen to be addressing the themes already delineated. Some of the speakers addressed issues fundamental to systems theory; an application of systems ideas in bringing about change was described; the fundamentals of critical systems thinking were defined; and finally important aspects of information systems were examined.

The 'Systems Thinking in Europe' conference is the second international conference of the United Kingdom Systems Society. Proceedings of the first international conference are available, entitled "Systems Prospects-The Next Ten Years of Systems Research" edited by R L Flood, M C Jackson and P Keys, also from Plenum.

M.C.Jackson

CONTENTS

## APPLICATIONS OF METHODOLOGY (BOTH HARD AND SOFT)

## PROBLEM STRUCTURING AND CRITICAL SYSTEMS THINKING

## INFORMATION SYSTEMS

## INFORMATION SYSTEMS DEVELOPMENT

PLENARY PAPERS

# TRANSFORMATION THEORY

Rafael Rodríguez-Delgado

National Institute of Public Administration (INAP)
Spain; President of the Spanish Society of General
Systems

## INTRODUCTION

The scientific and technological revolutions that
characterize our times have not reached our collective ways of
thinking. Our educational systems have not received yet the
full impact of modern physics, biology or sociology. They
belong, with a few exceptions, to the pre-atomic age.

By manipulating the atomic nucleus, modern Physics has
obtained the cosmic energy of the stars. By transforming the
electrons we have invented a new technological world (radio,
television, informatics, robotics). But we live, at the same
time, submerged in unchanging cultural and conceptual worlds
that have their roots in the far away past. We live,
therefore, in a dichotomic world: science and technology point
out to the future; ideologies, values and cultures point back
to an obscure past. A dangerous schizophrenia is the result.

The new powers acquired by mankind have changed everything
except mankind. The most urgent task for the final period of
the XX Century is the change of the human mind  transforming
opposing beliefs into a planetary conception of
interdependence, solidarity and intellectual synthesis,
compatible with individual freedom and cultural variety.

## UNIVERSAL TRANSFORMATION

Transformation appears as an universal phenomenon that
develops from the physical to the ideological levels.

In the Physicochemical level atoms are transformed by
processes of fission and fusion. Transmutations of chemical
elements can be accomplished by nature or by technology. The
same molecular structure may be transformed into different
states: solid, liquid, gas or plasma, depending on relations
among environmental changes and internal structures.

*Systems Thinking in Europe*, Edited by M.C. Jackson *et al.*
Plenum Press, New York, 1991

Qualitative transformation of systems can appear, as Prigogine (1980) states, in conditions far away from equilibrium.

Each level in the simplified AMPO scale: Atom - Molecule - Protist - Organism, can be conceived as a more complex form of organization in which emergent qualities appear. The same physico-chemical substance, through the fuzzy limit of the organic molecules, constitute the basis of all living systems, including human beings.

BIOLOGICAL LEVEL

At the Biological level transformations of the basic universal entity display the astounding variety of myriads of living forms appearing and disappearing in our planet during millions of years.

Biochemical transformations of incredible complexity take place in all internal components of the living systems. Biolectric currents in the nervous flow are transformed into biochemical flows at the synaptic level, to be transformed again in cyclic transduced currents. An electrons impact on the retina is changed into a series of bio-electro-chemical flows that arrive to certain brain locations where they are transformed into psychological perceptions and emotional responses. These inputs go to memories, from internal ways, and to muscular effectors to the external environment as sounds, gestures or concrete actions. Feed-back links close complex circuits in the internal and external worlds.

At the bio-ontogenetic level, each complex individual goes through a series of radical transformations. From two genetic cells, in the adequate environment, frenetic multiplication, differentiation and structural change produce a living being. The transformation of ova-spermatozoid conjunctions into living beings are cyclic reproductive processes (Foetus-Infant-Young-Adult(genetic cells)-Old).

However, in certain instance, philogenetic transformations occur. The chain of similar living beings is broken by the sudden apparition of living systems with new emergent properties. Philogenetic transformations are represented by the emergence of new species produced by natural mutation, or -at present- by means of genetic engineering.

Natural mutations -more complex processes than the natural selection mechanism postulated by Darwin- have produced the scale of living systems. Protists -unicellular beings-, branched out into complex organisms, vegetal and animal, conquerors of water, earth and air.

The same system is capable of internal transformations by using alternative possibilities of response -multistability of Ashby (1954)- in face of changing environments. Living systems may also develop original responses. They are creative, autopoietic in the terminology of Maturana and Varela (1980).

Philogeny -transformation of species- and ontogeny -genetic individual cycles- appear as complementary aspects of life transformation processes.

4

## CONCEPTUAL TRANSFORMATIONS

Dichotomies characterize the analytical thinking of the Western industrial civilizations, in contraposition with the global thinking developed by many oriental cultures.

Dichotomic thinking has produced violent conceptual oppositions in science and philosophy where fierce intellectual battles have been fought between materialists - corpuscular mechanics, determinism, dialectical materialism, positivism- and idealists -wave mechanics, indeterminism; dialectical idealism, spiritualism-.

Now, from the generalization of the basic principles of science emerge an integrated concept of Nature and Man in their different levels, scales and forms.

Einstein (1938, 1948) has transformed the opposition between the concepts of "matter" and "energy", "space" and "time" into unitary wholes: matter-energy, space-time. Bohr's complementarity principle (1957) has dissolved the opposition between "wave" and "corpuscle", that appeared as irreducible and paradoxical, because both concepts were based in well controlled experiments. At present, instead of opposite concepts, wave and corpuscle are conceived as "aspects" of the same basic element of the Universe, an element that does not conform with our classical concepts of "matter" and "energy".

The complementarity principle, born in Physics, has been extended, for instance, to neurophysicology and psychology -brain/mind-; to sociology -individual/society-; to methodology -analysis/synthesis- and, in general, to all aspects of nature and human thinking.

## MATERGON, A CONCEPTUAL SYNTHESIS

From a global viewpoint, a new name should be given to the cosmic entity that develops from the scale of physicochemical elements to the living scale that culminates in human beings.

I have suggested some time ago (Rodríguez Delgado, 1954, p. 43) "matergon" as a double hybrid term composed of: mat-ergon (matter-energy) or matter-gono (maternal angle, creative perspective from which to conceive Universe and Mankind as an unitary entity). This etymological interplay conveys a global perception of "reality" that supersedes the opposite conceptions, without eliminating them. The dichotomic perceptions appear as complementary aspects of an universal entity.

The general basic formula of matergon is the well known Einstein's equation: $E = mc^2$, that means, Energy (E) is equal to mass (m) multiplied by a constant (c) or velocity of light square.

If we integrate this basic formula with the concepts of

space-time and wave-corpuscle -born also in modern Physics- we would have a more complete definition of "matergon" that would take this form:

Matergon is: mass-energy, of wave-corpuscular aspects, developing in space-time as differentiated systems.

Matergon appears, therefore, as an intuitive concept represented by and integrated in a series of analytical concepts: mass-energy, space-time, wave-corpuscle.

This definition of matergon constitutes a first attempt to develop a global system theory of the Universe, based on a generalized entity divided into several levels of growing complexity that should correspond, more than the previous dichotomic concepts, to the kind of "reality" discovered by our sciences and technologies.

The matergon concept offers a way to reach an integrated idea of natural, conceptual and artificial systems, connecting the philosophical abstractions with scientific and humanistic models.

The continually changing frontiers or interfaces among systems, or among systems and environments, appear relativized and show different grades of fuzziness, depending on the cultural, scientific or philosophical viewpoints chosen by the observer, the researcher or the actor.


IDEOLOGICAL LEVEL

The ideological level (systems of ideas) is built upon the conceptual one (product of the human brain-mind), that in turn is built upon the psychological level (mind aspect of living beings with a brain). All depend on physiological structures (nervous, sensory, hormonal, blood, sex, liver, lungs, bones, etc.), that are based on the physicochemical level (organic and non organic molecules, atoms).

The ideological structures are innumerable. However, for the sake of simplicity, we reduce them to four fundamental models: materialism, idealism, dualism· and systemic synthesis.

Now we know that not only dichotomic thinking, but even the scale of observation, change our perceptions. Einstein (1938) perceived the Universe as a "continuum" of four dimensions because he was seeing the whole at a global scale in which the gravitatory fields of the celestial bodies had no clear limits. Following the quantic theory, he was also seeing at the subatomic scale a discontinuous Universe.

"Reality", perceived by different observers or by the same observer in different space-time scales, offers different, and even opposite, perspectives.

Integration of these perspectives is difficult. However an effort should be made to obtain a glimpse of the very complex whole of an Universe in perpetual change and transformation, whose unity and diversity appear before the human mind as multiple complementary aspects of a global reality.

IDEOLOGICAL MODELS

Idealism and Materialism are monist perspectives that aspire to intellectual hegemony, minimizing or excluding the other competitive conception. Dualism attempts to preserve both conceptions as totally different domains: the material and the spiritual, body and mind.

The systemic synthesis transforms these three ideological perspectives into complementary perceptions from a different epistemological position.

A loose correlation among the ideological, the conceptual and the physiological levels can be established through the following hypothesis:

(a) materialist monism is based in the concrete direct perception of sensory stimuli. As a mirror it reflects the external world, as Lenin (1948) repeatedly asserted.

(b) idealist monism would correspond to an intuitive global elaboration of the sensory data looking for a deeper vision of nature and man. Matter appears as "maya", delusion, to the oriental vision; or as "eidos", form, to the Platonic approach; or as "idea" to the Hegelian logic.

(c) dualism would emerge from an analytical dissociation of perception and intuition, trying to find a common point of linkage between both, or reflecting the moral dichotomy between good and evil. It should correspond to reason.

(d) systemic synthesis, more than an ideology, is a method to understand the three main ideologies mentioned above. It would correspond to intelligence, as a global unifying function.

An intuitive image of the correspondences among the four ideologies -equivalent to several explanatory pages- is offered in Figure 1.

The symbols chosen for representing each ideology have been, respectively, a full circle, a circumference, the circle and the circumference, and an abstraction of the yin-yan taoist representation of the unity of the contraries. The systemic symbol represents here a perspectivist method. An observer placed in the left side of the graph would see only a black hemisphere - materialism- and would extrapolate his/her perception to the other hemisphere, that he/she does not see. From the right side, on the contrary, only the white hemisphere -idealism- is perceived. A cautious observer that would rotate 180 degrees his/her points of vision, would see a white and a black sphere. Only a more careful observer who would rotate slowly the whole sphere would perceive a more complete reality -systemic perspectivism-. In this way the three classical perspectives are not denied, but considered as complementary.

As we have seen, in human system a series of transformations convert mass-energy inputs into complex outputs of differentiated matergon by means of physicochemical-biochemical-psychological-ideological almost unknown processes.

7

| CONCEPTUAL SYSTEMS | 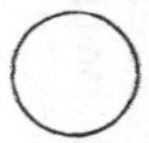 ○ | ● | 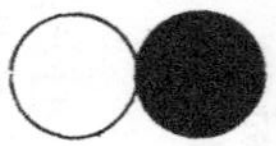 ◑ |  ☯ | REAL SYSTEMS |
|---|---|---|---|---|---|
| Philosophy | Idealism | Materialism | Dualism | Integration | ------ |
| Cosmology | Energy | Matter | Energy and Matter | Matergon | Universe |
| Physics | Wave (Physical energy) | Corpuscle (Physical matter) | Wave and Corpuscle | Physical matergon (Wave-Corpuscle) | Atom |
| Chemistry | Chemical energy | Chemical body | Chemical energy and Chemical body | Chemical matergon | Molecule |
| Biology | Vital energy | Living body | Vital energy and Living body | Biosoma. Living matergon | Protist |
| Psychology | Spiritual energy | Living brain | Psyche and Soma | Psychosoma. Orga-nismic matergon | Organism |
| HumanSociology | Idea Social energy | Group Social body | Ideas and Groups | Sociosoma Social matergon | Society |

Figure 1

From a global perspective, all these transformations appear as the result of the activity of our brain-minds, where ideological systems and cultural changes begin.


DIALECTIC MODELS

Dialectic models are, in fact, ideological models centered in the attempt to explain movement and change. Dialectics was born in the minds of classical Greek philosophers. Since the beginning it was object of violent rejection or enthusiastic acceptance.

In its modern version, Hegel (1816) explained everything by a logic triad: Thesis - Antithesis - Synthesis. The interplay of oppositions -Thesis and Anthitesis, Affirmation and Negation- was transformed into Synthesis that produced movement -Becoming-. Hegelian dialectics explained Becoming as the movement of Ideas.

Materialist dialectics, on the contrary, explained change by oppositions of material factors in Nature and Society. In human societies the main opposing social forces (proletariat and capitalism), would produce the synthesis of a classless society. Dialectical materialism was born with Marx (1892, 1947) and Engels (1946, 1950). For Lenin (1948) it had an absolute value, it was historical truth. Both dialectic ontologies have been submitted to sharp criticism -not always objective- especially by Bunge (1981, Chapter 4).

- Scientific dialectics

Bachelard (1938, 1940) and Gonseth (1944), editor since 1947 of the Journal "Dialectica", are the authors of the Scientific Dialectics school that tried to resolve the main opossitions by means of "dialogues" between philosophical and scientific dichotomies, as "object and subject", through the application of the extended principle of complementarity. A concrete objective was also to reduce the opposition between dialect idealism and materialist dialectics.

- Systemic dialectics and multilectics

Rodríguez Delgado (1986, 1987, 1988, 1989) is developing System Dialectics by means of the integration of the dialectical concept into a wider framework of methodologies to describe systems.

Each system is considered from different space-time perspectives: static (time = 0), dynamic (homeostasis and cyclic changes) and dialectic (qualitative transformations, including emergence and disapparition of the individual or collective system).

A dialectic methodology presupposes the interaction of two systems, an useful simplification when the aim is to reduce oppositions between two conflicting systems.

However, in "real" situations, when we study qualitative transformation processes in which there are three or more systems interacting synchronically, a new concept, that can be expressed by the term "<u>multilectics</u>", is necessary.

Processes of change and qualitative transformation are not linear. In complex systems -individual and social-, many inputs enter by different ways and are elaborated in many forms. Some inputs are preserved in memories. Multiple outputs impact on different environments. Therefore, the triadic logical structure of dialectics appears as a methodology that can be useful only in some extreme cases.

## TRANSFORMATIONS IN THE HUMAN WORLD

Through communication, ideas are diffused at the social level where they are accepted or rejected. If the impact is almost imperceptible, the new idea is quickly forgotten, or "rediscovered" in more favourable circumstances. When the idea -or the paradigm- finds a receptive atmosphere, reaching a "critical mass", evolutive social changes or sudden violent revolutions can be produced.

Transformation occurs at the individual and collective levels, as part of a process. Invention of new institutions or socio-political systems begin with an idea that is accepted by a group. Afterwards it is expressed in by-laws and regulations. Finally it reaches the social world. The emergence of Ministries of Defense, instead of the ancient Ministries of the Army, the Navy and the Air, represent, first, an ideological change, followed by a structural transformation. But this process is submitted to social restrictions and stimuli. Technologies, artifacts, new products, all kind of inventions, are, initially, individual ideas induced frequently by collective situations.

Technological changes produce successive "generations" of artifacts, as cars or computers, that are qualitatively different of the previous models.

The transformation of small groups of gatherers and hunters into shepherds and agricultural settlers initiated a series of qualitative social changes that, in about 10.000 years, culminated in the modern industrial revolutions.

In the human world, conceptual innovations bring about new cultures and civilizations. Sociotechnical systems, for instance, are new creations that emerged in the XXth. Century, linked to communications and information technologies. (See, Toffler, 1980, 1990; Friedrich and Schaff, 1982; Naisbitt, 1984, etc.)

Each historical period is a product of convergent trans-formations into ideas that produce qualitative changes in the structures, functions, relations and technologies of human action systems.

The most frequent explanations about social transformation processes point to one determinant factor: political want of power that originate international wars; economic opposition among social classes that unleash civil wars; attempt of one race to conquer hegemony over other races; ideological battles among religions; cultural clashes.

A systemic explanation should take into account the interaction of may factors: conflicts among interests and ideas in the social, economic, political, educational, ecological, strategic, aesthetic, moral, ludic, ideological, or other domains. The weight of each factor in each circumstance is difficult to evaluate. The components may change. Alliances can be suddenly reversed.

In some cases, an individual -a fierce warrior, a wise politician, a philosopher, a religious leader, a despotic autocrat- has been the initial instrument of transformation, in a positive or a negative way. In other cases, more or less coherent masses, activated by small groups of thinkers or agitators, have produced deep social changes.

Present changes have been increasingly global, planetary and accelerated, as if space-time had contracted. Future transformations would need more human variety than the previous ones: thinkers, educators, sociologists, historians, natural scientists, artists, should intervene more actively in the process of change. The growing gap between underdeveloped and industrialized countries can only be filled by means of an enormous and coherent effort by both groups.

The multiplication of problems that characterize the transformation of industrial societies into societies of knowledge needs "new ways of thinking", and new methodologies.

SYSTEMIC METHODOLOGIES FOR TRANSFORMATION

There is a panoply of systemic methodologies-Systems Analysis and Synthesis, Operational Research, System Engineering, System Dynamics, System Dialectics, System Educational Design, etc.- that could be applied to research and action for intelligent societal transformation. Among them the application of the generalized principle of complementarity is frequently ignored. (See Rodríguez-Delgado, 1950, 1957 a) and b). This method opens the way to understand and eliminate many apparent conceptual oppositions, to take adequate decisions, and to solve complex problems. When the opposite points of view are transformed into complementary concepts it is possible to find win-win solutions to conflictual situations.

More serious problems arise when the conceptual oppositions cover up economic or political interests. The conquest of land, gold, oil, strategic positions or any other valuable asset has been frequently disguised as ideology; as bringing civilization to savages; or as offering the true religion to unbelievers

Attempts to conquer hegemony by a faith, a race, a nation or an ideology have many irrational components that are considered more important than the possession of goods. Human life is voluntarily sacrified to non economic values. Lost prestige induced the samurais to kill themselves. Defy of the Roman emperors cult ended in voluntarily accepted martyrdom. Our societies are divided by social inequalities, lack of freedom, mutual ignorance and other ills that constitute a complex of factors in dangerous synergy.

SYSTEMS RESEARCH

Research in complementarity and conflict in each concrete situation is urgent in order to find ways and means capable of transforming positively our present societies. For instance, Europe was divided during centuries by may kinds of economic and ideological oppositions. Recent History has demonstrated that the factors of dissension can be progressively dissolved and that complementarity and collaboration have resulted more efficient than fighting. Research should aim to know how complex conflictive situations change in space-time and how this change can be oriented towards convergence or collaboration.

Research in transformation processes is so important as research in homeostasis, although so far it has not arised equal attention. A Theory of Transformation should be developed in order to understand qualitative changes in Nature and Human Societies, as a basis for intelligent action.

CONCLUSIONS

It would be very useful the creation of new institutions in which specialists and generalists could met to do research in fields such as: biological, psychological and conceptual basis of opposing ideologies and interests; unified theory and methodologies for transformation; applications of the generalized principle of complementarity to conflicts resolution; possibilities of implementing principles and values acceptable by the majority of our world -like the Universal Declaration of Human Rights-; design of new models of organization, capable of reducing the socio-economic and educational gap that exist between the affluent and the destitute social groups and nations of our world.

We are faced with enormous common tasks: reconstruction of our damaged environment; fight against hunger and disease; wide diffusion of useful knowledge; drastic reduction of all kind of weapons; planetary peace; development of human solidarity; transformation of our conflictive ways of thinking. If we apply intelligently our sciences and technologies it would be possible the creation of a New World for the XXI Century.

REFERENCES

Bachelard, G., 1938, "La Formation de l'Esprit Scientifique", Vrin, París.
Bachelard, G., 1940, "La Philosophie du non", P.U.F., París.
Bohr, N., 1957, "Atomic Physics and Human Knowledge", J. Wiley, New York.
Einstein, A., 1938, "The Evolution of Physics", Simon and Schuster, New York.
Einstein, A., 1950, 3rd. ed. "The Meaning of Relativity", Princeton University Press.
Engels, F, 1946, "Ludwig Feuerbach", Ed. Sociales, París.
Engels, F., 1950, "Anti-Düring", Ed. Sociales, París.
Friedrichs, G., and Schaff, A., (eds.) 1982, "Microelectronics and Society for Better or for Worse", The Club of Rome.

Gonseth, F., 1944, "Déterminisme et libre Arbitre", Entretien
    presidé para Gonseth, Editions du Griffon.
Hegel, The Philosophy of History, The Science of Logic, etc., in
    Friedrich, C. (ed.), "The Philosophy of Hegel", Random
    House, 1953, 1954, New York.
Lenin, V. I., 1948, "Matérialisme et Empiriocriticism", Editions
    Sociales, París.
Marx, K., 1892, "The Capital", Foreword to the 2nd. edition.
    Many editions and translations.
Marx, K., and Engels, F., 1947, "Etudes philosophiques",
    Editions Sociales, París.
Maturana H.R. and Varela, F.J., 1980, "Autopoiesis and
    Cognition", Reidel Publishing Co.
Naisbitt, J., 1984, "Megatrends", 2nd. edition. Warner Books.
Prigogine, I., and Stengers. I, 1979, "La Nouvelle
    Alliance,Metamorphose de la Science", Editions Gallimard,
    París.
Prigogine, I., 1980, "From Being to Becoming", W. H. Freeman,
    San Francisco.
Rodríguez-Delgado, R., 1950, "Introducción a una Filosofía de la
    Era Atómica", Editorial Lex, Havana.
Rodríguez-Delgado, R., 1954, Esquema del Nuevo Pensamiento, in
    Revista Venezolana de Síntesis, 1:25.
Rodríguez-Delgado, R., 1957, "Synthesis in a divided world", in
    Main Currents in Modern Thought, 13:25, New York.
Rodríguez-Delgado, R., 1957, A possible model for ideas, in
    Phil, of Sci., 24,3:253.
Rodríguez-Delgado, R., 1986, Systems Dynamics in a GST
    Framework, in The 1986 Internl. Conference of the Systems
    Dynamics Society, I:511, Sevilla, Spain.
Rodríguez-Delgado, R., 1987, System Dialectics: Applications to
    Integrated Development, in Intl. Conf. on Problems of
    Constancy and Change, 31st. Ann. Meet. of the ISGSR,
    II:966, Budapest.
Rodríguez-Delgado, R., 1988, Applications of Systems Dialectics
    to Integrated Development, in Systems Practice, I-3:259
Rodríguez-Delgado, R., 1989, System Dialectics: A new approach,
    in Congrés Européen de Systémique, Lausanne.
Toffler, A., 1980, "The Third Wave", William Morrow, New York.
Toffler, A., 1990, "Power Shift", Bantam Books, New York.

SYSTEMS APPROACH AS A STYLE:

A HERMENEUTICS OF SYSTEMS

Wojciech W Gasparski

Polish Academy of Sciences
Warsaw
Poland

"When we traded the results of our fantasies, it seemed to us - and
rightly - that we had proceeded by unwarranted associations, by
shortcuts so extraordinary that, if anyone had accused us of really
believing them, we would have been ashamed.  We consoled ourselves
with the realization - unspoken,.now, respecting the etiquette of
irony - that we were parodying the logic of our Diabolicals.  But
during the long intervals in which each of us collected evidence
to produce at the plenary meetings, and with the clear conscience
of those who accumulate material for a medley of burlesques, our
brains grew accustomed to connecting, connecting, connecting
everything with everything else, until we did it automatically,
out of habit.  I believe that you can reach the point where there
is no longer any difference between developing the habit of
pretending to believe and developing the habit of believing."

Umberto Eco : Foucault's Pendulum

ORIGIN OF THE PROBLEMS

This paper develops the subject-matter I initiated with an
article presented in Washington in 1982 at the Annual Conference on
Systems Methodology organized by the Society for General Systems
Research (now International System Science Society).

The paper, nota bene, with the same title as this work
(Gasparski 1982) stressed that even though the so-called systems
approach has unquestionably existed for more than three decades, its
methodological status is not yet clear.  It is still disputed by
the epistemologists.  Some methodologists tend to grant the systems
approach the status of a method, in the understanding in which the
word "method" can be found in the title of Rene Descartes' Discours.
It would be, according to them, a method completely different from
the one advocated by Descartes, a method not of segmenting the
studied subject but of deriving the properties of a part from the
features ascribed to the object as a whole, called a system.

Other philosophers of science treat the systems approach as a
heuristic or a metaphor, fostering scientific, technological,
organizational, etc. output.  Still others treat the systems approach
as one of particular methods, ie techniques which are specific to

Systems Thinking in Europe, Edited by M.C. Jackson et al.
Plenum Press, New York, 1991

15

cybernetics.    They identify the systems approach with treating the
discussed subjects as the so-called "black box", ie without dealing
with the internal mechanism transforming inputs into outputs of
these objects.    Among students of the systems approach are also
those who treat it as a set of convictions, beliefs, or even an
ideology (see Lilienfeld, 1978).

There have been many objections to the systems approach.    It is
charged, for what many others treat as its advantage, namely that it
is not a scientific method in the Cartesian sense (Monod 1974).    It
is also said that the systems approach unfairly competes with
cybernetics (Mazur 1976).

Each of the praises of the systems approach and all of the
objections are to some extent justified.    For the systems approach
is not one of those orders which are independent from the authors to
whose output one can ascribe this approach or who declare having
adopted this approach.    This statement makes us take up a different
attitude to systems approach, ie to consider it as a style.

THE NOTION OF STYLE

Style is generally associated with art.    It is thus said,
eg that impressionism is a style specific for painting, especially
French painting of the 1870s and later years of the past century.
It is said that the Gothic is a style typical for the architecture
of the middle and late Mediaeval period.    In literature one speaks
about styles of different writers.    The same is the case in music;
style is connected with a composer, eg the style of Chopin or
Schuman.

Associating the notion of style with painting, sculpture,
architecture, literature, etc. one often forgets that before the
fine arts were distinguished all ability to produce things was
considered an art.    This ability was identified with the knowledge
of principles of making things;   its notion implied that it uses
methods, rules and prescriptions;   art was understood as ability to
produce according to principles and rules.    This understanding was
presented by the classics:  Plato and Aristotle, who defined art as
'permanent disposition to produce according to accurate reasoning'
(Ethica Nicomach, 1140a 9).    Similarly, the Stoic, Cleanthes
(Quintilian, Inst. or, II. 17.41), defined art as an 'ability
delineating the path', ie one that allows for methodical production
of things.    Among the later ancient authors, rhetor Quintilian
defined art as producing things via et ordine, ie according to the
rules.

The Stoics stressed particularly the fact that art is underlied
by a whole system of rules and that is why they called it concisely a
'system'.    It was also defined as a system by Chrysippus (according
to Sextus Empiricus) as well as by Zeno of Citium (according to
Olimpiodorus), ie by the two initiators of Stoicism.

Thus, what the ancients called 'art' referred not to what we
mean by the use of this term but to what we call ability, skill,
technique.    The scope of the notion of art was thus broader, for it
embraced not only the skill of a painter or a sculptor, but also of
a carpenter or a weaver:  it concerned also these skills, included
the crafts.    Moreover, it embraced sciences, geometry or grammar,
in which also systems of accurate methods, rules of actions were
perceived and they were, together with sculpture or weaving,

included in arts.   Only later Cicero distinguished among arts the
ones which do not produce things but embrace with thought, animo
cernunt, ie the ones which we call not arts but sciences
(Tatarkiewicz 1988, pp. 62-63).

As regards style, for the ancients stilus was simply a stylus to
write, a handle (the traces of which we find in Polish stylisko - a
handle of a hammer, pick or spade).   Later this name was transferred
to the results of drawing this stylus on an appropriate slate,
ie practising writing, and later to writing itself.

Owing to Giovanni Paolo Lomazzo (1538-1600), who after the loss of
sight was restricted to study the theory of sculpture, architecture
and painting instead of performing the last mentioned one, the notion
of style gained a different meaning.   For - as Lomazzo remarked -
together with lasting orders in art (eg the Doric, Ionian, and
Corinthian) there exists a multitude of forms, which are not arbitrary,
depending on the choices made by the man dealing with art, but ones
depending on the epoch, milieu, and, finally, the man himself as such.
These forms are, according to Lomazzo, styles.

Styles are not forms handed down by the earlier generations to
the later ones, they are not like orders - unchanging - but change
together with life and culture under the influence of social, economic,
and psychological factors;  they reflect the times.   Often they change
radically, switching from one extreme to the other.
(Tatarkiewicz 1988, p. 206).

The notion of style is a twofold one, continues Tatarkiewicz,
quoting Julius von Schlosser (1866-1938) - on the one hand the style
of one artist, always to some extent own, individual, original, and on
the other hand the style of the epoch - the commonly used in it
artistic language (Tatarkiewicz 1988, P. 206).   Let us add to it the
sphere of creation to which the style is made relative.   Thus style
is the totality of features characteristic to art, epoch, the sphere
of creation, and the artist.

Style plays with respect to art a role similar to that played by
method in science or technology.   The difference between them lies
in the degree of freedom which the artist has on the one hand and the
scientist on the other one.   The former has a considerable freedom
in choosing the ways of performing his art;  in many cases he is the
author of his style - manner, as it was earlier said.   The scientist,
however, has to create within the framework of the standing
methodological paradigm or else the results of his studies would be
rejected by the normal science as non-scientific ones or he would be
made to wait in the anterooms of science until the paradigm changes.

Style is the soft or fuzzy method.   And this is the real systems
approach, ie the approach actually applied by authors of scientific
works, designs, technological and social solutions, etc.

STYLISTIC INTERPRETATION OF SYSTEMS APPROACH

When considering the systems approach as a style it is easy to
notice that it fulfils the requirements imposed by the definition of
style.   Thus systems approach is characteristic for art, namely, the
art of problem-solving (Gasparski 1981).   Systems approach is
doubtlessly specific for the epoch the beginnings of which go back to
the turn of the 1940s and 1950s.   This epoch, due to the great science
(Price 1963) and accompanying it big technology, is called the period

of scientific and technological revolution.  Systems approach has different variants depending on the sphere to which the solved problems belong, be this various scientific disciplines, branches of technology, kinds of design-making, education, social practice, etc. Individual character is given to systems approach by respective authors (manner).  This character is connected mainly with the mother disciplines of the authors (background) and microparadigms accepted by the teams to which the authors belong or by the recognized by the author reference group.

The analysis of the real systems approach requires empirical research, because referring to the works of the initiators of the systems trend is only of historical interest for this issue.  It may only serve to assess how the predictions of the classics of systems research have been fulfilled.  Empirical research, on the other hand, may be conducted in a twofold way.  The first one is to study the real systems approach as one studies scientific activeness on the basis of empirical methodology.  This methodology makes use of description based on systematic reconstruction of scientific procedure.  The aim of the reconstruction is not so much to reproduce the actions of respective problem-solvers but to identify the invariant features of the investigated action connected with solving problems of a certain type (scientific, pertaining to design-making, technological, scientific, etc).  As a result, a generalized, idealized description of this action is obtained.  The analysis of systems approach based on this method leads thus to reconstructing the features distinguishing this approach as a style of solving problems.

The second way of studying real systems approach is to study the convictions and ways of acting by those who apply this approach, with the use of methods of behavioural sciences.  Such research has been first carried out by Anna Lewicka-Strzalecka (1987).  Lewicka-Strzalecka used the method developed by G A Kelly for other purposes, one given in the English literature the name of repertory grid, which so far has no Polish equivalent.  A. Lewicka-Strzalecka studied not so much the discriminants of systems approach as the methodological approach specific for the agent (investigator, design-maker) systemically oriented in contrast to the approach of agents (investigators, design-makers) solving analogous problems but not systemically oriented.  She studied two groups of so-defined agents: (a) in situations determined by the problems they were currently solving, (b) in situations they postulated, ie views of the agents concerning the desired method - we would say style - in which, according to them, well- and ill-constructed problems should be solved.

Methodological systems approach was determined by the differences in assessments of ways of problem-solving between the above-presented groups of people.  The ways were defined with the use of an estimation table containing 18 of the so-called constructs defined as quasi-continua extending between clearly defined poles.  The continua were divided into seven places in which the respondents' answers were to be located depending on their convictions (assessments) concerning the accuracy of locating the way of acting with respect to the poles. For example, one of the constructs was determined by the poles: 'one-aspected' - 'multi-aspected', another one was located in the space determined by the poles: 'individualization of solution' - 'universalization of solution'.  The reader interested in the other constructs and the course as well as results of the research should refer, due to lack of space, to the source.  Here, only summary of the results shall be presented.

In the case of solving well-structured problems, the determinants of systems approach are, eg attaching great importance to the analysis of the final situation as well as the analysis of formulating the problem (goal and criterion of assessment) and the analysis of the original situation and, moreover, multi-aspected character, tendency to analyse the veracity of statements used in the process of problem-solving, meanings of the notions and effectiveness of the methods, as well as universalization of the solution. In the case of solving ill-structured problems, however, the issue of careful analysis of the way of implementing the solution came to the fore.

An important result of the research is confirmation of the validity of differences between persons oriented and not oriented systemically, and the differences concern more the postulated than the actual problem-solving.

SYSTEMS APPROACH, SYSTEMING, SYSTEMS THINKING, SYSTEMISM

Enthusiasts of systems approach tend to put into one bag systems engineering, cybernetics, dynamics of progress, general systems theory, soft systems thinking, critical systems thinking, etc and give to it one name, eg of systems thinking (Flood and Jackson 1989, p. 151). Other advocates of this approach try to explain clearly what is systems thinking (Emery 1969; Kramer, Smit 1977; Weinberg 1975). Still others try to characterize systems approach as such (Blauberg, Sadowski, Judin 1969) or under the name of systeming (Gasparski 1978) or systemism (Bunge 1979, 1983). The last-mentioned name seems to be the best because it locates the systems approach - as befits a style - among the other 'isms'.

F E Emery, in the Introduction to a collection of works edited by this author and entitled Systems Thinking, says that without the possibility of giving a positive answer to the question about the features distinguishing systems thinking the origination of such a set would be impossible. It turns out that giving such an answer is possible and the answer reflects - according to Emery - the essence of systems approach. This confirms, I believe, the synonymic character of the two names: systems thinking and systems approach.

According to Emery, systems approach is the only one which makes revealing the figural (Gestalt) features of objects possible, and the features refer to various levels of living organisms: from bacteria to human societies. At the same time, Emery makes a reservation that his remarks concern only biological and social systems and not issues of systems engineering which he omits due to the differences in language used to describe complex technological systems (Emery 1969, p. 7)

In a way different from that of Emery a pair of Dutch authors, Nic J T A Kramer and Jacob de Smit, attempt at distinguishing the constitutive features of systems thinking in all varieties of intellectual undertakings which are given the adjective 'systems'. They present examples both from the sphere of sociology, psychology, business, management, economy, technology, and from the natural sciences. These authors treat systems approach and systems thinking as synonyms, but only to a certain degree. According to them systems thinking serves the development of multidisciplinary means of communication and reflects the essence of systems approach (Kramer, de Smit 1977, p. 6).

The way method is thus discussed by Gerald M Weinberg for whom, unlike for Emery, Kramer and de Smit (and other authors who, due to lack of space are not quoted), systemic thinking is thinking in categories of general systems.  It is thus not 'discipline' thinking, as he calls it, eg scientific, electrotechnical, medical, etc but one concerning problems located beyond the disciplines.

The contained in the title word 'general' has a double meaning, wrote Weinberg;  it contains thoughts which have the most general application and may be available to the broadest public.  Putting thoughts occurring in respective disciplines into one structure and one common language, we make certain ideas from respective disciplines available to everybody.  If the ideas are well-chosen, in such a way that they can be generally applied, then the approach will spare man studying discipline thinking because he will not have to go once again along the path already covered in other disciplines (Weinberg 1975, p. 15).

The author of this book proposed more than a dozen years ago a name of systeming (analogously to scouting, cracking etc) as a name of acting which connects systems analysis and synthesis.  This term was connected with solving practical problems (in design-making), Fig. 1 (Gasparski 1978, pp. 16-18).

However, systeming has not caught on, which I admit without remorse, supporting the proposal of Mario Bunge to call the systems approach systemism (Bunge 1979, 1983).

At first Bunge considered systemism as a philosophical doctrine, competing with atomism and holism.  To recall: atomism is a standpoint according to which the whole is completely defined by its parts.  The outcome of this standpoint is the belief that knowledge of the properties of the parts is sufficient to understand the whole.  Holism occupies the opposite pole putting the whole to the fore, and stating that it is original with respect to the parts, that it influences the parts, is something more than a sum of the parts, cannot be explained in an analytical way and, finally, that a whole is better than any of the parts.  Holism according to Bunge is as false as atomism (the latter is often connected with reductionism), although each for a different reason and despite the fact that each contains true elements. The true elements of each of them considered jointly constitute systemism.

In the next volume of the Treatise on Basic Philosophy published several years later (1983), Mario Bunge does not speak about systemism as a philosophical doctrine but treats systemism as a synonym of systems approach (Bunge 1983, p. 260).

An approach is, according to the quoted author an ordered quadruple, A = (B, P, C, M), where B - background knowledge, P - set of problems, C - set of goals, M - set of methods (methodics). Depending on what background knowledge is involved in solving problems and on what kind of problems they are, eg mathematical approach (formal knowledge, such problems, conceptual methods), is distinguished, as well as humanistic approach (knowledge of culture, goals connected with understanding behaviour), etc.  If, however, the background knowledge is general philosophic knowledge then we have to do with three types of approaches:  the atomistic (or individualistic, or analytic), holistic (or synthetic) and systems (or analytic and synthetical) one.  These approaches differ in the type of ontology adopted as background knowledge, in epistemology, goals, and methods. And thus, the atomistic approach bases on atomistic ontology, according to which the world is an aggregate composed of individuals of several kinds and on reductionistic epistemology limiting the knowledge of the whole to the knowledge about its elements.  Holistic approach bases on organismic ontology according to which the world is composed, as an organic whole, of smaller organic wholes which cannot be further divided;  it is difficult to speak about a method here for intuition is mainly used.

Systems approach, Bunge continues, adopts as background knowledge basic systems ontology formulated on the basis of general systems theory.  The epistemology it uses is realism connected with rationalism, and its goal is to understand, predict, and control, while its methods are both analysis and synthesis as well as generalizing, systematizing, and empirical testing (Bunge 1983, pp. 259-260).

The definition of systems approach formulated by Bunge is a projecting definition, determining the approach in the way Bunge wished it to be understood and used.  I shall not deny that I approve of this approach, but one cannot fail to notice that real systems approach is rather a mixture of what Bunge calls holistic approach and systems approach in his understanding.  Bunge's treatment of systems approach, and he is a follower of the so-called exact philosophy, is an outcome of systems approach to systems approach, for Mario Bunge is one of the systems stylists, and a high class one. In this systems style his Treatise on Basic Philosophy is written, a ten-volume work, no less!

Let us finally quote the apology of systemism so formulated by the author of the Treatise:  systemism, or systems approach has the positive features of atomism and holism and is an approach comparable to the scientific one.  It differs from the scientific approach only in that it assumes knowledge of considerably lesser validity and methodological foundations.

The benefits derived from using the systems approach can be best seen in comparison with the competing approaches.  While systemism treats each system as a whole composed of individuals held together by borders (which are part of the structure of the system) and immersed in the environment, atomism leaves out both the environment and the structure;  holism leaves out the contents both of the system and the environment, and does not analyse the structure; environmentalism focuses on the environment at the expense of content and structure; structuralism leaves out contents and environment, let alone the history of the system.  Only systemism gives justice to all the three aspects and their transformations (Bunge 1983, p. 260).

Regardless of the name given to the systemic approach to problems, regardless of whether it is called systeming or systemism, or whether one will speak about systems approach or systems thinking, each of them contains as invariants certain aspects which are continually stressed by the advocates of this approach.  These aspects are: (1)  treating the considered object as a whole (the holistic aspect), (2)  treating the whole as a complex one (complexity), (3)  seeking what makes up the essence of the considered object (essentialist aspect), (4)  revealing elements and links between them both inside and outside the considered object (the structuralist aspect), (5)  sometimes dealing with the external links separately (contextual aspect), (6)  considering the object as a dynamic one due to the goal (teleological aspect).  These aspects are the features of the system style of problem-solving.

WHY HERMENEUTICS?

The main feature of cognition, wrote in his recently published book on H G Gadamer's philosophical hermeneutics Andrzej Bronk, responsible for its superstitionlike character, is according to philosophical hermeneutics its circular structure.   Introducing the notion of circular movement of cognition Gadamer fosters a specific form of dialectics.   He refers to the old hermeneutic rule that the whole should be understood on the basis of details and details on the basis of the whole.   This is derived from ancient rhetorics which compared perfect speech to a living organism, composed of a head and members.   Modern hermeneutics transferred this principle from the art of speaking to the art of understanding.   In both cases, namely, there occurs a specific circularity.   The anticipated meaning of a certain whole is made more and more prominent owing to the fact that parts, additionally defined by the whole, themselves define the whole. This is a process which on the one hand develops in time, for in the course of understanding of respective parts the image of the whole becomes more and more clear, and on the other one, the process is immediate because understanding of the part and the whole is constantly co-determined.   (Bronk 1988, p. 244).

Systems approach to various issues is in a way similar to hermeneutics, namely it is also circular.   It is discussed by the quoted author, who says that the modern attempt at solving the problem of relation of the part to the whole has been made in Ludvig von Bertalanffy's general systems theory (Bronk 1988, p. 246).   Similar is the case with the attempt at explaining the systems style, for it is also of the circular character.   Systems style is certainly a whole internalized by respective authors - systems stylists, the understanding of which requires studying its parts - constitutive features.   Studying these features is connected with understanding the style, and understanding style is enriched by better knowledge of its features.

All attempts at explaining systems style through its analytical interpretation lead to misunderstandings or end up with criticism (see Berlinski 1977) which only confirms that various superstitions are not tolerant one of another.   This lack of tolerance with respect to systemism is also the outcome of lack of clarity about its goal (teleological aspect of systemic approach):

(a)   fostering greater operational effectiveness (course, organization) of a process of problem solving?

(b)   increasing methodological effectiveness, ie obtaining cognitively better results in comparison to the ones obtained with the use of a different style, or perhaps obtaining results impossible to achieve in a different way?

(c)   improving the synthesis of mono-disciplinary results?

(d)   all above-mentioned goals together?

One of the tentative answers to some of these questions, mainly, however, to question (a) is the paper by M C Jackson and P Keys (1986) in which they transfer the discourse to a two-dimensional space. This space is defined by a four-field matrix the rows of which are types of problems, or rather the connected with them types of systems. They include, on the one hand, relatively simple systems with a small number of elements or either small number of interactions between elements or a limited repertoire of these interactions. On the other hand, they include very complex systems with a large number of interdependent elements, connected, moreover, with the environment in which they appear, and goal oriented. The columns of the defined matrix are types of decision-makers as the quoted authors called the persons solving the problems. They are either decision-makers with parallel goals (a unitary team) or decision-makers with different goals (a pluralistic system).

The quoted authors locate in respective fields of the matrix different methods of solving problems specific for various teams of decision-makers. The word "methodology" which they used as the name of these methods, even though common in literature is in fact only "a more impressive-sounding synonym for methods" (Blaug 1982, p. xi).

The quoted as an example attempt, while ordering the catalogue of methods and techniques, does not help to answer the questions nagging the methodologists. But are such answers at all possible? Probably A D Hall III was right when he wrote "... is of course, only a dream. But it exists amorphously in the aggregate of its parts (...). It is not yet whole, but it is undergoing a process of progressive systematization - like a star coalescing from cosmic dust" (Hall 1989, p. xii).

Hall's standpoint seems to justify our claim that systems approaches should be treated as a style, while the so-called systems methods and techniques as sui generis intellectual tools for solving problems, used by those who consider this style as their own and prefer a particular tool. This has always been done by artists par excellence. Thus style is intersubjective whereas the choice of the tool is subjective; attempts at objectivizing the choice of tool ignore the differences between authors using the same systems style. Understanding of similarities despite differences, as earlier understanding of texts, is helped by the hermeneutics of systems.

PRAXIOLOGY AS PART OF SYSTEMISM

The presentation of the specific for European thinking systems style would not be complete if it left out praxiology. I wrote about it as the French and Polish contribution to systemism itself (Gasparski 1984). Basically this is a broader contribution - not only French and Polish - for among the authors of praxiology, besides the French (Bourdeau 1882; Espinas 1890; Caude and Moles 1964; Kaufmann 1983) - there is an Englishman (Mercier 1911), a Russian (Bulhakov 1912), an Ukrainian (Sluck 1926), an Austrian (Mises 1949), Norwegians (Skirbekk 1983), an Italian (Ferrari 1988), a Czech (Zdimal 1990) and others represented at the International Conference "Praxiologies and the Philosophy of Economics" held in Warsaw in 1988. The fact that among the participants of this conference there were systems theorists (Boulding, Klir, Simon and others; see Praxiology (1991 Vol 1) indicates that elements of systems style are perceived in praxiology.

Affiliation of praxiology to the systems trend was declared by Kotarbinski (1977) who considered the general systems theory as a realization of his postulate of construing a general theory of complexes which he had formulated before the systems theory was presented to the scientific milieu. This postulate was the consequence of Kotarbinski's ontology formulated as early as the late twenties (see Kotarbinski 1966) under the influence of Lesniewski's ontology. The latter author, connected like Kotarbinski with the Lvov-Warsaw school of philosophy, formulated a theory of the relation of the part to the whole known as mereology (see Luschei 1962).

"Tadeusz Kotarbinski's ideas were related to our physicalism. He maintained conception which he called 'reism' and 'pansomatism', ie the conception that all names are names of things and that all objects are material things. Both Lesniewski and Kotarbinski had worked for many years on semantical problems. I expressed my regret that this comprehensive research work of Lesniewski and Kotarbinski was inaccessible to us and to most philosophers in the world, because it was published only in the Polish language ...", wrote Carnap in his Intellectual Autobiography (Carnap 1963, pp. 30-31).

Mario Bunge, the author of Ontology II: A World of Systems (1979), acknowledging the systems provenence of praxiology has recently written: "The solution to (...) praxtical problems calls for a number of specialized knowledge concerning the system of interest. Nevertheless, from such specialized studies a certain number of descriptive and normative principles can be formed. The field of knowledge that seeks to find such generalizations is called praxiology and it may be regarded as a branch of general technology and, more particularly, of general theory of systems." (Bunge 1989, p. 327).

Praxiological approach, in contrast to other branches of the general systems theory is not limited to the technical perspective - to use Linstone's words (1989). Praxiological approach is connected with personal and social (organizations) perspectives. Bunge wrote "... action theorists, with the sole exception of the Polish school of praxiology, have systematically ignored work. Why may this be so: general lack of academic interest in important problems, or aristocratic contempt for praxis?" (Bunge 1989, P. 334). This is one more question, this time related to the content not to the methodological context of systems approach, which should be added to the earlier asked questions of methodological character.

CONCLUSION

This paper is an attempt at proving that systems approach is a style, the dream envisioned by the authors in the recent decades. These authors, called systemists, worried by the atomization of the world, loneliness of man in the crowd, domination of particularisms over the global problems, thought: "Let's try another way". These attempts and positive results achieved from time to time, and, first and foremost, the increasing number of authors fostering this "another" way formed the systems style. One may believe that it was shaped under the influence of a more or less conscious ideal, which, as it seems, was best expressed by David Bohm. In his book Wholeness and the Implicate Order, this author remarked: "Destruction of the unity of cities, religions, political systems, conflicts in the form of wars, general violence, fratricide, etc are the reality." The whole is an ideal to which, may be, we should aspire. But this is not what we are driving at here. It should be rather said that the whole is real and the fragmentation is an answer of the whole to human activeness directed by the delusive perception shaped by fragmented thought. In other words, since the reality is a whole approaches taking into consideration only its section are given an adequate for such an approach fragmentary answers. For that reason we should be aware of fragmentary thinking, so that we know about it and are able to get rid of it. Human approach to reality may be comprehensive and then the answer will also be comprehensive. For this to happen it is important that man be aware of the activeness of his thought as such, ie thought as a form of perception, way of looking and not a true copy of the reality itself (Bohm 1980, p. 19). Systems style comes closer to such an insight into thinking and such a perception of reality, which is the belief of those who use it and for that reason encourage the others to do the same.

Finally, it should be remembered that the notion of style in science was introduced by a Polish scientist and specialist in science of science, Ludwik Fleck, as early as 1935 when the first issue of his book Genesis and Development of a Scientific Fact: An Introduction to the Science of a Style of Thinking and a Thinking Collective (Fleck 1979) was published. It was owing to his inspiration, with the mediation of Reichenbach, that Thomas Kuhn construed his theory of scientific revolutions. In his book Fleck wrote:

"We have defined the style of thinking as readiness to directed perception and appropriate treatment of what is perceived. We have also mentioned the specific mood which creates this readiness for adopting different styles of thinking. (...) It is expressed in common appreciation of a certain ideal, the ideal of objective truth, precision and exactitude. It consists of the belief that what is appreciated shall become available only, may be, in a very, or perhaps infinitely, distant future as an outcome of glorification of devotion to work. As a result of a certain cult of heroes and a certain tradition (...). And thus, step by step a certain construction is formed which from individual event in thought and history (discovery) through particularity of collective thinking becomes of necessity repetitive and thus objective and real cognition. An orderly common mood for scientific thinking (...) creates a specific, scientific style of thinking."

An interpretation of the above in relation to systems style is not difficult either as regards the readiness of the systemists' collective for directed perception and appropriate treatment of what is perceived as well as the mood, ie appreciation of a certain ideal (regardless of whether one is a follower of hard or soft systems methodology). It is thus possible to say – using Fleck's language – that systems style has become a fact of science.

REFERENCES

Blauberg, I, Sadovsky, V, Judin, E, eds, 1969, "Systems approach: Assumptions, Problems, Difficulties", in Russian, Moscow.
Blaug, M, 1982, "The Methodology of Economics", Cambridge University Press, Cambridge.
Bohm, D, 1980, "Wholeness and the Implicate Order", Routledge and Kegan Paul, London.
Bourdeau, L, 1882, "Theorie des sciences: Plan de science integrale", Libraire Germex Baillere et Cie, Paris.
Bronk, A, 1988, "Verstehen, Geschichte, Spache: Die Philosophische Hermeneutic von H G Gadamer", in Polish, Catholic University Press, Lublin.
Bunge, M, 1979, Ontology II: A World of Systems, in: "Treatise on Basic Philosophy", Vol 4, Reidel, Dordrecht.
Bunge, M, 1983, Epistemology and Methodology I: Exploring the World, in: "Treatise on Basic Philosophy", Vol 5, Reidel, Dordrecht.
Bunge, M, 1989, Ethics: The Good and the Right, in "Treatise on Basic Philosophy", Vol 8, Reidel, Dordrecht.
Bulhakov, S, 1912, "Philosophy of the Economy", in Russian, Moscow.
Carnap, R, 1963, Intellectual Authobiography, in: "The Philosophy of Rudolf Carnap", P Schlipp, ed, Cambridge University Press, La Salle.

Caude, R, and Moles, A, 1964, "Methodologie, vers une science de l'action", Gauthier-Villars, Paris.

Emery, F, E, ed, "Systems Thinking: Selected Readings", Penguin, Harmondsworth.

Espinas, A V, 1890, Les origines de la technologie, Revue Philosophique, t. XXX, Aout, 113-135.

Ferrari, G, 1988, "II Diritto tra Regola e Azione: Elementi D'Analisi Praeologica del Normativo", Casa Editrice Dott Antonio Milani, Padova.

Fleck, L, 1979, "Genesis and Development of a Scientific Fact", The University of Chicago Press, Chicago and London.

Flood, R I, Jackson, M C, 1989, Editorial, Systems Practice, Vol 2, No 2, p. 151-153.

Gadamer, H G, 1967, "Kleine Schriften I Philosophie Hermeneutic", Tubingen.

Gasparski, W, 1978, Designing and Systems, in Polish, in: W Gasparski and D Miller, eds, Projektowanie i Systemy, Vol I, pp. 11-20.

Gasparski, W, 1982, Systems Approach as a Style, in: "Proceedings of the International Conference on Systems Methodology", SGSR, Washington DC.

Gasparski, W, 1984, Praxiology: A French-Polish Contribution to Systemism, in: "Proceedings of the 6th International Congress of Cybernetics and Systems", Paris.

Gasparski, W, 1988, Theoretical and Methodological Tradition for Systems Science and Engineering in Poland, in: "Proceedings of the International Conference on Systems Science and Engineering", Beijing.

Gasparski, W, and Pszczolowski, T, eds, 1983, "Praxiological Studies", Reidel, Dordrecht.

Hall, A D, III, 1989, "Metasystems Methodology", Pergamon, New York.

Jackson, M C and Keys, P, 1986, Towards a System of System-Based Problem-Solving Methodologies, in Polish, in: W Gasparski and D Miller, eds, Projektowanie i Systemy, Design and Systems, Vol VIII, pp. 111-124.

Kaufmann, A, 1968, "The Science of Decision-making: An Introduction to Praxeology", Weidenfeld and Nicolson, London.

Kotarbinski, T, 1965, "Praxiology", Pergamon, Oxford.

Kotarbinski, T, 1966, "Gnosiology", Pergamon, Oxford.

Kotarbinski, T, 1977, Concepts and Problems in General Methodology and Methodology of the Practical Sciences, in: M Przelecki and R Wojcicki, eds, "Twenty-Five Years of Logical Methodology in Poland", Reidel, Dordrecht, pp. 279-289.

Kramer, N J T A, and Smit, J, de, "Systems Thinking: Concepts and Notions", Martinus Nijhoff, Leiden.

Lewicka-Strzalecka, A, 1987, "Identification of the Systems Approach in the Process of Solving Practical Problems", in Polish, Institute of Philosophy and Sociology Press, Warsaw.

Lilienfeld, R, 1978, "The Rise of Systems Theory: An Ideological Analysis", Wiley, New York.

Linstone, H A, 1989, Multiple Perspectives: Concept, Applications, and User Guidelines, Systems Practice, Vol 2, No 3, pp. 307-331.

Luschei, E C, 1962, "The Logical Systems of Lesniewski", North-Holland, Amsterdam.

Mazur, M, 1976, "Cybernetics and Character", in Polish, State Publishing Institute, Warsaw.

Mercier, Ch A, 1911, "Conduct and Its Disorders: Biologically Considered", Macmillan, London.

Mises von, L, 1949, "Human Action: A Treatise on Economics", William Hodge, London.

Monod, J, 1974, Chance and Necessity, <u>in</u>: F J Ayala and T Dobzhansky, eds, "Studies in the Philosophy of Biology", Macmillan, London.

Price, Solla de, D J, 1962, "Little Science - Big Science", Columbia University Press, New York.

Skirbekk, G, ed, 1983, "Praxeology: An Anthology", Universitetsfolaget, Bergen.

Slucky, E, 1926, Ein Beitrag zur formal-praxeologischen Grunlegung der Okonomik, <u>in</u>: "Annales de la classe des sciences sociales-economiques", Academie Oukrainienne des Sciences, Kiev.

Taterkiewicz, W, 1988, "The History of Six Ideas", in Polish, Polish Scientific Publishers, Warsaw.

Weinbeg, G M, 1975, "An Introduction to General Systems Thinking", Wiley, New York.

ACTION AND STRUCTURE IN PROBLEM SOLVING

Raul Espejo

Aston Management Centre
Aston University
Birmingham, B4 7ET

INTRODUCTION

The concern with problem situations is a response to their pervasive nature. A hallmark of human activities is that they are experienced as
problematic. In the quest for stability people in general find that they
need to bring about some kind of change in the situations of their concern. This quest is much more than countering perceived threats or ap-
prehending opportunities of one kind or another. The constant flux of
emotions, the outcomes of on-going interactions, the buffeting of new
stimuli, the emergence of new ideas, the fears of the unknown, and a
wealth of other factors are responsible for changing expectations about
what experiences are acceptable and desirable. As these expectations are
not fulfilled by their actions, people start to recognise problem situations; instabilities, that may be construed as opportunities or threats,
requiring their creative capacity to deal with them.

This paper offers a discussion of the nature of problem situations and an
approach to deal with them. The approach is based on cybernetic principles, hence it is referred to as a cybernetic methodology for problems
solving.

The cybernetic framework provides not only powerful insights about
problems of control and communications in complex situations, but also a
methodological grounding to support the work of problem solvers. These
problem solvers are the relevant viewpoints concerned with the situational stability. Hence, problem solving is understood as the production
of changes in the relations constituting the situation in order to bring
forth stability in the interactions of the relevant viewpoints.
Methodologically, this focus highlights that in problem solving it is essential to establish the appropriate viewpoints and the nature of their
communication mechanisms.

To help understanding this approach to problem solving, this paper offers
a wider discussion about the nature of problem situations in human activities and discusses alternative methodological approaches to deal with
them. The cybernetic methodology is offered as an alternative to the so
called hard and soft methodologies for problem solving.

*Systems Thinking in Europe*, Edited by M.C. Jackson *et al.*
Plenum Press, New York, 1991

The hard and soft methodologies are two extremes in the spectrum of possible methodologies. It will be argued that while the former may be inadequate to handle situations where the participants constitute different realities, the latter may be inadequate in situations where the participants are forced into constituting a single reality by an inflexible organisation structure. Hence, to overcome the shortcomings of both approaches the case will be argued for a different methodology based on cybernetic principles.

## ABOUT PROBLEMS

### The Problem Solver

This paper is concerned with problems that emerge when people need/want to work together in a co-ordinated fashion. In these situations an individual's ability to solve well defined problems, may be desirable but not essential. Thus, a person who knows precisely how to deal with a particular problem, but fails to communicate with others, is likely to render this capability ineffective.

Human activities are stability oriented rather than goal oriented. It is a fact that people are often ready to adjust or change their goals in order to achieve stability in their interactions. This view suggests that problem solving is much more than about finding the means to achieve certain goals, it is about finding stability in interpersonal interactions.

Moreover, since organisational activities are highly interconnected, achieving stability in one situation may prove to be the trigger of instabilities elsewhere. Finding a workable balance between them is the delicate "art of problem solving". Ackoff coined the term "mess" (Ackoff 1978) to describe this kind of situation; a mess is a web of problems. This point makes apparent the systemic nature of problems.

Therefore, effective problem solvers are those individuals who succeed in discovering and producing feasible and desirable changes in the multiple situations they participate. Indeed, problem solving is seldom the outcome of an individual's unilateral actions, more likely, it is the outcome of his effective participation in the constitution of an acceptable, and hopefully desirable, situation.

### Problems and Problem Situations

The problematic nature of human activities relates not only to perceived mismatches between expectations and actuality but, and most importantly, to the natural diversity of individual experiences in a given situation. While the former type of problems relates to one viewpoint, the one perceiving the mismatches, the second type relates to multiple viewpoints, those experiencing the same situation from different perspectives. These are the most intractable and difficult problems to tackle. Following Vickers' idea of appreciative systems, each of us, based on our individual histories, develops different readinesses to notice particular aspects of a situation. These readinesses in turn help to organise our further experience, thus creating unique and diverse appreciations about a situation (Vickers 1965, 1972). Though diversity of appreciations is likely to be an asset in any situation, it may also be the source of perceived problems, particularly when communications between the situational participants are inadequate. For a group of people to co-ordinate their actions and operate together they need to share these appreciations. But, if they are not aware of these differences and act unilaterally, most likely they will trigger all kinds of problems. Hence, problem solving entails creating the conditions for sharing experiences. Participants need to develop, gradually, richer and richer understandings of each others viewpoints.

Thus, there is a great difference between a multiple viewpoints and a single viewpoint problem solving situation.

For individual viewpoints (i.e. for individuals that express either their personal views or the views of a group of people they represent) problems emerge when they perceive their reality out of control. These perceptions may be triggered by mismatches between their preferred outcomes for a situation and the perceived present or anticipated future outcomes. Whether these mismatches appear to be well defined or not is important, but not central, to this definition. The point is simply that whenever individuals anticipate, or just feel anxiety, about the outcomes of a situation, they are perceiving a control problem. This anxiety may relate to current situations or to future situations. Individuals may recognise that something is out of hand today or may anticipate that something might be out of hand tomorrow; in either case they are perceiving a problem.

Thus, individual viewpoints construe problems as perceptions, with different degrees of uncertainty, of mismatches between expectations and actuality.

On the other hand, problems with multiple viewpoints are centred in their different appreciations about the same situation, that is, in the likelihood that they may be experiencing different equally valid, and not intersecting, mismatches for the same situation. Each viewpoint is essentially perceiving a different problem, though all of them are participating in the same situation.

Multiple viewpoint or stability (soft) problems emerge from the natural diversity of human activities, where individuals are not contributing to a given reality, out there independent of them, but are constituting their realities through interactions with others. Therefore, the problem is not improving the information of the participants for them to be closer to the reality of a situation. The problem is creating the conditions for the participants to share their experiences in a common operational domain. If sharing these different realities is hindered by poor communications, then, most likely viewpoints will have difficulties in creating a consensual domain, thus, possibly making the problem situation intractable. Problems are often rooted in these two facts; the different appreciations that people develop of the domains of common concern, and also, the inadequate communication mechanisms supporting their interactions.

## Control Problems

From the perspective of an individual viewpoint the emphasis in problem solving is in managing the complexity of its reality. The problem is finding the appropriate responses to counter the disturbances responsible for the perceived mismatch. Cybernetically, the viewpoint is aiming to control the outcomes of a black box. This is the case even if the viewpoint is not aware of the black box. The mismatch may be perceived anywhere in between a well defined and a totally undefined mismatch.

A viewpoint has to deal with at least two fundamental dimensions of uncertainty. Firstly, the viewpoint may have some degree of uncertainty (from low to high) as to what is the problem, secondly it may have some degree of uncertainty (from low to high) as to how to tackle the problem situation. Indeed, in this case, establishing what is the problem does not entail the process of creating a shared reality as much as distinguishing a focus of common concern. How to tackle the problem entails producing the appropriate responses to produce acceptable behaviours.

In global terms these two dimensions define four types of problems (Figure 1):

Operational problems are those where the viewpoint knows both, what the problem is, and how to handle it. These are simple problems that may lend themselves to mathematical modelling and computer applications. The real problem is not in the content itself, but perhaps in the fact that the capacity to handle all instances of the problem may not be available, or that the participants may lack motivation to tackle it.

Learning problems are those where the viewpoint "knows" what the problem is but does not know how to tackle it. The solution is part of a learning process.

Compromise or choice problems are those where the viewpoint may know how to tackle the problem, but may be uncertain as to which option to take; the viewpoint is unclear about its preferences. Indeed, a person may be unclear as to which option, among those available, to take, or the participants (in the one viewpoint) may have different preferences.

In these three types of problems the viewpoint has a clear structure for the problem, even though, in the case of choice problems, it may be unsure about its preferences.

However, the fourth type of problems, inspirational problems, are those where the viewpoint is not clear about what is that it wants to tackle, let alone how to tackle it. Dealing with these problems depend on the viewpoint's ability to make useful, insightful, distinctions in its operational domains. The viewpoint may only know that there is a feeling of uneasiness. These problems are particularly challenging. The viewpoint has no signposts and indeed, it requires of creativity to find out which are the issues or tasks of interest.

Yet, these problems are comparatively simple to tackle, they all assume one reality, with only one relevant viewpoint, experiencing a larger or lesser degree of uncertainty.

Stability (Soft) Problems

However, in all the above four cases, most likely the viewpoint will depend on other viewpoints to discover possibilities and produce any defined solution. In other words, the viewpoint will need to establish those changes that are both desirable and feasible in the given situational context. Inadequate grasp of the appreciations of the other participants may contribute to perceiving stability (soft) problems, that is, to the perception of instabilities in interpersonal interactions.

Therefore, stability problems are those emerging in situations where there are multiple viewpoints, and where problem solving is focused in the communications between the viewpoints rather than in the control of the outcomes of a situation.

It is important to reinforce that any problem, regardless of whether it is perceived as well structured or not, is potentially a soft problem as soon as its solution implies the participation of other viewpoints. For instance, the implementation of a response to a well structured organisational problem, such as the implementation of a stock control system, may

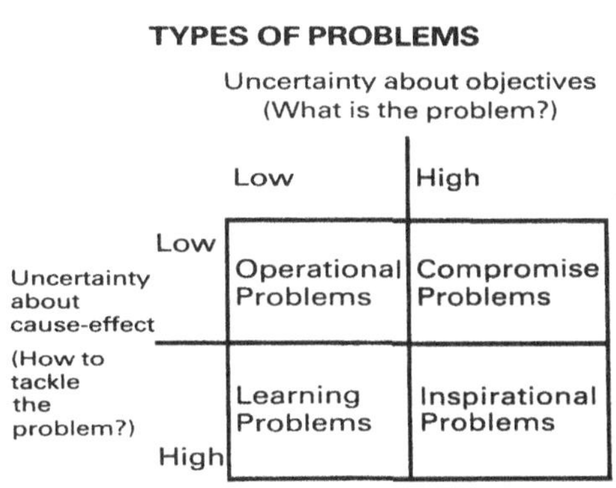

**TYPES OF PROBLEMS**

Uncertainty about objectives
(What is the problem?)

|  | Low | High |
|---|---|---|
| **Low**<br>Uncertainty about cause-effect | Operational Problems | Compromise Problems |
| (How to tackle the problem?)<br>**High** | Learning Problems | Inspirational Problems |

Fig.1 Types of Problems : One Viewpoint

33

be of the soft kind if it requires the participation of people who do not share the problem owner's appreciation of the situation. Analysts often encounter this kind of problems; they are inherent to the production of change in the real world. However, as a shared appreciation about necessary action grows, -something that does not imply that all viewpoints should think alike- the problem situation moves from softer to harder grounds. The emphasis in the latter type of problems is the production of a defined transformations and are mainly of a technical kind.


ABOUT METHODOLOGY

The problem is how do we improve our abilities to handle soft problems both, as problem owners, and as system analysts. In one form or another all of us deal intuitively with soft problems and solve them more or less effectively. But, this is not good enough.

We need an approach to deal with them; this approach, broadly speaking is a methodology. Thus, a methodology for organisational problem solving could be understood as a set of interrelated activities facilitating, and making more effective, the problem solving process.


Traditional Approaches to Problem Solving

Traditional approaches to problem solving, like some forms of Operational Research (Kidd 1985) and Applied Systems Analysis (Miser and Quade 1986) have centred their efforts in single viewpoint problems, where the control of some outcomes is paramount. These approaches are concerned with the definition of strategies and methods to achieve defined goals. They put the emphasis in applying scientific methods to those problems arising in organisations rather than in managing the processes leading to the perception of these problems.

In spite of the difference in names, both systems analysis and operational research focus their concerns basically in the same issue, that is, in the use of scientific methods to support the processes of problem solving in human systems.

These methods handle the technical aspects of a situation, rather than its human aspects. They seem to be more concerned with explaining, simulating or controlling a situation, rather than with communication processes to achieve the commitment of the concerned people. This technical focus has been greatly successful in a number of human endeavours, like design of production systems, computer systems, the space programme, etc, however, this focus, as has been made abundantly apparent by those propounding soft approaches, has not succeeded in the more unstructured situations, where interpersonal communications are not dominated by common goals.

The worldview of these methodologies is that problem solving is about finding responses to achieve well defined goals or objectives.

Unfortunately for those who put their hopes in them, these problem solving approaches are inadequate when the complexity of human activities is dominated by the need to maintain stability in interpersonal interactions. Among others, information analysts have felt this inadequacy; they refer to the so called "problems of implementation" (Boland and Hirschheim 1987), that is, the problems of transferring abstract designs into the real world. It is apparent that the diversity natural to the world we live in is not well captured by most of these approaches, it is also apparent that the flexibility required in the solutions is not there once they are implemented.

Of course, in spite of all the above distinctions, whether a situation is perceived as soft or hard, or softer or harder, depends on the context and the individuals concerned. The same situation may evolve from one extreme to the other, making appropriate the use of different methodologies, for a harder problem it is necessary a harder methodology, one that assumes that the participants share objectives, and where the inquire is about the means to achieve these objectives (i.e. a control methodology).

Quite naturally while soft problems do not lend themselves to precise techniques, individual viewpoints may use these techniques to do independent analyses of the situation.

Soft Methodologies

The recognition of the above situation has led to the development of new approaches like the Soft Systems Methodology (Checkland 1981), and several others, focused in the process of problem solving, rather than in their specific technical content (Eden et al 1984, Kling 1987).

In Checkland's view human activity systems are not objective entities in the world. People are simultaneously creating and participating in real world activities, as such they are naturally embedded in multiple realities. As a result of their personal histories people develop different appreciations about a situation rather than of a situation (Vickers 1972). An appreciation of the situation would imply that there is something objective outside in the world to be appreciated. In Checkland's view, problem solving is related to conversational processes in which people develop new insights relevant to the situations of concern. As such, problems are continuously being formulated and reformulated as a result of on going debates or conversations.

An analyst using the Soft Systems Methodology (SSM) starts his inquiry finding out about the problem situation by describing relevant structures and processes i.e. by building up a rich picture of the situation. This knowledge permits to hypothesise a few root definitions, relevant to the situation; these are concise, tightly constructed descriptions of human activity systems, which state the systems perceived as relevant to the situation by the analyst. What these systems do, that is their named transformations, is then structured in the form of conceptual models. The core of this methodology is comparing these models, which show the logical activities necessary to produce the transformations, with the real world situation (as described in the first activity). This comparison takes place in the real world (i.e. in the world of the clients). From this comparison it should be possible, for the clients, to derive systemically desirable and culturally feasible changes, that is, the directions for taking action in the problem situation.

The Soft Systems Methodology (SSM) makes apparent that the scope for change in human activities depends upon changes in the appreciations of the clients involved in the problem situation rather than on the merits of technical options.

This methodology would appear to take for granted the nature and quality of the communication channels relating the participants in the situation. In cybernetic terms it would appear to take for granted the structure of the situation. The emphasis of SSM is in phenomena, in the debates and action necessary to produce desirable and feasible changes, and not in the structure underlying these actions. SSM advocates debates as essential to the problem solving effort, yet it does not pay attention to whether they are taking place in a context that supports such processes. Moreover, SSM appears to be focused in the debates of the clients and not in the communication mechanisms (i.e. operational domain) relating those affected by the changes emerging from these debates. Both these aspects, the context supporting debates, and the communication mechanisms supporting the interactions of those implied by the 'culturally feasible' change, are essential to effective problem solving. In fact, while some structures may create the conditions for the effective participation of multiple viewpoints, others may inhibit this participation (Espejo 1991). Also, while some structures may increase the capacity for people to solve problems at the local level, others may inhibit this capacity, thus making systemically unfeasible changes that may be culturally feasible.

In other words, the scope for 'making available' particular appreciations, necessary to enrich the problem solving process, may be severely restricted by the organisational contexts in which the communications take place. The organisational structure may have inadequate communication channels to allow the participants to focus their different appreciations in the same concern. On the other hand, it is not enough to achieve this focus and recognise a 'culturally feasible' change to produce this change. If the existing capabilities and processes do not permit the organisation to produce a culturally feasible change, then this change will not happen however much the concerned managers may wish it to happen. If this is the case, however much people may communicate at certain levels, their problem solving capabilities will not be adequate.


THE CYBERNETIC METHODOLOGY

The above comments about the SSM suggests that both action and structure should be taken into account in problem solving. Of course if the situation requires a short term response, then the likelihood is that the overriding concern is going to be action, however, if the problem situation can be seen in a longer term perspective then, a focus in action and structure is more likely to produce effective responses. Moreover, even if the concern is the short term, appreciating the situational structure is likely to help in producing changes not only culturally but also systemically feasible.

It is this concern with action and structure in problem solving that needs of a methodological underpinning.

Cybernetics, or "the science of communications and control in the animal and the machine" (Wiener 1961), offers a way to deal with structure. To talk about the cybernetics of a situation is to talk about the mechanisms supporting communications, both in the operational and informational domains, between the people involved in the situation. This is a point that needs great emphasis. Communication is a far deeper phenomenon than information transmission. Communication relates to the structural coupling of the participants, that is, to their structural adjustments, in a history of recurrent interactions, and not to the transmission of information between them (Maturana 1987).

People's interactions, whether they are pursued explicitly or not, as they take place bring about transformations of one kind or another in an operational domain (i.e. the so called 'real world'). Whether these transformations are perceived as threats or opportunities is not the point, the point is that they trigger an adjustment, of one kind or another, in the stability of interactions and with that the need for problem solving.

The cybernetic framework provides not only powerful insights about the nature of communications in complex situations, but also provides methodological guidelines to tackle problem situations. These two aspects are often conflated by those who have a superficial understanding of cybernetics. One aspect is the models provided by cybernetics, like the Viable Systems Model (Beer 1979), another is the ontology of the observer as offered by second order cybernetics (Von Foester 1979, Maturana 1987) This second aspect supports the unfolding of methodological frameworks such as that introduced by this paper, and developed at length elsewhere (Espejo 1989).

Second order cybernetics helps understanding processes based on their on-tological grounding and, with that it offers a way to appreciate their complexity, or in the terms of this paper, a way to grasp their structure or the cybernetics of the situation (Espejo 1987). Problem solving, as implied above, is understood as the discovery and production of feasible and desirable changes to maintain and/or achieve stability in interper-sonal interactions, stability as perceived by the participants them-selves. Whatever IS a situation is what the participants in a situation make of it. Methodologically, this focus highlights that in problem solv-ing it is essential to establish the appropriate viewpoints, creating or constituting the situation, and the nature of their communication mechanisms. This concern is much much deeper than that of establishing the viewpoints talking about the situation.

At its deepest level, this paper advocates the need for a methodology to support the interplay of action and structure in the operational domain of the participants. It also advocates the need to support the interplay of content and context in the informational domain of the participants, and the interplay of the observed and the observing systems in the opera-tional domain of an observer. I call such a methodology the Cybernetic Methodology, (Figure 2). The emphasis of the cybernetic methodology is in the communication mechanisms between the participants in the problem situation. It is argued that inadequate mechanisms lead to inadequate ap-preciations about the situation, and also to inadequate actions. It is also argued that improvements in the situation depend, among other fac-tors, upon structural changes.

The cybernetic methodology highlights the fact that the creation (i.e. constitution) of human activities is strongly influenced by the channel capacity of the communication mechanisms underlying the recurrent inter-actions of individuals. The cybernetic view is that these individuals are constrained to different degrees by the organisation structures in which they are embedded, and therefore, that by changes and modifications in these structures, it is possible for them to develop different apprecia-tions of a problem situation. Moreover, while some structures are likely to inhibit their action and produce poor appreciations, others are likely to liberate their views and make more likely an effective action. There-fore, in the cybernetic approach it is argued that effective problem solving implies the creation of an effective structure, as effective as it is culturally feasible (for the creation of such an organisational structure must acknowledge the constraints dictated by the cultural environment).

The cybernetic methodology entails both a learning loop and a cybernetic loop. The learning loop is focused in action, the cybernetic loop in structure. While it may be helpful to think in both of them as separate loops, in practice they are braided in an operational domain. The opera-

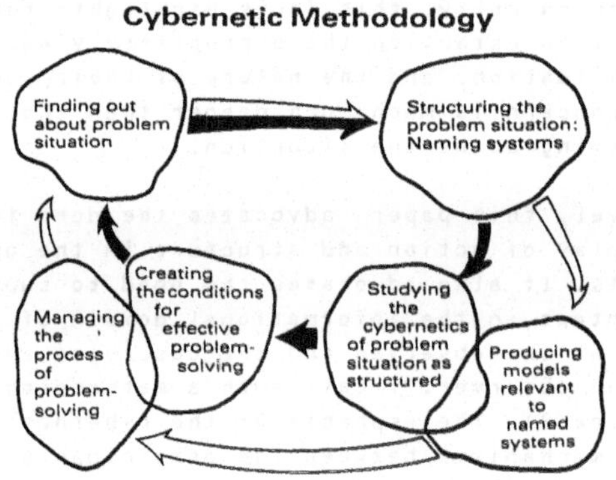

**Cybernetic Methodology**

Fig.2 Cybernetic Methodology, Inner Loop (in black) : Cybernetic Loop,
Outer Loop (in white) : Learning Loop

tional domain of particular interest in this paper, but by no means the only one, is that of action in organisations.

Making distinctions by naming systems and producing models (activities in the right hand side of the learning loop) take place in the informational domain of those constituting the situation. Finding out about the situation and managing the process of problem solving (activities in the left hand side of the learning loop) take place in the inter-active space of those constituting the problem situation. In this way problem solving is offered as an interplay between the fully fledge complexity of "real world" interactions and the much simplified, but also useful, world of models and abstraction.

Studying the cybernetics of the problem situation, that is, studying the control and communication mechanisms underlying the situation, implies studying the organisation(s) brought forward as relevant to the problem situation. Therefore, one (or in some cases more than one) of the systems named by the problem structuring activity of the methodology (Figure 2) may be the name of an organisational system. These studies produce models of communication and control mechanisms as perceived in the current situation. These models are then compared with (cybernetic) criteria of effectiveness. Mismatches between the perceived current situation and the abstract model define possible areas for improvement. Thus, the outcome of the modelling activities is an input to the debates among clients in the situation. These inputs are aimed at supporting the discovery of feasible and desirable changes in the cybernetics of the situation, thus creating the conditions for effective problem solving. Naturally changes of this nature affect the situation itself, thus closing the cybernetic (inner) loop for problem solving.

While the cybernetic improvements might not deal directly with the particular symptoms of the problem situation, they are intended to create the structural conditions for effective problem solving, i.e. for effective appreciation and participation. Adequate regulatory mechanisms reduce the chances of dealing with self inflicted problems. It is in these conditions that the participants are more likely to focus their problem solving capabilities in their different appreciations and preferences, rather than in conflicts triggered by poor organisational - communication- processes.

Perhaps the most relevant of the activities of the learning (outer) loop is managing the process of problem solving. It is at this stage that the management of complexity takes place. Debates should permit to establish what sort of improvements are desirable, and political negotiations should permit to establish their feasibility. Since producing feasible changes will require most likely the contributions of other people, success in problem solving relates to success in implementing the agreed transformations. However, while this implementation may be facilitated by an effective use of the cybernetic loop, most likely, it will produce stability problems to other participants operating at higher levels of resolution, for whom the same methodological approach may now be useful.

FINAL REMARKS

The purpose of this paper has been to argue the need for a problem solving methodology that takes into account both action and structure. The so called soft methodologies have been particularly concerned with the level of action. The cybernetic methodology raises the issue of braiding action and structure in an integrated framework for problem solving. This concern is not unique to cybernetics, recent work in sociology, in particular structuration theory (Giddens 1984) as reported by Walsham and

Han (1990) appears to be saying something similar. Understanding this methodological framework requires the grounding of epistemology in ontology, something that, by and large, has been ignored by management scientists. The seminal work of Ashby (1964) in the area of complexity, and of Maturana (1980) in the biology of cognition, have openned the way for methodologies such as the Cybernetic Methodology.

This paper was intended to make more clear the distinction between hard (control) and soft (stability) problems. The latter problems are inherent to human activities as people constitute different realities in their natural drift. It was argued that tackling stability problems can be improved by improving the communication mechanism between the participants (the cybernetic loop) and by focusing their different appreciations about the situation (the learning loop). The aim of the methodology is to work out desirable and feasible changes in the situation. However, feasibility is seen not only as defined by cultural concerns, but also by structural concerns.

## REFERENCES

| Ackoff R | 1978 | The Art of Problem Solving, NY, Wiley |
| Ashby R | 1964 | An Introduction to Cybernetics, London, Methuen |
| Beer S | 1979 | The Heart of Enterprise, Chichester, Wiley |
| RJ Boland | 1987 | Critical Issues in Information Systems |
| RA Hirschheim | | Research, Chichester, Wiley |
| Checkland P | 1981 | Systems Thinking, Systems Practice, Chichester, Wiley |
| Eden C , S Jones & D Sims | 1983 | Messing About in Problems, Oxford, Pergamon Press |
| Espejo R | 1987 | From Machines to People and Organisations, in New Directions in Management Science Eds. M Jackson and P Keys, U.K. Gower |
| | 1989 | Cybernetic Methodology for Problem Solving, Unpublished Monograph, Aston Business School, Birmingham, U.K. |
| | 1991 | Strategy, Structure and Information Management, in Espejo R. and M. Schwaninger eds. Organisational Fitness: Corporate Effectiveness through Management Cybernetics, Frankfurt/New York, Campus (Forthcoming) |
| Giddens A | 1984 | The Constitution of Society, Cambridge, Polity Press |
| Kidd, J. (ed) | 1985 | Managing with O.R., London, Philip Allen |
| Kling R | 1987 | Defining the Boundaries of Computing Across Complex Organisations in Critical Issues in Information Systems Research Eds RJ Boland and RA Hirschheim, Chichester, Wiley |
| Maturana H | 1987 | Everything Is Said by an Observer, in Gaia: A way of knowing, ed. by WI Thompson, California, Lindisfarne Press |
| Maturana H & F Varela | 1980 | Autopoesis and Cognition: the Realisation of the Living, Boston, Reidel |
| Miser H & E Quade | 1985 | Handbook of Systems Analysis: Overview of Uses, Procedures and Practice, Chichester, Wiley |

Vickers G            1965  The Art of Judgement, London, Chapman and
                           Hall
                     1972  Freedom in a Rocking Boat, Middlesex U.K.
                           Pelican Books
Von Foester H        1979  Cybernetics of Cybernetics in Klaus
                           Krippendorff (ed) Communications and
                           Control in Society, New York, Gordon and
                           Breach
Walsham G &          1990  Structuration Theory and Information Systems
Chun-Kwong Han             Research, Proceedings of the 11th International
                           Conference on Information Systems, IFIP,
                           Copenhagen, December
Wiener N             1961  Cybernetics, Cambridge, Mass. MIT Press

IMPLEMENTING TOTAL QUALITY MANAGEMENT THROUGH TOTAL
SYSTEMS INTERVENTION: A CREATIVE APPROACH TO PROBLEM
SOLVING IN DIAGNOSTIC BIOTECHNOLOGY (PTE) LTD

Robert L. Flood

Department of Management Systems and Sciences
University of Hull
Hull, HU6 7RX, U.K.

INTRODUCTION

Total Systems Intervention is a creative approach to problem
solving. It will be introduced in this article by describing and
discussing a practical example of its use. First of all we will
introduce the company, second the approach and then we will
describe and discuss intervention in the company employing the
approach.

BACKGROUND INFORMATION

After an introduction to Total Systems Intervention (TSI, see
Flood and Jackson, 1991a)[1,2] the Managing Director of Diagnostic
Biotechnology (Pte) Ltd (DB) in Singapore, Mr Lim Jiu Kok[3], invited
me in May 1990 to consult in his company and to apply our

---

[1]   Based on critical systems thinking (see Flood, 1990; Flood and
      Jackson, 1991b; and Jackson, 1991).

[2]   In which details of all the approaches discussed in this article can be
      found, thus relieving me of endless referencing.

[3]   To whom I am most grateful for allowing me to record my
      experiences in his company and for providing the technical input
      necessary to enable this paper to be completed.

*Systems Thinking in Europe*, Edited by M.C. Jackson *et al.*
Plenum Press, New York, 1991

creative problem solving approach. As the company's name suggests, its business is diagnostics using biotechnology. In June 1990 at a further meeting in Taiwan we agreed that I should spend two days in Singapore in September that year. I was to meet people from all functions of DB and to grasp a broad understanding of the main issues that they face. The second stage of involvement would be a two day seminar in November 1990 in the company to explore with the staff organisational difficulties and possible ways forward using systems ideas[4]. A third phase was also discussed, possibly to be an analysis of large projects undertaken by the company, to assess their effectiveness and efficiency.

As is always the case in consulting and as we shall find out below, things do not go exactly according to plan. What was to happen, however, was exciting enough and well worth while recounting. But we need to know more about DB and TSI before the story can be told.


INTRODUCING DIAGNOSTIC BIOTECHNOLOGY (PTE) LTD

Diagnostic Biotechnology (Pte) Ltd (DB) was founded in 1984 on Sing$6M venture capital. Today it has about Sing$20M authorised capital. Investors in the company include local and foreign venture capitalists. Biotech Research Laboratories Inc., (BRL), in Maryland, U.S.A., holds the majority of the foreign equity. It has a technology transfer agreement for a period of 7 years with BRL, which manufactures diagnostics, monoclonal antibodies and other research products for the biotechnology industry. Under this agreement, DB receives technologies in immunology and immunochemistry; cell culture; hybridomamonoclonal antibody production; and recombinant DNA-genetic engineering techniques. DB also has a contract with BRL to produce products for them subject to U.S.A. Governmental approval.

DB researches into and produces diagnostic kits and offers laboratory services using the latest biotechnology.

Biotechnology is any technique that uses living organisms, or parts of organisms, to make or modify products, to improve plants or animals, or to develop micro-organisms for specific uses. Biological

---

4    Not unprecedented in the biotechnology industry. For example, Fairtlough (1990) describes how systems thinking was used from the start in Celltech, another biotechnology company.

systems and organisms are applied to technical and industrial processes. Pharmaceuticals, speciality chemicals, foods, agriculture, commodity chemicals, are dealt with.

Current product lines in DB are the ELISA-HTLV II Assay kit and the Western Blot-HTLV III Assay kit for detection of antibodies against AIDS. The first kit is used for screening, while the latter is used for conducting confirmatory tests. Other products being introduced are HTLVI Assay kits, Western Blot-HTLVI Assay kits and a range of Hepatitis B and EBV diagnostics. For the longer term, DB plans to move into recombinant DNA technology, adding fusion and enzyme-engineering techniques to its existing tissue culture and monoclonal antibody technologies, for production of diagnostic kits.

Let us now turn our attention to Total Systems Intervention, a creative approach to problem solving employed to drive consultation in DB.


## INTRODUCING TOTAL SYSTEMS INTERVENTION: A CREATIVE APPROACH TO PROBLEM SOLVING

Total Systems Intervention (TSI) has been developed over the last 3 or more years by M.C. Jackson and myself. We have used, developed and refined the meta-methodology in a wide range of consultancies at both competitive consultancy rates in commercial and governmental bodies and without fees in charities, etc. The official account of TSI can be found in Flood and Jackson (1991a), from which the following is "reproduced."

It is the argument of TSI that the search for some super-method that can address the world's diverse mess of interacting problems is mistaken and must quickly lead to disenchantment. It would be equally wrong, however, to revert to a heuristic, trial and error approach and to tackle the mess in that way. We need to retain rigorous and formalised thinking, while admitting the need for a range of "problem-solving" methodologies, and accepting the challenge which that brings. It is our view that the future prospects of management science will be much enhanced if (a) the diversity of "messes" confronting managers is accepted, (b) if work on developing a rich variety of methodologies is undertaken, and (c) if we continually ask the question: 'What kind of problem situation can be "managed" with which sort of methodology?' TSI is an approach which accepts the triple challenge just detailed, so let us move on to consider its logic and process.

## The Logic and Process of TSI

The logic and process of TSI can be explained as comprising 3 parallel phases. These initially have to be explained in sequence. The 3 phases of TSI are labelled "creativity," "choice," and "implementation." They require maximum participation of involved and affected people. We shall consider each phase in turn, looking at the task to be accomplished during that phase, the tools provided by TSI[5] to realise that task, and the outcome or results expected from the phase.

## Creativity Phase

The task during the creativity phase is to use systems metaphors[6] as organising structures to help managers think creatively about their enterprises[7]. The sort of questions asked are: 'Which metaphors reflect current thinking about organisational strategies, structures, and control and information systems (including past, present and future concerns)?', 'Which alternative metaphors might capture better what more desirably could be achieved with this organisation?,' 'Which metaphors make sense of this organisation's difficulties and concerns?.' Of course, metaphor may

---

5    These are described very briefly in this paper.

6    The five metaphors used as a "top level set" reflect the development of organisation and management theory, and the systems and management sciences. They are complex networks organised as machine, organic, brain (or neuro-cybernetic, relating to Beer's viable system model), culture or political systems. Obviously there will be many affiliated metaphor. Each can be understood as flavoured versions of a general conception of "system". Explicit use of metaphor simplifies the task for the manager or problem solver because it makes sophisticated use of some understandings that each one of us already possesses. What we suggest is that loosely organised ideas relating to scientific knowledge are already there in our world. We regularly refer to organisations, societies etc., as operating like machines, or as teams, as evolving like species, as learning like brains, as acting as one or many cultures, or even as imprisoning our existence. At times, each of these, and other perceptions, represent a meaningful way of considering a situation. All we suggest with the metaphors is to step back and to recognise a pattern underlying these thoughts.

7    Morgan (1986) inspired this part of TSI.

46

be mixed, for example, giving rise to coerced-machines or brain-cultures.

The tools provided by TSI to assist this process evidently are systems metaphors. Different metaphors focus on different aspects of an organisation's functioning. Some concentrate on organisational structure, others highlight human and political aspects of an organisation. The main aspects of organisation highlighted, and those aspects neglected, by each metaphor will be disclosed in order to enhance discussion and debate.

The outcome (what is expected to emerge) from the creativity phase is a "dominant" metaphor which highlights the main interests and concerns and can become the basis for choice of an appropriate methodology. There may be other metaphors which it is also sensible to pursue into the next phase. The relative position of dominant and these "dependent" metaphors may indeed be altered by later work. If all the metaphors reveal serious problems then the organisation is obviously in a crisis state.

## Choice Phase

The task during the "choice" phase is to choose an appropriate systems-based intervention methodology (or set of methodologies) to suit particular characteristics of the organisation's situation as revealed by the examination conducted in the creativity phase.

The tools provided by TSI to help with this stage are guide-lines of what has been labelled a "system of systems methodologies." It is an idealised 2D framework relating the relationship between the participants (unitary, pluralist or coercive) to degrees of complexity (simple or complex). It helps to get inside methodologies and to assess the fundamental assumptions that they hold about the nature of social reality. Knowledge of the underlying metaphors employed by systems methodologies is derived from it. Systems methodologies may be linked directly to systems metaphors or the link may be secured through the system of systems methodologies.

The most probable outcome of the "choice phase is selection of a "dominant" methodology, to be tempered in use by the imperatives highlighted by "dependent" methodologies.

be mixed, for example, given rise to control machines or brain cultures.

The tools provided by TSI to assist this process evidently include systems metaphors. Different metaphors focus on different aspects of an organization's functioning. Some concentrate on communications structures, others highlight human and political aspects of an organization. The main aspects of organization highlighted and those aspects neglected, by each metaphor will be discussed in order to enhance discussion and debate.

The participants are expected to emerge from the creativity phase, having "metaphorically" highlighted the main concerns and issues, and with a favourite or basic root choice of metaphors which captures their predominant concerns, which in turn will mould the way they see the organization.

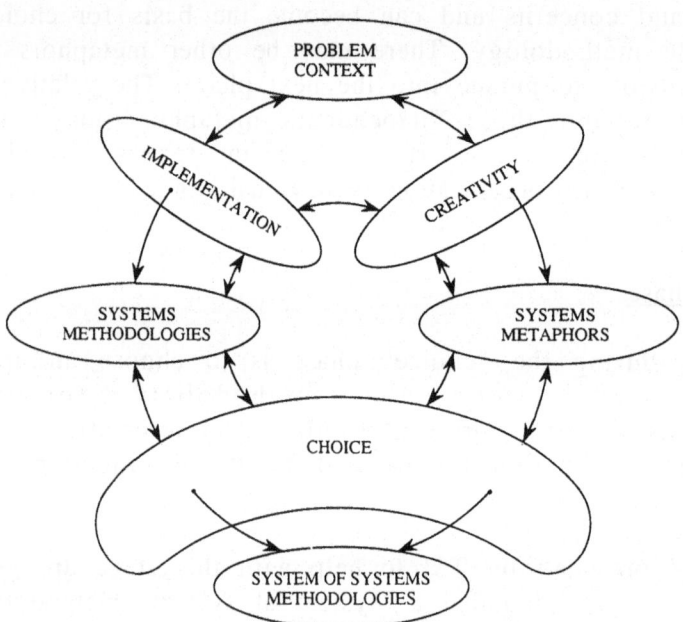

Figure 1. The process of Total Systems Intervention

The choice phase is underpinned by the system of systems methodologies. Systems methodologies may be linked, through systems metaphors, or the link may be severed, through the systems-based relationships.

The most important outcome of the choice phase is the selection of a dominant methodology to be instrumental in use, by the participants highlighting appropriate systems methodologies.

## Implementation  Phase

The task during the implementation phase is to employ a particular systems methodology (methodologies) to translate the dominant vision of the organisation, its structure, and the general orientation adopted to concerns and problems, into specific proposals for change.

The tools provided by TSI are the specific systems methodologies used according to the logic of TSI. The dominant methodology operationalises the vision of the organisation contained in the dominant metaphor. The logic of TSI demands, however, that consideration continues to be given to the imperatives of other methodologies. For example, the key difficulties in an organisation suffering from structural collapse may be best highlighted using the metaphors of "organism" and "brain" but the cultural metaphor might also appear illuminating, if in a subordinate way given the immediate crisis. In these circumstances a cybernetic methodology would be chosen to guide the intervention, but perhaps tempered by some ideas from a soft methodology.

The outcome of the implementation stage is coordinated change made in those aspects of the organisation currently most vital for its effective and efficient functioning.

TSI is an iterative meta-methodology. It asks, during each phase, that continual reference be made, back and forth, to the likely conclusion of other phases. This idea is captured in the circular representation shown in Figure 1.

We have now introduced DB and TSI and are therefore in a position to recount a meaningful story of how TSI was employed to intervene in DB.

## CONSULTING USING TOTAL SYSTEMS INTERVENTION IN DIAGNOSTIC BIOTECHNOLOGY (PTE) LTD

### TSI: Task, Tool and Outcome

Now, let us pick up the story line again. The initial visit to the company in September 1990 went ahead and a report of the first phase discussions was presented to Mr. Lim, and later to many other staff. The original idea to hold a two day seminar to educate the staff about TSI was reworked following the September findings. A participatory workshop was set up instead. The aim of

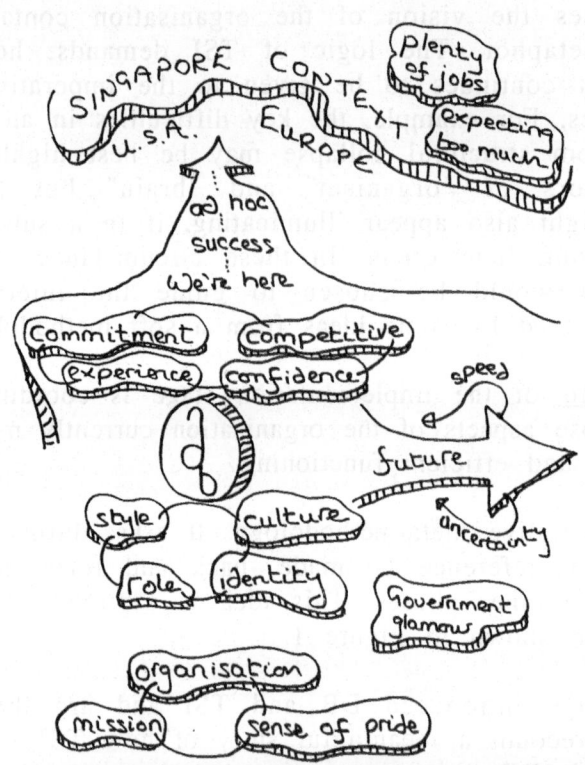

Figure 2. An expression of the main issues debated by
40 participants of a workshop at Diagnostic
Biotechnology (Pte) Ltd.

the workshop was to bring together personnel from all functions and many layers of the organisation and to enable a "technology transfer" to happen. The task was to delve into the organisation's main difficulties. The tool was TSI. The outcome that we expected would be recommendations for new ways forward for DB.

## Creativity

The two day session was extended to a three day one to allow the participants to be introduced to systems and management ideas. 40 employees attended. The first day was a difficult one for all involved. Most attendees were specialist natural scientists who had no experience of management concepts. Participation was consequently negligible, despite plenty of opportunity being given. A significant cultural difference which I had not considered explained the initial lack of success. As many participants later informed me, the Chinese way urges never to lose face in front of others. Since everyone started with a complete lack of management knowledge, and not one of them could admit this, day 1 was doomed from the start. On the following days however things came alive.

On day 2 TSI was employed in earnest. A very frank and open discussion about the main issues to which all felt confident to contribute[8] removed the barriers. In summary people felt: (a) a lack of mission; (b) no sense of pride, commitment, confidence, role or identity of individuals; (c) that the company is young and so are many of the staff, giving rise to a lack of overall experience; (d) control, communication and organisation are weak; (e) DB is a success story but 'Is it competitive?'; and (f) 'The future seems to hit us before we have has a chance to prepare for it.' Findings are also expressed in Figure 2.

Whilst this discussion was happening, attributes of the main metaphors used in TSI were "slipped in" and debated in the face of the issues[9]. There were two dominant concerns; that the company had undergone substantial expansion in 1989 but had not put in place proper communication, coordination and control

---

8    That is, except one senior manager who was very suspicious of the whole event and found ways of not being able to participate. Politics at play!

9    TSI does not have to be used "up front". It is often more effective when employed in the background.

procedures; and that the company was young and had not developed a corporate culture. The first concern was convincingly captured by ideas of viability portrayed by the brain metaphor, whilst the second fell in place with ideas of quality which is normally assumed to resemble the culture metaphor[10]. Political and machine metaphor were held in support. Participants supposed that viability and quality implied each other.


Choice

Total Quality Management[11] was introduced to the participants. We employ it as a systemic approach to problem solving acknowledging that it assumes all problems stem from a lack of quality. It was supposed that the ideas of this management belief would be highly relevant in the face DB's current difficulties. Quality was especially fit for the task because South East Asia is the home of quality management, the ideas being strongly promoted in Singapore by its Government, at that time led by Lee Kuan Yew.

There was one main concern at that stage. Traditionally quality is implemented top-down using the organisational hierarchical tree. This is hopeless because it merely implements quality according to a formal power structure. A coerced-culture if you like. We therefore began to conceive of the organisation in different ways. Mr. Lim had already been working on a highly participatory auditing and decision-making process. He had created a series of committees which would be used as "think-tanks" about the organisation. Each committee focussed on a particular area of the organisation. The committees were; Manpower Planning, Corporate Strategy, Customer Service, Control Systems, Research and Development, and Technical. Every member of one committee would also participate in one other committee. Thus each

---

10    Actually, the concept of quality can be thoroughly explored itself through the five main metaphors. This reveals that many assumed characteristics of quality management are only cloudy surfaces which obscure the real mechanical-coercive-cultural nature of the philosophy in most applications.

11    A brief definition follows. Quality is the meeting of agreed customers' requirements at lowest cost first time, every time. Customers are all those people who we supply products, resources or information to, and may be internal or external to the organisation. Management simply means that there is a management responsibility. And total means that all people across all functions and at every level must be involved.

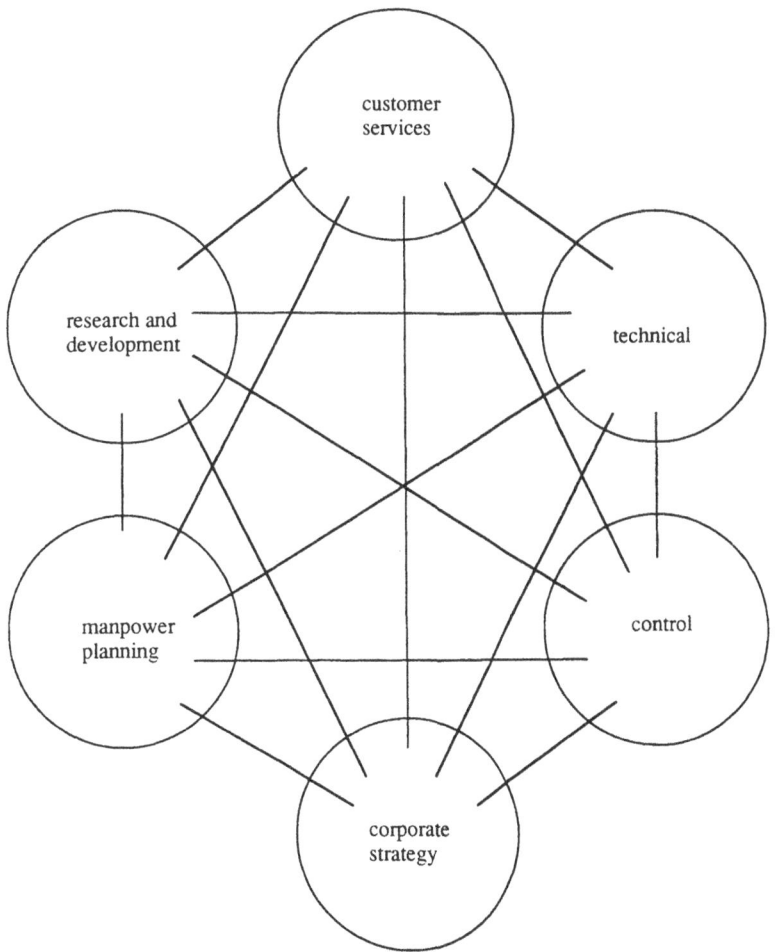

Figure 3. The participatory committee structure of Diagnostic
Biotechnology (Pte) Ltd., used to implement the first
stages of Total Quality Management.

committee was capable of looking in a focussed way at its main area of responsibility, and at the same time would see the organisation as a whole. My own task was to ensure that Lim's innovative arrangement covered all aspects of viability and so I worked on these committees, tightening up their outlook where possible with an on going viable systems diagnosis. The arrangement of the committees is shown in Figure 3. This proved to be an ideal way of implementing Total Quality Management, employing a top-down approach.

## Implementation

Total Quality Management was initiated in the following way. First of all a company quality mission was set by all 40 participants. This took a whole morning, but the value of the exercise was in creating a company image to which all participants felt they could relate and had ownership of. Also, many staff noted later that this was the first time they had grasped a genuine understanding of DB as a whole. Following this, committees were reformed with specialist and non-specialist participants in each group. Committees were then asked to define their mission within the spirit of the company mission, enabling the new identity to filter down through the company. A viable perspective was extremely useful here because it encourages identity, or mission, to spill down recursive levels. Groups then identified their customers (both internal and external), customers' requirements (as far as possible at this stage), and were asked to identify weaknesses in satisfying those requirements. With this information each committee was asked to define a number of projects which aimed to improve customer service. Some means of measuring the effectiveness of each project was requested. The concept of a vital few and useful many projects was adhered to. On completion of project definition committees were brought back together to present and share their findings. Main themes of projects were analysed and discussed. Let us now consider a few outputs from the application of the above quality seeking methodology.

The following mission for DB was agreed upon:

> *A quality biotechnology company excelling in manufacturing and distribution of diagnostic products across the world, striving to facilitate better health standards by continuously improving the product range through research and development.*

Ideally all 6 committees would have been reformed, but because of shortage of people we were only able to bring together 4. These were Manpower Planning, Technical, Research and Development, and Customer Services. Let us briefly look at the activities of each one.

Manpower Planning set the following committee mission:

> *To cultivate the skills of people who provide services and products to our customers*

The customers were identified internally as all DB staff; and externally as suppliers and users, distributors and agents. Weaknesses in meeting their needs were identified under two broad headings; attitudinal and technical. Attitudinal weaknesses included a lack of safety and quality consciousness, poor work intelligence and team commitment, no global outlook, closed mindedness and limited initiative. Technical weaknesses included a poor level of technical training and required skills to do jobs properly, and a limited general knowledge of techniques and products. The main project that was identified was a massive training program too large to document here. Performance measures were given. This project is now under way.

Technical set the following committee mission:

> *To enhance technical competence, coordination, and the implementation of all technical activities with a view to continuously improving the overall efficiency and to achieve customer satisfaction.*

The customers were identified internally as Research and Development, Production, Quality Control, Technical, Administration and Finance, and Marketing; and externally as end users, distributors and dealers. Weaknesses in meeting their needs were identified in technology transfer to production. The main project that was defined naturally aimed to tackle this "lack of reliability." The committee focused on the following shortfall areas: documentation on the stages of development related to the product, i.e., product specification, techniques involved, difficulties encountered, technical data and evaluation, and standard operating procedures; monitoring and reviewing the documentation and implementation process; involvement of Quality Control, Technical and Production during the final stage of research and development; introduction of process design into production; overall review of data; change when necessary; and final product evaluation. Project success was to be measured by product yield and customer feedback. This project is now under way.

Research and Development set the following committee mission:

> *To realise a commitment to achieve the company mission through improvement, innovation and diversification of product range by upgrading technology, forming strategic collaborations and supporting the smooth functioning of other departments.*

The customers were identified internally as all DB staff and externally as end users. The need internally was to provide customers with quality and friendly service, to continuously upgrade skills and ensure technical transfer; and externally, to develop quality products for accurate diagnosis and ease of use. Weaknesses in meeting their needs were identified as poor communication among colleagues, insufficient involvement of representatives from other functional areas in research and development seminars, no quality control in the latter phase of development of a product. The main project aimed to improve company wide participation and communication. For example, seminars and lectures were to be arranged to update staff on the latest technological and product developments. This project is now under way.

Customer Services set the following committee mission:

> *To provide and improve quality services to meet customers' needs.*

The customers were identified internally as all departments, the Board of Directors and shareholders; and externally as distributors, subsidiaries, end users and competitors. The main need internally was thought to be feedback; and externally, information on the company, the product range, the image, product information, response to complaints, shipping information, education about the product and technology. Weaknesses in meeting those needs were identified as poor communication and a lack of continuous updating on planning and control. A project was devised for the market information function to make the following improvements; greater teamwork and active participation, continuous review of procedures and proper documentation to help to keep abreast of market change, and to increase emphasis on follow-up. Information was to be gathered, categorised and circulated to all departments. The main measures were formulated in terms of the amount of useful information channelled into the organisation and adaptability achieved. This project is now under way.


## Achievements and Failures

In what ways did the first stages of intervention in DB tackle the main issues identified earlier? A major achievement was overcoming the lack of identity by formulating a company mission. The working out of the mission not only gave direction to DB, but had the added value of sealing employee commitment through mass participation in its construction. Setting committee missions also built into individuals an identity and a way of understanding their role in the whole organisation. This did help to develop pride, as did the sense of continuous quality improvement. DB as a success story became a "reality" for all rather than a probable myth that was being told along the corridors. Control, communication and organisation were being targeted by many of the defined projects and hence should improve and support preparation for the future.

The sessions were generally accepted as a break-through for the company by the participants, although in some ways the efforts remain bootless. For example: a senior member of staff stayed outside the intervention; the company had to keep functioning over the duration of the workshop and so very little contact could be made with "shop floor" staff; the amount of time available was hardly enough to "routinise" the new attitude and quality culture with the staff; the Managing Director has a powerful character and hold over the company which, despite his revolutionary management style, would probably continue to largely shape DB; etc.

These imperfections however did not represent a checkmate. The following ways forward were recommended and are now under discussion: (a) to plan and design management seminars and a management resource centre in DB; (b) to help to set up the committees with carefully worked out procedures, accountability, quality aims, customer identification and requirements etc., and thus to implement quality management throughout the company; (c) to work at least initially with the Managing Director to develop a clear understanding of how the committee structure, which promotes participation, can clearly complement the organisation of the various management functions.

We have now learned about Total Systems Intervention as a creative approach to problem solving, Diagnostic Biotechnology (Pte) Ltd., as a company suffering from the effects of its success, and have seen how TSI was employed in DB to manage the mess. Let us now draw out some points to conclude this article.

## CONCLUSION

The aim of this article is to inform readers about a creative approach to problem solving that we have called Total Systems Intervention (TSI). The aim has been met by presenting TSI in action in a biotechnology company that specialises in diagnostics. First we explored the business of Diagnostic Biotechnology (Pte) Ltd., (DB), and then we introduced TSI. Following this, consulting in DB employing TSI was discussed. Without a doubt, this report shows only one convoluted trajectory driven by TSI, of which there must be an infinite number. Respecting this never ending diversity requires us to introduce a diversity of ways of dealing with "it." This was evident in the case study discussed above. Overall, the details of the consultancy offer general lessons for consulting practice. Let us consider these.

Consultancy, or problem solving, is the management of messes. Our understanding of messes changes and therefore so should our management of "it." DB is a case at hand that shows how perceptions and needs change and how the methodologies that we choose for implementation may be appropriate in some contexts or at particular times but not others. For example, a "learning" methodology for company training was slowly and effectively replaced by intervention that employed both quality management methodology as dominant and viable systems diagnosis in support. Now I find it hard to conceive of this relationship between methodology changing in the DB consultancy. But only by continually putting into practice the logic and process of TSI will we be able to "guarantee" that the current or some other relevant and informed intervention is happening.

General principles of TSI have emerged from the DB and other consultancy practice. If nothing else, it is our belief that these should be adhered to by would be problem solvers. They are: (a) organisations are too complicated to understand using one management "model" and their problems too complex to tackle with the "quick fix"; (b) organisations, their strategies and the difficulties they face should be investigated using a range of systems metaphors; (c) systems metaphors which seem appropriate for high-lighting organisational strategies and problems can be linked to appropriate systems methodologies to guide intervention; (d) different systems metaphors and methodologies can be used in a complementary way to address different aspects of organisations and the difficulties they confront; (e) it is possible to appreciate the strengths and weaknesses of different systems methodologies and to relate each to organisational and business concerns; (f) TSI sets out a systemic cycle of inquiry with iteration back and forth between the three phases; (g) facilitators, clients and others are engaged at all stages of the TSI process. It is our argument, as already stated, that abiding by these principles will lead to relevant and informed intervention.

## REFERENCES

Flood, R.L. (1990). Liberating Systems Theory, Plenum, New York.

Flood, R.L., and Jackson, M.C. (1991a). Creative Problem Solving: Total Systems Intervention, Wiley, Chichester.

Flood, R.L., and Jackson, M.C. (1991b). Critical Systems Thinking: Directed Readings, Wiley, Chichester.

Fairtlough, G. (1990). Systems practice from the start: Some experiences in a biotechnology company. Systems Practice, 2(4), 397-412.

Jackson, M.C. (1991). Systems Methodology for the Management Sciences, Plenum, New York.

Morgan, G. (1986). Images of Organisation, Sage, Beverley Hills.

# FIVE COMMITMENTS OF CRITICAL SYSTEMS THINKING

Michael C. Jackson

Department of Management
Systems and Sciences
University of Hull
Hull, HU6 7RX

## INTRODUCTION

Critical systems thinking is a relative newcomer to the systems tradition of thought.  Yet it is now developing more quickly than any other part of systems thinking.  It is argued that, as it has evolved, critical systems thinking has taken on five commitments which distinguish it from other types of systems approach.  These five commitments are detailed below with comment on their origins and on the theoretical and practical work that has established them as pillars upon which critical systems thinking can be built and can be further refined.

## THE FIVE COMMITMENTS

Critical systems thinking embraces five major commitments. It seeks to demonstrate critical awareness; it shows social awareness; it is committed to the complementary and informed use of systems methodologies; to the complementary and informed development of all the different strands of systems thinking at the theoretical level; and it is dedicated to human emancipation.

### Critical Awareness

Critical awareness comes in two forms.  The first concerns understanding the strengths and weaknesses and the theoretical underpinnings of available systems methods, techniques and methodologies.  This is encouraged both at the level of particular methodologies and at the level of systems thinking as a whole.  Related to specific methodologies are the soft systems thinkers' assaults on hard systems thinking (Checkland, 1978; Ackoff, 1979; Churchman, 1979); Jackson (1982) and Mingers (1984) on soft systems thinking; Ulrich (1981), Jackson (1986, 1988a) and Flood and Jackson (1988) on cybernetics; and Jackson (1989) on strategic assumption surfacing and testing.  At the general level are Jackson's (1988b) review of systems methods for organizational analysis and design, Oliga's (1988) look at the methodological foundations of systems methodologies, Ulrich's (1988) programme for systems research, and Flood and Ulrich's (1990) examination of the epistemological bases of different systems approaches.

We can take as an example of an examination of a particular approach Jackson's (1982) critique of the ambitions of soft systems thinking as expressed in the work of Churchman, Ackoff and Checkland. It was argued that the assumptions made by these authors, about the nature of systems thinking and social systems, constrained the ability of their methodologies to intervene, in the manner intended, in many problem situations. Soft systems thinking had a quite limited domain of application. Using Burrell and Morgan's (1979) framework, it was shown that soft systems thinking is based upon interpretive assumptions. With Churchman, Ackoff and Checkland, systems thinking becomes much more 'subjective', the emphasis shifts from attempting to model systems 'out-there' in the world, towards using systems models to capture possible perceptions of the world. In Checkland's methodology, for example, systems models of possible 'human activity systems' are used to structure and enhance debate among stakeholders so that a consensus or accommodation about action to be taken can emerge. The recommendations of soft systems thinking are 'regulative' because no attempt is made to ensure that the conditions for 'genuine' debate are provided. The kind of open, participative debate which is essential for the success of the soft systems approach, and is the only justification for the results obtained, is impossible to obtain in problem situations where there is fundamental conflict between interest groups which have access to unequal power resources. Soft systems thinking either has to walk away from these problem situations or it has to fly in the face of its own philosophical principles and acquiesce in proposed changes emerging from limited debates characterised by distorted communication.

The other form of critical awareness involves closely examining the assumptions and values entering into actually existing systems designs or any proposals for a systems design. Critical systems thinking aims to provide the tools for enhancing this type of critical awareness as, for example, in Ulrich's (1983) 'critical systems heuristics'. Ulrich uses each of these three words in the sense given to them by Kant. To be critical one must reflect upon the presuppositions that enter into both the search for knowledge and rational action. A critical approach to systems design means planners making transparent to themselves and others the normative content of designs. All designs and proposed designs must be submitted to critical inspection and not presented scientistically as the only 'objective' possibility. Ulrich takes the systems idea in Kant to refer to the totality of the relevant conditions upon which theoretical or practical judgements depend. These include metaphysical, ethical, political and ideological aspects. In attempting to grasp the 'whole system' we are inevitably highly selective in the presuppositions we make. Ulrich follows Churchman in seeing Kant's systems ideas as an admonition to critically reflect on the inevitable lack of comprehensiveness and partiality of all systems designs. It is by reference to the whole systems concepts entering into these partial presuppositions that

critique becomes possible.  Finally, heuristics refers to a
process of uncovering 'objectivist' deceptions and of helping
planners and concerned participants to 'unfold' problems
through critical reflection.  It also signals that Ulrich does
not attempt to ground critical reflection theoretically, but
to provide a method by which presuppositions and their
inevitable partiality can be kept constantly under review.

## Social Awareness

Social awareness involves, as one of its aspects,
considering the organizational and societal 'climate' which
determines the popularity, or otherwise, of particular systems
approaches at particular times.  For example, it was
inconceivable that soft systems thinking could ever flourish
in Eastern European countries dominated by the bureaucratic,
'rational' dictates of the one party system.  With the move
towards free-market capitalism and political pluralism,
however, the circumstances which allowed hard and cybernetic
approaches to 'succeed' are changing, and softer methodologies
are likely to be more useful and used.

Contributions to this aspect of social awareness have fed
into critical systems thinking from 'critical management
science'.  Hales' (1974) and Rosenhead and Thunhurst's (1982)
analyses of the evolution of management science in terms of
the historical and material development of the capitalist mode
of production are particularly worthy of mention.

From within critical systems thinking itself, Flood and
Gregory (1989; Flood 1990a, 1990b) have argued for a
'genealogical' perspective on why certain systems theories and
methodologies become popular and others fall out of favour.
They set out four ideas on the nature of the history and
progress of knowledge - linear sequential, structuralism,
world-viewism, and genealogy, and relate these to accounts of
the development of systems thinking.  The linear sequential
model sees knowledge building chronologically and
cumulatively.  Structuralism represents 'deep' processes as
being at work in history, and uses the 'scientific' approach
to unearth these and build knowledge of them.  World-viewism
rejects the unilinear perspective and accepts the existence of
contrasting and even contradictory knowledges, although there
may be periods of settled or 'normal' science (Kuhn, 1970).
Genealogy, deriving from Foucault's writings, puts emphasis on
the effect power at the microlevel can have on the formation
and development of knowledges.  Localized power relations
outside of discourse can effect the success or lead to the
subjugation of knowledges.  In Flood and Gregory's opinion the
first three ideas on the history of knowledge are well
represented in accounts of progress in systems thinking, but
the genealogical view has not yet been exploited.  The result
has been a neglect of the effect of power at the micro-level
on the way the subject has unfolded.

Following Foucault's logic, Flood (1990a) has argued that
the project of 'liberating systems theory' must include the
emancipation of suppressed knowledges in systems theory
itself.  As an example of a subjugated knowledge in systems
thinking, Flood and Robinson (1989) provide General System
Theory (GST).  GST has lost favour in the systems movement but
the reasons can hardly be entirely scientific, they argue,

because the criticisms levelled against GST, and which have
become generally accepted, simply do not stand up to close
examination. Obviously a properly conducted genealogical
study of the systems field could contribute significantly to
the 'social awareness' of critical systems thinking.

Another side of social awareness is giving full
consideration to the social consequences of use of different
systems methodologies. For example, the choice of a hard or
cybernetic methodology often implies that one goal or
objective is being privileged at the expense of other
possibilities. Is this goal general to all organizational
stakeholders, or is it simply that of the most powerful?
Similarly, the use of soft systems methodologies, which are
dependent upon open and free debate to justify their results,
might have deleterious social consequences if the conditions
for such debate are absent. This aspect of social awareness
has been central to Jackson's research (1982, 1985a, 1988a, b,
1989), has led Oliga (1989) to seek to unmask the ideological
foundations of the different systems approaches, and provides
the rationale for Ulrich's (1983) demand that the systems
rationality of planners always be exposed to the social
rationality of the affected.

Complementarism at the Level of Methodology

This commitment originates from Jackson and Keys' (1984)
'system of systems methodologies', since taken in different
directions by the two authors (Keys, 1988; Jackson, 1990). It
was made explicit in Jackson's (1987a) argument for
'pluralism' in management science; an argument elaborated on
by Flood (1989).

The 'system of systems methodologies' was a classification
of systems approaches which allowed for their complementary
and informed use. It attempted to reveal what was being taken
for granted in terms of 'systems' and 'decision-makers' (later
'participants') in using each type of systems methodology.
This, it was felt, would enable potential users of systems
methodologies to assess their relative strengths and
weaknesses for the task at hand and to be fully aware of the
consequences of employing each approach. The dimensions of
'systems' and 'participants' were chosen because they seemed
to bring the greatest insight to the matter of distinguishing
systems approaches. And, conveniently for the argument,
systems methodologies did seem to make up a 'system of systems
methodologies' when allocated according to the matrix of
problem-contexts produced by combining these dimensions.

The 'system of systems methodologies' opened up a new
perspective on the development of systems thinking and
management science. Previously it had seemed as if these
disciplines were undergoing a 'Kuhnian crisis' as hard systems
thinking encountered increasing anomalies and was challenged
by other approaches (Dando and Bennett, 1981). By questioning
one of the underlying assumptions of this analysis – that
systems thinking has a well defined and somewhat uniform
subject matter – an alternative future was opened up. Instead
of seeing different strands of systems thinking as competing
for exactly the same area of concern, alternative approaches
can be presented as being appropriate to the different types
of situation in which management scientists are required to

act.  If this perspective is adopted, then the diversity of approaches heralds not a crisis but increased competence and effectiveness in a variety of problem situations.  Thus, the 'system of systems methodologies' represented the relationship between different systems methodologies as being complementary in nature and provided informed guidance about the assumptions that were necessarily being made in using any one systems approach.

The 'system of systems methodologies' also prepared the ground for an appropriate welcome to be given to Ulrich's (1983) 'emancipatory systems thinking'.  As was stated in the original 1984 article, the 'decision-makers' dimension could be extended to embrace coercive as well as unitary and pluralist contexts (an extension later made by Jackson, 1987b, 1988b, c).  At the time the authors did not know of any systems methodologies which assumed and acted as though problem-contexts might be coercive.  From the critical point of view this was obviously a weakness in the capabilities of systems thinking and made the construction of such approaches imperative.  When Ulrich's 'critical systems heuristics' became known in the UK it was like the discovery of an element which filled a gap in the periodic table.  Critical systems heuristics was arguably capable, where soft systems thinking was not, of providing guidelines for action in certain kinds of coercive situation.  It enabled systems designs or proposed designs to be carefully interrogated as to their 'partiality' and set down criteria for 'genuine' debates between stakeholders, which had to include both those involved in systems design and those affected by the designs but not involved.

Recently, Flood and Jackson (1991) have constructed a methodology (or perhaps meta-methodology), called Total Systems Intervention (TSI), describing procedures which critical systems practitioners can follow in trying to translate their complementarist thinking into action in the real-world

Complementarism at the Level of Theory

The commitment to complementarism at the methodological level requires an equal commitment to the complementary and informed development of all varieties of the systems approach at the theoretical level.  Different strands of the systems movement express different rationalities stemming from alternative theoretical positions.  These alternative positions must be respected, and the different theoretical underpinnings and the methodologies to which they give rise developed in partnership.  Further, the claim of any one theoretical rationality, whether functionalist, structuralist, interpretive or emancipatory, to absorb all others, must be resisted.  This should not lead the systems community to fragment.  The existence of a range of systems approaches, each driven by a different theoretical position, can be seen as a strength rather than a weakness of the systems movement.  All that is required is the guidance offered by complementarism, so that each systems approach is put to work only on problem types for which its theoretical rationality is appropriate.

Critical systems thinking's adherence to complementarism at the theoretical level rests upon its acceptance of Habermas' argument for human-species-dependent knowledge constitutive interests. According to Habermas (1972, 1974) there are two fundamental conditions underpinning the socio-cultural form of life of the human species. These he calls 'work' and 'interaction'. Work enables human beings to achieve goals and to bring about material well-being through social labour. The importance of work leads us to have a 'technical interest' in the prediction and control of natural and social affairs. Interaction enables human beings to secure and expand the possibilities for mutual understanding among those involved in social systems. The importance of interaction leads us to have a 'practical interest' in the progress of intersubjective communication. Disagreements between different groups can be just as much a threat to the reproduction of the socio-cultural form of life as a failure to predict and control natural and social processes.

Work and interaction have for Habermas pre-eminent anthropological status, but the analysis of 'power' and the way it is exercised is also important, he believes, if we are to understand past and present social arrangements. The exercise of power in the social process can prevent the open and free discussion necessary for the success of interaction. Human beings have therefore an 'emancipatory interest' in freeing themselves from constraints imposed by power relations and in learning, through a process of genuine participatory democracy, involving discursive will-formation, to control their own destiny.

There is a remarkable convergence in the way that three critical systems thinkers have used Habermas' ideas in developing their own approaches. Jackson (1985b, 1988b) has linked the technical interest to the concern systems methodologies show for predicting and controlling the systems with which they deal; and the practical and emancipatory interests with the concern to manage pluralism and coercion. It follows that the two dimensions of the 'system of systems methodologies' can be justified from Habermas' work and the different systems methodologies represented as serving, in a complementary way, different human species imperatives. Oliga (1986, 1988) argues that Habermas' interest constitution theory is an important improvement over the interparadigmatic incommensurability position of Burrell and Morgan, since

"... whereas Burrell and Morgan merely explain the different paradigmatic categories, Habermas explains and reconciles the interest categories in terms of their being individually necessary (although insufficient) as human species, universal and invariant (ontological) forms of activity - namely labour, human interaction, and authority relations." (Oliga, 1988).

He then goes on to conduct his own survey of how well the technical, practical and emancipatory interests are served by systems methodologies. Ulrich (1988) similarly uses Habermas' taxonomy of types of action - instrumental, strategic and communicative - to specify three <u>complementary</u> levels of systems practice, roughly parallel to the requirements of operational (or tactical) strategic and normative planning. Different systems approaches can then be allocated as appropriate to service operational, strategic and normative systems management levels.

One consequence of firmly basing complementarism at the theoretical level upon Habermas' work is that it clearly opens up critical systems thinking to a post-modernist critique. Since Habermas is seen by post-modernists as guilty of most of the major sins of modernism, it follows that critical systems thinking can be accused of exactly the same crimes. In Liberating Systems Theory (1990a), Flood has attempted, from a critical systems perspective, to come to terms with post-modernism, at least as it is expressed in the works of Foucault.

Flood argues that, despite their differences, Habermas and Foucault can be seen as contributing to a position opposed to theoretical isolationism (especially of the technocratic kind) and in favour of theoretical pluralism. Habermas provides a basis for accepting three types of rationality, for promoting the development of each, and for criticising the limitations of each. However, he is naive in the way he conceptualises power, believing that power can be made to follow knowledge; to issue forth from the force of the better argument. Foucault sees power as immanent in all aspects of social life, and as intimately linked to knowledge so that, for example, it determines what the better argument is. Various localized forces (which cannot be grasped through some grand narrative such as Habermas' social theory) decide what discourses should be dominant and what knowledges subjugated. Flood argues, therefore, that in order to achieve the maximum diversity in systems approaches, so that the fullest support can be provided to Habermas' human interests, it is necessary first to follow Foucault's method to reveal subjugated knowledges. Foucault provides the understanding and the means necessary to 'liberate' suppressed knowledges so that a diversity of approaches is achieved. These can then be subject to 'critique' according to the principles set out by Habermas for assessing the theoretical and methodological legitimacies and limitations of different knowledges. An 'adequate epistemology for systems practice' (Flood and Ulrich, 1990, Flood, 1990a) can be established on essentially Habermasian foundations, but with support from Foucault's conceptualization of power.

## Human Emancipation

Critical systems thinking is dedicated to human emancipation and seeks to achieve for all individuals the maximum development of their potential. This is to be achieved by raising the quality of work and life in the organizations and societies in which they participate. Critical systems thinking recognises its overall emancipatory responsibility and seeks to fulfil this by adequately servicing, with appropriate systems methodologies, each of Habermas' human interests. Methodologies which serve the technical interest assist material well-being by improving the productive potential and the steering capacities of social systems. Methodologies which serve the practical interest aim to promote and expand mutual understanding among the individuals and groups participating in social systems.

Methodologies serving the emancipatory interest protect the domain of the practical interest from inroads by technical reason and ensure the proper operation of the practical interest by denouncing situations where the exercise of power, or other causes of distorted communication, are preventing the open and free discussion necessary for the success of interaction. All human beings have a technical, a practical and an emancipatory interest in the functioning of organizations and society. So a systems perspective which can support all these various interests has an important role to play in human well-being and emancipation. And this is exactly what critical systems thinking wants to achieve. It wants to put hard and cybernetic methodologies to work to support the technical interest, soft methodologies to work to assist the practical interest, and to employ emancipatory methodologies to aid the emancipatory interest.

At the same time as acknowledging its overall emancipatory responsibilities, critical systems thinking perceives a special need, because of previous neglect, to nurture the development of 'emancipatory systems thinking'. It is noticeable that those involved in the creation of critical systems thinking have also been influential in seeking to develop emancipatory systems approaches. Jackson, Oliga and Ulrich have all sought to facilitate the emergence of new, emancipatory methodologies to tackle problem situations where the operation of power prevents the proper use of soft systems thinking. In theory this means encouraging the development and use of specifically emancipatory systems methodologies suitable for 'coercive' contexts (Ulrich, 1983; Jackson, 1985a; Oliga, 1990). In practice it includes supporting initiatives such as Community Operational Research (Rosenhead, 1986; Jackson, 1988c).

CONCLUSION

By about 1990, therefore, critical systems thinking had begun to take on some sort of form, at least in this author's mind. It seemed to be built upon the five pillars of critical awareness, social awareness, complementarism at the methodological level, complementarism at the theoretical level and dedication to human emancipation. As well as helping to consolidate existing work, this codification of critical systems thinkings' commitments helps us to separate it out from other near relations such as 'interpretive systemology' and 'emancipatory systems thinking'.

Interpretive systemology (Fuenmayor, 1991) aims to promote critical awareness in both of the senses discussed in the above by providing an interpretive systems theory. It is precisely because it lacks such a theory, Fuenmayor argues, that SSM is non-critical and lends itself to instrumental and regulative usage. However, because interpretive systemology shows no commitment to the other four features of critical systems thinking it should not be regarded as part of that movement.

Emancipatory systems approaches (such as Ulrich's critical systems heuristics) embrace, at least in part, critical awareness, social awareness and a dedication to human emancipation, but they do not concern themselves with complementarism. Emancipatory systems thinking is, therefore, narrower than critical systems thinking. It has concentrated on providing methodologies which through critique, and the engineering of particular social arrangements, can assist with the emancipation of human actors; putting them more in control of their own destiny. The domain of effective application of emancipatory methodologies is 'coercive systems'. But not all problem situations are usefully seen as set in coercive contexts. Some are best seen as unitary or pluralist. Emancipatory systems thinking, therefore, just like the hard, cybernetic and soft approaches, possesses a limited domain for which it is the most appropriate approach. Critical systems thinking, as we now understand, is about putting all the different systems approaches to work, according to their strengths and weaknesses, and the social conditions prevailing, in the service of a more general emancipatory project.

The work that has gone into developing critical systems thinking this far is considerable. But the challenges which remain are immense. The invitation is open to all systems thinkers to help us to meet those challenges.

## REFERENCES

Ackoff, R. L., 1979, The future of operational research is past, J. Opl Res. Soc., 30:93.

Burrell, G., and Morgan, G., 1979, Sociological Paradigms and Organizational Analysis, Heinemann, London.

Checkland, P. B., 1978, The origins and nature of 'hard' systems thinking, J. Appl. Sys. Anal., 5(2):99.

Churchman, C.W., 1979, Paradise regained: A hope for the future of systems design education, in: Education in Systems Science (B.A. Bayraktar, H. Muller-Merbach, J. E. Roberts, and M. G. Simpson, eds.), Taylor and Francis, London, pp. 17-22.

Dando, M. R., and Bennett, P. G., 1981, A Kuhnian crisis in management science?, J. Opl Res. Soc., 32:91.

Flood, R. L., 1989, Six scenarios for the future of systems 'problem-solving', Sys. Pract., 2:75.

Flood, R. L., 1990a, Liberating Systems Theory, Plenum, New York.

Flood, R. L., 1990b, Liberating systems theory: Toward critical systems thinking, Human Relations, 43:49.

Flood, R. L., and Gregory, W., 1989, Systems: past, present and future, in: Systems Prospects, (R. L. Flood, M. C. Jackson, and P. Keys, eds.), Plenum, New York, pp. 55-60.

Flood, R. L., and Jackson, M. C., 1988, Cybernetics and organizatin theory: A critical review, Cybernetics and Systems, 19:13.

Flood, R. L., and Jackson, M. C., 1991, Creative Problem Solving: Total Systems Invervention, Wiley, Chichester.

Flood, R. L., and Robinson, S. A., 1989, Whatever happened to general systems theory?, in: Systems Prospects (R. L. Flood, M. C. Jackson, and P. Keys, eds.), Plenum, New York, pp. 61-66.

Flood, R. L., and Ulrich, W., 1990, Testament to conversations on critical systems thinking between two systems practitioners, Sys. Pract., 3:7.

Fuenmayor, R. L., 1991, Between systems thinking and systems practice, in: Critical Systems Thinking: Directed Readings (R. L. Flood, and M. C. Jackson, eds.), Wiley, Chichester.

Habermas, J., 1972, Knowledge and Human Interests, Heinemann, London.

Habermas, J., 1974, Theory and Practice, Heinemann, London.

Hales, M., 1974, Management science and the 'second industrial revolution', Radical Science Journal, 1:5.

Jackson, M. C., 1982, The nature of soft systems thinking: The work of Churchman, Ackoff and Checkland, J. Appl. Sys. Anal., 9:17.

Jackson, M. C., 1985a, Social systems theory and practice: The need for a critical approach, Int. J. of Gen. Sys., 10:135.

Jackson, M. C., 1985b, Systems inquiring competence and organizational analysis, in: Proceedings of the 1985 Meeting of the SGSR, SGSR, Louisville, pp. 522-530.

Jackson, M. C., 1986, The cybernetic model of the organization: An assessment, in Cybernetics and Systems '86 (R. Trappl ed.), D. Reidel, Dordrecht, pp. 189-196.

Jackson, M. C., 1987a, Present positions and future prospects in management science, Omega, 15:455.

Jackson, M. C., 1987b, New directions in management science, in: New Directions in Management Science (M. C. Jackson, and P. Keys, eds.), Gower, Aldershot, pp. 133-164.

Jackson, M. C., 1988a, An appreciation of Stafford Beer's 'viable system' viewpoint on managerial practice, J. Mgt. Stud., 25:557.

Jackson, M. C., 1988b, Systems methods for organizational analysis and design, Sys. Res., 5:201.

Jackson, M. C., 1988c, Some methodologies for Community OR, J. Opl Res. Soc., 39:715.

Jackson, M. C., 1989, Assumptional analysis: An elucidation and appraisal for systems practitioners, Sys. Prac. 2:11.

Jackson, M. C., 1990, Beyond a system of systems methodologies, J. Opl Res. Soc., 41:657.

Jackson, M. C., and Keys, P., 1984, Towards a system of systems methodologies, J. Opl Res. Soc., 35:473

Keys, P., 1988, A methodology for methodology choice, Sys. Res., 5:65.

Kuhn, T., 1970, The Structure of Scientific Revolutions, 2nd ed., University of Chicago Press, Chicago.

Mingers, J. C., 1984, Subjectivism and soft systems methodology - a critique, <u>J. Appl. Sys. Anal</u>., 11:85.

Oliga, J. C., 1986, Methodology in systems research: The need for a self-reflective commitment, in: <u>Mental Images, Values and Reality</u>, (J. A. Dillon, Jr., ed.), SGSR, Louisville, pp. B11-31.

Oliga, J. C., 1988, Methodological foundations of systems methodologies, <u>Sys. Pract</u>., 1:87.

Oliga, J. C., 1989, Ideology and Systems Emanciption, paper for the 33rd Annual Meeting of the ISSS, Edinburgh, Scotland.

Oliga, J. C., 1990, Power-ideology matrix in social systems control, <u>Sys. Pract</u>., 3:31.

Rosenhead, J., 1986, Custom and Practice, <u>J. Opl Res. Soc</u>., 37:335.

Rosenhead, J., and Thunhurst, C., 1982, A materialist analysis of operational research, <u>J. Opl Res. Soc</u>., 33:11.

Ulrich, W., 1981, A critique of pure cybernetic reason: The Chilean experience with cybernetics, <u>J. Appl. Sys. Anal</u>., 8:33.

Ulrich, W., 1983, <u>Critical Heuristics of Social Planning: A New Approach to Practical Philosophy</u>, Haupt, Bern.

Ulrich, W., 1988, Systems thinking, systems practice, and practical philosophy: A program of research, <u>Sys. Pract</u>., 1:137.

# A SCENARIO-BASED APPROACH TO STRATEGIC INFORMATION SYSTEMS PLANNING

R. D. Galliers

Warwick Business School
University of Warwick
Coventry CV4 7AL, England

## INTRODUCTION

It is becoming generally accepted that a key to successful information systems planning lies in the close linkage of the information systems plan with that of the business, whether this is at the corporate or business unit level. While there is general agreement that there *should* be a strong linkage between information systems planning and business planning, it is still the case that many organisations' business and information systems plans are only tenuously linked, at best.

This linkage is improved if:
- key stakeholders are involved in, and are committed to, the formulation and implementation of the plan;
- the process of strategy formulation, implementation and review is integrated into on-going management activities;
- there is a senior management 'champion' and/or 'sponsor', who is prepared to take responsibility to ensure that this process is taken seriously;
- there is a partnership based on mutual respect and trust between information systems staff and their business colleagues, and
- the information systems function is organised and staffed in such a way that it 'fits' with that of the organisation as a whole.

There is nothing very new in the above, but it is still the case that organisations find difficulty in achieving success in their information systems planning. This is partly due to a lack of awareness by managers and their information systems colleagues alike: the former happy in the mistaken belief that information technology can be left to the technologists, and the latter happier to have information systems planning more concerned with technology than the business itself and/or the people who run it.

More important, however, is the difficulty that managers and information systems professionals alike have in determining the key information requirements and flows throughout the organisation. This is so, not only from the perspective of what is currently required, but also what is likely to be required in the future ... a problem exacerbated by business environments that are characterised by constant change; the global nature of much of today's business activity and competition, and a growing focus on information requirements and flows outside organisational boundaries (with a view to improving supplier/customer relationships and the establishment of electronic communications between collaborating organisations, for example).

This paper focusses on a means by which managers can identify the changing nature of their organisation's information requirement, while at the same time improving mutual understanding about the confusing and very different worlds of business, information and information technology. The approach described is based on an extension of soft systems methodology (Checkland, 1972,1981), first articulated by Wilson (1980). The approach has been further refined in many applications throughout the 1980s. It has been found useful, not only in clarifying current and future information requirements and flows, but also in forging a closer relationship between business and information systems strategic thinking. As a result, a closer relationship (and a much improved understanding) between business unit management and information systems management has ensued.

## ON INFORMATION REQUIREMENTS AND FLOWS

Before turning to the approach that is being advocated to determine information requirements, it is necessary first to consider the subject of information itself. This is because the degree of difficulty associated with developing *information* systems (*ie.*, systems that *inform* the recipient) is considerably greater than that associated with developing *data processing* systems (*ie.*, systems that automate an operational task). The latter may be accomplished by replicating observable actions (Yadav, 1983), while the former requires considerable awareness of the context in which information may be required and the manner in which it is likely to be interpreted to enable a required activity or decision to be made, *viz.*:

> ... information is that collection of data, which, when presented in a particular manner and at an appropriate time, improves the knowledge of the person receiving it in such a way that he/she is better able to undertake a [required] activity or make a [required] decision (Galliers, 1987a, p.4).

In other words, information may be understood as being both *enabling* and *contextual*, while data is context-free and simply the raw material from which information (meaning) may be *attributed* (*ibid.*, p.4; Checkland & Scholes, 1990, pp.54-56).

> From these considerations ... two consequences flow. Firstly, the boundary of an [information system] ... will always have to include the attribution of meaning ... [and] will consist of both data manipulation, which machines do, and the transformation of data into information [which humans do] ... Secondly, designing an [information system] will require explicit attention to the purposeful action which [it] serves... (*ibid.*, p.55)

It follows that a reasonable approach to information systems planning must take into account the context in which necessary actions and decisions are to take place, and the manner in which information is to be inferred from the data provided. Soft systems methodology is therefore a sound basis for information systems planning, given its emphasis on obtaining a shared understanding of an otherwise fuzzy situation, and its ability to clarify the activities that are required to meet objectives which may be only dimly perceived.

However, we should not be lulled into a false sense of security regarding the ease by which information requirements can be determined. We could easily form the mistaken belief that this process is simply a matter of rational analysis, based on an understanding of required activity. There is, however, an element of subjectivity about information, given, for example, the power that can come with it; the use of information for propaganda purposes (see, for example, Argyris, 1980; Feldman & March, 1981; Hedberg & Jönsson, 1978; Land & Kennedy-McGregor, 1981), and the difficulties associated with communicating needs between the analyst and the potential information system user (see, for example, Oliver & Langford, 1984; Valusek & Fryback, 1985; Mittermeir, *et al.*, 1982). Add to this the changing nature of information requirements (Land, 1982) - due to changes in the business environment, changes in job content/role, and changes in what is actually informative (brought about by the development in individual managers' capabilities/knowledge) - and it soon becomes evident that the process requires a complex mixture of rational analysis and subjective enquiry, synthesised into conclusions that are acceptable and recognisable by the key stakeholders involved.

## THE BASIC APPROACH

Much has been written on the basic approach to, and application of, soft systems methodology and the way in which it has developed over the years (*eg.*, Checkland, 1972,1981; Checkland & Scholes, 1990). It is therefore not my intention to describe the basic approach here. What may be useful though is to describe in outline an extension of the approach which has been developed by Wilson (1980,1984,1990) with a view to identifying information to support required organisational activities, and the manner in which this should flow between these activities.

Wilson's approach is summarised in Figure 1 below. Noteworthy features include the following:

- the identification of information requirements and flows is based on modelling a 'primary task' (as opposed to 'issue based') root definition of the organisation (see Checkland & Wilson, 1980);
- taking the existing organisational structure, 'activity-to-activity' information flows are converted to 'role-to-role' information flows in order to identify individual managers' information requirements, based upon an analysis of the activities for which each is responsible;
- required information systems are then determined, based on an analysis of the performance needs of each of the activities identified (Wilson, 1990, pp.231-232).

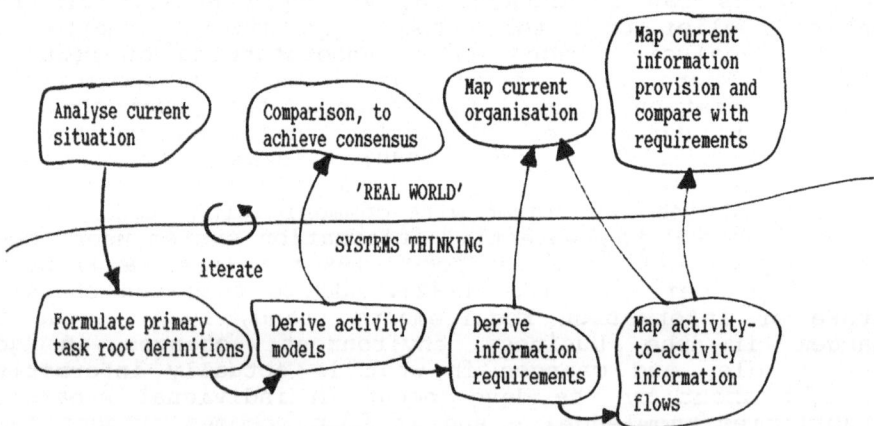

Figure 1. Information Requirements Analysis based on Soft Systems Methodology (Amended from Wilson, 1990, p.233)

## A SCENARIO-BASED APPROACH

In work undertaken over the past decade or so in the area of strategic information systems planning, the basic approach described above has been further developed to take account of the particular issues associated with linking business and information systems plans described in the Introduction to this paper. Examples of the manner in which these ideas have developed can be found in Galliers (1984,1985,1987ab); LeFevre and Pattison (1986), and Watson and Smith (1988).

The variation on the approach which is being advocated here attempts to take account of the following situations which executives often face when attempting to develop appropriate business (information) strategies:

- published mission statements (where these exist) may not reflect actual management activities and decisions, and may not be sufficiently detailed to assist in identifying key performance indicators let alone information requirements;
- key executives may have deeply held assumptions about the nature (both current and future) of the business and its environment which cloud their own perceptions, which are not debated with their colleagues, and which may not be shared by them;
- in line with the above, perfectly plausible (but sometimes quite radically different), views of the future may be held by different executives, which could lead to quite different requirements in terms of organisational arrangements, activities ... and information;
- organisational boundaries and managerial roles often arise almost as accidents of history, and may be inappropriate for current and future circumstances, and a change management strategy has therefore to be incorporated into the strategy approach (cf., arguments for 'change analysis' in Lundeberg, et al., 1981, and the socio-technical framework for information systems strategy advocated by Galliers, 1991, p.60).

The 'scenario-based' approach is illustrated in outline in Figure 2 and an example of the manner in which it can be applied is provided in the section after next. First, though, some of its key features are highlighted.


THE SCENARIO-BASED APPROACH: SOME KEY FEATURES

Perhaps the most significant feature of the approach – especially in the context of this paper and in distingishing it from Wilson's (1980,1984,1990) extension of soft systems methodology – is its acceptance of the likelihood that there will be a range of views expressed by senior management regarding an organisation's primary purpose and goals. One could go further: as a result of an application of the scenario-based approach, open diagreement regarding the primary purpose of the organisation can often result.

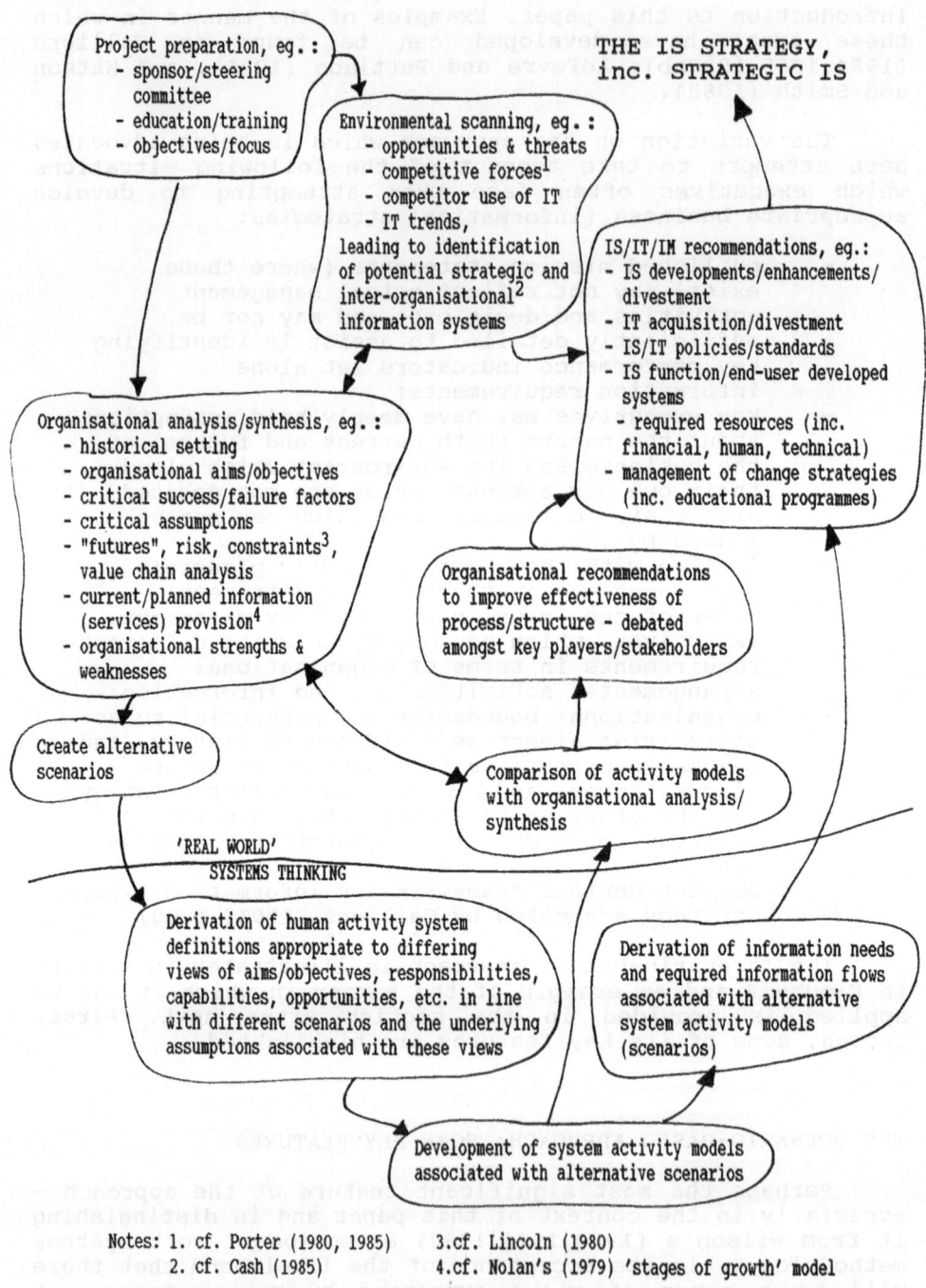

Project preparation, eg.:
- sponsor/steering committee
- education/training
- objectives/focus

THE IS STRATEGY, inc. STRATEGIC IS

Environmental scanning, eg.:
- opportunities & threats
- competitive forces[1]
- competitor use of IT
- IT trends,
leading to identification of potential strategic and inter-organisational[2] information systems

IS/IT/IM recommendations, eg.:
- IS developments/enhancements/ divestment
- IT acquisition/divestment
- IS/IT policies/standards
- IS function/end-user developed systems
- required resources (inc. financial, human, technical)
- management of change strategies (inc. educational programmes)

Organisational analysis/synthesis, eg.:
- historical setting
- organisational aims/objectives
- critical success/failure factors
- critical assumptions
- "futures", risk, constraints[3], value chain analysis
- current/planned information (services) provision[4]
- organisational strengths & weaknesses

Organisational recommendations to improve effectiveness of process/structure - debated amongst key players/stakeholders

Create alternative scenarios

Comparison of activity models with organisational analysis/ synthesis

'REAL WORLD'
SYSTEMS THINKING

Derivation of human activity system definitions appropriate to differing views of aims/objectives, responsibilities, capabilities, opportunities, etc. in line with different scenarios and the underlying assumptions associated with these views

Derivation of information needs and required information flows associated with alternative system activity models (scenarios)

Development of system activity models associated with alternative scenarios

Notes: 1. cf. Porter (1980, 1985)    3.cf. Lincoln (1980)
2. cf. Cash (1985)    4.cf. Nolan's (1979) 'stages of growth' model

Source: Amended from Wilson, 1984 (p.208); 1990 (p.233); Galliers, 1985 (p.5); 1987b (p.10) & Galliers, et al., 1988 (p.92)

Figure 2. The Scenario-Based Approach to the Formulation and Implementation of Information Systems Strategy

Certainly, it is most likely that a range of opinion will be expressed as to the business environment that is expected and the most appropriate responses to changing business imperatives in the light of these changed circumstances. Critical assumptions upon which business views are expressed, and upon which business decisions would be taken, are brought to the surface as a result of the debate which takes place in the organisational analysis/ synthesis phase. While this requires careful handling (*cf.*, the role of the facilitator, as described in Galliers, *et al.*, 1991), it is essential that 'taken-for-granted' views of key players in relation to both the current and future business environment are brought to the surface so that a shared understanding and shared vision of alternative futures can be obtained. Without this, the *implementation* of the resultant strategy may well be problematic, but also, key information requirements of a strategic nature may well not be identified.

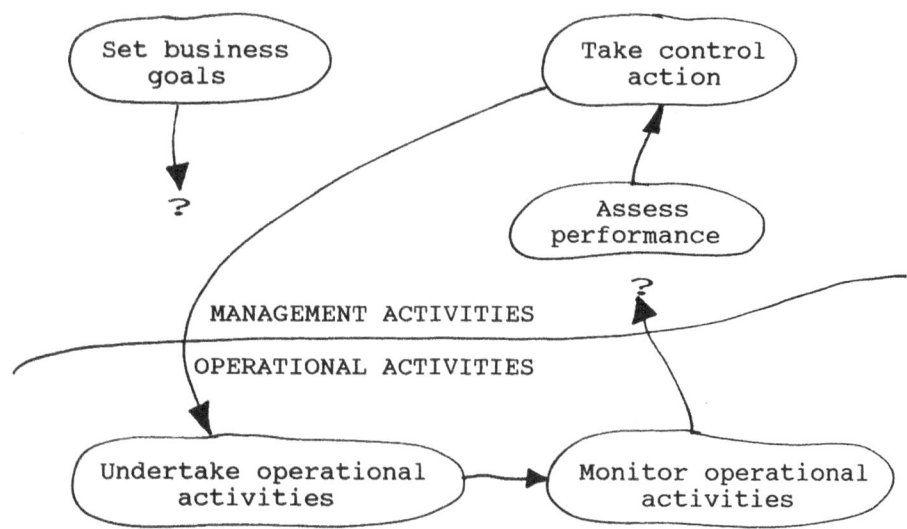

Figure 3. Management Control Activity: A Typical Process (Amended from Morien, *et al.*, 1988, p.5).

It is often the case that management control information relies almost entirely on the measurement of output from operational activities. The resultant control action is focussed on those operational activities as a result. This stems in part from a lack of connectivity between the planning process and the on-going activities of the organisation, and a lack of the kind of questioning described above which would add confidence in management that their planning was reasonably appropriate given the changing business conditions. A process lacking this connectivity is illustrated in Figure 3.

A more appropriate process is illustrated in Figure 4. Here, performance is assessed by not only monitoring operational activities, but also the business environment. Performance is assessed in this light, with control action being directed towards the setting of business goals; the assumptions underlying these goals, and the operational targets that arise from them, in addition to control action associated with carrying out the operational activities themselves. This approach not only helps to tighten the linkage between the planning process and the on-going activities of the business, but also assists in the identification of key information from the environment which might lead to a reassessment of the business plans and associated performance measures.

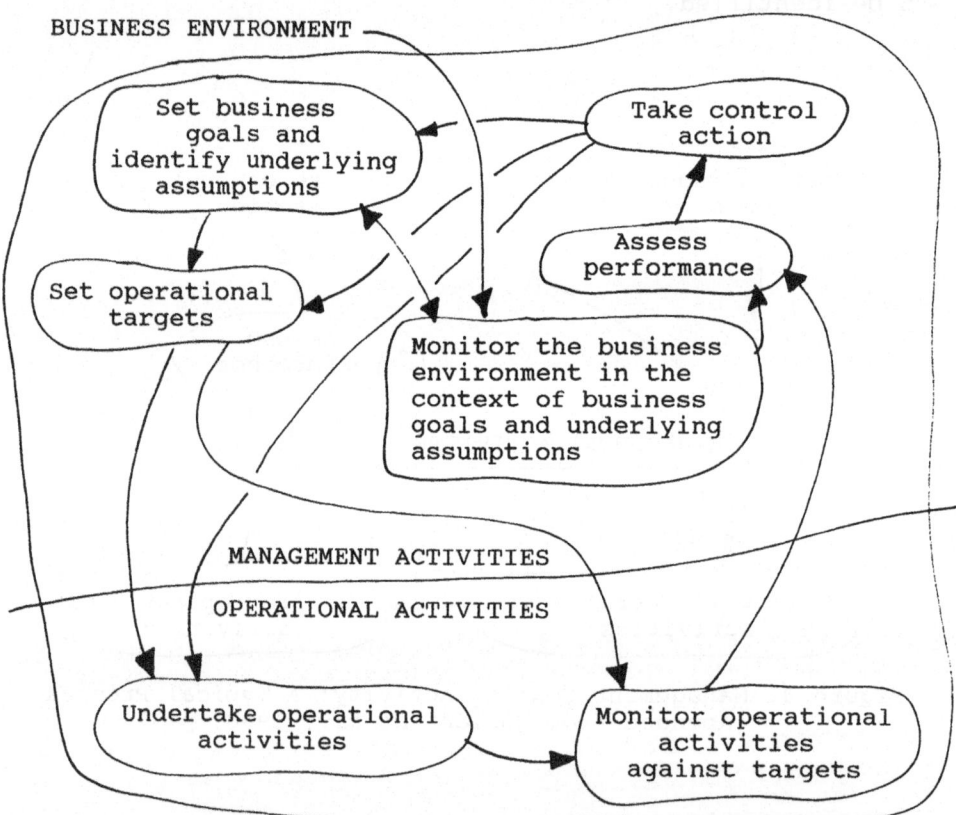

Figure 4. Management Control Activity: A more appropriate process (Amended from Galliers, 1984, p.625; 1987a, p.301; Morien, et al., 1988, p.6).

Different "futures" for the organisation can be established by identifying those environmental and organisational factors which appear likely to be fairly constant over the planning period ('facts'); those which appear to be subject to perceptable trends ('trends'), and

those over which there may be (considerable) debate
('issues'). Once these different "futures" have been debated
- and an appropriate organisational response to each agreed
- systems activity models can be built in the usual manner.
Similarly, required information and information flows
associated with each "future" can be identified.

It is highly likely that the information requirement
associated with a particular "future" will overlap to a
lesser or greater degree with the required information
associated with other "futures". For argument's sake, let us
assume that we have identified three "futures" for our
business, and that associated information requirements are
represented by the areas $F^1$, $F^2$ and $F^3$ as shown in Figure 5.
It is then a matter of judgement as to the extent to which
all the required information is collected.

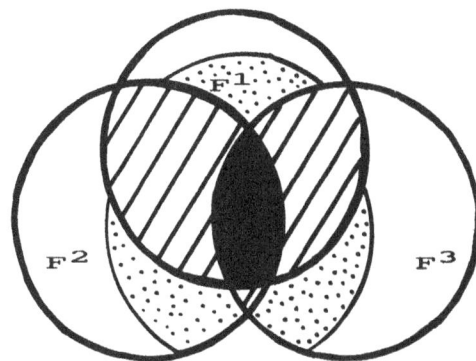

Figure 5. Required Information Associated with
Alternative "Futures" (Galliers, 1987c, p.361)

It is most likely that provision should be made to
collect the information represented by the area shaded
black, since this appears to be required no matter what
"future" emerges. In addition, we may decide to make
provision for the collection of information represented by
the hashed areas, since this appears to be required in two
cases out of three. Further, there may be some absolutely
critical information associated with a single "future"
(represented by the dotted areas) which we also decide to
collect. In this way, while not being absolutely certain
that we have prepared ourselves for every eventuality, we
have at least placed ourselves in a position whereby we can
develop reasonably flexible information systems, capable of
being adapted in line with changed business imperatives. In
addition, given that we constantly monitor the environment
so as to reassess the underlying assumptions upon which our
business strategy is built (cf., Figure 4), we can equip
ourselves with strategic information systems which provide
early warning of changed circumstances requiring control
action.

Other features of the scenario-based approach include the identification of critical success factors (Rockart, 1979) to assist in the determination of appropriate targets/performance measures (*cf.*, Figure 4). In addition, it is often useful to look at this from the opposite perspective and ask the question 'What key issues are likely to lead to disaster?!' While such 'critical failure factors' are often simply the opposite of the critical success factors that would have been identified anyway, this changed perspective often leads to new insights and a clearer understanding of the kind of warning signals that managers would prefer to have available if things are begining to take a turn for the worse.

## THE SCENARIO-BASED APPROACH: AN ILLUSTRATION

The following simple example will help to illustrate the application and utility of the approach.

Let us assume that we are running a small company which prints T-shirts and that our analysis has shown us that there are just two alternative scenarios that we would wish to consider for our business, *viz.*:

<u>Scenario 1</u>: *"Maintain the Status Quo"*
We will remain a small company, buying plain T-shirts from a single supplier and selling on printed T-shirts to two local retail stores. Ideas for slogans will continue to come from within the company, supplemented by informal, *ad hoc*, feedback from contacts within the retail stores. Key assumptions include our view that the business will continue to be viable and that we can be assured of on-going satisfactory relationships with our suppliers and retail outlets.

<u>Scenario 2</u>: *"Expansion"*
We wish to grow in size and see our profits increase. We should therefore develop our market by opening up our own retail outlet and selling customised printed T-shirts in addition to our normal lines. Printing capacity will increase and we will actively seek additional suppliers and retail outlets. Key assumptions include our view that there is a buoyant market for our staple product lines, together with an as-yet untapped market for customised T-shirts. In addition, we can be assured of our ability to forge satisfactory relationships with new suppliers/retail outlets, while maintaining our existing relationships.

Simple activity models for the two scenarios are given below in Figure 6. These are illustrative only and are not meant to be comprehensive in their treatment of all the likely activities necessary. Clearly, in both cases, more

detailed root definitions of relevant systems associated
with each of our scenarios would have to be devised were we
to be undertaking a formal information requirements study.

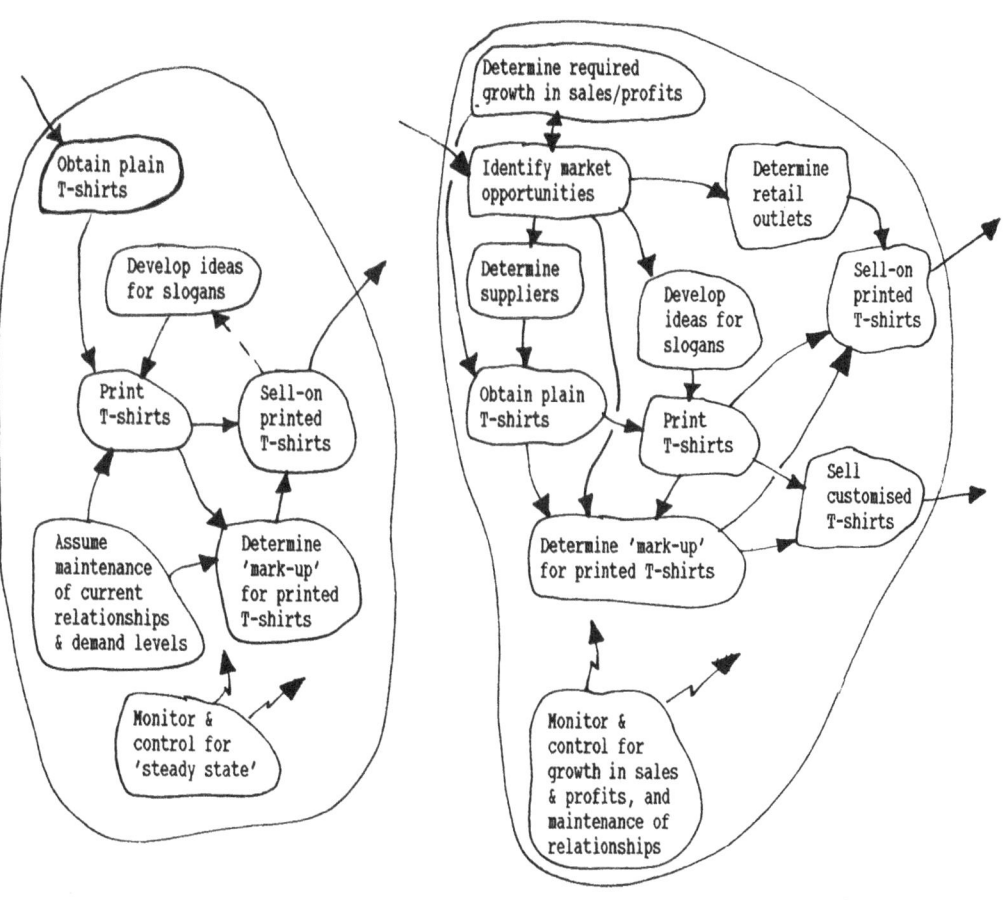

Scenario 1: "Maintain the Status Quo"          Scenario 2: "Expansion"

Figure 6. Activity models associated with two alternative
scenarios - an illustration

It is not necessary here to identify all the
information needed to manage and operate the business in the
above case. What I shall attempt to do is give examples of
the *differences* in information requirements associated with
the two scenarios and a discussion on the kind of strategic
information necessary to query the appropriateness of the
assumptions on which each is based. For Scenario 2, there is
a need to obtain additional information on, for example:

- alternative suppliers of T-shirts

- alternative retail outlets for T-shirts
- market opportunities/demand
- alternative slogans
- elasticity in demand (according to price) of 'normal'/customised T-shirts
- relative costs associated with printing and selling 'normal'/customised T-shirts.

Let us now turn the information required to test the assumptions underpinning the business strategies associated with each scenario. For Scenario 1, information will be feedback will be required from the suppliers and retailers regarding their on-going commitment to our business, and from the market place regarding their ability to remain in business. Economic and market information will also be required as to the continued market for our products. In the case of Scenario 2, similar economic and market information will be required in relation to the likely on-going demand for both the 'normal' and customised T-shirts. Additionally, market intelligence will have to be gathered and maintained in relation to the kind of slogans that are likely to be popular. Feedback information from the existing supplier and the two retail outlets of their reactions to our expansion plans will also have to be gathered and monitored on an on-going basis.

CONCLUDING REMARKS

Clearly, the T-shirt printing company is an all-too-simple example of the way in which the sneario-based approach can be used to identify requirements for strategic information systems. I hope it does serve to illustrate the different information requirements associated with different scenarios and the means by which underlying assumptions can be tested. As a result of this process, business strategies can be evaluated on an on-going basis and informed decisions made as to the extent of information provision likely to be needed given the alternative scenarios that are identified and studied. As key information is obtained, the scenarios themselves can be tested and altered, so that a flexible information systems strategy is maintained in the light of changing circumstances.

A key point is that, almost as a by-product of the approach, senior executive involvement and on-going commitment is maintained, and the strategy formulation process becomes an integral part of on-going management activities and concerns. While specific targets (performance measures) were not identified as part of the T-shirt printing company example, this would of course be an integral part of the process in a real world situation. This, as has been argued, helps to integrate the planning and management activities with on-going operational activities, thus cementing the relationship even further.

In addition, given that the process is seen to be highly relevant from a senior executive perspective (in that it helps to formulate appropriate business strategies and to provide useful information systems for senior executives), it tends to be championed by either Chief Executive Officers or highly influencial Board Members, thus increasing the chances of successful implementation of the information systems strategy. Further, and for the same reasons, it tends to raise the profile of the information systems function, and helps to forge partnerships based on mutual respect and understanding between information systems and business unit executives. All in all, however, it is likely that the shared understanding and visions that emerge amongst key stakeholders is the single, most important contribution that the approach offers.

## REFERENCES

Argyris. C (1980) "Some Inner Contradictions in Management Information Systems" In H C Lucas, *et al.* (Eds.) *The Information Systems Environment*, Amsterdam: North-Holland. Reproduced in R D Galliers (Ed.) (1987a), *op cit.*, pp. 99-111.

Cash, Jr., J I (1985) "Interorganizational Systems: An Information Society Opportunity or Threat?" *The Information Society*, 3(3). Reproduced in Somogyi & Galliers (1987), *op cit.*, pp. 200-220.

Checkland, P B (1972) "Towards a Systems-Based Methodology for Real-World Problem-Solving" *Journal of Systems Engineering*, 3(2).

Checkland, P (1981). *Systems Thinking, Systems Practice* Chichester: Wiley.

Checkland, P & Scholes, J (1990) *Soft Systems Methodology in Action* Chichester: Wiley.

Checkland, P B & Wilson, B (1980) "Primary Task and Issue-Based Root Definitions in Systems Studies" *Journal of Applied Systems Analysis*, 7.

Feldman, M S & March, J G (1981) "Information in Organizations as Signal and Symbol" *Administrative Science Quarterly*, 26(2), June, pp. 171-186. Reproduced in R D Galliers (Ed.) (1987a), *op cit.*, pp. 45-61.

Galliers, R D (1984) "An Approach to Information Analysis" *Proceedings of the First IFIP International Conference on Human-Computer Interaction*, London, 4-7 September. Reproduced in B Shackel (Ed.), *Human-Computer Interaction - INTERACT '84*, Amsterdam: North-Holland, 1985. Also in R D Galliers (Ed.) (1987a) *op cit.*, pp. 291-304.

Galliers, R D (1985) "Providing a Coherent Information Planning Environment to Meet Changing Organisational and Individual Information Needs" WAIT Computing and Quantitative Studies Working Paper, Perth, Western Australia, January.

Galliers, R D (Ed.) (1987a) *Information Analysis: Selected Readings* Wokingham: Addison-Wesley.

Galliers, R D (1987b) "Applied Research in Information Systems Planning" *Proceedings: Database '87*, Annual Conference of the British Computer Society Database Specialist Group, Edinburgh, April.

Galliers, R D (1987c) "Information Systems Planning in Britain and Australia in the Mid-1980s: Key Success Factors", Unpublished PhD Thesis, London School of Economics, June.

Galliers, R D (1991) "Strategic Information Systems Planning: Myths, Reality and Guidelines for Successful Implementation" *European Journal of Information Systems*, 1(1), January, pp.55-64.

Galliers, R D, Klass, D J, Levy, M & Pattison E M (1991) "Effective Strategy Formulation Using Decision Conferencing and Soft Systems Methodology" In P Kerola, R Lee, K Lyytinen & R Stamper (Eds.) *Collaborative Work, Social Communications and Information Systems*, Proceedings of the IFIP TC8 Conference, Helsinki, Finland, 27-29 August, Amsterdam: Elsevier Science Publishers BV.

Galliers, R D, Marshall, P H, Pervan, G P & Klass, D J (1988) "DSS Development within an Information Systems Planning Framework", *DSS-88 Transactions*, Eighth International Conference on Decision Support Systems, Boston, MA. 6-9 June, pp. 85-94. To appear in P Gray (Ed.), *Readings in Decision Support and Executive Information Systems* Englewood Cliffs, NJ: Prentice-Hall, 1991.

Hedberg, B & Jönsson, S (1978) " Designing Semi-Confusing Information Systems for Organizations in Changing Environments" *Accounting, Organizations and Society*, 3(1), pp. 47-64. Reproduced in R D Galliers (Ed.) (1987a), *op cit.*, pp. 179-202.

Land, F F (1982) "Adapting to Changing User Requirements" *Information and Management*, 5, pp. 59-75. Reproduced in R D Galliers (Ed.) (1987a), *op cit.*, pp. 203-229.

Land, F F & Kennedy-McGregor, M (1981) "Effective Use of Internal Information" *Proceedings of the First European Workshop on Information Systems Teaching (FEWIST)*, Aix-en-Provence, April. Reproduced under the title "Information and Information Systems: Concepts and Perspectives" in R D Galliers (Ed.) (1987a), *op cit.*, pp. 63-91.

LeFevre, A M & Pattison, E M (1986) "Planning for Hospital Information Systems Using the Lancaster Soft Systems Methodology" *Australian Computer Journal*, 18(4), November, pp. 180-185.

Lincoln, T J (1980) "Information Systems Constraints - A Strategic Review". In S H Lavington (Ed.) *Information Processing 80*, Amsterdam: North-Holland.

Mittermeir, R T, Hsia, P & Yeh, R T (1982) "Alternatives to Overcome the Communication Problem of Formal Requirements Analysis" In Ohno (Ed.) *Requirements Engineering Environments*, Amsterdam: North-Holland. Reproduced in R D Galliers (Ed.) (1987a), *op cit.*, pp. 153-165.

Morien, R I, Galliers, R D, Marshall, P H & Pattison, E M (1988) "Corporate Planning, Information Systems Planning and Performance Evaluation" Fifth National Evaluation Conference, Australian Evaluation Society, Melbourne, 27-29 July.

Nolan, R (1979) "Managing the Crises in Data Processing" *Harvard Business Review*, 57(2), March-April.

Oliver, I & Langford, H (1984) "Myths of Demons and Users" *Proceedings of the Australian Computer Conference*, Australian Computer Society Inc., Sydney, Australia, November. Reproduced in R D Galliers (Ed.) (1987a), *op cit.*, pp. 113-123.

Porter, M E (1980) *Competitive Strategy* New York: The
    Free Press
Porter, M E (1985) *Competitive Advantage* New York: The
    Free Press
Rockart, J F (1979) "Chief Executives Define Their Own
    Data Needs" *Harvard Business Review*, 57(2), March-
    April. Reproduced in R D Galliers (Ed.) (1987a), *op
    cit.*, pp. 267-289.
Somogyi, E K & Galliers, R D (1987) *Towards Strategic
    Information Systems* Tunbridge Wells: Abacus Press.
Watson, R & Smith, R (1988) "Applications of the
    Lancaster Soft Systems Methodology in Australia"
    *Journal of Applied Systems Analysis*, 15, pp. 3-26.
Wilson, B (1980) "The Maltese Cross - A Tool for
    Information Systems Analysis and Design" *Journal of
    Applied Systems Analysis*, 7.
Wilson, B (1984) *Systems: Concepts, Methodologies and
    Applications* Chichester: Wiley.
Wilson, B (1990) *Systems: Concepts, Methodologies and
    Applications* (Second Edition) Chichester: Wiley.
Valusek, J R & Fryback, D G (1985) "Information
    Requirements Determination: Obstacles Within, Among
    and Between Participants" *Proceedings of the End-
    User Computing Conference*, Association of Computing
    Machinery Inc., Minnesota. Reproduced in R D
    Galliers (Ed.) (1987a), *op cit.*, pp. 139-151.
Yadav, S B (1983) "Determining an Organization's
    Information Requirements" *Data Base*, 14(3), Spring.

# INFORMATION MANAGEMENT

Brian Wilson

Department of Systems and Information Management
University of Lancaster
Bailrigg, Lancaster

## 1. INTRODUCTION

The topic of Information Management is not one that is characterised by the clarity of its meaning or interpretation. Some of the confusion is due to the multiple interpretations of the word 'information' itself, but some is due to the origins of the role being in the world of data-processing. The role has grown out of a general desire, on the part of those organisations who are sufficiently concerned, to benefit as much as possible from the opportunities provided by developments in the technology. Some organisations see the technology as a way of improving the efficiency of what they are already doing in relation to information provision and communications in general, while others take these developments as an opportunity to re-think what they are doing and to plan and develop information provision to effectively and efficiently serve their needs. Although it is generally the desire to use modern technology that has awakened the interest in the role of information management, information has always needed to be managed even in the days of card-index driven systems or perhaps as far back as the quill pen.

It will be argued later that the management of information is as crucial to the performance of an organisation as the management of any other resource. Prior to this discussion, however, it is necessary to be clear about what we mean by information. A number of interpretations exist but to avoid confusion here I need to make my own interpretation clear. A simple but powerful definition and one which can actually be used to advantage is one that takes:

**Information to be data plus the meaning attributed to it.**

This definition makes the assumption that data is itself neutral. One might wish to argue that 'data' is equivalent to 'fact plus interpretation', which is no doubt the case. However, to be useful, I would wish to combine the interpretation with the overall meaning associated with the use of the data to form information. Thus a piece of data like, for example, 'the cumulative sales over a prescribed period for a particular geographical area' will take on one kind of meaning when used by a Salaries Clerk in calculating the amount of bonus due to the appropriate salesman; the same piece of data will take on a totally different meaning when used by a Sales Manager in assessing the performance of the sales team, or by the Marketing Manager when developing a particular marketing strategy. The piece of data, to avoid misinterpretation, will need to be bounded. Thus in the above example the 'prescribed period' and the 'geographical area' would need to be defined.

Hence it is important to differentiate between data and information. The route towards effective and efficient information support for an organisation is, first of all, to determine information needs (on the basis of what it is used for) followed by the specification of the basic data to be collected and the data processing that is then required. Given the above-defined interpretation, it is **information** that is the valuable corporate resource rather than **data** . The statement is often heard, usually accompanied by some gesture towards a large stack of computer print-out, 'we have too much information'. What is actually being said is 'we have too much data and not enough information'.

## 2. THE TASK OF INFORMATION MANAGEMENT

The view has been taken that information is a resource of considerable value to an organisation. Like personnel, finance etc., it is a resource which has corporate value and should be managed at an equivalent high level. The question then needs to be answered, 'what is information management?' At this stage it is probably easier to state what it is not, i.e. it is not the management of the data-processing technology and associated expertise. Information management needs an orientation to the business and not to the technology though, traditionally, this is where it has been assumed to reside.

The precise description of the role of information management, in terms of the activities to be undertaken, will need to be derived in such a way that makes it particular to the particular organisation concerned. However, the overall task that it is trying to achieve, can be meaningfully discussed in general terms. Information management is essentially attempting to bridge two areas of concern. One area is related to the particular business of the organisation and the other is related to the technology relevant to data processing and communication. As indicated above , a business orientation is required hence the view is also taken that the bridge must be crossed from the business or organisation side. The information and its management must be driven by the needs of the organisation and **not** by the technology available. To operate as a bridge the set of activities relevant to the information management task must therefore contain the communication processes to link business needs with the technological opportunities for satisfying these needs.

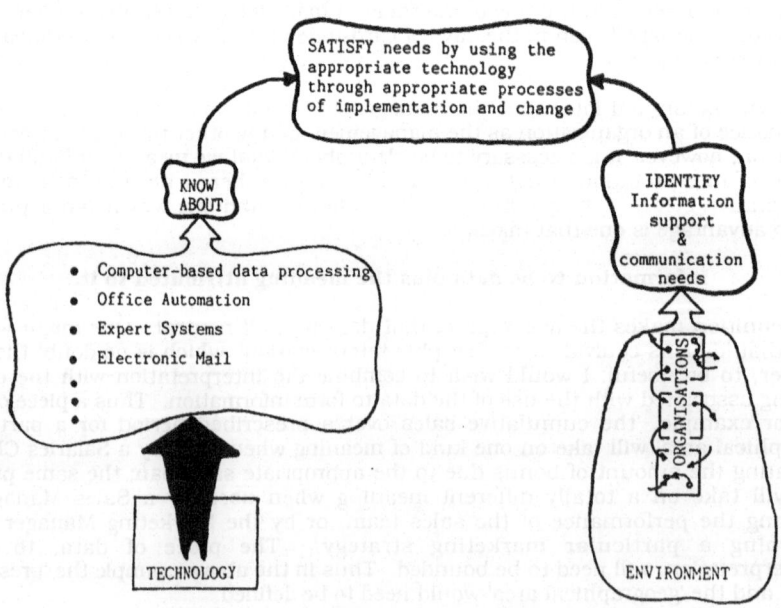

Figure 1. A Concept for Information Management

Figure 1. attempts to illustrate the concept of information management outlined above. Three major activities are seen to exist which describe the primary task. These are, to IDENTIFY the information support and communication needs of the particular organisation; to SATISFY these needs by using the appropriate technology through appropriate processes of implementation and change and, as a support to that activity, to KNOW ABOUT the opportunities provided by technological developments.

If the task of information management, illustrated in figure 1, is accepted, then it is implied that there are wide-ranging expertise requirements, which must be provided if the primary activities are to be realised. The 'identification' activity requires expertise in the use of concepts to describe the business processes undertaken by the organisation and their application in some form of methodology for information requirements analysis. There are a number of approaches utilising a variety of concepts (Wilson, 1990; Galliers, 1987; Avison & Fitzgerald, 1988). Whoever is doing this activity needs enough familiarity with the field to be able to select and use the approach appropriate to the organisation (and to personal preference). The central, (or bridge building) activity of satisfying needs is complex requiring a variety of expertise and skills. Firstly, enough knowledge about technological capabilities and developments must be acquired. This requires an understanding of the language of this area to be able to communicate with the experts and to extract realistic opportunities. Secondly, having identified **what** information requirements are to be satisfied, the appropriate technology needs to be selected in order to determine **how** the required processed-data is to be provided. Thirdly, inter-personal skills are required to liaise with users in order to facilitate and/or manage the participation necessary to ensure effective implementation. Fourthly, the ability to manage the changes accompanying the introduction of new technology is essential. It is unlikely that developments in an information network, either through modifications in requirements or through changes in the technology, can take place without changing the mode of working or, in some cases without changing or introducing new roles. (Here, an information network is taken to be the total set of formal information systems; computer-based and manual.)

Finally, the perceptive ability to appreciate the culture and social characteristics of the people in the relevant parts of the organisation is a skill which enables all of the above assessments and selection of courses of action to be **appropriate** to the organisation. Apart from the specific business activities, the organisation (or the part of interest) is determined by the values and characteristics of the people who operate within it.

It is unlikely that any one individual will have all of the above attributes and therefore it is to be expected that the introduction of information management would have a minimum organisation structure associated with it. The precise allocation of responsibilities can only be specified in relation to the people appointed. However, an overall responsibility can be defined by examining the set of activities that describe the role of information management.

## 3. THE ROLE OF INFORMATION MANAGEMENT

It was stated in the previous section that the role of information management can only be described as a set of activities by relating the analysis to the particular organisation. Certain activities may already be in place which modify the actual role requirements and replace some, otherwise necessary, activities by communication processes. However, within a recent project such a role was derived for a particular company and a generalised version of it is presented here. Significant aspects of it may well be transferable.

The particular problem faced by the company arose because of the introduction and widespread use of personal computers (PCs). A brief description of the approach to the problem is given in order to provide the context in which the total task of information management was explored.

Within the Engineering and Technical Centre of this company the use of PCs. had proliferated to the extent that each Engineer had his own. This was used to produce, analyse and store data which effectively became 'owned' by the individual Engineer. It was the case however, that the data was not the personal property of the originator, it belonged to the organisation and was shared by others. The individual ownership of data produced particular structures and interpretations which lead to inconsistencies and misuse when the data was shared. The company concerned believed that a change in attitude was required which would result in a shift from individual to corporate ownership. Their response was to make a structural alteration and appoint a 'data administrator'. The intention was that the appointment of a central responsibility for data would bring about the desired change. The big question of course was what could the role and responsibility of this data administrator be within the current environment within the Company. It is current philosophy in many organisations, as it was here, that 'each manager should have a PC on his/her desk' and if that is the case then this problem of individual data ownership is one that must be faced generally.

Within the situation described above, the role of data administrator could lie somewhere on a spectrum which extended from a mere 'store-keeper' (with responsibility for the capture, storage and availability of data) to the other extreme of 'information manager' (with a responsibility for the planning, progressing, maintenance and control of an information network).

We investigated this problem by developing a model of the task of information management (see Fig.2) and then used it to explore the implications of taking different responsibility boundaries for the role of data administrator.

The model of Fig.2 consists of a set of activities connected together on the basis of their logical relationships and the whole model could be looked at as a job specification for the role of data administrator (if it were to reside at the end of the spectrum represented by the task of information management in total). Each boundary which is drawn on the model represents a reduction in the area of responsibility. The set of activities within the boundary could be seen as a reduced job specification and the interactions between these activities and those outside the boundary represent the communication processes and procedures that would need to exist to link the data administrator with those other Managers who are undertaking these external activities. This would be the same kind of analysis required to explore the role of information management in a situation where a number of the activities were already in place. It is the total model which is of relevance here and a number of boundaries are mapped on to it in order to identify a number of sub-functions within the overall role. These demand particular expertise and could form the basis of the minimum organisation structure referred to in section 2. The sub-functions are defined as follows:

(1) Information Analysis function - with a concern for the assessment of methodologies for information requirements analysis and the use of the approach selected.

(2) Planning function - with a concern for the derivation of an information strategy and the conversion of this into a short-term plan of action for information system design.

(3) System Design and Implementation function - with a concern for assembling and using appropriate design and implementation skills so that the resulting designed information systems can be used effectively.

(4) Skill acquisition and Training function - with a concern for ensuring that necessary personnel and skills are available for the effective and efficient management and provision of information.

(5) Network Performance Assessment function - with a concern for the overall performance of information management within the particular organisation.

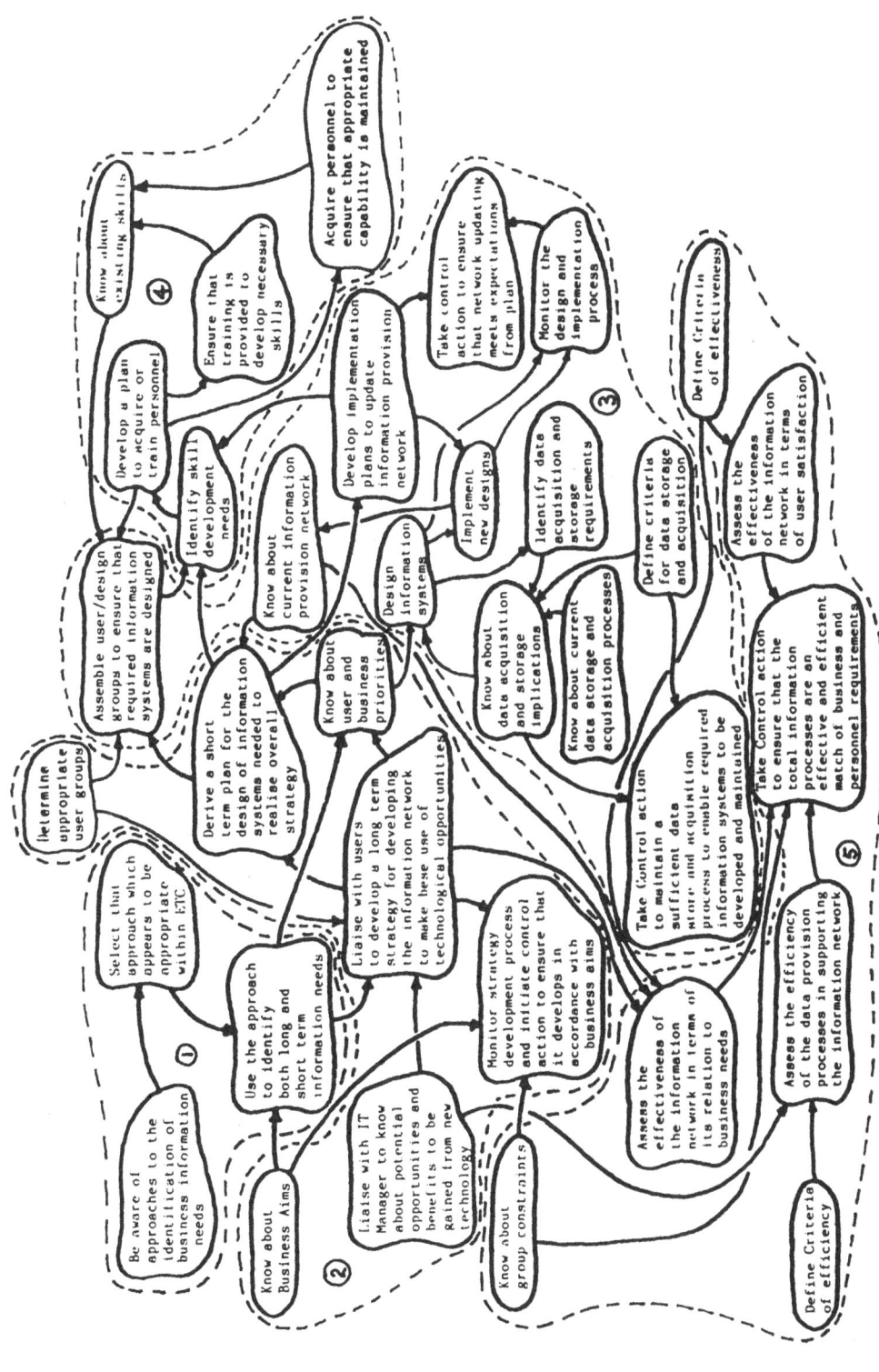

Figure 2. Activities appropriate to a role of "Information Management"

The model presented in figure 2 represents an expanded concept for the primary task illustrated by figure 1. It is therefore still a concept rather than an intended design. It specifics 'what' needs to be done in order to realise information management as described by figure 1. It does not specify 'how' to do it. How to do each activity would need to be explored in relation to the specific organisation, before such a concept (however desirable) could be converted into a feasible design. The model of figure 2 however could be used to derive the desired structure and the information requirements of the information management function itself. A detailed description of this process is given in (Wilson, 1990) . This represents a particular use of Soft Systems Methodology (Checkland, 1981; Wilson 1990) and could be one of the approaches adopted by sub-function (1) in figure 2.

## 4. ORGANISATIONAL INFORMATION SUPPORT

It is the case that information management requires an orientation to the information needs of an organisation in total, rather than to any specific business activity or function. It is not intended to remove the authority of the line manager and other service function managers in terms of information provision but to aid that provision and to ensure that where data is shared by a number of roles it can be made available in a way that assists its appropriate interpretation and can hence become meaningful information. A typical problem arises when data that is generated in one area of an organisation and not used by that area is required by another. Reluctance frequently exists to allocate effort to such data generation The removal of problems of that kind can be accomplished by an organisational commitment to the management of its formal information network. Informal communications will of course continue in an unmanaged way and will develop as a result of the style and culture of the particular organisation. They are a necessary part of maintaining the social structure.

The total information network may be illustrated by considering an organisation in transition. It is assumed that most organisations exist in a changing environment and must themselves change in order to remain viable. If this adaptation is not to be random some planning process will exist, however rudimentary, and it will make attempts to guide the organisation towards some desired future state This future state will also change as the assumptions about environmental impact and organisational capability will be re-defined and updated. It is therefore to be expected that proper information management must itself plan and develop the information network if the network is to become, and remain, effective in its support to the business of the organisation.

Consistency between business planning and information planning requires distinctions to be made between the various kinds of organisation information that are to be managed. Figure 3 attempts to summarise these distinctions.

For support (a), i.e. that information to support the current business, an information audit is required in which comparison is made between the existing information network and an explicit standard. Such a standard may be derived using a purposeful activity description relevant to the particular organisation and comparison made using a device called a 'maltese cross', (Wilson, 1990). Advantages of this type of analysis are that the standard is unique to the particular organisation and the activity description is independent of organisation structure. As this is something that is constantly changing the resultant information systems derived using the above approach are fairly robust.

Support (b) relies on being able to derive likely scenarios on the basis of the business plan. These may be in the form of activity descriptions, as mentioned above, in which case they are a useful source of organisation-structure-independent information requirements . Relating these to existing information support (as derived in (a)) and identifying development needs will provide the basis for an information strategy. This will also have to recognise user priorities before the corresponding I.T. strategy can be developed.

The information support to evaluate implementation (c), depends upon the identification of those measures of performance that will indicate whether or not the movement of the organisation away from its current state is in a desirable direction. What is meant by 'desirable' will need to be regularly re-defined as the environmental changes and their impact affect performance.

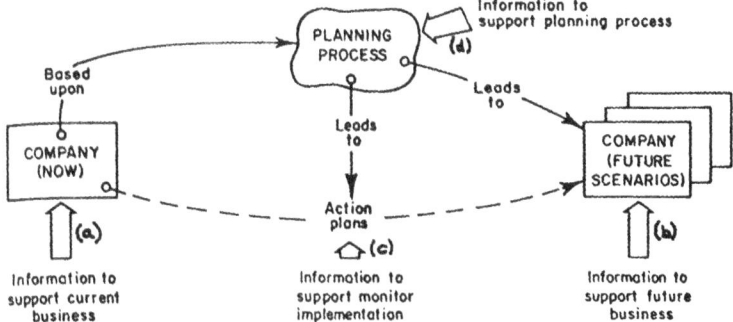

o Summation of Development needed in (a) ..... (d) proves an information Strategy Plan which leads to IT Strategy Plan (in terms of processed data requirements).

o Identification, Development and Maintenance of an Information Production Resource (IT & Manual) to satisfy needs of (a) ..... (d) is the task of Information Management.

o <u>Role</u> of Information Manager is to produce Information Strategy Plan ....... and manage its implementation.

Figure 3. Overview of Information Support

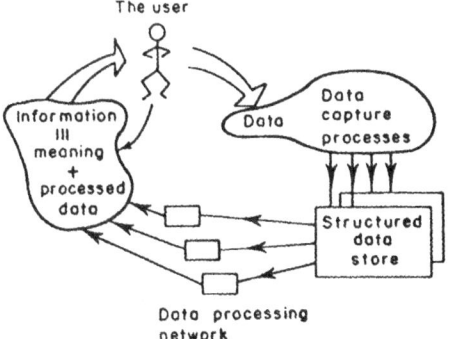

Figure 4. Total Information System

Information support for the planning process (d) is, of course, totally dependent upon the particular process adopted by the particular organisation. Again it can be specified using a purposeful activity description as the source of information requirements.

Support information (a) ..... (d) represents the totality of information to be managed and, to do this task effectively, the appropriate expertise needs to be available as indicated by the concept of Figure 1.

## 5. INFORMATION SYSTEMS

The discussion of information management has assumed a particular interpretation of the label 'information system'. In the introduction, information was defined as 'data plus the meaning attributed to it' and this definition leads to the particular interpretation illustrated by Figure 4.

The user must be included in the picture since it is the use of the data that leads to the meaning. Hence what are usually referred to as information systems are in fact 'processed-data systems'. They only become information systems if their outputs are derived from an analysis of use and the outputs actually get used. This implies that any analysis of information requirements must proceed anti-clockwise around the cycle illustrated by Figure 4.

## 6. CONCLUSION

The paper has attempted to present a concept for the management of information within organisations. The concept is consistent with particular definitions of the words 'information', 'data' and 'information systems' and, I hope, brings some clarity to what is currently a rather confused area.

Implementing such a concept within an organisation however, is not a straightforward process. There is still a widespread belief that information management is something that should be done by the data-processing department, if there happens to be one. Yet it is demonstrably the case that the major concerns of such a department are oriented towards the data and its associated technology and that their preferred starting point is a requirements specification. Thus any information management that does occur is the result of individual initiatives by some of the managers though usually from an individual rather than a corporate perspective.

Arguing the case for information management is also not straightforward. Requests for cost-benefit analysis are difficult to satisfy as it usually indicates a lack of appreciation of the concept and a desire for numbers.

Traditional cost-benefit analysis is not really appropriate in this area as there is unlikely to be a simple relationship between cost and benefit . Take, for example, the costs of processing the basic data to provide career profiles for the employees in a marketing department and the costs of processing the basic data to forecast potential revenue income from a particular market segment. The costs may be very similar but how do you evaluate the benefits? The latter can be provided as numerical estimates and may have considerable benefit to the company in terms of its future orientation and performance. The benefit from the provision of career profiles cannot be evaluated in the same way. At best it is intangible and its absence may have some long term effect on morale. Some organisations may not value this aspect particularly highly as evidenced by the lack of explicit attention given to employee career aspirations within their personnel or human resources departments.

**In general the benefits arise from better performance in doing the task (supported by the information) but the costs of supplying that information are unrelated to the task.**

Some appreciation of benefit may be obtained by examining the costs of **not** implementing the activities associated with information management. A recent example is of a company who wished to develop an I.T. strategy. They already had a large data-processing organisation and hence the in-house skills missing were those of business analysis leading to information requirement identification and the long -term planning of information needs leading to an information strategy. Consultants were employed to do this at a cost of some £300,000 and the I.T. strategy produced will cost some £76m over the next five years to implement. The task has been done but no in-house skills remain. Of course the task is not one-off and in maybe five years time the exercise will need to be repeated. Changes do not occur at five yearly intervals and some appreciation of the effect of continuous changes on the business and its information requirements needs to be present otherwise what actually gets implemented may well turn out to be irrelevant to the business by the time it exists.

There is some cost associated with acquiring or developing the necessary in-house capability for information management but this cost is not high and can be evaluated. The benefits, as exemplified above, are difficult to evaluate but they may be itemised as follows:

(1) The information that managers actually need will be properly identified and made available.
(2) The management of this process will ensure that the information is timely.

(3) Effective linking of information needs to the appropriate technology for supplying them will be a managed process.

(4) Current waste, accruing from processing data that does not actually get used will be identified and removed.

These are all difficult to quantify but in any situation some feel for their magnitude may already be held by the appropriate managers.

REFERENCES

Avison, D.E. & Fitzgerald, G. (1988). Information systems Development, Blackwell Scientific Publications, Oxford.

Checkland, P.B. (1981). Systems Thinking, Systems Practice, John Wiley, Chichester.

Galliers, R. (1987). (Ed.). Information Analysis, Addison-Wesley, Sydney.

Wilson, B. (1990). Systems: Concepts, Methodologies and Applications, John Wiley, Chichester.

APPLICATIONS OF SYSTEMS THINKING

APPLICATIONS OF SYSTEMS THINKING: INTRODUCTION


For the systems thinker, the emergence of a new European order is an unparalleled opportunity.   Quite suddenly, economic and social structures have become more complex as old boundaries have collapsed. Herein lies the rationale for the systems theme behind many of the papers in this section.

Some of the contributors have identified new opportunities for a systems approach ranging from the structure of entire subsystems of the Economic Community to the economic modelling of particular countries. Increasingly, the new Europe which viewed itself as more open and free from previous restrictions may now appear to be creating its own new barriers against the rest of the world.   The challenge for all organisations in Europe is to create a balance between autonomy and cooperation.   Other authors have considered in detail, questions of organisational structure and the implications of a change in system configuration.   Fundamental to these questions is the very nature of systems development:   should we expect systems to evolve or is it necessary to envisage step changes to configuration and behaviour.

One of the most interesting features of systems thinking is the breadth of application.   Some papers have presented applications within specific disciplines and derived wider insight into organisational activities.   Disciplines such as geography, education, health care and management science are represented but all these papers are concerned with creating understanding and awareness among participants.

This process of understanding exemplifies the nature of systems thinking and some of the authors have taken this as their theme.   In Europe, change has been a central experience and the study of this change is leading to deeper understanding of how systems learn about themselves.   The concepts of cognitive science have been explored in a number of papers.

It is apparent that different countries within Europe have adopted different routes to their current state of systems thinking.   Within this diversity, alternative unifying principles are emerging and are presented as new sets of systems images and definitions.   Even so, the essential cohesiveness of systems thinking has been observed and enhanced in these papers.   They are all concerned with a structured approach to complexity and the development of understanding between participants in the new European environment.

THE DESIGN OF EUROPEAN INDUSTRIAL SYSTEMS: A SYSTEMIC

INVESTIGATION

R.K. ELLIS (1) AND A. DELGADO-FERNANDEZ (2)

(1) THE UNIVERSITY OF BUCKINGHAM
SCHOOL OF ACCOUNTING, BUSINESS AND ECONOMICS
BUCKINGHAM MK18 1EG, ENGLAND
(2) CITY UNIVERSITY
DEPARTMENT OF SYSTEMS SCIENCE
NORTHAMPTON SQUARE, LONDON EC1V 0HB
ENGLAND

INTRODUCTION

Industrial systems capable of developing, surviving and contributing to society beyond the year 2000 are emerging in many parts of the world, and particularly in Europe. Such an emergence will provide an opportunity for systems thinkers to make an international contribution of significant proportions in the design, implementation and operation of a European Industrial System comparable in size and power to both the Japanese and American Industrial Systems.

There is a rich source of, largely untapped, intellect and ideas which are available to systems thinkers to play a leading role in examining the potential for the development of an integrated European Industrial System. This will, of necessity, include the economic, political and cultural aspects of the European Community.

This paper, which is based on an ongoing research programme, is attempting to answer a set of linked and complex questions.

BASIS OF THE RESEARCH PROGRAMME

This research programme is associated with the exploration of systems thinking in the analysis and design of an integrated European Industrial System. This programme is attempting to address a set of questions which include:-

(1) What are the European industrial objectives post 1992?
(2) Who, if anyone, is giving any consideration to evaluating and operating a European Industrial System capable of competing as a unity on the international stage?

*Systems Thinking in Europe*, Edited by M.C. Jackson *et al.*
Plenum Press, New York, 1991

(3) Where, in Europe, will the technical, political, social and economic leadership come from?

(4) How will Europe cope with the increasing pace of technological change?

(5) What will be the impact of organisational socio technical systems on a potential European Industrial System?

We are witnessing the development of industrial systems in both the developed and the developing world. A significant example of such a development is currently underway in Europe. It is also apparent that this development is piecemeal in nature, with all the attendant problems which are associated with such a development. The countries which embarked on the development of the European Community (EC) by a process of unification in the 1950's were, in the main, already industrialised. However, these countries were also recovering from a disastrous war and, therefore lacked any form of a unified systems approach. At this initial point there was little concern over external competition and the EC commenced with the development of organisations including EURATOM and the EUROPEAN IRON AND STEEL COMMUNITY.

Despite the apparent success in achieving a degree of integration in the agricultural, financial and political fields the EC is embroiled in problems associated with the diverse industrial schemas of operation. If the EC is to compete with the existing and developing industrial systems of Japan, USA and the countries of the Pacific rim, then there is a clear need for further integration in Europe. In the search for a unified industrial system the EC needs to consider the development of a social Europe without geo political borders which will permit and encourage the mixing of political, economic, social and cultural factors associated with advanced industrial systems. Even if a Pan European society can successfully mix the factors referred to there are still likely to be significant problems associated with the development of an integrated industrial system. It is to these problems that this research programme is addressed.

THE EUROPEAN COMMUNITY AS A SYSTEM

We need to consider the EC as a 'meta system' which comprises sub systems which, in themselves, are systems which are complex and contain high levels of variety.

The EC can be considered to be made up of (at least) the following sub systems:-

(1) A Political sub system.
(2) An Economic sub system.
(3) A highly diverse Cultural sub system.
(4) A Legal sub system.
(5) A Financial sub system.
(6) A Military sub system.
(7) An Industrial sub system.

The EC as a system is co-ordinated by the European Commission, with considerable interference from the national governments of the various member states. There are over 22 individual commissions involved in the regulation of factors which vary from farm produce to genetic engineering.

With respect to industrial organisations, there are EC directives concerning Health and Safety at Work, parental leave, information technology, training associated with technological change etc. The EC has also focussed on polit-ical adjustments with, for example, the manufacture of steel, the production of energy and the construction indus-tries of member states.

With respect to the development of industrial capability the EC has indicated that priority objectives are to consol-idate the promotion policy of both minor and major indus-tries, and to intensify industrial co-operation amongst major industrial organisations.(2)

The European production statistics (1) reveal a general-ly promising outlook for European industries, particularly in Telecommunications, Electronics, Aerospace, Pharmaceuti-cals, Insurance and Information Engineering. Less favourable are the Automotive, Construction, Transport and Oil/Gas sectors. The EC is encouraging industrial development in those EC countries which, as yet do not possess a highly developed industrial infrastructure.

Where technological development is concerned "The EC spends as much on R&D as Japan, yet the dispersion of effort and the sums spent on trying to match and adapt products to meet the myriad of differences in product standards dissi-pates energies. Member countries companies are therefore less competitive than their global competitors in terms of technological and production development." (3) Dahrendorf (4) argues that the EC has become an organisation which protects declining industries rather than promoting newly emerging industries. This is the clearest indication that the approach being taken by the EC with respect to the development of an integrated industrial system is non holis-tic.

In considering organisational systems, particularly in the context of an organisation in relation to its environment, Ellis (5) argues that there are five major sub systems that must be considered, and these are:-

(1) Strategic sub system.
(2) Technological sub system.
(3) Managerial sub system.
(4) Human sub system.
(5) Operational sub system.

Taking these sub systems as representative of industrial organisations, this model can be used as a basis for translating the system representation to the meta level exemplified by the EC Industrial System. The contained sub systems will interact with, and be interdependent upon systems in the environment. These environmental systems are those previously referred to, including for example the Political, Economic, Cultural sub systems.

THE CONTINUING INVESTIGATION

Carrying out a systemic investigation which involves Political, Economic, Cultural aspects is clearly no easy task, given the rapid processes of social and technological changes that are currently in process on a global scale. The questions that have been posed require a systems investigation on a large scale. That means that a methodological approach, which considers the EC in viable system terms, needs to be used.

Clearly the EC can be considered as viable system, and the investigation will proceed using, as a basis, the Viable System Model (VSM) of Beer (6). Such an approach is considered to be realistic given the need to investigate the following aspects:-

(1) The technical leadership necessary for the development of industrial infrastructure.
(2) The co-ordination of the Industrial sub system with the other EC sub systems.
(3) The development of a modern EC Industrial sub system which will exemplify a high quality philosophy.
(4) The development of a pan European training programme that will improve technical and managerial skills.
(5) To cope with rapid technological change by developing a European strategy for the introduction of an integrated approach to product and process innovation.
(6) The inclusion of post 1992 objectives on European Industrial Systems.

It is likely that this investigation will provide a novel application of the VSM, in that prior applications of the VSM have been confined to 'single' organisations with specific problem situations under investigation. This application will be directed at a meta organisational system which is political in nature and multi cultural. Therefore, this investigation will not only be using the VSM in the design mode, but will be testing the validity of the VSM in the associated enquiry process.

CONCLUSION

The investigation process to date has considered the use of the following methodological approaches:-

(1) Checkland's (7) Soft Systems Methodology (SSM).
(2) 'Hard' systems methodologies such as Systems Engineering (SE) and Systems Analysis (SA).

Both of these approaches have been rejected as being too narrowly focused. The VSM approach is considered to have sufficient width and rationality to provide a variable trajectory given the nature and scope of this investigation.

Further reports will be published in due course.

REFERENCES

1. Annual Report of the IFO, BIFE and PROMETEIA Institutes. "Europe in 1993". Brussels. (1989)

2. "Annuary of El Pais". Madrid. (1989)
3. J.Dudley, "1992 Strategies for the Single Market". Richard Clay Ltd. Suffolk, England. (1989)
4. R.Dahrendorf, "Whose Europe? The Future of Europe". Pro-Print Co Ltd. London. (1989)
5. R.K.Ellis, Modelling Technological Change: A Socio Technical Approach, in: "Proceedings of Second International Conference for Industrial Engineering". Nancy, France. (1988)
6. S.Beer, "Diagnosing the System for Organisations". Wiley. Chichester, England. (1985)
7. P.B.Checkland, "Systems Thinking, Systems Practice". Wiley. Chichester, England. (1981)

MODELLING MEDICA: A TECHNOLOGY TRANSFER SYSTEMS APPROACH TO THE
EVALUATION OF A EUROPEAN COMMUNITY FUNDED RESEARCH PROJECT IN
MEDICAL INFORMATICS

David Smith

Audio Visual Centre
University of East Anglia
Norwich NR4 7TJ UK

## 1. SUMMARY

This paper discusses the use of a 'Soft Systems' approach to the
evaluation of Project MEDICA, one of forty multinational and
interdisciplinary projects within the 'Exploratory Action' phase
of the European Commission's AIM (Advanced Informatics in
Medicine) Programme.  Led by Professor Julian Hilton, Director
of the Audio Visual Centre at the University of East Anglia,
MEDICA (Multimedial Medical Diagnostic Assistant) was based on
collaboration between industrial, clinical and academic workers
from three EC countries and from Sweden, and involved specifying,
designing and partially implementing a 'Cognitive Support
System' for psychiatry.

A fundamental objective of the AIM Programme Exploratory Action
(which ran from July 1989 until December 1990) was the transfer
of advanced information technologies from research and development environments into routine healthcare practice throughout
the European Community and to achieve this in ways which might
effectively benefit the New Information and Communication
Technology (NICT) industries within the Community.  This focus
of the AIM Programme led the MEDICA team to conclude that the
most profitable approach to evaluation would be to consider the
Programme and its constituent projects as comprising a Technology Transfer System.

The development of a systems model of AIM, with MEDICA as one
of its subsystems, focussed attention onto the general features
of technology transfer systems.  The basic function of such a
system could be defined in terms of bringing about specific
changes in one system or subsystem through the managed importation of devices and/or cultures from another system or subsystem.  This definition laid particular emphasis on the complex
nature of the innovation process and pointed up the significance
of the management of change as an interface process in technology transfer, and thus provided a formal framework for the
identification of suitable criteria for evaluation of many
aspects of Project MEDICA.

Technology Transfer System modelling was found to be appropriate
to the needs of Project MEDICA, and to offer a coherent and cost
effective integrating framework on which larger scale and longer
term evaluations might be built. It was concluded that the
systems-based approach offered greater intellectual coherence
than many alternatives, and that its holistic perspective could
deal effectively with the complexities associated with the
development and implementation of 'pre-normative' systems.

## 2. INTRODUCTION: WHY EVALUATE MEDICA ANYWAY?

This may seem at first sight to be a rather strange question.
Of course we should evaluate - we always do! Evaluation is just
taken for granted! It was certainly a condition of funding in
the AIM Programme, as it is for any other, that there should be
a process of evaluation. The only problem was in clarifying
what the evaluation was expected to achieve. This is a quandary
which affects most project managers at one time or another. All
accept and recognise the needs for rigour and accountability,
towards which evaluation is notionally addressed, but it is
sometimes difficult to reconcile the actual process of evaluation
with the satisfaction of those needs.

The MEDICA team faced this issue as one of its first priorities.
An early decision was taken to adopt the so-called "illuminative
evaluation" approach, which would be concerned with description
and interpretation, rather than measurement and prediction. The
role of the evaluator would notbe the usual subtle blend of
judge, examiner and recording angel, but the evaluation process
would be decision-oriented, serving three basic functions:

* A management information tool specific to MEDICA
* A management information tool in the overall context
  of the AIM Programme
* A factor in a wider process of academic legitimation
  within relevant peer groups

These functions were taken as the basis for the formulation of
general objectives for the evaluation of MEDICA. However, they
were far too diffuse, even ideological, to direct the processes
of evaluation. What was sought, therefore, was a framework of
theory which was formally sound yet which permitted sufficient
flexibility in its application to cope with the pragmatics of
real-life evaluation. For example, the fact that relatively few
factors in human activity systems are amenable to exact quant-
ification, and many of the techniques currently availavle for
the estimation of potentially significant variables (such as
'attitude change') are both imprecise and difficult to validate.

The complexity and diversity of purpose of human activity systems such as education and medicine often come as a surprise to
researchers from more 'engineering' oriented fields. There are
many problems of evaluation which cannot in fact easily be
solved by the convention of using strictly controlled experiment
supported by complex inferential statistical procedures: indeed,
there are severe ethical and practical constraints to this
research paradigm even in fields such as the trialling of new
drugs. Where inputs cannot be clearly identified or outputs
precisely quantified, statistically-oriented approaches can all
too easily degenerate into expensive 'data-trawling' operations,

with attention focussed onto the apparently measurable rather than what is likely to be clinically significant.

Awareness of typical systems functions and interactions, and a holistic approach to real-life complexity is essential to plan and evaluate changes in working life and professional practice, such as might be expected from the development and widespread application of a new tool such as the sort of system which the MEDICA prject was developing. 'Systems Thinking' is a valuable antidote to any tendency to make simplistic assumptions about the relationship between interventions and their outcomes, and it was consequently decided that the framework for the evaluation of Project MEDICA should be founded on this principle, and more specifically on the "Soft Systems Methodology" of Checkland (1981).

The fact that, as stated above, the transfer of advanced NICTs from research and development into routine healthcare practice was a clearly stated fundamental objective of AIM, supported the conclusion that the most profitable theoretical framework for evaluation would be to treat AIM as a technology transfer system with MEDICA as one of its component subsystems, whose specific function was the transfer of relevant techniques and technologies from academia and industry into psychiatry.

3. "SYSTEMS THINKING" IN PROJECT EVALUATION

A technology transfer system involves the management of an INTERFACE between the SUPPLY SIDE and the DEMAND SIDE of some technology system. Interface management will involve activities on both sides of the system. The interface is not a passive boundary, but rather an active MEDIATING SUBSYSTEM, with inputs from and outputs to both the supply side and the demand side, and supporting various TRANSFORMATIONS, both at the interface itself and within each subsystem. The interface subsystem can be assessed in terms of the relationships between inputs and out-puts and the efficiency and effectiveness of the transformation processes. These relationships will, however, inevitably be very complex. Some interface activities will have many characterist-ics of supply-side factors, whilst others will appear to act like demand-side factors. Thus, concentration on the assessment of pure demand side factors cannot give more than a fragmentary picture of the functioning of the system as a whole.

Furthermore, a systematically managed interface subsystem will not represent the whole of the interaction between supply side and demand side. Other factors will almost certainly impinge on the relationship between supply and demand. The true pattern of relationships is extremely complicated, and the intervention of indirect relationships between supply and demand increases the (already high!) inherent difficulty of ascribing outcomes to specific inputs. It is likely that any changes within the total system will have multiple antecedents (avaoiding for the moment the word 'causes').

MEDICA occupied a position at the interface between the supply and demand sides in a system which could be labelled "COGNITIVE SUPPORT IN PSYCHISTRY". This was a subsystem of "ADVANCED INFORMATICS IN PSYCHIATRY", which was inturn a subsystem of "ADVANCED INFORMATICS IN MEDICINE". All of these also inter-faced with a variety of more general systems environments, summed

up in the following systems description (Root Definition),
derived by applying Checkland's "CATWOE" inventory:

> "a system for transferring multimedia cognitive support
> technologies from the research and development field
> into routine psychiatric practice, through the
> specification, construction and dissemination of a
> practical healthcare system"

In specifying and developing a cognitive support system for
clinical psychiatry, MEDICA was acting as a supply side agent.
But there was no point at which the project functioned solely on
one or other side of the system.  It would not, for example,
have been consistent with the scale of the project if it had
been directly concerned with attempts to originate all of the
tools and techniques fundamental to the root definition.  The
need for 'enabling technologies' therefore placed aspects of
MEDICA firmly on the deamnd side of the equation.  On the other
hand, MEDICA was not the ultimate end-user of these enabling
technologies, but aimed to add value to them on behalf of its
intended clientele.  This implied effective mediation between
supply and demand, as discussed previously.  Criteria for
evaluation were therefore located within each of the three subsystem domains: <u>supply</u>, <u>demand</u> and <u>interface</u>.

## 4. TARGETTING EVALUATION ACTIVITIES

There is a body of opinion that evaluation is properly concerned
with the exact determination of quantifiable dimensions of the
system under scrutiny, and with the accurate and precise
measurement of changes in those dimensions associated with the
satisfaction (or otherwise) of specific objectives of the system.
To work in this way often requires us to ignore factors which
may not (yet) be amenable to being operationalised or quantified,
irrespective of their potential clinical significance.  In the
evaluation of aspects of healthcare which do not involve
'biological intervention', this can often reduce to concentration
on financial aspects of the system:

> "...Let us establish what they cost us and how much they
> bring in, and leave alone all the things we don't know
> enough about either to praise or to blame, and which are
> probably neither good nor evil but only necessary..."
> Denis Diderot, Rameau's Nephew

Whilst the dynamic complexity of the MEDICA Project was the main
consideration leading to the rejection of the reductionist model,
it was necessary at the same time to be aware that the 'holistic'
conception of evaluation is potentially subject to a 'combina-
torial explosion' of possibly significant factors.  With so many
features of the system and its components being capable of being
accommodated in the soft systems model, there is a great danger
of the evaluation process becoming diffuse and over-general.
The definition of evaluation as a management information tool,
rather than as a process of arbitrary judgement, has helped to
circumvent this problem by requiring that evaluation and
communication activities should be targetted in terms of:

> * selecting information which is most likely to be useful
> * ensuring it reaches people who are likely to act on it

There were in fact two major foci for attention within the evaluation process, namely the project team and the MEDICA system (in its usual narrow IT sense). The evaluation of the role and actions of the project team was in many ways the less problematic of the two, since relevant events took place within the immediate time-scale of the project. The evaluation of the advanced technological products of MEDICA was far more difficult, sincemany of the expected impacts of the system would only happen long after the first exploratory phase of the project was completed. The important thing from the MEDICA point of view was, however, to maintain the integrity of the evaluation, and not to allow it to degenerate into a series of isolated actions, selected solely to target the small number of measurable factors which were immediately apparent.

This is where the systems modelling approach more than proved its worth. The basic function of a technology transfer system is, as stated previously, to bring about specific desired changes in one system or subsystem through the managed import of technological devices and/or cultures from another. This clearly points up the significance of change management as an interface process. This is something which has been extensively studied for several years, and there is a substantial body of literature on the subject as it relates to a wide variety of domains. It was possibly to construct an evaluation strategy on the basis of current theory and practice in the field of innovation and the management of change, as well as the detailed consideration of purely technical factors.

A number of qualitative approaches to innovation proved helpful in this context, including Gosling's rather trivialised product cycle model (1981), as well as the vast body of information on the pragmatics of change management. Project MEDICA was thus able to avaoid many of the common 'evaluation traps', such as the well-known 'learning-curve' effect, whereby the adoption of an innovative technology is often associated with a short-term deterioration of performance, which is often distorted by the compressed time-scale of conventional pretest/post-test models of evaluation (These almost always favour established methods - or technologies over innovations during a critical learning curver phase).

It was possible to avoid these and similar pitfalls by making evaluation a continual and iterative part of project management. The basic MEDICA systems development strategy involved early prototyping and the close involvement of relevant clinical practitioners as full members of the project team. Evaluation was embedded in every stage of the development process, and there was no need for artificial distinctions between 'formative' and 'summative' aspects. In conformity with the preferred strategy of 'illuminative evaluation', there was no intention to draw globally valid conclusions from the MEDICA evaluation; at least in this exploratory action phase of AIM. Most reports were based on case studies, from which only limited generalisations were attempted.

## 5. CONCLUSIONS

Technology transfer is not an instantaneous process. It can take many years for a new device or procedure to become firmly and effectively 'grafted' into the routine of an established practice culture: and even longer before the full implications of transfer for the evolution of practice become apparent. And during the adjustment period, both the technology and the practice culture are likely to continue evolving (usually at quite different rates and in different directions). This is the well-known 'moving target' paradox. Whatever a project team may be hoping for, it is pointless for them to attempt to over-determine the nature and extent of the convergence of technology and practice in the short term: or indeed even to assume that they are well-determined in the first place.

Evaluation based on a-priori objectives can become the route for the imposition of a stultifyingly rigid bureacratic and unimaginative management style on what should be a creative enterprise. Our systems-based approach combined considerable flexibility with intellectual coherence, whilst its holistic approach dealt effectively with the real-life complexities associated with the development and implementation of basically 'pre-normative' experimental systems such as those within the AIM Exploratory Action.

The experience of Project MEDICA supports the view that a soft systems approach offers a coherent and cost-effective integrating framework on which large-scale evaluations may be based. Furthermore, the theoretical bases of many evaluation techniques are rather poorly developed, and this model also provides the foundations for the construction of a more solid theory base for the evaluation of complex multimedia systems in real-life applications.

## 6. REFERENCES·

Checkland P (1981). Systems Thinking, Systems Practice. Chichester, John Wiley.

Gosling W (1981). The Kingdom of Sand. London, CET.

THE SOCIAL SYSTEM OF NATIONAL HEALTH CARE AND SYSTEM OF HUMAN SOCIETY

Ferenc Kun

Department of Social Medicine
University Medical School Debrecen
Debrecen, Hungary

In this paper I continue the considerations of a former one, published
in the Proceedings of the Conference on the  Operational Research and the
Social Sciences /Kun 1989/.

In this abovementioned paper, and also in our former researches, I arrived at the conclusions that in the developed societies such an organisation process is taking place which regards the health care of the full
population as a task of the whole society and for this end it forms it to
a more and more organic and unified system. This unifying process first
appeared in the form of a health service to insure the health care of the
whole population. This mode of service got in the Constitution of the World
Health organisation the name of "National Health Service". (When a system
of personal health service has been extended as a right to cover an entire
national population or nearly so, it may be defined as a national health
service.)

This transformation, regarding its outset, started with the direct
connection between the physician and the patient the exclusion of which to
the whole population with intermittent character received crucial importance
several times in the history when plagues occured. These late measures,
however, significantly differed from the modern systems of health care.
These aformentioned measures mainly aimed to inhibit the spreading of
plagues and were characterized by the application of drastic means /quar-
antine etc./. The intention /and the practice/ is nowadays to combine the
humanitarian character - adapted to the individual - of the doctor-patient
- relationship and to stamp out epidemics in the bud. /The general activities of the Stations of Public Health and Epidemics, e.g. protective
inoculations./

This change was made possible by the new results of the social - and first
of all - the scientific and technical progress. It follows from the fact
that the present standard of the activities of the national health service
greatly depend on the scientific-technological level of the given country.

However, the scientific-technological progress resulted in certain
negative effects besides the obvious positive ones in the health state of
the population. The scientific-technological revolution basically changed
the conditions and way of life of human beings. The first conclusion which
should be made is that the scope of duties concerning the health care have

to be extended to factors determined not only by biological or natural matters but socially as well. The conditions and way of life are formed during interpersonal activities, in the process of conscious connections for determined purposes. The root of the problems is, that in many times these purposes are not in harmony with the aims and interest of protection of health. This results in a dichotomy: the physician diagnoses and supposes the social sources but he /she has neither possibility nor competence to the exact identification of matter. Even less can he/ she influence them to an advantageous direction. Another aspect of the question is that the persons -managers- dealing with these problems are not directly interested in enforcing the aspects of health-protection.

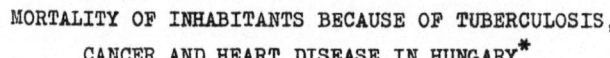

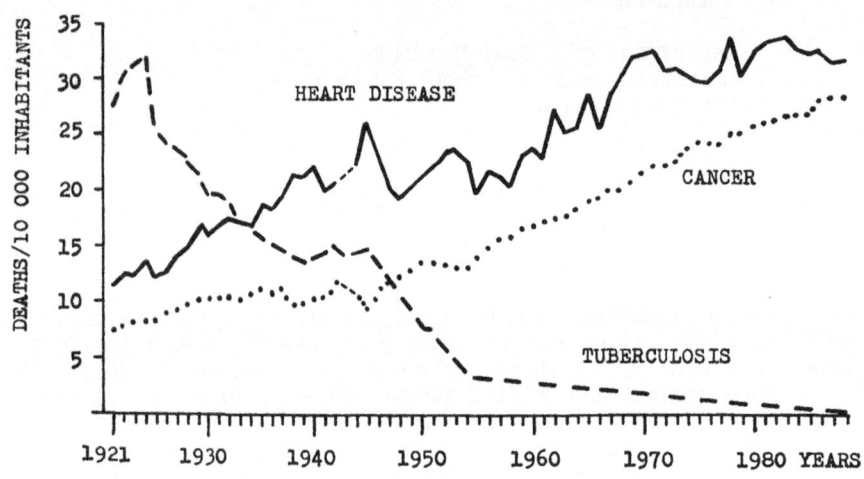

*On the basis of the data of the Central Bureau of Statistics

FIGURE 1.

I would like to point out that I exaggerated earlier the problem, because the countries take into consideration - in different extents - not only the risk factors of the diseases, but make significant economic-, technical-, organisatory- etc. steps to influence them advantageously. Nevertheless this arrangements - in certain countries, e.g. in Hungary - do not achieve the wanted results. This statement can be supported by the following two diagrams, the first of which is taken from my cited paper /KUN 198./ since it is the exact basis of our further conclusions too. It shows the mortality rate in the case of the most frequent diseases /heart-disease, cancer and tuberculosis/ during the past 60 years /from 1921 to 1980/ in Hungary. As it was referred to in our former paper, in the last years the mortality due to each of these two non-infectious- chronic diseases, cancer and heart disease, exceeds the TB-mortality in its most fatal period in the early twenties (30 case/10.000 inhabitants), meanwhile these ratios were 10/10.000 for cancer and 5/10.000 for heart diseases. This phenomenon is caused by the not-optimally regulated scientific-technical revolution.

The increasing role of non-infectious diseases in the mortality in Hungary can be better demonstrated if they are discussed as cardiovascular diseases.

The second figure shows that the mortality due to the total cardiovascular group is higher <u>in itself</u> /about the double/ than that of the TB-once named as morbus Hungaricus. In 1988 this rate reached 70,22! If we add the cancer mortality rate to this, it becomes clear, that this sum is the treble of deaths caused by TB in the worst post-war years of poverty. The cumulated number is 70.22 + 28.34 -approximately one hundred! This means that yearly 1% of the population becomes the victim of these two diseases.

MORTALITY OF INHABITANTS BECAUSE OF TUBERCULOSIS
AND CARDIOVASCULAR DISEASE IN HUNGARY[*]

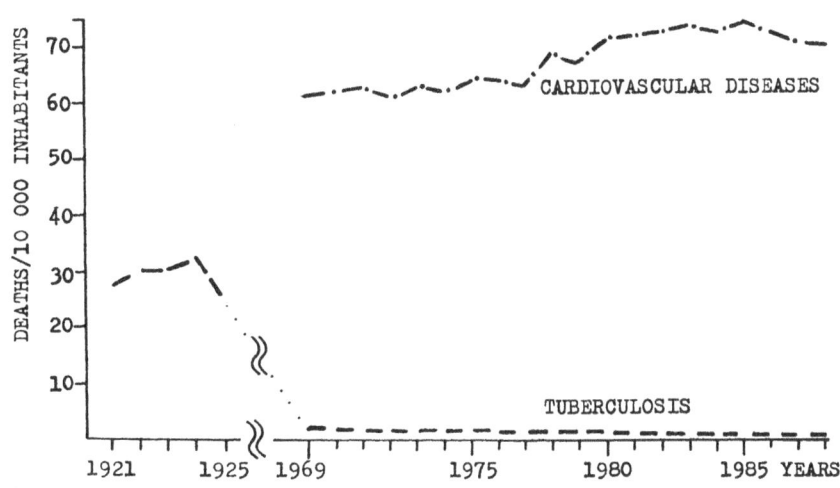

*On the basis of the data of the Central Bureau of Statistics

FIGURE 2.

According to the data of the international comparison the restructuring of mortality until 1960 took place similarly to the developed countries "the contribution of epidemic diseases is insignificant (about 1%)" /Boján 1990/. The differentation became from the midsixties more and more obvious. Namely the mortality due to the non-infectious-chronic diseases in the East-European countries from this year increased further, while in other countries improving tendencies appeared.

## Conclusions

This concept is realized by us by elaborating a theory of the formation of a global social subsystem based on prevention of diseases, preservation of health. As a matter of course these considerations mainly deal with certain branches of medical sciences, however, they embrace certain aspects of social sciences too.

The realization of our aims consists of three stages:

1. The definition and systematization of the research. Our starting point is Checkland's theory and we apply the theoretical systems theory, in which logic, semantics and philosophy - with the help of modelling techniques - plays primary role (Checkland, 1988).

2. The difinition and systematization of objects of the research. First of all the different spheres of different social activities and human relations of different levels should be considered.

The tasks will be completely diverse when the problem of preservation of health is regarded in the economic or technical sphere of production and eg. in the field of cultural and intellectual life. Our efforts can be fruitful only if the decision-making is adapted to the specific fields of the given social activities.

Our tasks will be quite different when dealing with human relations of different levels. The problems are entirely diverse when the object of research is the human relation on micro level or when - on medium level - the relations of institutions are regarded or when global relations /eg. international political conflicts; environmental problems or questions of peace-or-war/ must be taken into consideration.

3. Third stage of our activity: Permanent and continuous control of the justness and efficiency of practical applications of the decisions and -if necessary- their modification. The criteria of the efficiency are mainly the demographic indices -first of all the data of the average life expectancy- and the differentiation of them according to the different types of letal diseases.

The constant interaction among the three stages, the exchange of information and the adequate decisions based on these form the criteria of the efficiency.

## REFERENCES

Baján, F., 1990, "Demográfia," Medical School, Debrecen.

Checkland, P.B., 1980, "Systems Theory, Systems Practice," Willey, New York-London.

Kun F., 1989, The Social System of National Health Care, in: "Operational Research and the Social Sciences," M.C. Jackson, P. Keys and S.A. Copper, ed., Plenum Publishing Co., New York-London.

# SYSTEMS THINKING IN ORGANIZATIONAL ANALYSIS AND DESIGN

Petr Gerlich

The Department of Management
The Technical University of Ostrava
Czechoslovakia

## INTRODUCTION

The Technical University of Ostrava was the first university in Czechoslovakia that widened its curriculum by introducing lessons of management science, and initiated research in that area. The Department of Systems Engineering was established in 1967 already. Orientation to the application of operational research and systems engineering, especially in technical "hard" systems in the sixties and seventies was not only the reflection of the then approach to the operational research in the world. That approach was also associated with a negative attitude of power structures to the attempts of studying behavioural aspects of management systems in the Czechoslovak specific social and political conditions.

In the conception of operational research, the mathematical models, projects of data processing computer systems, simulation, modelling, regulation and control were dominating. The practical applications often failed and even in the field of "hard systems" they faced broader social and economic circumstances, which imposed restraints on the implementation and efficient making use of the results. The situation can be illustrated by stating the example of a failure in a transport system optimalisation in an enterprise; the system failed in operation because of lack of direct financial incentives of the drivers as for the consumption of fuels was concerned.

In the eighties, relatively more favourable conditions were created for so-called "rationalization of management", to which the Department of Systems Engineering is focusing too. When I started my carrier in the Department of Systems Engineering in 1983, the area of organization analysis and design methodology became a subject I was greatly interested in, though the research was pursued rather theoretically, quite lacking an approprivate  interest of practice.

Changes in the political regime and a transition to a market economy after November 1989 have brought quite new possibilities into that field of activity. The emergency of new Department of Management is therefore associated with the need of putting special emphasis on the problems of real world that are prior to the methodology. The problems of management, the department is dealing with in the frame of its research work and

advisory services, cannot be structured, they are complex, it is difficult to define them and they exert subjective impacts. To solve them, it
is necessary to employ appropriate systems of thinking, especially new
directions denoted as "soft thinking" and "organizational cybernetics"
(M.C. Jackson and P. Keys, 1987).

MANAGEMENT AND ORGANIZATIONS

## Organizations nad Systems

The human organizations are complex objects of real world. The
existence of organizations is a prerequisite to management. Therefore, if
we want to solve the problem associated with the management, it is
necessary to be familiar with the methods of analysis and design of
organizations.

The complexity of organizational problems does not rest exclusively
in the very quantity of elements and links, in the openess of the
organization and its contacts with the environment, in organizational
parts. The complexity ensues from a variety and interactions of viewpoints,
by means of which the participants consider the organizational problems
(R. Espejo, 1987). As far as the organization is concerned, a number of
systems can be defined as concepts aimed at their better understanding.

There is a way how to overcome the complexity and to understand the
essence of organizational phenomena lying deep under the surface - it is
an application of organizational cybernetics. But doing so, it is necessary, however, to avoid a limitation of  traditional cybernetics schemes
conceiving the subject of exploration in a rather mechanical way.

The article is aimed at indicating the potentialities of organizational cybernetics for organizational analysis and design; these potentialities rest above all in looking for the basic regularities of organizational arrangement and behaviour. The article draws from the knowledge
comprised in works of S. Beer (1979, 1985) carried out in that field and
further developed by M.C. Jackson (1987, 1988) and R. Espejo (1987). The
author of the structural arrangement from the cybernetics point of view
(especially co-operation) exerted on behaviour of the system in a given
type of organizational environment (turbulence).

## Organization Componnents

If we identify ourselves with the idea that an organization can be
seen as a system, we identify the organization with such qualities that
characterize it as a whole, indicate the form of its internal structure
and behaviour (V.N. Sadovskij, 1974).

One of the starting issues is the question what the basic component
of organizational system consists of S. Beer (1979) distinquishes
5 systems - implementation, coordination, control, intelligence and
policy within his "Viable system model".

Through a more detailed decomposition, a conception of human activity
as the basis of an organization can be attained. Even P.B. Checkland (1981)
uses the words defining human activity when creating "conceptual models" in
the frame of "soft system methodology".

These activities are focused on the attainment of individual aims
A together with the real results R. Potential possibilites to reach these
aims are given by resources Z (time, material resources, money, information) Efficiency of the transformation depends on motivation,

qualification and rate of freedom (variety of action) the bearer of activity holds.

The human activities seem to be in a mutual interaction, which is directed by means of:

1. Co-operation - that is by interaction among bearers of activities by means of which the aims, the ways of resource exploitation and activity performance are set more precisely with regard to the current aims of the organization. Co-operation can be directed by horizontal coordination.

2. Control - that is the execution of the power in the framework of a hierarchic power system, by means of which the aims and results are keeping in harmony inside a feedback loop.

## Organization Structures

The components of the organization and their interaction create hierarchic structures. The way of their arrangement influences organization behaviour towards the environment. The increase in rate of organization reduces the rate of human action variety in the organization as well as the variety of reactions of the organization as a whole to the stimuli from the environment. The rate of organization is given by specialization, coordination, configuration, delegation of powers and formalization (Kieser and Kubitzek, 1983). These five variables, each of them getting n-values, would define $5^n$ types of arrangement.

From the cybernetics point of view, the things will be considerably simplified if we focus on one variable only, the one that indicates the way of coordination, that is the way of arrangement of vertical and horizontal linkages in the system. We are going to specify the following types of organization (Fig. 1).
  I. The stable type based on dominant vertical linkages and a low rate of division of work (line type or a charismatic organization).
  II. The ultrastable typ (Ashby, 1964) representing a system with firm linkages among particular activities that are predominantly conrolled in feedback loops (functional or bureaucratic type).
  III. The multistable type arising form relatively autonomous subsystems with internal horizontal linkages inside of those subsystems (divisional organization).
  IV. Co-operative organization is the expression of synergic effect as a result of co-operation among particular subsytems functioning at a limit of stableness. The limits of stableness can be reached only by increased supply of energy-by adequate motivation influencing the human behaviour at the limit of availability (project, matrix organization).

In the direction from type I to IV, the share of co-operation among people is growing at the expense of veertical feedback coordination. The rate of coordination is thus the property that differenciates the types I. - IV.

## Organization and Environment

Organization is an open system (L. Bertalanffy, 1968). In order to survive, it has to adapt itself to the environmental changes. The view of organizational performance as well as of rationality of arrangement depends on the environment - contingency approach.

Environment that influences an organization by means of long-termly determined stable aims and at the same time it does not provide the organization either with opportunities or hard economic restrictions which are typical conditions for a planned economy, such an environment demands reactive behaviour, where stability of performance is a primary criterion. Turbulent dynamic environment creates on the other hand new opportunities, but demands entrepreneurial behaviour, the effort to take full advantage of organizational resources potential, the ability to adapt itself, eager looking for new opportunities and innovation.

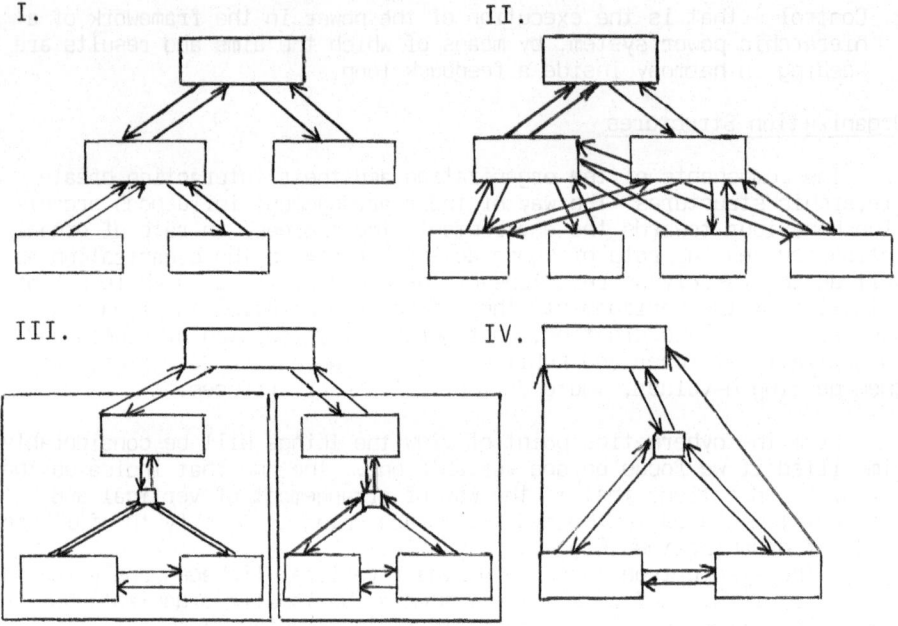

Figure 1. Diagram illustrating the types of
organization components arrangement

The external environment of the organization differs according to branches (P. Lawrence and Lorch, 1970), technologies (J. Woodward, 1971), and innovation.

There is another external factor - a stage of enterprise development and its size. Dependence of enviromental characteristics, structure and behaviour is depicted in Fig. 2.

The study of relations among the structure, functioning and external factors of the organization have resulted in the following conslusions:
- with the growth of dynamism, rigidity and turbulence of environment, the organization is changing from an executive into an entrepreneurial one,
- with the growth of innovative efficiency the organization is changing from a bureaucratic and administrative into a specialized or a pioneer one,

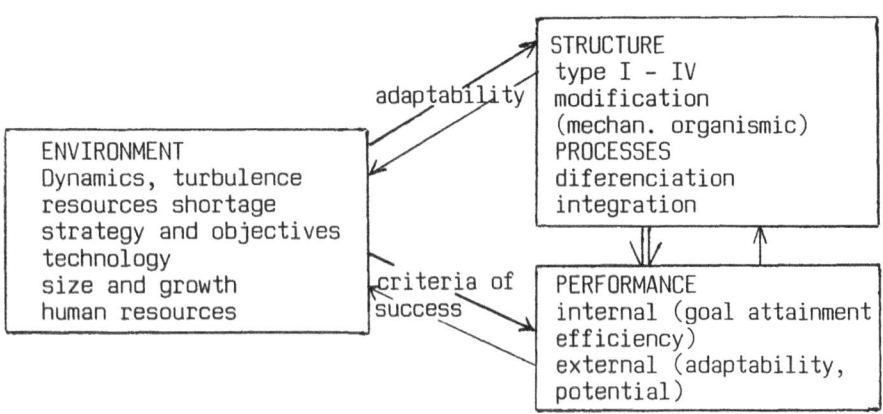

Figure 2. Relations among System Characteristics
of Organizations

- for a series production the organization is mechanic, for a piece
  production an organic one,
- with the development and growth of the organization, the arrangement
  is changing from a visionary and pioneer form to the top - it is a
  synergic top, then a capability fo react to the changes of environment
  is falling down across the administrative and bureaucratic stages.

There is one common cybernetic variable into which all factors of
environment can be converted; it is a variety. According to the law of
requisitive variety (R. Ashby, 1964), figuratively espressed by words
"only variety can destroy variety", it is possible to derive that only
an increase in variety of organization behaviour can cope with an environmental variety. Behaviour variety is deremined by the type of
arrangement - it is growing from the type I to IV. At the same time the
rate of co-operation isgrowing. The conclusion is then as follows:
"Growing variety of environment can be destroyed by human co-operation".

A higher rate of co-operation, however, implies an increase in
costs and energy for the maintenance of the system, growth of complexity
and proable unstability and irregularity in the behaviour of the system.
Therefore, it is suitable to involve a type of arrangement appropriate to
the variety of environment.

The growth of autonomy (from type I to IV) connected with divergency
of partial aims and attributes of individual organizational units is
inseparably associated with competition. Competition and Conflict is an
incidental phenomenon of co-operation. In the past, especially in a
planned economy, an opinion often dominated saying that competition
should be excluded even at the expense of a limited co-operation. On the
other hand, there are authors, e.g. Peters (1984), who put special
emphasis on entrepreneurship and competition and their influence upon
organization motivation and innovative efficiency. To reach enterpre-
neurial behaviour in a turbulent environment full of unpredictable states
so that the organization could be sucessfully adapted and survive, it is
necessary to bring co-operation and competition into balance.

A turbulent environment with a higher variety requires an adequate
organizational culture as a set of forms of human behaviour, norms,
values, visions, supporting a wll balanced rate of co-operation and
individuality (behaviour autonomy) of people.

ORGANIZATIONAL ANALYSIS AND DESIGN

Organizational design is concerned with the proposal of organization arrangement in order to reach a desider behaviour in a given environment. It is necessary to take into consideration that a design of social systems is in question, that is a plurastic-systemic problematic situation (M.C. Jackson, 1987). It is suitable to take the advantage of a "soft system methodology" of P.B. Checkland (1981).

The methodology was applied in the course of implementation of a number of projects recently. The experience tells, however, that not all applications were successfull in our cicrumstances. The reasons are as follows:
- unwillingness or fears of the participants to be involved in a discussion and express frankly their attitudes to a problem situation,
- dislike of power structures in the enterprises to tolerate or accept different viewpoints,
- the strenght of inertia and the effort to preserve status quo and a low efficiency,
- unfamiliarities and novelty of a "soft" method, customer´s expectations from a designer that he will find an optimal solution and put it into practice himself.

The need to create organizational culture and introduce "methods of organizational changes management" is arising. These methods would encourage the acceptance of changes as a logical factor of the enterprise life as well as that of a man, working in that enterprise. (Herakleitos: "There is nothing more stable than a change.").

CONCLUSION

Dynamic changes in the Czechoslovak economy on its path to a market economy,a creation of turbulent environment bring about a change in the behaviour of an organization as a whole as well as in the behaviour of individualis.

New opportunities for the development and application of new direct-ions of Management Science in the area of organizational analysis and design are being offered currently. An immense space is available to verify experimentally the methods used in conditions of advanced market economies. In that field the achievements of British Management Science are of a prominent significance and they are on the world level. Thus, there is a chance to initiate joint Czechoslovak and British research projects in the area of Management Science.

REFERENCES

Ansoff, H.J., 1984, Implementing Strategic Management. Prentice Hall, Englewood Cliffs.
Ashby, W.R., 1964, An Introduction to Cybernetics, Methuen, London.
Beer, S., 1979, The Heart of Enterprise, Wiley, Chichester.
Beer, S., 1981, Diagnosing the System for Organizations, John Willey, Chichester.
Bertalanffy, L., 1968, General Systems Theory, George Braziler, New York.
Espejo, R., 1987, From machines to people and organizations: A cybernetics insight to management, in: "new Directions in Management Science", Gower House, Hants.
Gerlich, P., 1989, Systems Characteristics of organization and their utili-zation in organization project (research thesis), VŠB Ostrava.
Checkland, P., 1981, Systems Thinking, Systems Practice, Wiley, Chichester.

Jackson, M.C. and Keys, P., 1987, New  Directions in Management Science,
        Gower, House, Hants.
Jackson, M.C., 1988, Systems Methods for Organizational Analysis and Design,
        <u>Systems Research</u> Vol 5, No3, pp 201-210.

THEORIES OF AUTONOMOUS SYSTEMS:

A COMPARATIVE ANALYSIS

Georg von Krogh

SDA BOCCONI

The Graduate School of
Business Administration
Milano, Italy

Olav Solem

The Norwegian Inst. of
Technology
The University of
Trondheim,
Trondheim

INTRODUCTION

In an attempt to develop a theory for strategy implementation in a newly acquired company, von Krogh (1990) analyzed and applied theories of sociocybernetics. As part of the work an extencive literature review of acquisition strategy studies was performed.

Based on this work the main purpose of the present paper is to give an overview, to compare and to discuss recent development in theories of autonomous systems. This is to be done especially with respect to how to combine two autonomous systems.

The interest for theories of autonomous systems was stimulated through a parallel empirical study of acquisitions to the extensive literature review mentioned above. Analyzing the findings from the empirical study, it became evident that the term autonomy was of great importance for the success of the acquisition. Furthermore, success seemed to be related to autonomy in strategic decision making, realization of synergies and various forms of communication. There were indications that problems of misunderstandings, misfit, resistance, lack of task coordination, stress and lack of follow up were avoided by the granting of autonomy to the acquired organization. This suggested that the theory of autonomous systems could be useful in the work of developing new theories of strategy implementation in a newly acquired organization.

In elaborating theories on autonomous systems, the theory of autopoiese is especially central. Three exponents of the discourse on autopoietic systems will here be considered: Maturana and Varela, Luhman and Bråten.

AUTOPOIESE AFTER MATURANA AND VARELA

The theory of autopoietic systems was originally developed within the field of neurobiology by the two Chileans Maturana and Varela. Their ideas can be traced back to earlier work by Maturana (1958, 1960), Maturana and Frenk (1965) and Varela (1979). The ideas were developed as a consequence of the researchers confrontation with questions concerning the autonomy of biological systems.

Maturana and Varela (1987) view autopoietic systems as self-producing systems. There is no output or input relationships with the environment. Everything produced is produced by the system itself. Living systems, for instance, engage themselves in self-formation through a set of recursive operations. Instead of input-output system transformation these recursive operations account for the transformation in a system (Maturana, 1981). Every state in a system unit produces another in the same unit, thus creating circularity (Maturana and Varela, 1987) - referred to as circular organization. In this theory all living systems with respect to mind or consiousness are closed entities, and thus autonomous. Another important aspect of autopoietic systems is self-reference, i.e. a state in an autopoietic system is internally determined by earlier states in the same system.

As underlined above self-production, circularity, autonomy and self-reference are four main principles in the theory of autopoietic systems making them closed systems. But also within the theory of autopoiese, the term "openness" occurs. According to the ideas of Varela an autopoietic system description is merely a complementary description of a system exiting together with the open-system description.

According to Maturana and Varela (1987) certain kinds of interaction and input/output description are accounted for by what they term "structural coupling", i.e. leading to structural coherence, which means coherence between components and relations in two systems. Autopoietic systems are not closed in terms of interaction, but in terms of organization. Organization is the relation between system components which make the system belong to a certain class. The pattern of relations that define the systems organization is in turn defining the interactions that the system can go into.

Maturana and Varela (1987) invite an understanding of social systems not as autopoetic systems, but as systems of interaction or "structural coupling". They introduce levels which might distinguish the level of autopoiese from the level of interaction. A human being, for instance, is a part of the social system insofar as it interacts in the network which defines the social system. As such the social system is seen as a medium in which the units and individuals realize their own autopoieses (Maturana and Varela, 1980).

Interaction in a social system, which is the concern of this paper, can in the theory of Maturana and Varela be seen as communication. Communication in turn can be interpreted as a kind of action.

AUTOPOIESE AFTER LUHMAN

Niklas Luhman is a German sociologist who has taken an interest in the theory of autopoietic systems attempting to generalize the theory. Luhman (1986, 1984,1984/2) distinguishes between communication and consciousness. The last term, consciousness, implies the possibility to explain the autonomy of the individual while communication implies the possibility to explain the autonomy of the social system, which is seen as a meaning-processing system.

Luhman (1986) suggests that social systems consist of and use communication as their mode of reproduction. Information, utterance and understanding are seen as the three basic entities which when integrated forms communication. Information, utterance and understanding are recreated from situation to situation, a cycle that requires self-reference and consciousness. To be able to create communication one has to have had previous communication. This previous communication creates consciousness. When consciousness is present, new communication can in turn be created. This view implies that all communication is encased within the system, and it fulfills the self-production idea of autopoiese.

On the individual level Luhman treats the mind as a closed system. Closure of the mind can be understood as the control of ones own negation possibilities, - the possibility to say "no"! (Luhman, 1988, p. 15).

The question about self-reference is also treated by Luhman. In Luhman (1984) he suggests that to refer is an operation. System-reference labels an operation which uses the terms system and environment to identify the system. Self-reference is an operation in which the one who performs the operation identifies himself with this operation.

## DIALOGICAL SYSTEMS THEORY AFTER BRÅTEN

Stein Bråten, a Norwegian sociologist, is the third exponent of the discourse on autonomous systems to be considered. Bråten is especially interesting because he addresses the problem of combining two autonomous systems, which is accomplished through the principles of complementary autonomy. More specifically he chooses to use communication, not in Luhman's way, but in the sense of a dialogical circle. Communication is taking place between systems possessing different perspectives and being autonomous in the control of their own perspective.

Bråten mainly uses the individual actor as the unit of analysis. In Bråten (1986, 1988), however, he generalizes his theories to also encompass the socio-cultural systems that are systems sharing the same worldview.

Bråten (1986/2, p. 11) suggests that meaning develops in the dialogue between individuals "within the reality which they create and in which they exist." There must be a primary dialogical circle, both within the individual and between individuals. Consciousness or awareness assumes the presence of more than one perspective, and is established through the means of dialogue. Within the individual the dialogue is to take place with a virtual other until the dialogue with the actual other starts. In the introduction of the virtual other Bråten (op. cit.) deviates from the two other works regarding self-reference, by also introducing other reference.

Important elements in the dialogical systems theory are the terms model monopoly (based on Bråten 1973, 1981) and discourse. In a given social setting involving discourse there could be model strong and model weak actors, having their names after the model resources they possess.

## A COMPARATIVE ANALYSIS OF THE PRESENTED THEORIES

This chapter is based on Bråten (1986/2) and extends his work. The three described positions are compared on the following points:

- Constitution of the social system
- Two autonomous systems combining

### 1.   Constitution of the social system

Maturana and Varela (1987) suggested that social systems are constituted of human beings who individually realize their autopoiese. The social systems after this scheme only gives the conditions which are necessary for the realization of the "individual autopoiese". The language in this sense is important because it shows how the society exists for the members of the society due to the fact that language is invented by individuals.

Social life as said above can be understood as individuals interacting. These interactions only arise as a result of coordinated behavior between individuals, who in other respects are seen as independent. Coordinated behavior is a result of communication. Communication, however, is in turn not seen as transferred information. Maturana and Varela (1987) suggest that communication does not depend on what is transferred in interactions, but rather on the one who receives it.

Luhman's (1986) suggestions that social systems are systems of communication, and that autopoiese happens on a societal level depart from Maturana and Varela's position. A society or an organization realizes its autopoiese rather than the individual. In this sense Luhman (1986) does not follow the "biological roots" of Maturana and Varela, but merely seeks to participate in the development of a "general theory of autopoiese" (Luhman, 1986).

It might thus also be said that Luhman and Maturana/Varela differ in their way of regarding coordinated behavior. Luhman (1986) however shares the view on communication as closed held by Maturana and Varela (1987), but on a different level, not limited to the effects of communication within "the individual", but rather within society since communication in itself is used to produce and reproduce society.

Bråten (1986/2) suggests that communication (the dialogue, not to be confused with Luhmans term communication) is a characteristic of the social systems, but contrary to the two other positions he states that consciousness comes into being when there are more than one perspective. The social system maintains its creativity and thus its capacity for development by allowing a dialogue between two complementary perspectives.

Maturana and Varela (1987) could open for the presence of rival perspectives as they understand communication in subjectivistic terms being only dependent on the reactions of the individual receiver. Different patterns of reactions would relate to different perspectives. But consciousness does not presuppose interaction in the terms of Bråten. An autopoietic system does not have the ability to take another autopoietic systems perspective because it is closed. A perspective cannot transcend from one system. Varela's (1979) suggestions, however, deviates somewhat from this strict view by allowing for conversation between two autonomous systems through a "mediating common ground" like, for instance, a common language.

Further, treating Luhman (1986) on the "autopoietic level", in this case meaning society, it is found that communication in terms of utterance, information and understanding does not extend to the other autopoietic system. The "other system" is accounted for in terms of "interaction" and described as the environment of the system. But he recognizes the need for a position in which the shift of perspectives are allowed for. This already indicates that Bråten's theories can be seen as complementary to those of Luhman. Perspectives can serve as an explanation of why communication is understood or not understood, or why it is rejected.

Luhman's theories do not account for, at least explicitly, the presence of perspectives. Hence, in comparison with Bråten, certain values are not explicitly present in his theories. Purpose and steering are other topics to be commented. In the theory of Maturana and Varela (1987) the purpose of the system seems to be to reproduce itself. Luhman (1986), however, states the following (Luhman, 1986, pp. 183):

"The theory of autopoietic systems formulate a situation of binary choice. A system either continues its autopoiesis or it does not."

This can be interpreted as a choice within the system. Thus continuation is a result of choice, and before choice consciousness (the ability to make a choice) and purpose (reflected in a criterion for choice) must be present. Thus the ideas of Maturana and Varela (1987) and those of Luhman (1986) seem to be incompatible on this point. Purpose, if assuming correct interpretation of Luhman (1986) could also be to break the autopoiese with respect to communication, for instance, through rejection.

Luhman (1986) states that "society" needs a simplified model of itself. This corresponds to Bråten's work, but the latter is focusing on models held by the system of the other system, on which own actions and subsequent effects under conditions of sufficient model resources could be simulated. A system which has such a model which could account for own actions and predict exactly the outcome of own actions in a discourse would be model strong. This means that Bråten also to a certain extent includes the system's description of itself in such a model. Luhman (1986) does not in his work account for the possibility of asymmetric distribution of model resources. Bråten (1986) suggests, based on the Conant-Ashby theorem, that the model-strong actor can control the model weak actor. And this is at the core of the differences; Bråten states that control or monological thinking can be present, but not wanted because it prevents creativity and development. On the other hand Luhman and Maturana (Varela) do not understand external steering in terms of models or perspectives. This is not unnatural since, as seen above, information cannot be taken in from the environment. Then there is no external steering.

## 2. Two autonomous systems combining

How does the three theories account for the situation where two autonomous systems combine? This question is especially relevant for the case of acquisitions, reflecting a situation where two organizations are to combine.

For Bråten (1986/2) the answer to the question derives from his suggestion of the dialogical circle and the cyclic presence of the virtual and actual other. He notes in Bråten (1988,pp. 218):

"By virtue of this postulated intersubjective circle in the participant, involving the primary participation of the virtual others viewpoint, complementing the subjective viewpoint of the participant, no qualitative shift need be entailed when any actual other steps into the circle and fills the space of the participants virtual other."

Thus a recognizable event of shift between actual and virtual other can be identified when two autonomous systems begin their interaction. Two systems thus have a common "systems border" even though they maintain their respective autonomy meaning that they maintain their perspectives.

Varela (1979) suggests an approach with a dialogical circle. Building on Pask's theory of conversation (see for instance Pask, 1978 and Pask, 1975), he is not deviating much from Bråten's (1986/2) position. Maturana, however, maintains the notion that complementarity of perspectives does not exist other than for an external observer because none of the two systems are able to take the other system's perspective (Bråten, 1986). Thus two systems "combining" is understood as different degrees or maturity of "structural coupling". A certain culture, corresponding to coordinated behavior, does develop as a result of these couplings (Maturana and Varela, 1987).

One has a mediating common ground, a language or "life world" (see Bråten, 1988, on the term "lifeworld", see for instance Habermas, 1984, in the theory of Maturana and Varela this is referred to as a "consensual domain"). This ground, however, has not been created through a process of "stepping outside the system's boundaries" by taking another perspective. Any social interaction is thus dependent on how the system "sees" the other system and not on the other system's perspectives, or on how it is "seen" from a third observer. At the outset any other system B is only a source of disturbance in broad terms for system A (Maturana and Varela, 1987, Kennealy, 1988). When two such systems start to interact they can form structural couplings and at the same time maintain their self-production.

Where then does Luhman (1986) stand with respect to the question of origin of two systems combining? According to him an autopoietic system has a binary choice of continuing its self-production. From this it follows that every system either communicates or it does not communicate, according to their choice of whether to reproduce own components or not. As mentioned, Luhman (1986) suggests the distinction between "interactions" and "societies". He notes that systems in interaction cannot import "ready-made" communication from their environment, they have to be introduced to the topic, maybe to learn about the utterances, the way of understanding and to gain information about the topic, in other words to communicate. In this case a new boundary encompasses a combined enlarged system. The result is a "society" in Luhman's terminology. This is different from Bråten's view. In Luhman's theories communication forms autonomous systems and there is no dialogical circle inherent in the system which eventually prepares the system for approaching the other system. According to Luhman's perspective the autonomy of the system dissolves when communication starts whereas in Bråten's perspective it is maintained even when the systems communicate.


FINAL REMARKS

To summarize, the tree positions differ in their view on autonomy as a result of two systems "combining". Maturana and Varela assume complete autonomy for the individual. Luhman assumes that there is no autonomy at the point of communication and Bråten assumes that autonomy is preserved even in communication. This leads one to ask if there is not another and intermediate position, namely the possibility that a system enjoys different degrees of autonomy?

By including perspectives in the description in the combination of two systems, various degrees of autonomy can be applied rather than the binary constitution of autonomy/not autonomy. The degree of coherence in perspectives or the presence of system specific perspectives could, for instance, account for the different degrees of autonomy. This in turn indicates how Bråten's theory is complementary to Luhman's theory of autopoiese.

Based on the above discussion, a model for combining two autonomous system has been established (von Krogh, 1990). Altogether 10 propositions was formulated indicating how to implement strategic decisions in a newly acquired organization.


REFERENCES

Bråten, S. "Between dialogical mind and monological reason: postulating the virtual other" in Campanella, M. (ed.) "Between rationality and cognition: policy making under conditions of uncertainty, complexity and turbulence", Albert Meynier, Torino, 1988

Bråten, S. "The third position; beyond artificial and autopoetic reduction"; in Geyer, F. and van der Zouwen, J. (eds.) "sociocybernetic paradoxes", Sage, Beverly Hills, 1986

Bråten, S. "Paradigms of autonomy", Teubner, G. "Autopoiesis in law and society", De Greuyter, New York, (forthcomming) paper dated 1986/2

Bråten, S. "Modeller av menneske og samfunn", Universitetsforlaget, Oslo, 1981

Bråten, S. "Model monopoly and Communication", Acta Sociologica, 2, 1973, pp. 98-107

Kennealy, P. "Talking about autopoiesis - order from noise?" in Teubner, G. (ed.) "Autopoietic law: a new approach to law and society", Walter de Gruyter, Berlin, 1988

v. Krogh, G. "A theoretical analycis of strategy implementation in a newly acquired organization", Dr.ing.-thesis, UNIT - NTH, Trondheim, 1990

Habermas, J. "The theory of communicative action", Beacon Press, MA, 1984

Luhman, N. "The unity of the legal system", in Teubner, G. (ed.) "Autopoietic law: a new approach to law and society", Walter de Gruyter, Berlin, 1988

Luhman, N. "The autopoiesis of social systems" in Geyer, F. and van der Zouwen, J.(eds.) "sociocybernetic paradoxes", Sage, Beverly Hills, 1986

Luhman, N. "Soziale systeme. Grundriss einer allgemeinen Theoire." Suhrkamp, Frankfurt, 1984

Luhman, N. "Die Wirtschaft der Gesellschaft als autopoietisches System", Zeitscrift fur Sozologie, 13, 1984/2, pp. 308-310

Maturana, H., Varela, F. "Kunnskapens tre", Ask, Aarhus, 1987

Maturana, H. "Autopoiesis" in Zeleny, M. (ed.), "Autopoiesis", North-Holland, New York, 1981

Maturana, H., Varela, F. "Autopoiesis and Cognition; the realization of the living", Heidl, London, 1980

Maturana, H. "The fine anatomy of the optic nerve of anurans - An electronm microscope study", Journal og Biophysical and Biochemical Cytology, 7, 1960, pp. 107-120

Maturana, H. "Efferent Fibres in the Optic Nerve of the Toad", Journal of Anatomy, 92, 1958, pp. 92-21

Maturana, H., Frenk, S. "Synaptic Connections of the Centrifugal Fibres in the Pigeon Retina", Science, 150, 1965, pp. 359-361

Pask, G. "A conversation theoretic approach to social systems", in Geyer, F., Van der Zouwen, J. (eds.) "Sociocybernetics", Martinius Nijhof, Leiden, 1978

Pask, G. "Conversation, cognition and learning", Elsevier, Amsterdam, 1975

Varela, F. "Principles of biological autonomy", North-Holland, New York, 1979

# CYBERNETIC ASPECTS OF THE MANAGEMENT SYSTEMS

# ENGINEERING

Piotr Sienkiewicz

Polish Cybernetics Society
Warsaw

## INTRODUCTION

Besides the technological progress, one of the elements of the universal progress of civilization is the progress in the field of organization. It consists in the development of the organizational structures as well as of the methods of controlling the processes. Both "the technological gap" and "the organizational gap" can be regarded as the systemic restraints which inhibit the development of national economy, the expansion of a company or an office.

Cybernetics has brought about an interest in models of systems and controlling the processes while systems engineering has caused the progress in methods of systems and their structures design. The development of cybernetics and systems engineering has formed the systemic trend in the contemporary scientific research on organization and management.

In conventional studies on organization and management the attention has been focussed on such notions as values, behaviour, styles, structures, organizational climate, organizational culture etc.

In case of the systemic trend its basic notions are: social system, organizational structures, development, information, decisions, efficiency, optimization etc.
Two domains can be distinguished in the structure of the contemporary science of management:
- management systems theory, and
- management systems engineering.

In the first domain, models of the management systems are created, in the second - methods of the optimum designing of structures and processes.

## BASIC PROBLEMS

### Organization

Organization is a social system in which all the functional sub-systems cooperate, due to certain co-ordinating links (control), to achieve local and global objectives of the system. Organization is made up of the

*Systems Thinking in Europe*, Edited by M.C. Jackson *et al.*
Plenum Press, New York, 1991

people belonging to the hierarchic organizational structures who, thanks to their knowledge and skills as well as to their informational, material and technological resources, transform the environment of the system according to their objectives and values. Every organization operates in an active, dynamic and stochastic environment. Every organization strives for achieving desirable conditions and preventing those which are undesirable for the developmental objectives of a system (efficiency). Desirable conditions with regard to the effectiveness of the operation determine a strategic direction of the organizational and economical development of a system.

The operation of an organization is determined by the following parameters:
- external (social needs, market, political, legal and economical conditions, access to resources etc.);
- internal (people with their norms and patterns of individual and group behaviour, material, technical and informational resources, models and rules of decision making, information systems and structures).

Every organization is an autonomous system with homeostatic characteristics and with ability to be in control for its own benefit as well as ability to prevent the loss of control.

## General model

It has been assumed that organization as a social system is composed of two basic systems:
- management system containing decision making system and information system;
- executive system containing operational system and logistic system.

Process of control is accomplished on two levels: on "central level" through management system and on "local level" owing to the process which controls the operational system as well as logistic system.

Example: the objective of organization is to maximize the efficiency function

$$F(x) \longrightarrow max$$

under the conditions $x \in X = x:h'(x) \leqslant 0, \ h''(x) = 0$
Suppose that $x = (v,d)$ and
$$v \in V = \{v:h_0'(v) \leqslant 0, \ h''(v) = 0\} \quad -\text{"central" decisions;}$$
$$d \in D_v = \{d:h_x'(v,d) \leqslant 0, \ h_x''(v,d) = 0\} \quad -\text{"local" decisions;}$$
so the optimization problem takes the following form:
$$\max_{x \in X} F(x) = \max_{(v,d) \in X} F(v,d) = \max_v \max_d F(v,d)$$
The objective of the executive systems: for given $v \in V$ such a vector $d^*(v) \in D_v$ has to be determined that for each $d \in D_v$ the following condition is fulfilled:

$$F(v,d^*(v)) = \max_{d \in D_v} F(v,d)$$

The objective of the management system: to find such a $v^* \in V_0$, where $V_0 \subset V$, that for each $v \in V_0$ the following condition is fulfilled:
$$F(v^*,d^*(v^*)) = \max_{v \in V_0} F(v,d^*(v)).$$

Problems

The basic problems of the contemporary management systems engineering and theory which respond to general tendencies in the practice of management include:

1. Integration of micro-scale problems ("man in the universe of organization", elementary units management) and of macro-scale problems ("organization in the universe of systems", large systems management);

2. Decision making problems belonging to the "complex-dynamic--uncertain" class ("deterministic man in a stochastic world");

3. Problems of integration of tactical and strategic decision making processes (tactical decisions result from the selection of strategies of system development);

4. Problems of the development of organizational structures which are flexible and easily adaptible ("keeping pace with environmental changes") and those which are active ("provoke changes in the environment");

5. Problems of optimization of information systems: designing processes of collecting, transmitting, processing of information and complex information servicing of management ("optimization of the info-sphere of the system");

6. Problems of optimization of decision making systems: designing systems supporting decision making (current and developmental), designing early warning systems etc. ("optimization of decisions and choices");

7. Problems of organizational applications of system analysis to multi-variant and multi-criterion evaluation of system efficiency of various socio-political and economical situations (from "the exclusively right solution" to "alternative thinking");

8. Problems of humanizing of the functioning of organization and management ("primacy of humanistic values over the narrow technocracy");

9. Problem of the opening up of organization and systemic philosophy ("to think globally and to act locally");

10. Problem of adaptation of the system to the requirements of the informational society ("information as the most precious resource and access to information as the source of organizational power").

CONCLUSION

The influence of cybernetics and systems theory on the development of the problems of organization and management can be described as a passage "from art to systems engineering". The universality of problems, methods and techniques of the management systems engineering should establish propitious conditions for their effective use for the improvement of the functioning of economical, political, educational and other organizations under the emergence of a new European order.

BIBLIOGRAPHY
Sienkiewicz P., 1987, "Inżynieria systemów kierowania", PWE, Warszawa;
Sienkiewicz P., 1989, "Systemy kierowania", WP, Warszawa.

# SYSTEMS EVOLUTION IN MODERN SYSTEMS RESEARCH

# AND A FORMAL MODEL FOR EVOLVING SYSTEMS

S.-J. Gao  and   F. J. Charlwood

Department of Systems Science
City University
Northampton Square
London EC1V 0HB
England

## INTRODUCTION

Although the study of systems evolution originated in early 60's as a discussion of self-organizing systems (von Foerster 1960; Ashby, 1962), it gained general awareness and became a defined research field in systems science only after the Brussel school's work in non-equilibrium thermodynamics in the later 70's and early 80's (Nicolis et. al, 1977, Prigogine, 1980). Over the past 10 years, much work has been done and many papers published in the study of progressive change of systems, i.e. systems evolution, although much of the work may be under the heading of "self-organization" in systems which is regarded as a specific manifestation of evolutionary process. Among the more important work are Prigogine's "Dissipative Structure Theory" (Nicolis et.al, 1977, 1989; Prigogine 1980; Prigogine et. al, 1984), Haken's "Synergetics" (Haken, 1983a, 1983b, 1988), Eigen's "Hypercycle" (Eigen et. al,1977, 1978a, 1978b), the study of "Cellular Automata" (Wolfram, 1983, 1984), and above all, the synthesis of thermodynamics and Darwin's theory of evolution (Weber et. al, 1988, Wicken, 1987). It is a real multi- and inter- disciplinary study and this is consistent with the spirit of systems research. It has been argued that systems evolution has become a new chapter of general systems theory (Jdanko, 1987).

## SYSTEMS EVOLUTION

Generally speaking, systems evolution is the process when systems change their structures so that they can function more efficiently in a changing environment. In most literature, it is defined as the process when systems move to higher ordered states with the entropy within them declining. We adapt this interpretive definition accepted in the systems community rather than attempt to define it in more rigorous terms.

Systems evolution is closely connected with those concepts such as emergent property, structural stability (Thom 1975). Any evolutionary process, in which a system is either transformed from an incoherent state to a coherent one, or changed from a less efficient functioning structure to a more efficient one, is manifested by the emergence of a new macroscopic structure which transcending the old one. This process is marked by creativity and novelty. Structurally, it is characterized by the qualitative structural change of the systems. From this very point view, the mathematical Dynamical Systems Theory (DST), which is competent in dealing with structural stability and emergent phenomena, is regarded as one of the prospective tools for the study of systems evolution (Abraham, 1988).

*Systems Thinking in Europe*, Edited by M.C. Jackson *et al.*
Plenum Press, New York, 1991

## Dissipative Structural theory

The Brussel school's work on dissipative structure theory is still one of the most influential in the study of systems evolution, not to say it is the most well known one. Drawing from the non-equilibrium thermodynamics, the main conclusion is that "non-equilibrium is the source of order" (Prigogine et al. 1984). Extending to the general case, it has been argued that if a system is open to its environment, far from thermodynamic equilibrium, governed by nonlinear dynamics, and possessing sufficient microscopic fluctuations, the change of environment causes the change of flux (matter, energy and information) through the system and this , coordinating with the system's nonlinear dynamics, may give rise to a new state of the system and, the system evolves spontaneously to a new ordered state. Special dynamical equations, such as partial differential equations of the reaction-diffusion type, master equation, are employed to model the evolving systems, and many results and techniques of dynamical systems theory have been used (Nicolis et. al, 1977, Prigogine, 1980).

Prigogine's work has provided a general framework of systems evolution and this is widely accepted in most discussions of systems evolution.

## Synergetics

In Synergetics, an open system is away from thermal equilibrium and is composed of many different competing and cooperating subsystems. It can undergo self-organization (evolution) in the sense that new ordered state at macroscopic level can emerge spontaneously. Examples are various, from natural systems to society. The inner dynamics governing those systems and the particular evolution process might be different, but it is believed that there is a universal principle underlying these phenomena. This principle was named by Herman Haken as "Slaving principle" and this is what Synergetics is searching for (Haken, 1983a, 1983b). According to this principle, the behavior of a system in the neighborhood of critical points is dominated by a few collective degree of freedom which are called order parameters The order parameters salve subsystems and give the total system its specific structure or order. To understand the specific process of evolution of a particular system, the key step, according to synergetics, is to detect the order parameters at the critical point. Given dynamical equations describing the system, the mathematical techniques employed to find the order parameters is the "adiabatic approximation". The use of **DST** in synergetics is plentiful (even more so than in dissipative structure theory). It is even stated that the slaving principle contains some deep theorems of DST as its special cases, such as the center manifold theorem, the slow manifold theorem, and adiabatic elimination procedure (Haken 1983b, pp316).

## Hypercycle

In studying the mechanism underlying the process of which biological systems come into existence from the existing molecules capable of replication, Eigen et al. analyzed the effect of mutation and competitive selection on molecules' structure. They have discovered the necessity of cooperation among the closed, cyclic replication process (Eigen and Schuster, 1977, 1978a, 1978b). It has been stressed that auto-catalytic and cross-catalytic processes play the essential role in establishing the dynamic correlation between chemical components. This process, named as "Hypercycle" by Eigen, establishes the feedback loops and multi layer connections between the components and therefore leads to the emergence of macroscopic order --- a biological system with properties irreducible from the composing chemical molecules. It is believed that this self-organization process links the biological world with the inanimate world and hence makes the universe as an evolutionary continuum.

This investigation is based on impressive mathematical treatment. The simplest system capable of selective self-organization can be defined by a dynamical system in such a general form:

$$dX/dt = (A - \Lambda - \Phi) (X)$$

where $X = (x_1, x_2, .., x_n)$ is the vector of population variables; $A$ reflects the positive contribution (amplification) to $\mathbf{X}$, $\Lambda$ the negative one (decomposition), while $\Phi$ refers to the out flux of $\mathbf{X}$ from the system. The mathematics techniques employed here is of dynamical systems theory (fixed points analysis in particular), although not explicitly stated.

## Cellular Automata

Although the recent interest in cellular automata was inspired by computer simulation of complex spatial patterns evolved from random initial conditions, they were originally introduced some thirty years ago as mathematical models for complex natural systems containing large numbers of simple identical components with local interactions. They consist of a lattice of sites, each with a finite set of possible values. The value of the sites evolve over time in discrete steps according to simple deterministic rules relating the value of a particular site with previous values of a neighborhood of sites around it (Wolfram, 1983). It has been found in computer simulation that a wide range of complex behavior can arise from random initial conditions through simple construction rule. Various attractors can be detected as similar to those found in nonlinear dynamical systems, such as point attractor, cyclic attractor, and strange attractor (Wolfram, 1984). The emergence of organized structure from dis-organized structure is the defining characteristic of self-organization process and therefore the theory of cellular automata is regarded as one of the potential tools to study systems evolution.

Discrete dynamical systems equations are employed to describe cellular automata and can be tackled by using dynamical systems theory.

## DYNAMICAL SYSTEMS THEORY (DST) AND THE STUDY OF SYSTEMS EVOLUTION

In mathematical sense, a dynamical system is described as a flow on a manifold, it evolves over time (Hirsch, 1984). The study of nonlinear dynamics consists of the study of structural stability, bifurcation and global behavior of this flow and the recent development in mathematical treatment of dynamical systems has greatly enriched our understanding of the complex behaviors of systems governed by nonlinear dynamics. In principle, any systems evolving over time can be modeled by such dynamical equations. The long range behavior of the system is proscribed by various attractors the dynamical equations possess. The bifurcation process, through which one type of attractor is replaced by another can serve as a prototype of emergent phenomena found in any evolutionary process. As argued by Abraham, in the study of self-organizing systems, or evolving systems in general, Dynamical Systems Theory (DST) seems to provide with:

1) simple geometric models for systems' complex behavior;
2) a complete taxonomy of attractors possessed by dynamical systems;
3) a mathematical rationale for the complex systems to evolve along a particular evolutionary path among different choices (Abraham, 1988).

As mentioned in previous sections, DST also brings with the solid mathematical techniques for the study of evolutionary process of particular systems.

## FORMAL MODEL FOR EVOLVING SYSTEMS

Formally, we can view a system as a whole of its components $\mathbf{E}$ together with the relations among them $\mathbf{R}$, i.e. $\mathbf{S} = (\mathbf{E}, \mathbf{R})$. Here we tend to describe a system by its internal variables and interrelations among its elements (or subsystems). When the system is changing over time $\mathbf{t}$, we denote the system as $\mathbf{S_t}=(\mathbf{E_t}, \mathbf{R_t})$ and call it a dynamical system. $\mathbf{R_t}$ represents the dynamical relations between a system's components. If the system is open to its environment, we denote $\mathbf{S_t}=(\mathbf{E_t}, \mathbf{R_t}, \lambda )$. $\mathbf{S_t}$ is constrained and influenced by the environment through this matter/energy /information flux and the influence is denoted by a parameter (or a parameter vector) $\lambda$. If the change of environment is relatively slow compared with the life time of the system, $\lambda$ is considered as a constant when we are analyzing the structure of a system at certain stage of the evolutionary process. A dynamical system $\mathbf{S}$ can be described by a group, say $\mathbf{n}$, of "observables". These observables are represented by an n-dimensional variable vector $X = (x_1, x_2, .., x_n)$ which takes its value on an n-dimensional space M, the state space of this system. When M is such a space where each point is assigned a rule which specifies the time changing behavior of that system at that particular point, it is called a generalized manifold. The rule which governs the system's time changing behavior at X not only describes the interactions among the systems components, but also reflects the impact of environment at that moment, and hence can be denoted $F_\lambda$. We also call $F_\lambda$ the inner dynamics of the dynamical system. In general, the system $S_t = (E_t, R_t, \lambda)$ is then described by the dynamical equation:

$$dX/dt = F_\lambda (X).$$   (an autonomous system)

The behavior of the system proscribed by the dynamics $F_\lambda$ is what of the most interest to us and it can be depicted on the manifold. By looking into its qualitative characteristics like attractors, basins and structural stability, we can understand how this system can evolve over time.

*Attractor*: let A be a sub-set of M, $U \supset A$ is a neighbourhood of A with $A \supseteq \cap_{t>0} F_\lambda^t$ U. A is called an attractor of the system $S_t$, iff it satisfies the following properties:

(1) Attractivity: for every V with $V \supset A$ we have $V \supseteq F_\lambda^t U$ for all sufficiently large t.

(2) Invariance: $F_\lambda^t A = A$, for all t.

(3) Irreducibility: if there exist another attractor A', $A \supseteq A'$, then A' = A.(also see Ruelle, 1989).

Apparently, an attractor is a time-independent set which attracts initial conditions from some region around it during a time-dependant process (evolutionary process). It represents an asymptotic state of the real world system to which the system move to as $t \rightarrow +\infty$. Any system recognized as a system can be viewed as a macroscopic attractor which makes the system maintain certain recognizable properties although its subsystems may be at different states (described by different attractors at the lower level)."*Every object, or physical form, can be represented as an attractor C of a dynamical system on a space M of internal variables*" (Thom, 1975). The ultimate state of thermodynamic equilibrium can also be regarded as a "final attractor" which, as described by the second law of thermodynamics, is the doomed state of all systems. The actual process, which moves a system to some other non-equilibrium attractors, is attributed to the exchange of matter, energy and information with the environment which imports "neg-entropy" to the system.

There are four different types of attractors appearing in dynamical systems: i.e. *static (point) attractor, periodic attractor (limit cycle), pseudo-periodic attractor (torus), and chaotic attractor (strange attractor)*. One attractor is separated by separatrices from others on the phase space and each attractor is the only one in its attracting basin. A complex system may possess some or all different types of attractors in the phase space, but at any given time t, the system can only be attracted to one of those attractors and this is decided by history ---- the system's previous state or initial conditions.The type, number and relative position of the attractors gives us a global picture of the system's possible state.

There are two different factors which might affect the state of a system: the microscopic fluctuation within the system and the perturbations from the environment. Affected by the fluctuations and perturbations, the system can be described:

$$dX/dt = F_\lambda (X) + \delta (X)$$

where $\delta (X)$ denotes the fluctuations and perturbations. Denote $T_A^\lambda, N_A^\lambda, P_A^\lambda$ the type, number, and relative position of the system's attractors at $\lambda$ respectively. Let $\Omega^\lambda = (T_A^\lambda, N_A^\lambda, P_A^\lambda)$ and $\Omega^{\lambda 1} = \Omega^{\lambda 2}$ if and only $T_A^{\lambda 1} = T_A^{\lambda 2}, N_A^{\lambda 1} = N_A^{\lambda 2}, P_A^{\lambda 1} = P_A^{\lambda 2}$. If $\Omega^{\lambda'} = \Omega^{\lambda 0}$ when $\lambda'$ is in the neighborhood of $\lambda_0$, we say that the system is structurally stable at $\lambda_0$. Evolution occurs when the structural stability of the system has been lost. The loss of structural stability is called a bifurcation, and during the bifurcation the transformation of the system's state to a new one, which is qualitatively different from the previous, can always be observed. The system at the new state possesses some properties which cannot be deduced from the old one and this is called *emergence*. Emergence is one of the striking phenomena exhibited by systems evolution (Swenson, 1989).

For the evolving system **S** , there are many different ways that the bifurcation can occur. Among them, the following bifurcation patterns have been detected :

point attractor $\longrightarrow$ point attractors;

point attractor $\longrightarrow$ periodic attractor;

periodic attractor $\longrightarrow$ periodic attractors;

periodic attractor $\longrightarrow$ pseudo-periodic attractors;

point attractor $\longrightarrow$ periodic attractor $\longrightarrow$ chaotic attractors;

(Haken, 1983b). From the mathematical sense, more bifurcation patterns have been found including more complicated ones relating to attractors, repellors, and saddle points all together (Stewart, et. al, 1986).Although it is still far away from having established the general theorems for various bifurcation patterns, there are some solid mathematical techniques for tackling bifurcations in some specific cases: Elementary Catastrophe Theory for bifurcation from point attractor to point attractors (Poston et. al, 1978), Hopf Bifurcation Theory for bifurcation from a point attractor to a periodic attractor (Hassard et al., 1981), period-doubling cascade for bifurcation to chaotic attractor (Feigenbaum, 1978). Applications can be found in many fields (Nicolis et. al, 1989; Guckenheimer et. al, 1983).

Usually, the dynamical equations describing a real-world system are nonlinear, high dimensional and very complicated. In that case, we use numerical analysis techniques rather than analytical methods. By comparative numarical analysis, we can detect various attractors at each parameter value representing different evolutionary stage and then construct the bifurcation diagram of the system by plotting the points representing the structure of the system on the parameters plane. This gives us the picture of the global bifurcation behavior over the whole range of parameters.

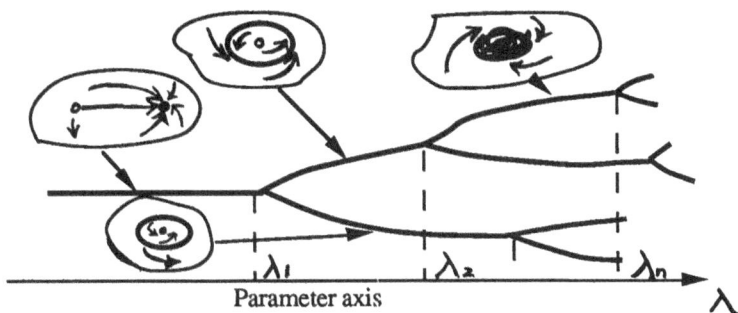

Parameter axis

*Figure. Big Picture of Bifurcation Casecade*

## Applications and Conclusions

The formal model and proposed study programme based on DST concentrates on the geometrical structure of systems and hence can be used widely to study systems evolution from the structure point view. It is complementary to those methods which stress the physical laws underlying the evolutionary process, such as Principle of Maximum Entropy Production (Swenson, 1989). Examples of this structure analysis can be found in many fields, physics, biology, ecology, just to name few (Haken, 1983a, 1983b; Zeeman, 1986).

Evolution is believed to be the behavior exhibited universally by open systems. It is a progressive change process in contrast to the change which drive systems to their thermodynamic equilibrium as described by the second law of thermodynamics. Darwin's theory of evolution and the second law of thermodynamics are no longer contradictory from the open systems theory point of view. They study different behavior exhibited by systems at different surroundings. In modern systems research, systems evolution has become a defined research field and many effects have been devoted to it over the last ten years. Among different methods, the use of mathematical dynamical systems theory has been discussed in this paper. It is believed that DST provides not only a useful model for systems evolution, but solid mathematical treatment as well. The mathematical techniques are especially useful and vital when we are trying to analyze the evolutionary process from the structural point of view. Our understanding of the being and becoming of the natural world as well as ourselves is surely enriched by the effort of the study of systems evolution, to which the DST plays its whole role.

References

Abraham, R.H. (1988) Dynamics and Self-organization. In F.E.Yates (eds) *Self-organizing Systems*: the Emergence of Order. Plenum Press: New York. pp599-614.
Ashby, R. (1962) Principles of the Self-organizing Systems. In H. von Foerster and Zopf, G.W. (eds.) *Principles of Self-organization*. Pergamon:New York.pp155-178.
Eigen, M. and Schuster, P. (1977) The Hypercycle, Part A: the Emergence of the Hypercycle. *Naturf.* 64:pp541-565.
------- (1978a) The Hypercycle, Part B: the Abatract Hypercycle. *Naturf.* 65:pp7-41.
------- (1978b) The Hypercycle, Part C: the Realistic Hypercycle. *Naturf.* 65: pp341-369.
Feigenbaum, M.J. (1978). Quantitative Universality for a Class of Nonlinear Transformations, *J.Stat.Phys.* 19. pp25-52.
Guckenheimer, J.A. and Holmes, P. (1983) *Nonlinear Oscillations, Dynamical Systems, and Bifurcations of Vector Fields*. Springer-Verlag: New York.
Haken, H. (1983a) *Synergetics*. Third edition. Springer-Verlag: New York etc..
------- (1983b) *Advanced Synergetics*. Springer-Verlag: Berlin etc..
------- (1988) *Information and Self-organization*. Springer-Verlag: New York etc..
Hassard, B.D., Kazarnoll, N.D. and Wan,Y. -H.(1981) *Theory and Application of Hopf Bifurcations*. Cambridge University Press: Cambridge:
Hirsch, M. (1985) The Dynamical Approach to Differential Equations. *Bull.Amer.Sco. Math.*, Vol.11 pp1-64.
Jdanko, A.V. (1988) Evolutionary Cybernetic Systems Theory Considered as a Chapter of General Systems Theory--- a Viewpoint. *Kybernetes* Vol.17, No.5, pp44-51.
Nicolis, G. and Prigogine, I. (1977) *Self-organization in Non- Equilibrium Systems*. Wiley: New York.
--------- (1989) *Exploring Complexity*. Wiley: New York.
Poston, P. and Stewart, I. (1978) *Catastrophe Theory and Its Application*. Pitman, London.
Prigogine, I. (1980) *From Being to Becoming*. Wiley: New York.
Prigogine, I. and Stenger, I. (1984) *Order out Chaos*. Bantam Books Inc.
Ruelle, D. (1989) *Chaotic Evolution and Strange attractors*.Cambridge University Press: Cambridge.
Stewart, H.B. and Thompson, J.W.T.(1986). Towards a Classification of Generic Bifurcations in Dissipative Dynamical Systems. *Dynamics and Stability of Systems*. Vol.1.No.1. pp 87-96.
Swenson, R. (1989) Emergent Attractors and the Law of Maximum Entropy Production: Foundations to a Theory of General Evolution. *Systems Research*, Vol.6, No.3:pp187-197.
Thom, R. (1975) *Structual Stability and Morphogenesis*. W.A.Benjamin: Reading, MA.
von Foerster, H. (1960) On Self-organizing Systems and their Environments. In M.C. Yovitz et al.(eds). *Self-organizing Systems*. Pergamon: New York.
Weber, B. and Depew,D.J. (eds) (1988) *Entropy, Information and Evolution*. MIT Press: Cambridge.
Wicken, J. (1987) *Evolution, Information, and Thermodynamics: Extending the Darwinian Program*. Oxford University Press, Oxford.
Wolfram, S. (1983) Statistical Mechanics of Cellular Automata. *Rev.Mod.Phys.* 55:pp601-642.
-------, (1984) Universality and Complexity in Cellular Automata. *Physica. 10D*. pp1-35.
Zeeman, E.C. (1986) Dynamics of Evolution. In S.Diner, Fargue, D. and Lochak, G. (eds.): *Dynamical Systems: a renewal of Mechanism*. World Scientific: Singapore.pp155-165.

# ORGANIZATIONAL CLOSURE AND THE QUANTUM VIEW OF ORGANIZATIONS

Gerrit Broekstra

Rotterdam School of Management
Erasmus University
P.O.Box 1738, 3000 DR Rotterdam
The Netherlands

## INTRODUCTION

Configuration theory is, I believe, one of the most promising trends of modern organization theory. Here, effective organizations are conceived of as configurations or, to make the parallel with quantum physics, to be in certain "quantum states" composed of tightly interdependent and mutually supportive elements such as strategy, structure, culture, technology, and human skills. As a consequence, a change of configuration may often entail a more or less revolutionary "quantum jump" (Miller and Friesen, 1984). The theoretical underpinnings of what appears to be emerging as a theory of configuration, however, are not well-established and not yet very strong. It is the general aim of this paper to contribute to this theory formation.

In contingency theory, which I regard as the logical precursor of configuration theory, the organismic open-systems or input-output/control paradigm became generally accepted as the underlying theoretical foundation and, indeed, is regarded as the standard view. The organization as an organism responding to the needs of the environment has thus become a prominent metaphor in organizational studies (Morgan, 1986).

As has been pointed out in a seminal paper by Goguen and Varela (1979), the dual, but complementary, system notion of control or input-output behavior is the autonomy or recursive view. Whereas in the control/autonomy distinction the control, or "other-law," perspective places the emphasis on the environment, the autonomy, or self-law (Varela, 1979), view arises when emphasis is placed on the system's internal constitution. Changing one's attention from control to autonomy is like exchanging figure and background, as in the well-known vase-two faces example. From the cognitive point of view, this may cause a dramatic Gestalt switch and completely change one's understanding of a system. In this way, when our cognitive focus switches to holistic notions like cooperative interaction, self-organization, and autonomy of a system, "then environmental influences become perturbations (rather than inputs) which are compensated for through the underlying recursive interdependence of the system's components" (Goguen and Varela, 1979).

*Systems Thinking in Europe*, Edited by M.C. Jackson *et al.*
Plenum Press, New York, 1991

In this autonomy realm of systems theory, Maturana and Varela (1980) pioneered the notion of autopoiesis which received a first wave of considerable attention cresting aroung 1980. More recently, the debate, particularly concerning the difficulties in applying the autopoiesis concept to social systems, appears to give rise to a second wave of interest (Mingers, 1989; Morgan, 1986; Broekstra, 1991). However, the generalized, though admittedly less exotic sounding, version of autopoiesis, *organizational closure*, developed at around the same time by Varela (1979), has received virtually no attention by organizational theorists. This, I believe, is unjustified.

Organizational closure may prove to be a powerful concept that may serve a central role in providing a sounder theoretical underpinning of the emerging configuration approach, just like the input-output open system model plays its central role in contingency theory. The purpose of this paper is to explore the connection and possible synthesis of these two important perspectives: the concept of organizational closure from the domain of systems thinking, and the holistic configuration approach from the realm of modern organization theory. It is of interest to note that both ideas originated at about the same time, the late seventies, early eighties, but, to my knowledge, have not been explored in synthesis.


SOCIAL-COGNITIVE STRUCTURE AND CONFIGURATION

An excellent and early example of the "total system" approach as they called it, is the typology of configurations as constructed by Miles and Snow (1978). They portrayed entire organizations "as integrated wholes in dynamic interaction with their environments." They were interested in the process of organizational adaptation, and were thus able to distinguish four organization archetypes, of which "each of these types has its own strategy for responding to the environment, and each has a particular configuration of technology, structure and process that is consistent with its strategy" (Miles and Snow, 1979). Three of these pure types, the Defender, the Analyzer, and the Prospector were considered stable, consistent, and effective in their respective environments. The fourth type, which they baptized the Reactor, was considered unstable, inconsistent, and ineffective.

For the purpose of this paper, it is important to note that Miles and Snow systematically based the description of their four configurations on a conceptual framework. This represented a model of organizational adaptation, called by them: the "adaptive cycle." In order to attain effectiveness in adaptation, from a top-management perspective, three major interrelated problems were identified by them: the entrepreneurial problem, the engineering problem, and the administrative problem. The four configurations then "represent alternative ways of moving through the adaptive cycle" (Miles and Snow, 1978). Although this point is not really emphasized by Miles and Snow, it is implicitly evident that these authors clearly sought their starting point in the (internal) cognitive structure of the manager grappling with the problem of effective organizational adaptation, not in the operational structure of the organization, "out there." As van Dongen would argue, this cognitive structure originates ultimately from the social interaction among managers, management scientists, etc. To emphasize this facet of cognitive mindsets, the adaptive cycle could therefore also be called a social-cognitive structure (van Dongen, 1991).

As a sidepoint, this focus on cognitive structure is, in my view, strongly reminiscent of the approach taken by Quinn and Rohrbaugh (1983) to the problem of effectiveness. They argued that the effectiveness debate, which raged in the seventies and left the effectiveness literature in disarray, had started at the wrong end. In accordance with the above, they argued that the focus should be "on the cognitive structure of the organizational theorist, not on the operational structure of the organization" (Quinn and Rohrbaugh, 1983). In other words, their method was based on making the "implicit and abstract notions of multiple theorists and researchers explicit and precise." The question they posed was really about the underlying social-cognitive structure, "How do individual theorists and researchers actually think about the construct of effectiveness?" Analogously, we could imagine that Miles and Snow might have posed the question, "How do top-managers actually think about the construct of (effective) organizational adaptation?" Their answer included the conceptualization of the adaptive cycle. Quinn and Rohrbaugh's findings suggested that three competing value dimensions underlie conceptualizations of organizational effectiveness. Most significantly, but not surprisingly given the central role of effectiveness in the organizational literature, these value dimensions turn out to be analogues of existing models of organizational analysis. As I have shown elsewhere, the four resulting models match perfectly with the four key systems of the Consistency Model (Broekstra, 1991).

Miles and Snow were well aware of the fact that a new approach of synthesis was needed in organization theory. They stated that "the behavior of organizations as total systems cannot be fully understood and predicted without concepts appropriate for this level of analysis. Typologies provide an excellent vehicle in this regard since their primary strengths are codification and prediction." A stronger case for what he called in the title of his paper "a new contingency approach: the search for organizational Gestalts" was made by Miller (1981), who observed that "the field of organizational theory seems to be reaching a crisis point: a state in which a central research approach is proving to be manifestly inadequate." Here he referred to the atomistic and linear method of analysis that has led to many conflicting findings in the field of organization theory. He therefore called for "a radically different approach to discovering predictive regularities in organizational data and a new view of organizations" (Miller, 1981).

In 1984, he and his colleague Friesen in collaboration with Mintzberg delivered (Miller and Friesen, 1984). They stated their preference for an approach of synthesis so that a "holistic, integrated image of reality" could be obtained. They suggested that organizational data on strategic, structural, and environmental variables tend to cluster tightly to produce distinct "quantum states," or configurations. The existence of stable and complex forms of interdependency among the variables, where the nature of the relationships vary from one to another Gestalt, has important implications for prediction and the perspective on change, which will not be discussed here. Clearly, their approach was empirical rather than conceptual. By means of a multivariate statistical procedure they were able to find a taxonomy of ten common configurations in a sample of 81 firms described along 31 variables of strategy, structure, information processing, environment and performance. Six of these archetypes were regarded as successful performers, the remaining four unsuccessful.

As to the justification of the choice of the main categories of variables selected by Miller and Friesen, they referred mainly to the general body of research on the strategy-structure and structure-environment literature and its relation to performance. For our purposes, it is of interest to note that their cognitive stance was somewhat vaguely rooted in the organismic metaphor. In fact, in selecting a broad variety of environmental, structural, and strategy-making variables, they were not guided by some well-defined conceptual framework. Instead, they "employed a rough biological analogy, viewing firms as organisms, which with their structural "anatomies" and strategic "behavioral repertoires" had the task of surviving a specific set of environmental challenges. Performance would ultimately reveal whether the match between anatomy, behavior, and environment was a successful one" (Miller and Friesen, 1984). They then continued to explain in general terms their definition of "some representative components within each of the three categories." As Morgan (1986) has argued, the application of a particular metaphor like the organismic one has the advantage of bringing out clearly certain features. A weakness is that some other aspects are underrated. This turns out to be also the case with Miller and Friesen's list of variables. Scrutinizing them with the help of a more fully integrated conceptual framework, the Consistency Model, developed by this author, certain missing variables can be detected particularly in the area's of technology, culture, politics, and human resources. This does underscore the importance of an integrated cognitive structure, and a balanced conceptual and empirical approach. In this respect, Miles and Snow's pioneering work was definitely superior. Nevertheless, Miller and Friesen's pathfinding study stands out as one of the best published in the eighties.

Whereas Miles and Snow implied that the "adaptive cycle" could be conceived of as the underlying configuring force putting organizational data into distinct clusters, Miller and Friesen discussed this driving force only in the more general terms of how under the umbrella of "the approach of synthesis as the objective of research...causation is viewed in the broadest possible terms." They state: "The approach of synthesis is really for *networks* of causation. Each configuration has to be considered as a system in which each attribute can influence many of the others by being an indispensible part of an integrated whole" (Miller and Friesen, 1984). This brings us to the next subject.

ORGANIZATIONAL CLOSURE AS THE DRIVING FORCE

Put into the context of cognitive structure, the "network of causation" could be given a far better foundation when we apply the concept of organizational closure as developed by

Varela (1979). First, however, one has to understand the meaning and difference between the terms *organization* and *structure*, as advocated by Maturana and Varela (1980). The relations between the components of a composite system that define it as an integrated whole of a particular kind, is called the organization. It is a network of relevant relations. The structure is then constituted by the actual components with all their properties and the actual relations between them that concretely realize a system as a particular member of a given class (see also Mingers, 1989). Although the term structure is still acceptable, the term organization used in this sense is bound to cause much confusion. I have therefore replaced Maturana and Varela's term organization by the term "deep structure" (Broekstra, 1991), but for reasons of tradition retained Varela's expression organizational closure.

Clearly, cognitive structures as represented by Miles and Snow's adaptive cycle, or this author's Consistency Model (Broekstra, 1984) are deep structures. The actual configuration, a Defender, for example, would then be the structure. To bring out more clearly the essence of integrated wholeness, or the self-referring nature of configurations I now suggest we consider applying Varela's notion of organizational closure. This is defined as follows (Varela, 1979; I replaced the term organization by deep structure).

> That is, their deep structure is characterized by processes such that (1) the processes are related as a network, so that they recursively depend on each other in the generation and realization of the processes themselves, and (2) they constitute the system as a unity recognizable in the space (domain) in which the processes exist.

The main point of this paper is thus that a configuration is fundamentally a "homeostatic machine" maintaining invariant its deep structure through a process of organizational closure as defined above. Since any configuration is here proposed to be, in fact, an autonomous system, we may paraphrase Varela's Closure Thesis (1979) in the following way: every configuration is organizationally closed. This means, as Varela stated, that if you are interested in the configuration aspects of an entity, go and look for the way in which its deep structure closes onto itself (cf.Varela, 1978). And, this opens some exciting new research perspectives, for example, in the area of (resistance to) change (Broekstra, 1991a).


CONCLUSIONS

Once the systemic notion of organizational closure is accepted as a powerful explanatory driving force underlying configurational structures, the attention switches to uncovering the self-referring deep structures that might serve as good candidates to which the organizational closure thesis can be applied. I purport that the Consistency Model, which builds on the work of Miles and Snow, is such a candidate, since it basically consists of a vital number of key processes interrelated in a recursive network supporting each other through consistency relations (Broekstra, 1984, 1991).

The model is also fundamentally a self-referring

cognitive structure through which managers or theorists may usefully view an organizational situation. I conclude that the notion of organizational closure has quite enriched the holistic conception of organizational models like the Consistency Model. Rather than viewing them as more or less truthful representations or abstractions of an organizational situation "out there," we should view them as cognitive structures "in here." From a philosophical point of view, I find it interesting to note, that a self-referring cognitive structure as represented by, for example, the Consistency Model, is congruent with the self-referring nature of consciousness in general. In my opinion, this is consistent with the metaphysical assumption of the holistic nature of reality. In this sense, systemic lenses may provide a more intelligent picture, or quantum view, of the organizational world than the more common fragmentary approaches we encounter still abundantly in daily managerial practice.

REFERENCES

Broekstra, G., 1984, MAMA: Management by Matching, in: "Cybernetics and Systems Resarch," R.Trappl,ed., Elsevier Science Publ., Holland.
Broekstra, G., 1991a, Consistency, configuration, closure, and change, in: "Steering, Autopoiesis, Configuration," R.J.in'tVeld, ed., Kluwer Ac.Publ.
Broekstra, G., 1991b, Effective management = Holomanagement, Proc. 35th Annual Meeting ISSS, Ostersund, Sweden.
Dongen, J.H. van, 1991, "Organizing Realities: Rethinking Organizing and Social Integration," to be publ.
Goguen, J.A., and Varela, F.J., 1979, Systems and distinctions; duality and complementarity, Int.J.General Systems, 5:31.
Maturana, H.R., and Varela, F.J., 1980, "Autopoiesis and cognition," Reidel Publ.Comp., Holland.
Miles, R.E., and Snow, C.C., 1978, "Organizational Strategy, Structure, and Process," McGraw-Hill Kogakusha, Tokyo.
Miller, D., 1981, Toward a new contingency approach: the search for organizational Gestalts, J.Management Studies, 18:1.
Miller, D., and Friesen, P.H., 1984, "Organizations: a Quantum View," Prentice-Hall, New Jersey.
Mingers, J., 1989, An introduction to autopoiesis-implications and applications, Systems Practice, 2:159.
Mintzberg, H., 1989, "Mintzberg on Management: Inside our Strange World of Organizations," The Free Press, New York.
Morgan, G., 1986, "Images of Organization," Sage Publ., London.
Quinn, R.E. and Rohrbaugh, J., 1983, A spatial model of effectiveness criteria: towards a competing value approach to organizational analysis, Mngt.Science, 29:363.
Varela, F.J., 1978, On being autonomous: the lessons of natural history for systems theory, in: "Applied General Systems Research," G.J.Klir, ed., Plenum Press, New York.
Varela, F.J., 1979, "Principles of Biological Autonomy," North Holland, New York.

# LANDSCAPE ECOLOGY AS AN OPERATIONAL FRAMEWORK FOR ENVIRONMENTAL

# GIS: ZDARSKE VRCHY, CZECHOSLOVAKIA

Ian Downey, Ian Heywood and James Petch

Geographic Information Systems Unit,
The Department of Geography
The University, Salford, M5 4WT, England

INTRODUCTION

The landscape is a finite but dynamic resource which requires careful management.  There are complex questions about its use and protection which range in scale from the local to the global. At a localised level the environmental manager may be required to evaluate, for example, whether a specific stand of trees should be felled or a given habitat protected.  On a regional scale he may need to determine which farming practices should be permitted or what type of recreational activity should be promoted. The problems of seeking a solution to such issues are complicated further by the range of individuals and organisations that may be affected by a chosen strategy, and who will want to be involved in decisions.  Landscape preservation, for example, affects individual farmers and regional agricultural economies whilst questions of air pollution are of concern to power generation managers as well as to private or public forest organisations.

This paper examines how two approaches to management are being used to formulate environmental policy in the Zdarske Vrchy protected area of Czechoslovakia.  Both have been used extensively to aid environmental policy making, but have evolved under quite different cultural and scientific influences. Jointly they provide a new perspective on environmental management.  The first approach originated in North America in direct response to the collection of large volumes of data for land inventory programmes during the late 1950's and early 1960's (Heywood, 1990).  The technology, now known widely as Geographic Information Systems (GIS) utilises computer mapping, digital image analysis and data management routines to help solve environmental problems. The second approach, expressed in the term 'landscape ecology' emerged from Central Europe and involves studying the landscape as a amalgam of natural and cultural conditions which has evolved over long periods.  Landscape stability can be assessed in terms of its territorial structure and relations between use and carrying capacity.

*Systems Thinking in Europe*, Edited by M.C. Jackson *et al.*
Plenum Press, New York, 1991

This study, is an exploratory step in using modern GIS techniques in the context of a central European approach to landscape management. Its purpose is, first, to provide a much needed conceptual basis to the environmental application of GIS technology and second to develop a working methodology for multi user landscape policy assesment.

THE GIS APPROACH

A GIS is a computerised means of storing, organising, analysing and presenting data which is geographically referenced. Conceptually it is convenient to visualise a GIS as storing data about a region as a series of map layers. Each map layer might represent a different element of the environment. Such as; natural resources geology, soils, hydrology, etc, or human related features such as population, agricultural productivity, pollution. The power of a GIS lies in the ability to integrate and analyse any number and combination of map layers to answer spatial questions such as:-

- What geographical patterns exist in data?
- Do these patterns relate to other (geographical) phenomena?
- Where is the best place to do something?
- What areas will be affected if this action is taken?

In answering these questions the GIS is used to build spatial models using the interaction of a given set of data layers about the environment. A GIS can make use of any geographically referenced data using a known co-ordinate system. The range of data which can be used in a GIS includes information taken from paper maps, satellite images, air photographs and statistical records.

THE LANDSCAPE ECOLOGY APPROACH

Landscape ecology is a relatively new discipline, developed by geographers and ecologists in Central Europe following the end of World War II. Its purpose is to provide a rational basis for landscape appraisal, planning, management, conservation and reclamation. Zonneveld (1982) describes the approach as being as much a state of mind and attitude as it is a scientific method. Landscape ecology draws much of its theoretical and conceptual foundation from General Systems Theory, Biocybernetics and Ecosystemology (Naveh and Lieberman, 1983) which support a holistic approach to the study of the landscape. Theoretically, landscape ecology offers a conceptual solution to bridging the gap between the study of natural, agricultural and urban systems which as, Naveh and Lieberman (1983) state, ".... is making landscape ecology increasingly the major scientific basis for the creation of more balanced and far sighted policies and decision making tools".

The landscape ecology approach uses the concept of ecological stability based on the premise that the landscape is made up of a mosaic of stabilising and destabilising segments. Stable segments of the landscape are often associated with natural

152

ecosystems which are rich in diversity. Unstable segments on the other hand tend to be associated with those areas where man's interaction with the land has been to maximise its productivity, for example areas of mono culture or industrial developments.

In Czechoslovakia, this theory of diversity and stability in ecosystems has been developed into a working methodology by Low, Bucek and co-workers (Bucek et al 1985). Their methodology considers landscape practices (destabilising) in the context of a framework of stabilising biocentres and connecting corridors. This framework is called the skeleton of ecological stability. This strategy, they argue, can be used to plan landscape management programmes on local, regional and national scales.

The design of skeletons of stability follows a defined procedure (Bucek and Lacina, 1979, 1981). First the landscape is divided into areas with specific ecological conditions, termed geobiocoenes. Then an assessment of the current state of vegetation is made within each of these areas. Statements about the functional possibilities of each area and groupings of bigeocoene types are drawn up. These are based on historic observations about landscape use and the current intensity of economic activity. This process leads to the production of a hierarchy of ecologically important elements of the landscape which provides the environmental manager with a set of working spatial units as a blue print from which to evaluate and develop a landscape plan.

SOME COMMON GROUND

Despite the differing cultural background from which GIS and landscape ecology have emerged they have much in common. First they both assume the need to take a holistic approach to a problem, second they both require that a given problem should be viewed from a systems perspective and finally both require the definition of a hierachical set of spatial units for data management and application purposes. Their connectivity is central to the flow of information through the system. In fact as, Naveh and Lieberman (1983) stress, despite the absence of the term landscape ecology from the literature on land use capability analysis and assessment out of which GIS technology developed in North America, it is in reality founded on some of the basic principles which underpin landscape ecology.

GIS AND LANDSCAPE ECOLOGY: THE ZDARSKE VRCHY PROJECT

Zdarske Vrchy (Zdar Hills) protected area covers some 715 square kilometres. It is underlaid by a granite massif audits topography, typical of Bohemia-Moravia, is that of a dissected plateau. Maximum relief is about 200m and slopes are rarely greater than 20 degrees. It is important in regional hydrology because it is the source area for several Bohemian and Moravian river. Forests occupy about sixty percent of the area and dominate the central ridges. State protection was established in 1970 to preserve the harmonious cultivated landscape which represented a balanced combination of natural and man made elements.

The current version of the Zdarske Vrchy GIS has been implemented using the SPANS GIS software. The SPANS GIS has a comprehensive modelling language which can be used to build both qualitative and quantitative landscape models. In association with the GIS a range of additional computer hardware and software has been used for both the capture and output of digital information from the GIS. This includes a range of image processing equipment, digitising facilities and plotters.

The base materials for the information system were contemporary maps and historical topographic and thematic maps at various scales from 1:10,000 to 1:200,000, aerial photographs from the 1970's; multispectral aerial photographs at 1:25,000 and 1:5,000 from 1986 and a Landsat TM image for May 1987 used at scales of approximately 1:25,000 and 1:50,000. Meteorological data and ecological survey data from the previous fifty years were also used. The range of paper maps for the region provided spatial information on forest extent, drainage, relief, transport networks, historical animal habitats, tourist paths, recreation areas, urban communities, administrative boundaries and the location of protected areas. The Landsat image was used to provide information on current land cover for the region. This was used to derive the skeleton of ecological stability and a range of other landscape indices including a forest edge index. The image processing procedures used to establish these data layers are discussed in more detail in Downey et al (1990).

In addition, data on regional demographics and animal habitat surveys were incorporated into the GIS. The final data layer incorporated was a map of the biogeocenoses made by Bucek and Lacina in 1975. This map, compiled from field and map data is a spatial representation of the natural units of the Zdarske landscape.

APPLICATIONS OF THE ZDARSKE VRCHY GIS

Each application of the GIS first requires a model of the object or property of interest to be constructed. The structure of this model is determined by the nature of the problem. Problems currently under investigation include potential forest damage from SO2 emissions, bird habitat analysis and recreation impact assessment.

For each problem a conceptual or scientific model is derived. This is expressed, as a flow diagram, statistically, or as a mathematical function. This model is then translated into what is termed a 'landscape model'. Five stages lead to the development of the landscape model:

-     defining the problem,
-     developing the conceptual or scientific model,
-     identifying the spatial and non spatial
      components of the model,
-     compiling the required data layers with in
      the GIS
-     establishing the nature and magnitude of the
      relationship between the layers in the model.

This process generates a new map which is a spatial expression of the conceptual model (Figure 1).

From a management perspective, it is essential to identify which of the model components are important in maintaining the stability of the landscape. The GIS uses the maps of biogeocenoses and the skeleton of ecological stability as "spatial filters" through which the results of the landscape model are passed. This allows the identification of areas which could experience adverse conditions or provide unstable landscape structures. Identification of these areas is essential if management priorities are to be targeted accurately and scarce financial resources spent efficiently. The point here is that any process which can be translated into landscape terms can be passed through these filters. The base maps describing landscape structure and limitations can be utilised as filters, in many different ways, to provide operational answers to functional and structural problems of landscape. They provide a currency for dealing with landscape problems related to forces and events which are essentially unrelated.

The conceptual and scientific models used within the GIS to construct the landscape model are designed with experts who have a substantial knowledge of the field. Future research will investigate how advances in the development of knowledge based systems could be used to capture knowledge from a range of experts and incorporate this within the GIS. Should this be possible it would allow for wider application of the technology, particularly in those areas where environmental decisions must be taken without access to a knowledge base for the problem.

HABITAT ANALYSIS AND MODELLING FOR ENDANGERED SPECIES

One problem being addressed by the integrated GIS and landscape ecology approach in Zdarche Vrchy is the development of a methodology for habitat assessment for an endangered species. Preliminary investigations have concentrated on the occurrence of Black Grouse (Tetrao tetrix) (Pauknerova et al, 1990).

The management of the protected landscape area has been monitoring local population of Black Grouse for many years. The decreasing number of breeding pairs and the reduction in habitat extent is of particular concern. The decline of the Black Grouse population is seen by ecologists to be a result of an increase in the number of destabalising activities introduced into the landscape by man, particularly tourism, agriculture and forestry.

The first stage in assessing the reasons for the decline of the bird and identifying areas for its reintroduction involved using the GIS to identify those areas in Zdaske Vrchy where the landscape structure corresponds to what is known about the species behaviour. It is known that the species favours a landscape with scattered tree groups and a mixture of different cover types. It is a pioneer type species which lives near the edge of the forest and non-forest areas. Therefore, forest maps were used in association with remotely sensed image analyis to produce a spatial classification of forest edge density (Pauknerova etal, 1990). In a similar way other landscape

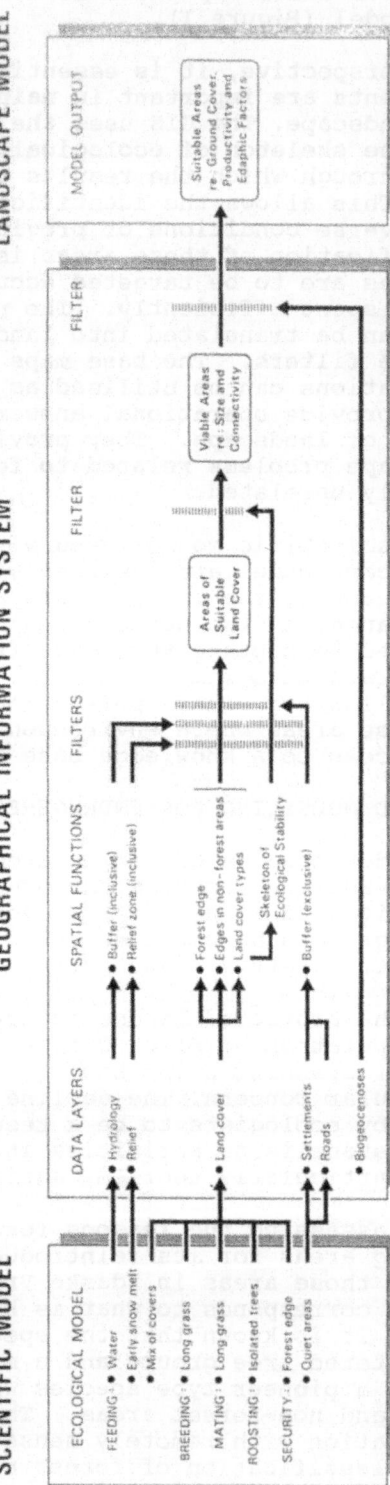

Fig 1.    Habitat Analysis for Black Grouse

elements were compiled in the GIS.  These are presented in Figure 1.

It is the interaction of these individual landscape elements which is important in determining which habitats are suitable for the bird, why existing habitats have disapeared and those areas favourable for colonisation.  The GIS modelling language was used to develop a series of qualitative and quantitative landscape models. Evidence about the importance of the various landscape elements was drawn from scientific literature, previous management experience and empirical observation in the field.

CONCLUSION

To date the Zdaske project has been an experiment in linking GIS technology with the theory and methods of landscape ecology. Preliminary work shows that the two approaches are complementary and can be developed mutually. At this stage in the project it is too early to assess their full potential as an operational tool. Future developments in the Zdaske project will concentrate on evaluating the different needs of the Zdarske managers, deriving landscape models to satisfy these needs and assessing their usefulness.


REFERENCES

Bucek, A. and Lacina, J. 1979, Biogeograficka diferenciace krajiny jako jeden z ekologictych podkladu pro unzemni planovani. Uzemni planovani a urbanismus 6:6: 382-387, Terplan, Praha.

Bucek, A. J. Lacina & J. Low. 1985, Vyhodnoceni prostorovych a funkcnich parametru USES, Brno, 16pp. (Evaluation of spatial and functional parameters of territorial systems of ecological stability).

Bucek, A, and Lacina, J. 1981, The use of biogeographical differentation in landscape protection and design.  Sbornik CSGS, Praha, 86 (1), 4-50.

Brokes, P., Downey, I., Pauknerova, E., Petch, J. and Corlyand, A. 1990, Habitat analysis and Modelling for endangered species. International Symposium on Ecology and Culture Ecological Development of the Svratka River Headwaters.  ICUN, Oct 9-11.

Downey, I., Heywood, D. I.,  Kless, P., Pauknerova, E. & Petch, J. 1990, GIS for Landscape Management, Zdarske Vrchy, Czechoslovakia in Proceedings of  GIS for the 90's, Second National Conference on Geographic Information Systems, Ottawa, Canada.

Heywood, D.I. 1990, Monitoring For Change: A Canadian Perspective on The Environmental Role for GIS, Mapping Awareness, Vol 4, No9, pp24-26.

Naveh, Z. and Lieberman, A.S. 1988, Landscape Ecology Theory and Application, Springer-Verlag, New York pp 356.

Petch J. 1990, The Tradition of Landscape Ecology in Czechoslovakia, in Cousins S. and R. Haines-Young, (eds) GIS in Landscape Management, Taylor and Francis, London (in press).

Zonneveld, I.S. 1972, Textbook of Photo-Interpretation, Vol 7. (Chapter 7: Use of aerial photo interpretation in geography and geomorphology). ITC, Enschede

# INTRODUCING GEOGRAPHICAL INFORMATION SYSTEMS INTO BRITISH

# LOCAL GOVERNMENT: DEVELOPMENTS IN KIRKLEES METROPOLITAN DISTRICT

Derek E. Reeve

Centre for Local and Regional Analysis
Huddersfield Polytechnic

'It is unlikely that any of our County, urban or
District authorities will be untouched within five
years by [the digital mapping] revolution which has
already started. Its a fair bet that even a few
Parish Councils will be dabbling with their home
computers' (Waters R S, 1989)

'Vendors exaggerate, most data sets contain errors,
most things are more difficult than experts say,
grafting GIS onto organisations is difficult'
(Rhind D, 1990)


## INTRODUCTION

British Local Governments are in a dilemma with regard to
Geographical Information Systems. There is presently a
powerful bandwagon pushing them towards implementing large
GIS databases. All the major computer vendors have GIS
systems, and are marketing them heavily into local
government; there have been influential central government
reports extolling the virtues of GIS; there are new journals
and glossy exhibitions; and academics, perhaps for their own
reasons, are ushering local government towards GIS.

On the other hand, there remain within local government
memories of the failures of a previous generation of
land-and-property databases. During the early 1970s a number
of British authorities embarked upon major exercises to
develop property databases which had objectives remarkably
similar to those being quoted today. These early systems
generally failed to achieve their objectives, leaving a
legacy of unfulfilled promises, cost overruns, half
implemented systems, and implemented systems for which there
proved to be little operational demand. Fingers were badly
burnt. Also, a careful reading of current GIS literature
would produce sufficient doubts to give a cautious Chief
Executive pause for thought.

*Systems Thinking in Europe*, Edited by M.C. Jackson *et al.*
Plenum Press, New York, 1991

This paper provides a brief of review of the current debate about GIS in local government, and secondly outlines the strategy being adopted by Kirklees Metropolitan District in to attempt to ensure that it can determine the suitability of GIS for its present requirements.


GIS IN LOCAL GOVERNMENT : ALTERNATIVE VIEWS

Local government officers charged with the responsibility of investigating the relevance of GIS for their authorities often will have had no direct experience of GIS, and thus are forced to rely upon evidence from sources such as vendors literature and demonstrations, academic papers, and reports of other local authorities' experiences. Such evidence can be both partial and confusing.

A very positive case can be built which argues that the bottlenecks which held back the 1970's land-and-property databases have now been removed. Whereas the computer technology of the Seventies struggled to meet the demands of land-and-property systems, the computing revolution has delivered the power needed at reasonable costs. In the Seventies, the software used for land-and-property databases was both difficult to use, and often inefficient in handling spatial data. Modern GIS systems are easier to use, and because of advances in spatial computing algorithms and spatial data-base structures, much more efficient. In the Seventies, a corporate database, was necessarily a centralized database. Now workstations and networks make it is possible to envisage corporate systems which appear to users as confederations of semi-independent stations, which might make them more amenable to local government organisational structures. Many of the early land databases floundered because they greatly underestimated the size of the database creation task, particularly digitizing property boundaries and georeferencing. Thanks to the Ordnance Survey digital mapping programme, and the emergence of private sector suppliers of geographic referencing systems, the creating a geographic database need no longer be such a daunting task. The benefits which it was argued, two decades ago, would flow from corporate land-and-property databases - data integrity, data currency, efficient data storage, digital mapping, and improved service to citizens at lower costs - are now practical propositions. Technology, it is argued, has caught up will ambition.

Cautious observers, however, might respond that so far current GIS has promised much, but little delivered. Most current local government initiatives are described as 'pilot' studies, and few would claim yet to have achieved a full implementation. Furthermore, comments emerging from the pilots suggest that some of the problems which beset the earlier databases may have changed their forms, but have not entirely disappeared. Modern GIS software is highly complex, many of the systems are still developing,and inevitably the systems contain bugs. It sometimes seems that the vendors are using their early customers as guinea-pigs. Despite the

emphasis on user-friendliness in the promotional literature,
GIS software is still not easy to learn or to use. For
example ARC/INFO, the largest selling GIS system, now has
over a thousand commands and could not be widely used
throughout an authority without considerable customisation.
Just as the 1970's LAMIS system required FIND2 gurus to
protect 'ordinary' users, today's systems require their high
priests of ARC. Despite the glossy literature, some of the
systems have small client bases, and one wonders at the level
of training and post sales support which such firms can
offer. Indeed, one wonders at the ability of some companies
to survive - pity the council which makes a wrong choice. The
inadequate quality of the Ordnance Survey digital data means
that creating 'geography' within systems is not as simple a
task as it might seem to external observers. For example,
Ordnance Survey only attach a single code to any feature,
which can have interesting effects. Famously, the Houses of
Parliament are shown as having been built with only three
sides, the fourth being depicted as the high water mark of
the Thames. Plymouth (Markham 1990) report that it took
twenty weeks of editing before their OS data were usable.
Most of the modern optimism for GIS has been fueled by
technological developments, but in local government the major
obstacles to successful implementation have been
organisational (Grimshaw, 1988). Again, several of the
current pilot studies have drawn attention to the continuing
truth of this. Consider, for example, Gault's (1989) remarks
on the 'political' status of GIS in Birmingham. Here,
apparently, rates of development within departments have
varied enormously 'due to combinations of factors such as
personalities, skills, vision, clout, ability to manipulate
budgets etc', and 'while a corporate ethos may be a laudable
goal to strive for, the reality of most local governments is
that there are empires to be built and defended'.

It is human nature to emphasize successes and to play down
ones failures. In recent months, however, there has been an
increasing openness about problems with GIS. Openshaw(1990)
in a commendably frank post-mortem of a failed GIS summarizes
the worst doubts which some observers have about the present
generation of GIS projects. He writes of GIS software so
badly written that it seemed purposely designed to punished
its user for mistakes; of technical (vendor) incompetence,
inadequate support, lack of commitment and professionalism,
and lack of adequate training. The vendor, he suggests,
effectively expected the user to pay to develop a skeletal
system into a commercial product that the vendor could then
sell.

There may be a further similarity between the 1970's
land-and- property systems and today's developments. Some of
the Seventies systems became victims of cut backs in local
government expenditure. Regardless of the technical
sophistications of modern GIS, the looming financial
restrictions on local government finance in the immediate
future might similarly curtail current initiatives.

The wave of enthusiasm for GIS has so far been fueled by

promises. As the current round of pilot local authority projects begin to make their reports, a new air of realism might emerge. Pessimists might sense a bubble about to burst.

## KIRKLEES: GIS INITIATIVES

Against this background, those local governments which have initiated GIS developments have done so on an experimental basis, authorizing limited pilot projects rather than full implementations. Kirklees Metropolitan District (KMC) certainly are pursuing this measured approach, having initiated two linked GIS pilot which are intended to provide KMC's decision takers with an informed basis upon which to determine whether to expand to a fuller implementation.

One of the pilot studies, the 'Town Centre GIS Pilot' is essentially a technical exercise, designed to answer questions about the adequacy of the software, and to give KMC's staff first hand experience of the realities of creating a property database. The pilot is limited both spatially and functionally. Spatially the database will cover only Huddersfield town centre (the area inside the Inner Ring road) and a small residential area adjacent to the centre. An area covered by nine OS 1:12550 map sheets. Functionally, the pilot is taking a number of data sets from key departments, Planning, Estates and Highways, and with the help of consultants intends to build a limited number of applications, including local land charges and a highways maintenance function.

A distinctive feature of the Pilot is its 'parallel running' arrangement. Most local authority's having been through a preliminary evaluation exercise, go ahead with pilot studies based upon a single software package. This compromises the exploratory nature of the exercises, since it is difficult at the end of such pilots to recommend that the pilot software should be ditched and a completely fresh start made with an alternative system. For single system pilots, the choice is realistically between continuing with the pilot software or cancelling the project. KMC, however, being aware of the considerable variety of systems on the market, were concerned that they should not become locked into a single system prematurely, and thus asked the Polytechnic to carry out a shadowing exercise using alternative software. At the end of the pilot, KMC will be in a position to compare the capabilities of two packages, measured using the same data. The two software systems run on the same hardware, and if KMC are dissatisfied with the performance of their pilot software, they will be able to port the alternative GIS built in the Polytechnic onto their machines.

As the Pilot exercise is presently only half-way through its course it is not possible to make any definitive remarks about its outcome. It is possible, however, to make some preliminary remarks. Despite the fact that the systems are among the top selling GIS packages in Europe, we have found bugs in both packages, and have had to rely heavily upon vendor support for assistance. The systems are complex to master and if the pilot goes ahead on either package, considerable customization and training will be required before the systems could be widely used within the authority. Our early experience of manipulating OS digital data echoes the comments coming from other pilot studies. To build property parcels from OS data will require considerable editing of the map data, and it is not clear whether it is wise to pursue this course. If we do decide to do so, the relative efficiency with which the two systems can achieve this task could be an important determinant of which system is preferred.

The second KMC exercise is intended to research the demand, and support, for GIS across the authority. To have adopted a formal method to arrive at a rigourous Statement of User Requirements would have been an enormous task, and any case KMC's systems analysis officers had not previously adopted formal methods. In lieu of formal analysis, the 'Kirklees Spatial Data User Needs Survey' was initiated, with a team of officers conducting extended informal interviews with officers in every operational group in the council. The interviews cover issues relating to how the group presently processes mapped and other spatially tagged data, flows of spatial data between departments, and whether officers can envisage that GIS facility being of assistance to them in carrying out their responsibilities. Chrisman(1987) notes that many GIS exercises in the USA have similarly substituted a social survey approach in place of more rigourous means of systems definition, and he makes some cogent criticisms of this approach, focusing particularly upon the danger that informal methods might end up 'automating chaos, without understanding it'.

Whilst acknowledging Christman's strictures, our survey will provide, at least, a top level understanding of the flows of maps and site information across the authority and will provide a gauge of perceived needs, and support, for GIS within the authority. If the technical exercise is successful and a decision is taken to expand into other areas of the council's activities, the survey should help prioritize the areas demanding further investigation.

Again, the exercise is only partially complete, but it is possible to make some preliminary remarks. There is within the authority a very large redundancy of mapped information, with numerous groups maintaining their own copies of Ordnance Survey maps, which often are out of date and in poor condition. Also the arrangements for passing OS maps from area to area are ad-hoc, depending on personal contacts. Considerable officer time is spent tracking down map sheets, xeroxing, and cutting and pasting to obtain required scales. If a GIS simply improves OS map use in the authority it will make a major contribution. There is a strong demand from officers for a corporate, site-orientated database. Officers are presently frustrated by the length of time it takes them to trace site details by searching manual records and making telephone enquiries. There is also a wide spread requirement for routing and vehicle scheduling across the authority (children to schools, meals-on-wheels, books to invalids, etc), which might be assisted by GIS facilities. Distinct from corporate needs, some sections have specific needs which might justify local GIS systems. For example, the officers concerned with providing services for the partially sighted, need to be able to maintain a database to record locations of accessible buildings and street hazards. Also, in addition to operational needs, there is a clear role for GIS to be used for research purposes within the authority. Research officers are aware of many worthwhile research exercises which they cannot presently undertake for lack of GIS facilities.

CONCLUSIONS

'A nice trick if you can do it'

At present the case for major investment in corporate land-and-property systems in local government is not proven. The advantages which would flow from successful implementations have long been recognised, and our survey of KMC officers suggests that there is presently a strong officer support for such systems. Whether the software systems are presently adequate, whether the necessary data can be assembled for a reasonable cost, and crucially whether the organisational changes which such systems require can be accommodated should become clear in the very near future.

Note : The opinions expressed here are the author's, and should not be taken to represent those of the Kirklees Metropolitan Authority.

References

Christman N.R., 1987, Fundemental principles of geographic information systems, in: "Proceedings AutoCarto 8", Baltimore.
Gault I., 1989, Land information systems in Birmingham, or the second city's stuttering steps towards geographic information systems, in: "Proceedings of the Mapping Awareness conference", Oxford.
Grimshaw D. J., 1988, The use of land and property information systems. International Journal of Geographical Information Systems, 2,1,57-65.
Markham R. et al 1990, GIS at Plymouth City Council - coming to terms with reality. Mapping Awareness, 4, 8, 10-13.
Openshaw S., 1990, Lessons learnt from a post mortem of a failed GIS, in: "Proceedings of the Association for Geographic Information conference", Brighton.
Rhind D., 1990, Geographic Systems in Local Government, in: ETC Newsletter No 1, SERRL, Birkbeck College.
Waters R. S., 1989, LIS and GIS - technology, organisation and investment, in: Proceedings of the Mapping Awareness conference, Oxford.

# TAKING SYSTEMS THINKING INTO SCHOOLS

Alida Bedford

School of Information Science
Portsmouth Polytechnic
Portsmouth, Hants, UK

## INTRODUCTION

This paper aims to explore the argument for taking elements of systems thinking into schools in the UK. It comes from research currently being carried out in Hampshire secondary schools. This research focusses on using systems thinking to aid the integration of Modern Languages and IT in the secondary sector of secondary education.

The act of taking systems thinking into schools is an act of integration in itself. This is because it necessitates combining established with new practices, from within the school environment (system) and from a philosophy which has been in evidence for some time outside that system.

## BACKGROUND TO RESEARCH

The idea for this research came from three major influences:

a) the experience of having to consider and initiate the use of the microcomputer in English Language Teaching (ELT), as part of my own work;

b) learning about systems thinking as part of a "conversion" course in Information Systems;

c) the realization that "1992" would bring with it increasing demands in the labour market for individuals with a combination of Modern Languages and IT skills.

Additionally, the recently developed "innovation" of the National Curriculum in UK state education includes in its requirements that IT should be integrated across the curriculum (Penfold 1987).

Unfortunately not enough has been done to date to enable all teachers to facilitate this integration. Far more practical in-service training is required in IT, especially in the use of the microcomputer for this is a sophisticated technology in both qualitative and quantitative terms. Integration has been more successful in the primary sector (ibid), mainly because these teachers are trained and well practised
in providing integrated education.

Secondary teachers, however, largely work as subject specialists. If we examine human activity in terms of everyday living, it would seem that primary schools offer a truer reflection of the way people behave, for everything we do, think and say is interconnected. We may do different things at different times, and in a variety of situations, but we do not live in a disorderly collection of isolated pockets of activity which somehow identify with one person. Additionally our ongoing life experiences, whether active or reflective, take us always into newer dimensions of ourselves. Vickers (1983) calls this continuous process "appreciation". Put simply, "today I am the result of all my previous experience, and I will never be the same again as I am now".

## WHY SYSTEMS THINKING IS NEEDED

Primarily, knowledge of systems thinking raises our awareness that all is interconnected in the universe. Its basic tenet: "the whole is greater than the sum of all its parts" is the principle argument which demonstrates connectivity and the strengths that can be gained from working cooperatively.

Also in systems thinking we know that organization is a driving force, governed for us through human behaviour. It is organization that contributes to the success or otherwise of a system. This is especially true of an information system, so volatile because of its reliance on (human) will.

If we now relate this to secondary schools, we can see that whilst dividing the curriculum into subject specialization might make life easier in some ways, it does prevent those within such a system from knowing the daily business of their colleagues in terms of curriculum application. Teaching one, maybe two subjects eases lesson preparation, but it can be terribly isolating. Such isolation can lead individuals to believe that their subject is the centre of the universe, not an interconnected part of it. It could be argued that the bringing of integration of skills and knowledge across the curriculum, in this case with IT, will bring people closer together. Working with others does facilitate one's own learning processes.

From both personal observation and teacher comment during the research, being carried out at a time when there is an overwhelming amount of change, it is evident that the enormous amount of extra work is further separating colleagues. Teachers noticed they rarely had a proper conversation with each other during a term. These teachers worked in the same department, and so we can see that separation of professionals is increasing between colleagues with the same interests. Yet it would be so beneficial if teachers could stand still long enough to organize themselves into cooperative groups. If we return to the idea that the whole is greater than the sum of all its parts, then cooperation is going to enable greater achievement than everyone running around separately but trying to go in the same direction. This is frequently self-evidently true, but rarely carried out in practice.

The more we look at the problem area of integrating Modern Languages and IT skills, the more evident it becomes that systems thinking needs to be used, and not just because of what has been previously explained. There is an additional complication.

The teaching and learning of Modern Languages has its own history. This includes a certain amount of paradigm shifting in terms of methodology. Many teachers, however, are too busy to take interest in a changing academic climate. Even bringing in schemes throughout a Local Education Authority will not change teachers' classroom behaviour overnight. If real educational change - whatever form it takes - is to be effective it needs time, money and retraining. None of these is in sufficient supply in UK state education. Teachers also need to be convinced that change is desirable.

What we currently have in schools in the area of Modern Languages is a group of people who have their own beliefs about the structure of language, how learning takes place, and who each have their own particular approach in the classroom. Many language teachers tend towards eclecticism rather than stay strictly within a particular methodology. This gives them the freedom to use a greater variety of techniques and methods which work for them and their learners.

The way that systems thinking can help is first of all that it does not deny that differences in perception and practice can happily co-exist. Instead it may raise people's awareness that there are already in use many different ways of helping learners which they are free to try out. Thus it should encourage the sharing of knowledge and skills. This is crucial if Modern Languages and IT are to be integrated effectively. It does not matter if individuals want to use microcomputers in different ways. What they have to learn about them is the same, so it would be more efficient to work cooperatively. In this way they could work on sensible organization of curriculum development that considers the use of technology as a possible tool (amongst many) for extending students' learning.

## SOME TECHNIQUES USED IN THE RESEARCH

This section discusses particular techniques chosen through personal decision, in the same way that classroom approach is eclectic. Why choices were made will be explained. If you are repelled by what it described here it will be assumed you have/ will find techniques which fit you. Some of the techniques used to date are:

a) Rich Picture;

b) Root Definition;

c) Conceptual Model.

Each will be considered here under the following headings:

(i)   reasons for using the technique;

(ii)  a brief explanation of how each was done;

(iii) the teachers' reactions;

(iv)  my reactions;

(v)   conclusions and suggestions for future use.

a) Rich Picture

(i) Language teachers are used to including pictorial representation in classroom activity, and this has been transferred during changes in methodology. They can therefore easily identify with techniques that focus on illustration.

(ii) Two schools were involved for this technique, but analyzed separately in terms of being different systems. It was clear, however, that they would both be subject to the same influences from the environment as regards central government intervention. eg the introduction of the National Curriculum. A basic understanding of these issues was therefore necessary.

Information and impressions were gained through classroom observation (quite structured), interviewing (highly structured), informal conversation (unstructured but often very deliberate on my part), and by occasionally looking for what could be useful information eg on the staff noticeboards.

Once all teachers willing to be observed had been, and a selection of interviews had been completed, some very strong elements in the systems concerned could be expressed through Rich Picture form.

(iii) The teachers' reactions were very illuminating, and so it was right to assume they could easily relate to this technique. First of all the Rich Pictures enabled them to focus on certain aspects of their craft. One striking example was one group's realization that they have a very strong tendency to use the target language (that being learnt) for 'safe' activities. ie where dialogue is pretty predictable and activity well practised before.

The first teachers presented with a Rich Picture were very helpful about improving the technique, and added extra information. We would like to share with you the idea of using a key. This idea was used in the Rich Pictures for the second group. This enabled me to use far more detail. The teachers in this group were also able to use the key to explore and discuss their Pictures very thoroughly. This was particularly useful as this group had recently undergone a major staffing upheaval.

(iv) It was personally very encouraging to find positive outcomes from using this technique in a real world situation, especially in a domain new to systems thinking.

It was also encouraging that one group felt able to suggest an improvement, as this demonstrated involvement and understanding.

(v) Using the Rich Picture technique has proved to be viable with Modern Languages teachers. It generates extra information, and promotes understanding of the system people work within.

Including a key is very useful in terms of facilitating understanding, and enabling the compiler to use greater detail. Additionally it helps with the act of creation. (It became a lot easier to create symbols once a key was used for interpretation.) Often it takes people new to something to bring a fresh perspective.

b) Root Definition

(i) Asking a group to form a Root Definition provides a focal activity in the early stages of problem solving in system development. It forces people to get rid of the clutter that often invades human thought patterns and prevents them from starting with a simple (not simplistic) view of a complex problem.

The idea was to make teachers compile their own Root Definition which they could work with for outlining a model to help them plan how to integrate Modern Languages and IT.

(ii) As teachers are always busy people, with little time for reflection, an example Root Definition was given. This was meant to be inappropriate so they would not use the original! The group had to tell me what to write.

(iii) The teachers rewrote my Root Definition into their own by following its linguistic pattern - this is hardly surprising considering their subject! This also removed the necessity of applying Checkland's (1988) CATWOE test, which did not feel appropriate for this context. The teachers decided the microcomputer could be useful to provide variety in language teaching, for themselves, whilst working with Lower secondary pupils on the Hampshire Modern Languages Skills Development Project (HMLSDP).

(iv) It was surprising the group were so teacher-oriented, as opposed to learner-oriented, but it was realized their honest statement is a true reflection of what it is like being in the classroom. Preventing one's self from becoming bored with one's job by aiming for variety does improve working relationships, attitudes and personal motivation.

(v) It was worthwhile using this technique as it removed the influence of my ideals, and replaced these with a clear, simple statement from the people engaged in the system being analyzed. They are the people who will remain in the system, so it needs to be theirs.

It is uncertain whether others will approve of not using the CATWOE test, but it was very apparent at the time that people were worried about spending time on themselves as a team. This is a disturbing trend in education, and something needs to be done to reverse it. Teachers are very aware of this.

c) Conceptual Model

(i) Outlining a Conceptual Model is the follow-up activity once a Root Definition has been agreed, according to Checkland (ibid) in SSM. It guides people towards developing a model of a system that they can base planning, design and activity around. It may not be necessary to compile a new model of the system if the old one works - it could be used as a focal activity to help team work.

Usually, the Conceptual Model is an abstract representation of a new model of a system. It has been drawn in a less abstract way (Avison & Fitzgerald 1989), and this is what it was decided to aim for in school. No-one could seriously believe that at the end of a school day, and in the penultimate week of the Winter term, any group would find mental abstraction a riveting activity. This also relates strongly to feelings people have about how they perceive themselves as spending their time.

(ii) Giving practical examples is the best way of presenting new ideas, so a large Conceptual Model was drawn on a blackboard - prepared prior to the meeting. It is suggested here not to make any diagram so well drawn people feel intimidated by it. Clarity and simplicity are all that are needed.

A brief explanation was given. The group were left to devise their own model. This did not take them very long. I drew the new model according to a series of suggestions, adding a few points (verbally) for them to consider, until they were satisfied. The Head of Department (HOD) made paper notes throughout this and the previous activity, and proved to be very talented at drawing Conceptual Models at great speed.

(iii) This activity promoted good, interactive discussion, although the two youngest teachers (both women) did not contribute as much.

(iv) It was surprising how little time this took. Although the young women did not say much, they do ask very relevant questions which more experienced staff are not so good at.

(v) It is proposed that being less abstract with Conceptual Models has its place in the teaching domain, especially in the present climate of too much change too soon. However, it is a worrying trend that teachers feel guilty about spending time on meetings which should be helping them. This was brought up later with the HOD. He said that once you get teachers outside the school environment they are far more relaxed. We came to the conclusion that the ideal situation would be for them to do in-service work in a different location where they could not go to others meetings or sports practices, sit in a meeting preoccupied by a mental shopping list of things that need doing.

## CONCLUSIONS

Many conclusions have already been covered in the previous section. Some do show a worrying trend in the wider field of UK education. Teachers are having to take on implementing major changes, and this is having a devastating effect on their professional development.

It is suggested that using systems thinking as part of a methodological lifecycle could bring people together synergistically so that one of the changes they are expected to cope with can be effected more efficiently. ie integrating IT and a specific subject in the curriculum.

Previous personal experience of working in a team situation where everyone was part of a cohesive whole, had an outcome which far exceeded anything we did as isolated parts, not just in terms of output, but also in the way we related to our work and each other, and how we felt about ourselves.

Systems thinking, and the already tried techniques that have developed from it offer an opportunity to education to start to determine its future in a systematic way. It is systematic cooperation between professionals from any discipline which effects change in the way that they want it to go.

## BIBLIOGRAPHY

Checkland,P, 1988, "Systems Thinking, Systems Practice", John Wiley & Sons
    Ltd, Chichester.
Land, F & Hirschheim, R, 1983, Participative Systems Design: Rationale,
    Tools and Techniques, Journal of Applied Systems Analysis, 10:91.
Penfold, B, 1987, IT in National Curriculum, Educational Computing, 8:7.
Schoderbek, PP, Schoderbek, CG, Kefalas, AG 1985, "Management Systems",
    Business Publications Inc, Plano, Texas.
Vickers, G, 1983, "Human Systems are Different", Harper & Row Ltd, London.

# HEALTH AND SAFETY LEGISLATION - THREAT OR OPPORTUNITY?

T.G. Gough

Division of Operational Research and Information Systems
School of Computer Studies
University of Leeds
Leeds, LS2 9JT

## ABSTRACT

This paper is designed to focus the attention of the systems community on health and safety issues. The paper seeks to show that, despite the fact that concern for health and safety in employment has already been reflected in legislation, little attempt appears to be made in current methodologies or in systems practice to incorporate health and safety into the systems design process. There is the impression that health and safety is someone else's responsibility, not that of the systems community. A brief review of present legislation and the new Article 118A of the Treaty of Rome will be used to explore the question of whether legislation is likely to prove a threat to good systems practice or encourage better design and more usable systems. An attempt will be made to assess the likelihood of legislation providing the impetus for turning good intentions on health and safety into effective action.

## INTRODUCTION

The importance of health and safety at work has been a subject of concern for generations. In fact, Hippocrates was concerned about lead toxicity in the mining industry in the fourth century BC ! Watson (1984) suggests "One of our highest priorities in public health is the protection of millions of Americans during their working lives". This concern for health and safety in employment has been reflected in legislation directed at that concern and reflecting the priorities at the time (see Equal Opportunities Commission 1979 for a brief historical review of UK legislation). More recently in the UK there has been introduced the Health and Safety at Work Act (in 1984) and the regulations for the Control of Substances Hazardous to Health (in 1988). With the rapid growth in the use of interactive systems and the subsequent attention given to the human-computer interface there has been an increase in interest in some aspects of health and safety, especially those associated with VDUs (see Willcocks and Mason 1988 for a reasonably succinct summary and International Labour Office 1989 for a more detailed review). However, whilst most systems designers would find it hard to disagree with the suggestion that the aim of systems design should be to produce systems which do not put the health and safety of their users at risk, it is not often obvious, in practice, that systems design activity is directed to the achievement of 'safe' systems.

HEALTH AND SAFETY

There are four main aspects of health and safety which need to be addressed. Firstly there are general workplace hazards associated with the physical office environment (see Damodaran, Simpson and Wilson 1980 for a suggested list). To their list 'Sick Building Syndrome" should perhaps be added (see, for example, Wilson 1987). Secondly, there are the hazards associated with fatigue. Muscular fatigue is most frequently associated with postural problems. The growth in the use of keyboard devices has resulted in cases of repetitive strain injury (RSI) which can cripple arms and hands and is classed as an industrial disease by the UK Health and Safety Executive. The second main type of fatigue is visual fatigue, the main causes of which, other than postural strain, include differences in viewing distance, frequent eye movements, prolonged near point focussing, and poor lighting. The hazards associated with fatigue are primarily attributed to failure to pay attention to 'ergonomics' during the design process. For a further examination of these issues, see, for example, Shackel 1974, Cakir, Hart and Stewart 1980, Grandjean and Vigliani 1980 and Grandjean 1984.

Getting the ergonomic factors right will reduce the hazards associated with fatigue but these hazards are in some cases only part of a third (and larger concern) about more serious (long term) risks associated with the regular use of visual display units. There are four main risks on which attention is focussed, radiation, photosensitive epilepsy, facial dermatitis and birth defects. The existence (or otherwise) of these risks has been discussed and evaluated in numerous conferences and papers over a number of years (see, for example, Bennett, Case, Sandelin and Smith 1984, Pearce 1984, and Fletcher 1985). A report from the World Health Organisation (1988) stated unequivocally that "visual discomfort experienced by many VDU users must be recognised as a health problem". There is no evidence from this two-year study of a link between adverse effects on pregnancy and the use of VDUs, but "this should not be interpreted as saying that working on VDUs is entirely safe". Stress has been identified as the fourth main area of concern (see, for example, Cohen 1984, section 5 and Work and Stress 1987). Stress is an issue that is affected both by the quality of the workplace and by the design of the job. The replacement of human by electronic communication coupled with increasing integration of systems reduces social contact and increases the pressure by removing the 'natural breaks' in the routine of the office.

SYSTEMS DESIGN

All four aspects of health and safety have implications for systems design but little attempt seems to have been made to incorporate consideration of health and safety into the systems design process. Discussion tends to be fragmentary or even non-existent. There is an impression that health and safety is some else's responsibility and not that of the systems designer, except in one or two special cases, for example, in dialogue design. This criticism of insufficient attention to health and safety is valid across the range of methodologies. It is ignored almost completely in the traditional systems analysis approach (see, London 1976, Bingham and Davies 1978, and Kilgannon 1980, for example). The structured approach directs attention to the structure of data and programs (for example, Yourdon and Constantine 1978, Gane and Sarson 1979) or even to the whole analysis and design process (Cutts 1987) but the structure of the workplace is largely ignored. A shift in emphasis to a more people-centred approach only leads to a marginal improvement (Mumford and Weir 1979), if any (Norman and Draper 1986). Despite the centrality of 'human activity' Soft Systems Methodology does not seem to address health and safety explicitly in either its original (Wilson 1990) or its contemporary form (Checkland and Scholes 1990). The increasing interest in human-computer interaction has probably narrowed rather than widened the health and safety focus, for example, in Shneiderman (1987) the emphasis is on performance rather than health and safety nor are health and safety issues addressed in Johnson and Cook (1985) nor in Harrison and Monk (1986). Health and safety is not regarded as a critical issue by Boland and Hirschheim (1987) nor as influencing productivity by Lieberman, Selig and Walsh (1982). It is not included in human factors by Christie (1985) nor is it discussed in the wide-ranging review of information systems development by Avison and Fitzgerald (1988).

## ARTICLE 118A

The new Article 118A (of the Treaty of Rome) calls on Members States of the European Community to bring about improvements in the working environment for the health and safety of workers. The proposals comprise a Framework Directive and five 'daughter' directives, specifying minimal health and safety requirements for: the workplace; the use of machinery, equipment and installations; the use of personal protective equipment; visual display units; and the handling of heavy loads. The purpose of the Directives is to ensure that all Member States have comprehensive and comparable health and safety legislation. The framework Directive establishes essential responsibilities of employers and employees. Some changes in British Law will be necessary to implement the new Article and it is likely that new regulations under the Health and Safety at Work Act 1974 will be used to achieve this. A British Standard (BS 7179:1990) has also been issued in which specific guidance is given on some of the matters covered by the Directives.

Particular attention has been drawn to the VDU Directive which will require employers: to analyse display screen workstations to evaluate safety and health conditions, taking appropriate measures to remedy any risks found; to ensure workstation entering service after 31 December 1992 meet the requirements contained in the Annex to the Directive, which sets standards for display screens, keyboards, furniture, lighting, working environment, task design and software; to ensure that workstations in service before 31 December 1992 are adapted to comply with the requirements of the Annex by 31 December 1996; to plan activities so that daily work on a display screen is periodically interrupted by breaks or changes of activity; to give workers an entitlement to an eye and eyesight test before starting display screen work, at regular intervals thereafter, and if they experience visual difficulties.

## THREAT OR OPPORTUNITY

The only reported reaction to the Directives from British employers has been to see them as a threat, an unnecessary and costly imposition which will raise operating costs unnecessarily. This is certainly the view of the CBI where the head of health and safety argues that the health risks in IT have not been proved to warrant such legislation. This view is echoed by individual employers. It is not yet clear whether the British Government shares the employers' view. It is possible, perhaps, that the CBI and others have yet to see a link between the standards set out in the Directives and the claim that the rapid growth in the number of people using computer keyboards has been matched by growth in the number of cases of repetitive strain injury (RSI). May 1990 saw the first out of court settlement in the UK for a claim against a bank for RSI injuries sustained by an office employee. Other claims are being pressed. In the USA Pacific Bell is reported as invest-ing $8 million in a "programme to help curb stress, strain and the costly effects of using a computer" as noted in a news report in Computer Weekly of 26 July 1990, where it was also reported that Pacific Bell had had 689 cases of 'RSI' between 1986 to 1989 at an average cost of $48,000 per case.

There are, therefore, two threats, the one perceived by UK employers of increased costs and restrictive practices resulting from the Directives, the other perceived (and experienced) by a number of VDU users of a continuing risk to their health and safety if changes in operational practice are not made.

Systems designers have the opportunity of making the best use of the legislative framework to eliminate the second threat in a way which also allays the fears of employers about the first threat. It is open to systems designers to address the risks to health and safety both perceived and actual by using the proposed standards to design workplaces and jobs which address all the issues raised earlier in this paper, for the benefit of both employers and employees. Better job design and better workplace design will lead to an increase in office productivity. To take a particular example, implementing the Directives could provide the first real chance to replace QWERTY with something more 'user friendly'! Article 118A could provide the impetus for turning good intentions on improving the human computer interface into good quality working systems, providing systems designers see it as an opportunity rather than a threat.

## CONCLUSION

Given the concern about health and safety in the workplace, already reflected to some extent in legislation, the absence of health and safety from the general systems design agenda is rather surprising. It is clear that if the health and safety aspects are to be handled effectively by systems designers then considerations of health and safety must become part of the mainstream of the systems design process. Impending legislation provides an excellent opportunity to make this happen.

In commenting on the publication of a new student's guide to Agriculture (Health and Safety Executive 1989), Carl Boswell, the Chief Agriculture Inspector, remarked "One of the most important conclusions which I hope the student will reach after studying the literature is that commitment to health and safety is of paramount importance and a prerequisite to efficient and effective farming". The theory and practice of systems design should encourage its students to come to similar conclusions on the key role of health and safety in systems work.

## REFERENCES

Avison D E and Fitzgerald G, 1988, "Information Systems Development - Methodology, Techniques and Tools", Blackwell Scientific.

Bennett J, Case D, Sandelin J and Smith M, (eds.), 1984, "Visual Display Terminals - Usability Issues and Health Concerns", Prentice-Hall.

Bingham J E and Davies G W P, 1978 "A Handbook of Systems Analysis", Second Edition, Macmillan.

Boland R J and Hirschheim R A, (eds.), 1987, "Critical Issues in Information Systems Research", John Wiley.

Cakir A, Hart D J and Stewart T F M, 1980, "Visual Display Terminals", John Wiley.

Checkland P and Scholes J, 1990, "Soft Systems Methodology in Action", John Wiley.

Christie B, (ed.), 1985, "Human Factors of Information Technology in the Office", John Wiley.

Cohen B G F, (ed.), 1984, "Human Aspects in Office Automation", Elsevier.

Cutts G, 1987, "SSADM - Structured Systems Analysis and Design Methodology", Paradigm.

Damodaron L, Simpson A and Wilson P, 1980, "Designing Systems for People", NCC.

Equal Opportunities Commission, 1979, "Health and Safety Legislation - Should We Distinguish Between Men and Women?"

Fletcher A C, 1985, "Reproductive Hazards of Work", Equal Opportunities Commission.

Gane C and Sarson T, 1979, 'Structured Systems Analysis: Tools and Techniques", Prentice-Hall.

Grandjean E, (ed.), 1984, "Ergonomics and Health in Modern Offices", Taylor and Francis.

Grandjean E and Vigliani E, (eds.), 1980, "Ergonomic Aspects of Visual Display Terminals", Taylor and Francis.

Harrison M D and Monk A F, (eds.), 1986, "People and Computers : Designing for Usability", CUP.

Health and Safety Executive, 1989, "Student's Guide to Agriculture".

International Labour Office, 1989, "Working with Visual Display Units", Occupational Safety and Health Series No. 61.

Johnson P and Cook S (eds.), 1985, "People and Computers: Designing the Interface", CUP.

Kilgannon P, 1980, "Business Data Processing and Systems Analysis", Edward Arnold.

Lieberman M A, Selig G J and Walsh J J, 1982, "Office Automation - A Manager's Guide of Improved Productivity", John Wiley.

London K, 1976, "The People Side of Systems", McGraw-Hill.

Mumford E and Weir M, 1979, "Computer Systems in Work Design - the ETHICS Method", Associated Business Press.

Norman D A and Draper S W, (Eds), 1986, "User Centred Systems Design - New Perspectives on Human-Computer Interaction", Lawrence Erlbaum Associates.

Pearce B G, (ed.), 1984, "Health Hazards of VDTs?", John Wiley.

Shackel B, (ed.), 1974, "Applied Ergonomics Handbook", Butterworths (reprinted from Applied Ergonomics, Vol. 1, Nos. 1-5 and Vol. 2, Nos. 1-3, 1974, Butterworths).

Shneiderman B, 1987, "Designing the User Interface - Strategies for Effective Human-Computer Interaction", Addison-Wesley.

Watson W C, 1984, Public Health and the Workplace, in Cohen, (ed.), "Human Aspects of Office Automation", Elsevier.

Willcocks L and Mason D, 1987, "Computerising Work - People, Systems Design and Workplace Relations", Paradigm.

Wilson B, "Systems : Concepts, Methodologies and Applications", Second Edition, John Wiley.

Wilson S, 1987, "The Office Environment Survey", Building Use Studies.

Work and Stress, 1987, Vol. 1, No. 3, July-September, Taylor and Francis.

World Health Organisation, 1988, "VDUs and Workers' Health".

Yourdon E and Constantine L L, 1978, "Structured Design : Fundamentals of a Discipline of Computer Program and Systems Design", Second Edition, Yourdon.

EXAMINATION OF THE CONCEPT OF 'KEY FACTOR FOR SUCCESS'

FROM A SYSTEMS THINKING PERSPECTIVE

Tatsuyuki Negoro

The Sanno Institute of Management
1573 Kamikasuya, Isehara, Japan, 259-11

## INTRODUCTION

As the resources in a company are limited, the company should concentrate on reinforcing some particular aspects of its activities if it wishes to succeed in the industry. One concern of the theories of competitive strategy is to establish a suitable idea for identifying with certainty what aspects should be strengthened. The concept of 'Key Factor for Success(KFS)' is an attempt to achieve this, but the concept seems to me to be vague or rough. The aim of this paper is to clarify the problems of the existing description of KFS and try to refine the concept from a systems thinking perspective. The next section will briefly describe the literature and the problems. Then, we discuss supplementary ideas which should be considered. In conclusion, a new description of KFS will be presented.

## THE CONCEPT OF 'KEY FACTOR FOR SUCCESS'

The concept of KFS was first described by K.Ohmae(1983). The term of 'Strategic Excellence Position(SEP)' which C.Pumpin (1989) uses is basically a similar concept.

There are many competitive activities in the work of a company; examples are the capability of production, of marketing and of R&D. A company can choose a specific policy in each of these fields. One may pursue low-cost production, whilst another may try to gain a good reputation for the quality in the field of production. In the field of marketing, a company may additionally choose from a number of policies. One may pursue increasing the advertisement, whilst another is trying to extend its own retail network.

A company cannot gain complete advantage over its competitors in every field, because its resources are limited. Moreover, not every possible policy will result in good sales or good profits. An example Taken from the microcomputer industry will illustrate this point. Will a company which achieves 'the lowest cost per unit', increase its sales? The answer will be 'No' if the company has only a narrow range of software available for use with its hardware. In other words, no company can avoid considering the range of software available as a important factor. In this example, the company should choose a policy in a specified suitable way in a particular field. The factor to be pursued in this way in the field is said to be 'the KFS' (Key Factor for Success). In the microcomputer industry, the availability or quality of softwares may be the KFS. In this case product design is an example of a selected field(activity), and trying to increase the range of softwares is one of a specified suitable way. Although the choice of KFS is limited, the alternatives of KFS are not only one in some industries. For example, in the air transportation industry, you can find companies pursuing 'quality of services' and 'punctuality' and other companies pursuing 'low price'.

Having decided on a clearly-defined KFS enables a full concentration of forces. The concept of KFS focuses on a skilful utilization of existing strengths and opportunities for synergy. Stating the future direction of the company in terms of KFS gives the staff a clear direction.

The advantage of concentration that KFS indicates is also emphasized in the well-known competitive theory of M.E.Porter(1980). He identifies three basic possible strategies : the overall cost leadership strategy, the differentiation strategy and the focus strategy. The overall cost leadership strategy may correspond to pursuing a low-cost product as a KFS. The differentiation strategy may be correspondent with trying to gain a good reputation for the quality as a KFS. Introducing a new high technology adopted product is also a differentiation strategy and to pursue product technology as a KFS. On the other hand the focus strategy is to pursue a KFS in selected segment(s) of an industry. Some companies do not participate in every segment in the market. They concentrate on a single one or a few segments. The company can choose either overall cost leadership strategy or differentiation strategy in the segment(s). Porter emphasizes the difficulty of pursuing both overall cost leadership strategy and differentiation strategy. He claims that '…[The] firm failing to develop its strategy in at least one of the three directions … is in an extremely poor strategic situation. … The firm … is almost guaranteed low profitability.' It may be said that Porter is illustrating the danger of pursuing plural KFSs.

The points of KFS are two: one is that a limited number of factors can lead a company to success in an industry, the other is the importance of concentration of forces. Porter's argument is to identify three patterns of concentration.

The problems we would like to present about the existing description of KFS concern whether its ability to establish an actual relevant strategy for the real world. The first problem is related to 'the limited number of factors' point. What kind of factor can be vital? How to define the factor? The second problem corresponds to 'the concentration' point. Is concentration on only one factor enough for success? How to deal with other factors besides approved KFS? The answers to these questions are important in the real world, but the existing description does not answer them adequately. The first problem is relevant to the second one because identifying a vital factor should be before concentration. We will try to answer these questions by presenting ideas of objective, attribute and activity which are confused in the existing description. Objective will be defined in the next section, and attribute will be argued in the following section.


OBJECTIVE AND DOMAIN OF BUSINESS

One important viewpoint of systems thinking is as follows. The sum of the best solutions obtained from the parts taken separately is not the best solution to the whole. Improvement of the parts does not always lead to the improvement of the whole. The thing to be explained should be treated as a part of 'a containing whole' (Ackoff, 1981).

Management tries to improve the status quo. But it is not effective or best policy to improve anything that comes its way. This is what the above viewpoint says. In this case, what is 'a containing whole'? It is the 'domain' of the business which is based on the consumer needs. Let us explain it.

A company produces a product, but the product itself is not an objective for which the consumer purchases or uses. What the consumer desires is to transform his state(State A) into another state(State B) using the product(see figure 1). For example, a person without Coca-Cola is transformed into the person with Coca-Cola using a Coca-Cola(product). In terms of the objective, the states can be interpreted in various ways which depend on the person in a certain circumstance. For example to ingest water can be an objective, and to take refreshments or to kill time can be others. The product is a means to enable the customer achieve the objective. Objective can be plural. It varies dependant on the person in a certain circumstance, and the person sometimes is conscious of the objective but sometimes unconscious. It is fundamentally an abstraction interpreted. The attributes of a product for satisfying objectives may be termed consumer needs. Ease to drink is an attribute for ingesting water, good taste is one for refreshment, etc. Different objectives require different attributes. As a product is a means, other products can satisfy the same objective. For instance, merely tap water can be another means to ingest water and coffee can be one for refreshment.

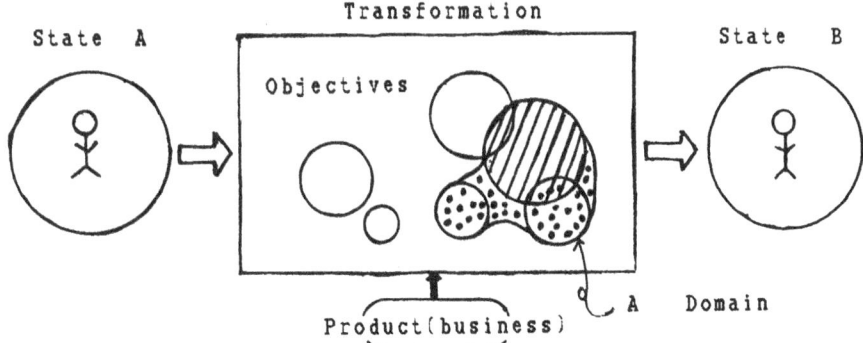

Fig. 1. Objective and Domain

A company should try to identify the objectives which its product(s) satisfy. Objectives should be interpreted and prioritized. It might be that some objectives should be ignored. A set of those prioritized objectives we can term 'domain of business'. For a business producing soft drink, refreshment can be the most important objective. On the other hand, for a business of drink for sportsmen ingesting water can be the important one. To look fashionable may be important for some other drink businesses.

A domain should be a 'containing whole' for a business which provides product(s). Every improvement must be evaluated in relation to the domain. The domain tells what attribute(s) of the business are vital. Considering the video machine business for example, consumers may have three major objectives; recording and replaying TV programmes, viewing movie tapes, producing their own recordings.

In a domain where viewing movie tapes is the primary objective, the range of movie software is a vital attribute in the business. But in another domain where producing their own recordings is the primary one, the ease of using a camera is the vital attribute in the business. Sony Corporation has been defeated in the former domain, and yet at present it is trying to reestablish itself in the latter domain. Concerning evaluated attributes in relation to a domain, management staffs of a company can debate about its competitive advantage. This is the topic of the next section.

## ATTRIBUTE AND COMPETITIVE ADVANTAGE

Models are not models of parts of the real world, only models of ways of perceiving the real world, that is to say, models 'relevant to debate about reality' (P.B.Checkland and J.Scholes 1990). It is another viewpoint in systems thinking. We present this kind of model about competitive advantage.

Actual companies sometimes cannot avoid pursuing more than one attribute in their business. The range and quality of the softwares in the microcomputer industry is the vital attribute because handling data on software is an important objective for the consumers and the range affects the scope of choice. But if a company with extensive range of of software does not have a suitable price (not necessarily the lowest but not too high) for the related hardware, it will not achieve good sales or good profits. There might be only one vital attribute, but this does not mean that the company need not pursue other attributes. Moreover, there are some cases where companies must try to raise their position on the plural attributes. For example, a VTR camera manufacturer who achieves low costs per unit but who does not succeed in introducing a lightweight model, will not enjoy good sales or good profits. On the other hand, a company which develops a lightweight model but an expensive one will not do so either. The problem is, to what extent should a company pursue each attribute? The model you can see in figure 2 is a device for management staffs to debate about this kind of question.

This model illustrates the relationship between physical gap of attribute against a competitor and the impact on the consumers. Strong impact leads to brand switching from the competitor to the company. Four zones in terms of attribute gap are displayed on the model. If improvement of a gap of an attribute leads to increase the impact sharply, the company is in the Effective Zone in terms of the attribute. A small plus gap against the competitor does not have a big impact. On the other hand a small minus gap does not lead to defeat automatically. The Tolerance Zone contains this small plus and minus gaps. If a company has a very big plus gap in an attribute, to maintain this big gap may be more costly rather than to increase the impact. This gap should be in the Insignificant Zone which shows too great emphasis on quality. ( Of course whether a gap is insignificant depends on the consumer or segment of the market because different consumers have different objectives or criteria.) In contrast there is a case where a company has a big minus gap. This gap accelerates brand switching from the company to the competitor. This case is the Avoidance Zone into which the company should avoid entering. In sum a company should maintain all attributes of the business which are required from the domain in the Effective Zone or the Tolerance Zone. If not, the company has to reconsider its domain. The attributes which are classified vital from the viewpoint of the domain should be in the Effective Zone. The secondary attributes are allowed to be in the Tolerance Zone.

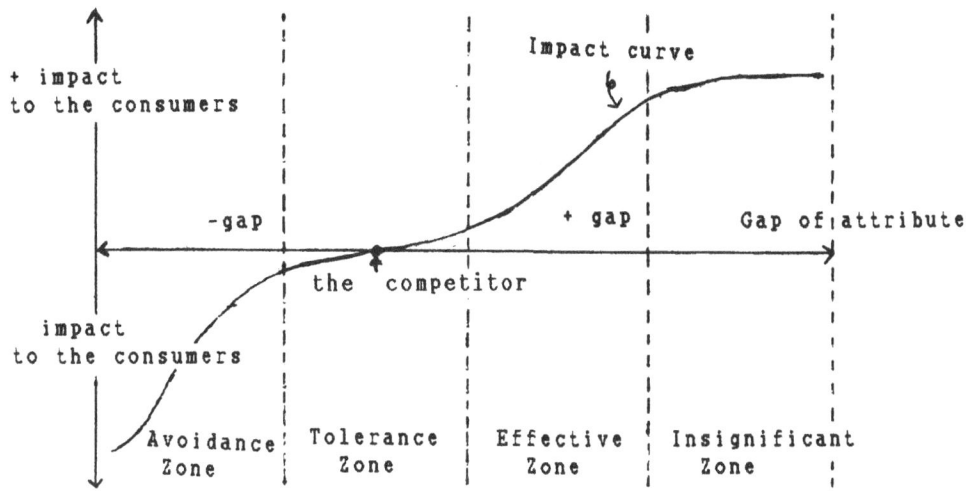

Fig. 2. Attribute and four zones

How do you use the model? First list the attributes for the objectives from the domain you identified and select the possible competitors. Then debate about the competitive effect of each attribute against each main competitor. Suppose you identify the domain for your video aplliance business as follows; producing consumer's own recordings as the primary object and viewing movie tapes and recording TV programmes as the secondary ones. To satisfy the primary objective, you need a camera as a product of which the attributes should be lightweight, high resolution capability, ease of operation, etc. You can debate whether these attributes are in the Effective Zone or not against the main competitors. On the other hand to satisfy viewing movie tapes a attribute that your business should have is the extensive range of movie tapes available with your hardware. You must place this attribute within the Tolerance Zone or you should abandon the objective from your domain. The price should be also reconsidered whether it is in the Effective Zone or the Tolerance Zone for the sake of the objectives. The zone model is a device for such a debate.

CONCLUSION

Based on the above supplementary ideas we would argue about the description of KFS. We have various activities to improve attributes which are R&D, Procurement, production, sales, finance, human resource management, etc. The relationship amongst objectives for the customers, attributes of the business and activities of the company is illustrated in figure 3. The existing description of KFS does not have the clear distinctions of the ideas of objective, attribute and activity.

Our first challenge against the existing description was ; What kind of factor can be vital? And how to define the factor? Our answer is that you should first identify your domain so that you can list the attributes which you should control, then to improve these attributes you can enforce related activities. Our second challenge was ; Is concentration on only one factor enough for success? And how to deal with other factors besides approved KFS? Our answer is that you should place every attribute that is required from the domain in the Effective Zone or Tolerance Zone and enforce your related activities.

Strategy based on KFS in our argument is to enforce the activities to improve the attributes which are derived from the domain. Concentration should be formed in this process. This argument indicates the danger within Porter's theory. Because choosing only one strategy makes a company not sensitive to the other strategies. As a result, the company might lose the necessary capability in a certain attribute which is not the most important but should not be ignored. A important thing is to identify the domain but not choose one of three strategies. In our argument, the three strategies are not exclusive. The number of domain to be interpreted and identified are many and focus strategy is merely one of them. The overall cost leadership strategy and the differentiation strategy are not exclusive because these are merely an exaggeration of particular attributes which should be pursued to some degree at the same time.

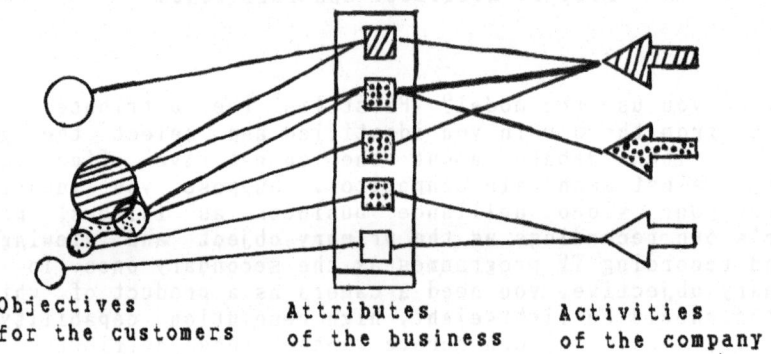

Objectives                Attributes              Activities
for the customers         of the business         of the company

Fig. 3. Objective, Attribute and Activity

REFERENCES

Ackoff, R.L.,1981, "Creating The Corporate Future," John Wiley, New
    York.
Checkland, P.B., and Scholes J.,1990, "Soft Systems Methodology In
    Action," John Wiley, Chichester.
Flood, R.L., and Jackson, M.C., 1991 , "Total Systems Intervention,"
    John Wiley, Chichester.
Krick, E.V.,1969, "An Introduction To Engineering & Engineering
    Design," John Wiley, New York.
Levit, T.,1960, Marketing myopia, Harvard Business Review, Sept.-Oct.
Ohmae, K.,1983,The Mind of the Strategist," Penguin, Harmondsworth.
Ono, K. and Negoro, T.,1990, "Management of Manufacturing company,"
    Kaiseisya, Tokyo(In Japanese).
Negoro, T.,1990, How to recover the disadvantage in Software,
    Diamond Harvard Business Review, Aug.-Sept.(In Japanese).
Porter, M. E.,1980, "Competitive Strategy," The Free Press, New
    York.
Pumpin. C.,1989, "The Essence of Corporate Strategy," Gower,
    Aldershot.

A SYSTEMATIC APPROACH TO LEARNING FROM PAST EXPERIENCES

IN DIAGNOSTIC PROBLEM-SOLVING ENVIRONMENT

Kambiz Badie

Djanbazan Biomedical & Rehabilitation Engineering
Research Center, Tehran, Iran

INTRODUCTION

Among the different learning approaches, a supervised learning from
the past problem-solving experiences is meaningful in the sense that it
guides the problem-solver to the ideas which may lead to an efficient
method of increasing the problem-solving performance, despite the fact
that the knowledge essential to problem-solving may not be identified
thoroughly.

In this paper, we propose an approach to learning from the past
experiences of diagnostic problem-solving in an environment where
various types of knowledge are utilized in a controlled manner. The
approach is believed to be a suitable start for conceptualizing
diagnostic patterns in human experts, and is particularly significant
when improvement of human judgemental skill is to be considered.

SOME CONCEPTS IN LEARNING

Learning can be defined operationally to mean the ability to
perform new tasks that can not be performed before or perform old tasks
better as a result of changes produced by the learning process. A
question arising is that "if a problem-solving environment is fully
known, in the sense that all the possible characteristics of a problem
are summerized into a definite number of sample patterns through which a
stable problem-solving structure can be constructed, do we still need to
think of learning as an essential process?". The answer is "YES", since
the problem-solving system may purposefully decide occasionally to
delearn a variety of items regarding its past problem-solving structure,
and instead retain the capability of learning for future cases of
problem-solving. Obviously, delearning process is to be started from the
part which has not been utilized so frequently in the recent
experiences.

The most widely studied paradigm for symbolic learning is the
inductive paradigm which is responsible for inducing general concept

description from a sequence of instances of the concept and known
counter examples of the concept (Michalski and Step, 1983; Gennari et
al., 1989).

Analytical learning paradigm has also been proposed to actuate
analytical learning from few exemplars in addition to a rich underlying
domain theory. Here, past problem-solving experiences which are the
exemplars, are utilized within the framework of a deductive process in
order to infer new search control rules enabling more efficient
utilization of domain knowledge. In contrast with inductive method of
learning, the aim of analytical learning is mostly to improve the
performance of a problem-solving system with no particular emphasis on
extending the library of concept description (Minton et al., 1989 ;
Dejong and Mooney, 1986).

## HUMAN DIAGNOSTIC SKILL IN TERMS OF KNOWLEDGE UTILIZATION

### Sources of Knowledge to be Utilized in Diagnostic Problem-Solving

It is believed that a human expert is accustomed to make use of a
variety of sources of knowledge in order to solve a diagnosis problem.
The success of the expert in a diagnostic problem-solving process,
therefore depends on the validity and adequecy of these sources. It is
also right to think that a time-efficient solution in diagnosis depends
on the way the sources of knowledge are efficiently utilized. The role
of experience is thus to realize the more reliable sources of knowledge
as well as to discover the more efficient methods of knowledge
utilization in diagnosis.

Different sources of knowledge are to be utilized by an expert,
based upon the nature of the diagnostic problem :
a) Source of knowledge describing the diagnosis environment
Examples can be mentioned for the characteristics of the failure pattern
to be diagnosed, existing facts regarding the peculiarities of the
diagnosis problem (e.g. diagnosis is to be performed under a set of
limitations in the cognitive capabilities of the expert as well as the
motivations in the objectives of diagnosis, and specifications of the
environment where the system under diagnosis has been operating.
b) Source of knowledge describing noise/distortion generation mechanism
For example the relationships between the sources of noise/distortion
and their configurations.
c) Source of knowledge describing failure behavior mechanism
When failures occur in a system, they somehow influence the behaviors of
the elements in that system. Production rules are suitable tools to
demonstrate the way such an influence can be described. Semantic
networks can also be utilized to show the interrelationships between the
behaviors of the faulty elements, when some failures have occured in the
system.
d) Source of knowledge describing the constraints
This source of knowledge is utilized to demonstrate whether the
conditions up to a certain stage of diagnostic problem-solving such as
the status of the classes of failures diagnosed up to that stage are
totally acceptable or not.

### Expert's Skill in Diagnostic Problem-Solving

The sources of knowledge mentioned above are incorporated together
and an appropriate strategy will then be responsible to utilize them in

a manner justifiable within the frame-work of the results obtained from the expert's past experiences. Such a strategy is believed to be developed in the expert's mind according to his/her former experiences of solving similar problems. The thing important is that the expert is to realize the proper sources of knowledge, as well as the efficient method to utilize them.

Suppose that a number of utility degrees have been assigned to the sources of knowledge based upon the results obtained from the past experiences, and the activity in problem-solving is to be started. It is reasonable for the activity to be started from the source for which the past diagnostic cases show the highest support. Because of the incertitude in expert's criteria for utilizing sources of knowledge, he/she should sometimes reconsider the way new facts are obtained on the basis of applying the existing knowledge to the existing facts. Regarding this, some sort of performance index is required to show how successful the process of utilizing a source of knowledge has been. The information concerning this index is contained in the source of knowledge describing constraints which was previously discussed. It is therefore seen that some sources of knowledge may be utilized more than once within the whole process of diagnostic problem-solving. To consider this point, utility degrees for the 2nd, the 3rd,... utilizations of the knowledge sources should also be taken into consideration (Badie, 1990).

## Learning From The Past Experiences of Diagnostic Problem-Solving

The following hypotheses are postulated for an expert who is at the mode of learning from the past diagnostic experiences :
1) Having accomplished a diagnosis process, the utility degrees of all sources of knowledge which have been utilized in this process are incremented by a predetermined value. This principle is valid not only for the 1st utilization, but also for the 2nd, the 3rd, the 4th,...times they have been utilized.
2) Whenever the utility degree of some sources of knowledge gets below a certain threshold, it would then be reasonable for the expert to delearn them at least for a short period of time in order to avoid extra computation belonging to the process of comparing the utility degrees. Regarding this, it is reasonable to make a list for the deleted sources of knowledge in order to show their priority in deletion. Considering the limitations in the cognitive capabilities of the expert, such a delearning process can provide him/her sufficient capacity for a probable modification of the diagnosis structure in future.
3) Whenever, the expert feels that the solution of his/her problem is obtained in a cost-consuming manner, he/she should do efforts to add some new sources of knowledge to the previous diagnosis structure. New sources of knowledge can be occasionally those which have already been deleted in a previous learning based on the mechanism discussed in 2).
4) Whenever, within the past diagnosis experiences, it is noticed that the knowledge describing the problem-environment has been utilized several times, with the successful results mostly belonging to the recent utilizations, it would then be reasonable to reconsider the criteria for the way this source of knowledge is to be utilized.

It is seen that learning in our approach is performed in three phases: a phase of statistical inductive learning for the utility degrees of the sources of knowledge as well as the threshold values, a phase of addition (or deletion) of the new (or old) sources , and finally a phase of adjusting the parameters which belong to the criteria responsible for utilizing the sources of knowledge. It is also interesting to see that the information on any success or failure observed within the process of diagnostic problem-solving can be fed into an analytical learning module which is responsible for justifying these effects. There, the treatment can be expressed in terms of adding or deleting some sources of knowledge.

## APPLICATION OF THE PROPOSED APPROACH TO DIAGNOSIS OF NEUROPSYCHIATRIC DISORDERS

Among the biological disorders, neuropsychiatric disorders are particularly complex in the sense that classes of disorders are in many cases similar in their signals and symptoms, and the way a symptom is assessed by human psychiatrist is uncertain. In the meantime, these disorders hold a multi-class nature with an uncertainty in their class formation mechanism which is mainly due to the ambiguity in their causal physiological or psychogenic relationships. Because of these reasons, it is worth approaching diagnosis of neuropsychiatric disorders from a knowledge-based problem-solving view-point like the one discussed in this paper. Such an approach is particularly helpful when the differential diagnosis between the candidates of the disorder can not be performed conveniently by the human psychiatrist. Regarding the application of our approach, following objectives are to be considered :
a) To study how human psychiatrists make use of different knowledge sources in order to solve their diagnostic problems.
b) To realize the sources of knowledge, out of the available sources, which have the most utility in ordinary diagnostic problems.
c)  To build an efficient strategy for differential diagnosis of the disorders.
d)  To look for new sources of knowledge in order to approach solving difficult diagnostic cases.

Following sources of knowledge can be used by an expert in psychiatry when it comes to diagnosis of neuropsychiatric disorders :

1) Factual knowledge regarding the reality of symptoms and signals acquired from a patient at different time stages.
2) Knowledge concerning the way environmental factors (either genetic or psychogenic) lead to appearance of disorders in patient's biosystem.
3) Knowledge describing disorder formation mechanism
Rules as well as causal networks can be used in order to demonstrate the interrelationship between disorders and their manifestations in terms of symptoms and signals
4) Knowledge describing the way artifacts or noises influence the symptoms and the signals.
5) Knowledge describing causes for artifact or noise generation in the environment.
6) Knowledge describing relevance of the set of classes of disorders already interpreted.

Diagnostic problem-solving structure for neuropsychiatric disorders is illustrated in Fig.1. As is seen from the figure, a central supervisory module is responsible for utilizing a source of knowledge according to the framework of the results obtained from its past experiences. It should be noted that a source of knowledge describing the relevance of the set of disorders can help the expert find out whether the set of interpreted classes can co-exist or not. If the classes of disorder can not co-exist, a number of them is eliminated and the process of knowledge utilization will then be repeated.

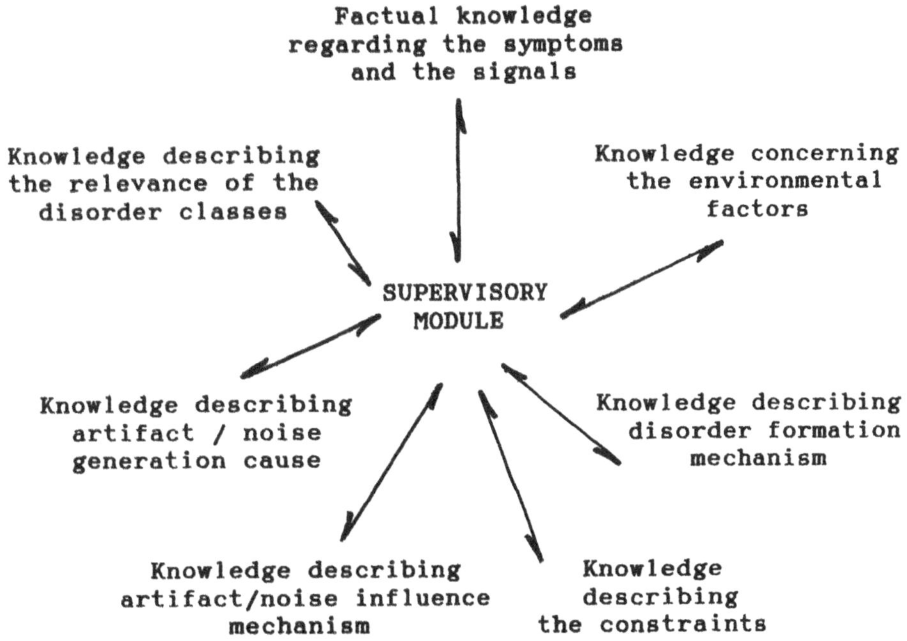

Factual knowledge
regarding the symptoms
and the signals

Knowledge describing
the relevance of the
disorder classes

Knowledge concerning
the environmental
factors

SUPERVISORY
MODULE

Knowledge describing
artifact / noise
generation cause

Knowledge describing
disorder formation
mechanism

Knowledge describing
artifact/noise influence
mechanism

Knowledge
describing
the constraints

SUPERVISORY   MODULE: Either for deciding the source
of knowledge to  be  utilized,
or responsible  for   learning

Fig. 1  Diagnostic problem-solving structure for
neuropsychiatric disorders

A problem in diagnosing neuro-psychiatric disorders is the way their manifestations such as the related symptoms and signals co-occur in context of time. In fact, this is a factor which can make the final differentiation between a variety of disorders with high similarity. Therefore one aspect of learning is to understand the temporal relationships which exist between the disorder's manifestations at different time stages. It should also be noted that knowledge concerning premorbid conditions of the patient can help identifying pathological chronicity of a disorder, and therefore can help regulating the treatment regime for the patient.

## CONCLUDING REMARKS AND FUTURE PROSPECTS

An approach was discussed for learning from the past diagnostic problem-solving experiences in an environment where various types of knowledge are to be utilized. The approach is based on the concept of knowledge utility which is for indicating how frequent the sources of knowledge have been utilized within the past problem-solving experiences. Since the knowledge utilization process may be occasionally unsuccessful, it is worth considering the utility degrees not only for the 1st utilization, but also for the 2nd, the 3rd,...times these sources are being utilized within the process of diagnostic problem-solving. As it is seen from the hypotheses mentioned for learning, although many items are conceptually acceptable, they are not however clear from a quantitative view-point. For example in the 3rd principle, it is seen that concepts like "difficultly", or "cost-consuming" are fuzzy, and therefore some thresholds are required in order to clarify the maneuver domain of the principle. Obviously, a systematic method of learning, mostly of statistical nature, is required to select suitable thresholds in this concern. It is thus seen that, to actuate a learning process, learning in different aspects is required. It is also interesting to see that using inductive learning techniques can help realizing which source of knowledge holds the most amount of utility at a certain mode of problem-solving. Analytical learning techniques can also be utilized to figure out the sources of knowledge which are to be added or deleted within the process of modifying the problem-solving structure. Elaborating on these points is a part of our future research work.

## REFERENCES

Michalski, R. S., and Stepp, R. E. (1983). Learning from observation : Conceptual Clustering, in : R. S. Michalski, J. G. Carbonell and T. M. Mitchell (Eds.), Machine Learning : An Artificial Intelligence Approach (Tioga, Palo Alto, CA, 1983).
Gennari, J. H. et al. (1989). Models of incremental concept formation, Artificial Intelligence, Vol.40, pp 11-61, Sept. 1989.
Dejong, G. F., and Mooney, R. (1986). Explanation-based learning : An alternative view, Machine Learning 1 (1986) 145-176.
Minton, S. et al. (1989). Explanation-based learning : A problem perspective, Artificial Intelligence, Vol.40, pp 63-118, Sept. 1989.
Badie, K. (1990). A knowledge utilization approach to solving recognition-type problems : Toward intelligent recognition, Proceedings of The Seminar on Artificial Intelligence sponsered by Informatic Society of Iran, Sept. 1990 (in Persian).

# Interdisciplinarity and Self-Organization in Computational Neuroepistemology

## An Alternative Methodological Approach to Cognitive Science

Markus F. PESCHL

Dept. for Philosophy of Science
Unit for Epistemology and Cognitive Science (University of Vienna)

## 1 Methodological Problems in Traditional Cognitive Science

As is well known, cognitive science is far from being a well defined discipline with an own methodology (as, for instance, physics or biology). A multitude of disciplines is involved in this fuzzy research program which has its roots in the early fifties – the development of *cybernetics* has brought about an interdisciplinary discourse concerning the questions of human mind, thinking, a mechanistic view of cognition, etc. (*F.Varela* [VARE 90]). Cognitive science has developed in the context of computer science, Artificial Intelligence (AI) and cognitive psychology. As will be shown its traditional form is dominated by the concepts and ideas of computer science. Normally the following disciplines are assumed to take part in the "interdisciplinary" discourse of traditional cognitive science (*Osherson* et al. [OSHE 90], *Posner* [POSN 89], *Stillings* et al. [STIL 87], *Winograd & Flores* [WINO 86], *Peschl* [PESC 90], etc.):

- *(Cognitive) Psychology.*
  The *information processing* revolution of the fifties and sixties brought about a paradigmatic shift in the goals, research methods, models and experiments in this discipline (*H.Simon* et al. [SIMO 89]). In this period the idea of understanding cognition as information processes has been developed. This shift has to be seen in the context of the developments of computer science (faster machines, higher programming languages, etc.) and of Artificial Intelligence. This concept is based on the assumption that information processes are operating on an internal representational structure (which is realized in most cases in symbols). *Newell & Simon*'s *Physical Symbol Systems Hypothesis* [NEWE 76, NEWE 80, NEWE 89] represents the most extreme position in this approach; in short it says that a symbol manipulation and symbolic knowledge representation is the necessary and sufficient means for "general intelligent action". I am stressing this hypothesis, because it dominates the concepts of cognitive psychology as well as of AI.

- *Artificial Intelligence & Computer Science.*
  As has been mentioned above computer science and its concepts are dominating the

theories of cognition. These concepts are very much influenced by the *von Neumann* computer architecture which is based on the *Turing* machine [TURI 36] and on the assumptions of formal logic (i.e. everything can be formalized and, thus, computed by a symbol manipulating algorithm). These assumptions are applied to the problem of simulating cognitive processes. Orthodox AI (i.e. symbol manipulating AI) even developed commercial products which are based on these concepts (i.e. expert systems, knowledge based systems, etc.).

- *Philosophy.*

  The questions being asked by cognitive science are very old and have been considered by philosophy and epistemology for centuries. In traditional cognitive science, the solutions and results of these investigations which are provided by philosophy are repressed by computer science issues. Of course, many theories, concepts, etc. are not adequate in the light of modern (natural) scientific knowledge, it has turned out, however, that epistemology can provide solutions to many problems which are discussed nowadays in cognitive science. *Descartes, Kant, Hume, Bacon,* etc. were working on the problem of knowledge and its representation already centuries ago. The interdisciplinary discourse in traditional cognitive science, however, is not capable of integrating this knowledge into the body of knowledge of AI and cognitive science. The reasons for this problem will be discussed in the following sections.

- *Linguistics.*

  *Language* is the most important means of communication between humans. It is also one of the last steps in our evolution and represents a very sophisticated system of (knowledge) representation; in many cases the latter fact is ignored and (natural) language is assumed to be the given and well proven basis for symbolic computation. Due to the importance of language in the process of knowledge transfer and knowledge processing it is no wonder that cognitive science and AI are interested in the investigation of this phenomenon in order to achieve artificial intelligent behavior. Linguistics reduces language to syntactic and symbolic structures in a system of formal logic. Most approaches investigate the phenomenon of language on this superficial level. Strong interaction between concepts of AI, cognitive science and linguistics can be observed in the domain of natural language processing and understanding.

- *Cybernetics.*

  Cybernetics is an interesting discipline because of its integrating and interdisciplinary function in the discourse of cognitive science. Its concepts would be very important for cognitive science (*Varela* [VARE 90]); they are, however, in most cases not considered. Especially second order cybernetics has brought forth interesting approaches in the field of cognition (*Maturana, v.Glasersfeld, v.Foerster, Varela,* etc.).

- *Neuroscience.*

  An increasing interest in this discipline can be observed in the discourse of traditional cognitive science. Neuroscience was of second order interest for many years, because the gap between the investigation of small circuits of neurons and so-called "higher" cognitive phenomena which were investigated by cognitive science was to big. The development of knowledge and methods in neuroscience as well as in the field of neural computing has brought about a possible convergence.

Results (i.e. publications, cognitive models, books, etc.) in cognitive science show that interdisciplinary cooperation is rather reduced to a slogan; having a closer look at these projects, results, etc. shows that interdisciplinarity is often understood only as quoting and comparing other discipline's results which fit in the own theory. It is no wonder, however, that "real" interdisciplinarity cannot take place in this structure of discourse in traditional cognitive science. The following reasons seem to be the main problems:

- *lacking consequent integration* of the results of the interdisciplinary discourse of the participating disciplines; i.e. the results which are "produced" in the course of the interdisciplinary discourse are not taken seriously or they are assumed to be irrelevant for one's own discipline. In many cases they are assumed to be irrelevant if they do not fit in one's own scientific understanding or paradigm. As there are no methodology and no experiences in interdisciplinary cooperation in traditional cognitive science the importance of integrating the other disciplines' knowledge (also if it does not fit) has not been realized and is misunderstood as a failure in one's own discipline. As a consequence such a scientist would have to leave the interdisciplinary discourse or he/she would have to change his/her paradigm (which is, of course, very difficult), if the aim is to establish "real" interdisciplinarity.

- *lacking reflection* of the results of one's own discipline; i.e. in most cases the scientist of discipline $x$ is not even aware of the scientific paradigm (in the sense of *T.S.Kuhn* [KUHN 67]) he/she is working in. Thus he/she is not capable of understanding and seeing the basic problems and implicit assumptions he/she is making.

- *different levels of discussion* are one of the most important reasons for misunderstandings, confusions and failures. In the case of traditional cognitive science we have seen that this great number of disciplines is working on very different levels of discourse. An AI researcher, for instance, who is searching for a fast reasoning mechanism will not be interested in a discussion about small circuits of neurons which could be interpreted as a very simple form of reasoning on a neural level. He/she could neither contribute nor learn very much from such a discussion, because he/she has completely *different aims*. This is only one example for the unuselessness of the constraint of bringing together as many disciplines as possible.

As an implication it is rather of interest to bring together a *well chosen small group* of disciplines which is capable of finding a *common level of discussion* and *reflection*.


# 2 Computational Neuroepistemology – a Methodological Alternative in Cognitive Science

The *computational neuroepistemology* (CNE) approach tries to "purify" traditional cognitive science by reducing the participating disciplines to the following three: epistemology, computer science and neuroscience. Cognitive science is understood as the continuation of traditional epistemology in the context of modern (natural) scientific knowledge. The aim is to establish a a common level of discussion in the interdisciplinary discourse. I am understanding such a discourse as a cybernetic feedback loop being a *dynamic process* of interactions ([KROH 87, KROH88, KROH 90]. This process can reach only a *stable equilibrium* if all constraints being made by each participating discipline are "satisfied". For our approach to cognitive science this means that an equilibrium can be found only if the participating disciplines are integrated in such a way that they can formulate common results coming up to the claims of each of them. In the following paragraphs I am going to show how such an equilibrium can be found by these three disciplines by discussing their roles and contributions in the interdisciplinary discourse:

- *Epistemology & Philosophy*:
  I restrict philosophy to *naturalistic epistemology* and *philosophy of science*; epistemology is understood as the discipline investigating the question of what is *knowledge*. This means that this conception of epistemology has always one root in the natural sciences

in order to avoid such misleading, inadequate, purely speculative and unsuitable developments (for natural sciences, as they are well known from literature). The *Churchlands* [CHUR 86, CHUR 89, CHUR 90a] and *E.Oeser* et al. [OESE 88], for instance, are exponents of this approach being called *neurophilosophy* and/or *neuroepistemology*. In this approach to cognitive science (CNE) of consequently integrating epistemology into cognitive science, philosophy/epistemology plays a *speculative* role which is *constrained*, however, *not* determined by the empirical results. It also plays the role of reflecting these results and the methods of computer science as well as of neuroscience in respect to their relevance and context; in other words it also puts constraints on these results. Hence, a system of mutually constraining each other is established in the process of interdisciplinarily developing knowledge.

- *Neuroscience*:
  Neuro science represents the *empirical* part of this approach. It provides the empirical evidence for epistemological investigations. As our brain is the substratum (material basis) of our cognitive processes, neuroscience investigates physical (and physiological) processes (and the behavior correlated with it); epistemology has to integrate these (empirical) results into its theories, hypotheses, etc. and has to interpret and examine the applied methods and empirical data in order to value and perhaps qualify or revise them. Neuro science, on the other hand, has to examine the plausibility of epistemologists' theories on cognition, mind, etc.. This mutual examination, revision and correction of methods, empirical data, theories, hypotheses, etc. ensures a well balanced cooperation between two approaches (philosophy and neuroscience) which have *one* goal, but completely *different* methods, approaches and means.

- *Computer Science*:
  Computer Science provides the *generative* aspect to *computational neuroepistemology*. This means that CNE is making use of computer science's simulation techniques and, thus, is capable of artificially generating cognitive phenomena. Computational neuroepistemology does *not* apply the traditional methods of symbol manipulation, but rather makes use of the alternative (quite young) concept of *artificial neural networks*. Computers are playing an important role as simulating instruments for artificial *neural networks* in order to achieve a deeper understanding of cognitive processes in an interdisciplinary context. The PDP approach provides "compatibility" to neuroscience and philosophy by having similar (and more realistic) assumptions concerning knowledge representation (e.g. distributed representation), learning, etc. and by providing similar process structures (e.g. parallel processing, spreading activations, etc.).

By restricting cognitive science to these three disciplines the basic *boundary conditions* for a successful dialogue are created. The level of discussion is well defined; epistemology, computer science (connectionism) as well as neuroscience are assuming the same level of investigation as well as of abstraction. Natural and artificial neural processes represent the basis for the (*bottom-up*) investigations and for the discussions about cognitive phenomena being made by computational neuroepistemology. The other disciplines, which are not participating in this discourse, will "*emerge*" in one or the other form out of this approach; i.e. psychological results, for instance, can be (re)interpreted under an alternative aspect – a kind of "neurally grounded" psychology could emerge from this approach.

## 3  Feedback and Developing Knowledge in Computational Neuroepistemology

The point being of interest for our considerations is the investigation of the *cybernetical flow of information & knowledge* in the process of developing knowledge. Let's have a closer look at the development of (scientific) knowledge in computational neuroepistemology. This process can be understood as a *circular structure* alternately applying deduction and induction. "...scientific method has the same *circular* structure as the *elementary* mechanism of *trial & error-elimination* that is observable already by the lowest animals" (*E.Oeser*, [OESE 90], p 151). What does this mean? A *self-correcting, self-organizing* and *self-regulating* mechanism being responsible for generating "true" knowledge is assumed. This trial and error-mechanism, having been mentioned in the quotation, represents nothing but such a self-organizing structure. "Trial and error" can be interpreted as alternately applying deductive and inductive actions. The created knowledge can be called "true", of course, only in the context of this mechanism. As I am assuming a *constructivist* perspective truth is understood as fitting into the environment.

This process is highly *theory-laden* and you have to be aware that the applied methods are in most cases "self-fulfilling prophecies"; i.e. the structure you are giving to the environment by applying methods will be in most cases very much the same structure the theories are giving to it. Problems will only arise if new or abnormal phenomena are observed. There is either need of a (inductive) *correction* of the current (set of) theory(-ies) or even of a – as *T.S.Kuhn* calls it – *scientific revolution* [KUHN 67]. The latter means, simply spoken, to inductively construct a completely different set of theories and approaches and deductively "test" them whether they are fitting better. This circular process, hence, has a *self-correcting* or *self-organizing* aspect *in se* which helps avoiding dead ends in research. If a *stable state* between the inductive and deductive part is found we speak of a "proven theory"; it is the "*Eigenvalue*" of this feedback system.

What does this mean for the computational neuroepistemology approach? As we have seen the "heart" of the organization of CNE is the tension between the characteristic properties of the participating disciplines: the *speculative, empirical & generative* aspect. The speculative and empirical part can be integrated in the model of developing knowledge having been discussed above in the following manner:

- the *empirical* part is realized by the *neuroscience*; i.e. it has a rather *deductive* character, as it is to a great extent applying methods and mostly doing empirical research.

- *Epistemology* is playing the more *inductive* or *speculative* role. On the one hand they are providing their mostly *speculative* concepts and ideas to neuroscience and computer science; they can make use of it either as a "stimulant" or as a constraint. This has to be decided in the common interdisciplinary discussion. On the other hand it has to *inductively construct* and *integrate* the results coming from neuroscience as well as from computer science. They are kind of *constraints* for the further development of (more or less speculative) concepts and theories.

This circular process can be understood as a spiral developing knowledge and searching for a stable state. What is the role *computer science* and its connectionist approach is playing? This third component (i.e. the *generative* function) in the concept of computational neuroepistemology plays an important role in the self-organizing system of developing knowledge as it is capable of *evaluating* as well as of *generating* knowledge in the circular process of bottom-up (induction) and top-down (deduction). The circular relationship between observed object and theories is expanded by adding the possibilities of *simulation* standing *inbetween* and being themselves such bottom-up/top-down processes. The aim is to find an *equilibrium* in the dynamics of the interactions between the three disciplines; these interactions are "realized", as already has been mentioned, by the "*interdisciplinary discourses*" in which all participants discuss and decide on the validity of the presented knowledge. This is the central "commitee" where the decisions on accepting or rejecting knowledge as "definite" are made.

## Concluding Remarks

This system of developing knowledge is *self-regulating* as each discipline has the same rights and constrains the other disciplines. Hence, a mutual system (network) of constraints is established in each interdisciplinary session. Each *decision* in this discourse implies a restriction of freedom of action – on the one hand this is a very important factor for finding an "Eigenvalue" ($\Rightarrow$ knowledge as "fixed point" in a mathematical sense); on the other hand this process runs the risk of getting in a dead end. For this reason the interdisciplinary group has to be flexible enough to recognize such dangers and to change parts of their theories in order to avoid such developments. Normally the risk of such a development will be quite small, however, because of the diametrically opposed approaches, assumptions, methods and concepts being provided by the disciplines. This ensures a *critical* dialog between the disciplines and protects the system of self-organization against rash decisions. Conclusions are made and controlled by a mechanism of *self-correction*. As an implication this organization of interdisciplinary cooperation is also open to *paradigmatic shifts*; i.e. if the argumentation against one's discipline's contribution is cogent this discipline has to revise its concepts or even change its scientific paradigm (or it has to leave the interdisciplinary research group). Otherwise it would "disturb" the dynamic process of self-regulation and would prevent the group from finding an equilibrium in the process of developing knowledge in this alternative approach to cognitive science.

From this perspective the restriction to these three disciplines becomes comprehensible. Other disciplines which are traditionally taking part in the discourse of cognitive science do not fit into this dynamic system of developing knowledge as they are discussing on different levels and have different aims; this implies that no closure can be established and the whole system would never reach a stable state, because the participants are – as in traditional cognitive science – talking in most cases at cross purposes. As has been mentioned a new kind of psychology or linguistics could emerge from this approach which is trying to "neurally grounded" (*Churchlands* [CHUR 90]) investigate the phenomenon of knowledge.

# References

[CHUR 86] Churchland P.S. (1986): Neurophilosophy. Toward a Unified Science of the Brain; *MIT Press, Cambridge, MA, 1986.*

[CHUR 89] Churchland P.M. (1989): A Neurocomputational Perspective – The Nature of Mind and the Structure of Science; *MIT Press, Cambridge, MA, 1989.*

[CHUR 90] Churchland P.M. & Churchland P.S. (1990): Could a Machine Think?; *Scientific American, January 1990, pp 26-31.*

[CHUR 90a] Churchland P.M. (1990): Cognitive Activity in Artificial Neural Networks; *in Osherson et al. (eds.), An Invitation to Cognitive Science, MIT Press, Massachusetts, Vol. 3, pp 199-227, 1990.*

[KROH 87] Krohn W., Küppers G. Paslack R. (1987): Selbstorganisation – Zur Genese und Entwicklung einer wissenschaftlichen Revolution; *in Schmidt, Der Diskurs des Radikalen Konstruktivismus, pp 441-465, Suhrkamp, stw 636 (1987).*

[KROH88] Krohn W. & Küppers W. (1988): Die Selbstorganisation der Wissenschaft; *Suhrkamp, Frankfurt/M., stw 776, 1988.*

[KROH 90] Krohn W. & Küppers G. (1990): Science as a Self-Organizing System. Outline of a Theoretical Model; *in Krohn et al (eds.), Selforganization. Portrait of a Scientific Revolution, Kluwer Academic Publishers, Netherlands, 1990, pp 208-222.*

[KUHN 67] Kuhn T.S. (1967): Die Struktur wissenschaftlicher Revolutionen; *Suhrkamp Taschenbuch, Frankfurt, stw 25.*

[NEWE 76] Newell A. & Simon H.A. (1976): Computer Science as Empirical Inquiry: Symbols and Search; *Communications of the ACM, March 1976, Vol. 19, Number 3, pp 113-126*

[NEWE 80] Newell A. (1980): Physical Symbol Systems; *Cognitive Science 4 (1980), pp 135-183.*

[NEWE 89] Newell A., Rosenbloom P.S. & Laird J.E. (1989): Symbolic Architectures for Cognition; *in Posner M.I. (ed.), The Foundations of Cognitive Science, MIT Press, Massachusetts, pp 93-131, 1989.*

[OESE 88] Oeser E. & Seitelberger F. (1988): Gehirn, Bewußtsein und Erkenntnis; *Wissenschaftliche Buchgesellschaft Darmstadt, 1988.*

[OESE 90] Oeser E. (1990): The evolution of scientific methods; *Fresenius' Journal of Analytical Chemistry (1990) 337, pp 150-154.*

[OSHE 90] Osherson D.N. (ed.) (1990): An Invitation to Cognitive Science; *MIT Press, Massachusetts, 1990.*

[PESC 90] Peschl M.F. (1990): Cognitive Modelling. Ein Beitrag zur Cognitive Science aus der Perspektive des Konstruktivismus und des Konnektionismus (Cognitive Modeling. Constructivist and Connectionist Aspects of Cognitive Science); *Deutscher Universitäts Verlag/Vieweg, Wiesbaden, 1990.*

[POSN 89] Posner M.I. (ed.) (1989): Foundations of Cognitive Science; *MIT Press, Massachusetts, 1989.*

[SIMO 89] Simon H.A. & Kaplan C.A. (1989): Foundations of Cognitive Science; *in Posner M.I. (ed.), Foundations of Cognitive Science, MIT Press, Massachusetts, 1989, pp 1-47.*

[STIL 87] Stillings N.A., Feinstein M.H., Garfield J.L. et al. (1987): Cognitive Science, An Introduction; *A Bradford Book, The MIT Press, Cambridge MA (1987).*

[TURI 36] Turing A. (1936): On Computable Numbers, with an Application to the Entscheidungsproblem; *in Proc. London Math. Soc., ser 2, 42(1936), pp 230-265.*

[VARE 90] Varela F.J. (1990): Kognitionswissenschaft – Kognitionstechnik. Eine Skizze aktueller Perspektiven (Cognitive Science); *Suhrkamp, stw 882, Frankfurt/M., 1990.*

[WINO 86] Winograd T. & Flores F. (1986): Understanding Computers and Cognition, A New Foundation for Design; *Addison-Wesely Publishing Company, Inc., (1986).*

# SYSTEMS THINKING IN POLAND

Piotr Sienkiewicz
Roman Wojtala

Polish Cybernetics Society
Warsaw

## INTRODUCTION

The history of the scientific thinking is a self-knowledge of people acting in the civilization based on science. It also protects the scientists from the absolutism of the contemporary scientific theories, it teaches how to perceive their variability and dependency from the current state of knowledge and, sometimes, from the various non-scientific factors (ideological or political).

So far the history of the systems thinking has not been written. In this paper only some events, concepts and persons have been presented - those particularly near and dear to the authors.

Perhaps the origin of the systems thinking in Poland should be searched in the works of the great astronomer M. Kopernik (N. Copernicus, 1473 - 1543), in the works of the very first central state school authority in Europe - National Education Commission (1773 - 1794) or in the development of the Lvov - Warsaw philosophical and logical school (1895) to which belonged, among others, K.Twardowski (1866 - 1939), J.Lukasiewicz (1878 - 1956), S.Lesniewski (1886 - 1939), A.Tatarski (1902 -      ), R.Ingarden (1893 - 1970) or in the works of the representatives of the mathematical centre in Lvov, such scientists as H.Steinhaus (1887 - 1972) and S.Banach (1892 - 1945).

## MAIN TRENDS

Philosophical trend. Polish cyberneticists readily recall philosopher B.Trentowski (1806 - 1869), who wrote on cybernetics as "an art of ruling over a people". He used that notion in his work entitled "Relation of philosophy to cybernetics" (1843), independently of A.M. Ampere. Certain features of the systems thinking should also be noticed in the works of philosophers and logicians belonging to the Lvov - Warsaw school and, particularly, in the holistic concept of mereology by S.Lesniewski (1916). Lesniewski logical systems are composed of three theories: prototetics, ontology and mereology. Mereology is a theory of the relation of a part to the whole, which constitutes the fundamental problem of the systems thinking, it can also be regarded as a set and class theory in a collective sense. Lesniewski systems exerted an influence on the works of T.Kotarbinski (1886 - 1981), on his philosophical idea, so called reism. Kotarbinski is the author of praxiology (Gr praxis - activity, logos - science), that is a science on efficient activity. In it, he stipulated for the establishment of a complex theory, as formulated by L. von Bertalanffy, K.Boulding and others. Kotarbinski incident theory is very close to the complex theory.

R.Ingarden, in his lecture on "Quelques remarques sur la relation de causalité" (Rome, 1946), pointed at the idea of a "relatively isolated system" in connection with the question of causality. This idea is related to the notion of a system used in biology by I. von Bertalanffy ("An outline of General Systems Theory", 1957).

Cybernetical trend. The concept of a "relatively isolated system" by Ingarden has been further developed in an original cybernetics of the logician - H.Greniewski (1903 - 1972). Using this notion, Greniewski formulated three principles of duality which allowed to formulate elements of general cybernetics (1959). He developed the cybernetic and system elements of economical planning as well as general models of human being, so called Golems.

Greniewski concepts has been further developed by the outstanding economist - O.Lange (1904 - 1965), who worded out elements of the economical cybernetics (1965) and formulated very original, mathematical concept of "entirety and the development of systems in light of cybernetics" (1962).

References to the Greniewski concept are made, among others, by M.Kempisty in a model of associative memory (1970) and by J.Konieczny in a cybernetic theory of combat (1970). Greniewski concept is a starting-point to the mathematical theory of Cybernetical Isolated Systems (CIS) of J.Jaron (1968). Formal description of a cybernetical system includes fundamental notions aggregated in the following sets: set of repertoirs, set of calendars, set of trajectories, set of boundary organs, set of assigners. Cybernetical systems described by means of the quintuple are called cybernetical systems of the CIS class. The description of five simple systems, whose names define their character are given as examples of the terminology and symbols presented above. These are the following: "negator", "alternator", "conjuctor", "retarder by one instant" and "bicopiator". These cybernetical systems will be used in the sequel to construct the examples of complex cybernetical systems.

The original concepts of the engineer and cyberneticist - M.Mazur (1909 - 1983) include: autonomous systems theory (1966), based on it cybernetical theory of human character - psychocybernetics (1975) and qualitative information theory (1970).

Mazur worked out a concept of "autonomous system" defined as a system: 1) able to control itself, and 2) able to preserve its ability to control itself. The first condition requires organs for reception and accumulation of both energy and information. The second condition requires an organ maintaining functional equilibrium of energetical and enformational processes (homeostasis). Thus, any statement resulting from the analysis of the autonomous system applies to the human being. Analysis of informational processes from the physical piont of view (potentials, conductances, energy flow) elucidates the physical nature of psychical phenomena: memory, emotions, reflections, intuition, consciousness, thinking, motivation, decision making etc. The basic concept in Mazur qualitative information theory is information (not amount of information). In his works, two kinds of degenerated processes of informing are analysed; one in which the code chains are full separate but not full (disinforming). In both cases, the following possibilities are considered: a) there is image information but no original information (simulation), b) there is iriginal information but no image information (dissimulation), c) the image information differs from the original information (confusion).

Mazur autonomous systems theory has found its application in the analysis of socio-economical systems and in the analysis of human characters.

Mathematical trend. Apart from the already mentioned CIS mathematical theory, other theories referring to praxeology are: a) M.Nowakowska activity theory (1973) which comprises models of activity systems, models of decisions in hazardous situation, theory of social change etc., b) K.Tchon theory of events systems as basis for interdyscyplinal communication (1981).

Among the mathematical system concepts the following should be mentioned: 1) mathematical theory of information system (1982) and Z.Pawlak mathematical theory of conflicts (1988); 2) R.Kulikowski mathematical models of the development of socio-economical systems (1970); 3) S.Piasecki mathematical theory of organization (1970); 4) S.Wegrzyn mathematical theory of the development processes (1988).

Engineering trend. Both W.Gasparski methodology of design (1985) and J.Konieczny activity systems engineering (1984) refer to praxeology. Large technical systems (as for example electric power systems) and socio-economical systems are the object of W.Bojarski analysis and systems engineering (1989) and theory of systems efficiency (1987) refer to cybernetics and systems analysis.

CONCLUSION

Systems thinking in Poland includes various trends derived from different sources. They are still being developed and Polish systems sciences have not lost touch with the systems research centres around the world. Prognosis covering the period up to the end of XX century is reasonably optimistic. The economic crisis has its influence on the development of research and application of systems analysis and systems engineering. This optimism comes from the fact that, at present, the majority of the polish organizational, economical and technical problems cannot be solved without modern high technology and modern systems methods.

BIBLIOGRAPHY

Greniewski H., 1969, "Cybernetyka niematematyczna", PWN, Warszawa;
Jaron J., 1976, "Podstawy cybernetyki", Wrocław,
Konieczny J., 1984, "Inżynieria systemów działania", WNT, Warszawa;
Kulikowski R., 1970, "Sterowanie w wielkich systemach, WNT,Warszawa;
Lange O., 1962, "Całość i rozwój w świetle cybernetyki", KiW,Warszawa;
Mazur M., 1975, "Cybernetyka i charakter", PIW, Warszawa;
Pawlak Z., 1988, "O konflikcie", Logos, Warszawa;
Sienkiewicz P., 1988, "Poszukiwanie Golema", KAW, Warszawa;
Sienkiewicz P., 1989, "Inżynieria systemów kierowania", PWE, Warszawa.

# OLD CONFUSION IN NEW EUROPEAN SYSTEMS THINKING

Otto Hansen

Department of General and Applied Linguistics
University of Copenhagen

## INTRODUCTION

The answer comes easily enough if we ask where systems originate? They originate in the human mind, or more specifically in the human language capacity. The notion is wrong that we discover systems. It is wrong even when we talk about nature with its systematic order among microorganisms, plants, and animals. Nature has no intention and makes no plans. It all happened by chance. Even the human central nervous system is a system only because we say so.

Of course there is that serious reservation Churchman (1988) voiced when claiming that the question if God exists is the most important one in systems thinking. If such "a perfect being exists", Churchman said, our planning must relate to this existence. Thomas Aquinas' (Summa contra Gentiles I, 13 (35)) expression for that perfect being is *aliquid quod est maxime ens*, and He could have created in accordance with a divine system. Philosophically, ontologically as it were, it has not all that weight when Churchman says that if God does not exist, "then we not only have a lot of explaining to do in terms of our values, but we also have to find a whole set of godless values to guide us". What does matter is firstly that God, in the words of Prigogine & Stengers (1985, p. 272), if he made use of his absolute knowledge "could get rid of all randomness", and secondly that people's values are intrinsically bound up with doctrines grounded in a faith their forefathers actually had.

## ARE THERE THEN NO SYSTEMS?

The genes of an organism have no interest in the species to which we systematically refer them. They care even not whether they belong to a species with asexual or one with sexual reproduction, i.e. one belonging at the lower side of the evolutionary jump of the introduction of sexuality or on the higher side. There is an eukaryotic alga *Chlamydomonas*, a unicellular organism which can reproduce sexually (with to our systematic view essential advantage), but ordinarily reproduce asexually. While this means simple division into two new identical individuals, sometimes an alga by successive divisions produces eight, sixteen, or thirty-two sex cells. They are called gametes and fuse in pairs. The gametes of these pairs usually come from different original individuals. The new organisms are diploid, i.e. they have double the chromosome number, and they are able to survive adverse conditions, even to remain dormant for a long time. Eventually the zygotes, as the diploid cells are called, divide into four haploid cells, cells with the same chromosome number as the mother cell. But they have become genetic variants. Of course when we have higher organism mutants that stop crossbreeding with their ancestors, we call the mutants a new species and say that this is the way we got the evolution of the species. Still, when we originally had the chemical mutation, the change in a DNA sequence became possible because one of 7,000 hydrogen atoms in a DNA base randomly, randomly theoretically and practically, has two protons instead of only one.

It should be noted that an original event by chance, when first there may necessitate an effect. Something may have been caused although no law determined it. Mathematically expressed, statistical vagaries can determine limiting values which are themselves random variables. Insufficient representation of such situations may yield illusions that have led to construction for example in biology of apparent systems misleading as to the understanding of evolutionary mechanisms.

## THE SYNTACTIC SYSTEM

Then still there is a system: the primal, universal, or innate human grammar. I dealt (Hansen, 1989) with the biology of the language faculty as that which in Humboldt's (1820) and Chomsky's (1976 a, pp. 207-209) meaning made man man, that in Chomsky's (1976 b) words "fixed function, characteristic of the species, one component of the human mind, a function which maps experience into grammar". Human language, not as all the different semantic apparatuses which human intelligence generally have spread over the globe, but the genetically determined logical capacity of syntactic analysis and construction, is a system. Thomas Aquinas (Summa contra Gentiles III, 104 (3) & 154 (7)) maintained that God necessarily had to provide man with the grace of speech, the very act of speaking being characteristic of rational nature. Søren Kierkegaard (1923, IV, p. 352) said that "it is not permissible to have man having invented language himself". And then I asked (Hansen, 1989) if speaking man is an ethical creature?

## THE CONFUSION

If I have now in a way taken us back to Churchman's (1988) involving the question of God's existence in systems thinking, we may note the theological perspective in the Fall. Kierkegaard (1923, IV, pp. 350-353) actually in a grand synthesis interconnected the acquisition of language, original sin, human sexuality, and the very existence of history.

In any case the problem is that our analytical and constructive capacity provides no guarantee concerning semantics. Whatever it was that made the Old Testament poet tell the story of the tower that was to have its top in the heavens, the single language became confused, and people no more understood what they said to one another (Genesis 11, 1-9).

202

Among the words that seem to have got different meanings to different people are those supposed to cover central systems concepts as right and wrong, good and evil, gain and loss, and efficient and inefficient. What I say is that the concepts are differently interpreted. That individuals will not always agree if some action is right is a common experience. If we mean to put the question if a business proposition is economically reasonable, there will be professional systems agencies with analytical techniques they could defend as equally applicable in the north of Britain and in the south of Italy. If, however, the business executives from the two areas had compunction concerning the morality of the economic plan, it might well turn out that right and good could sound quite differently in northern and southern Europe. This phenomenon may then also turn out to present difficulties to the European integration, to the technical co-operation within the EEC. We may find that the roots lie deep. My contention is that the differences must be sought in different philosophies created some centuries ago by the great European religious upheaval. It could be that even not always apparently dissimilar European subcultures could argue systems science and its practical application fundamentally differently because some earlier Europeans clashed about faith and reason.

## THE REFORMATION

Actually the faith-reason conflict is older than the Reformation. Bernard of Clairvaux was deeply sceptical about reason, and he vehemently fought the Aristotelian Abelard. In the words of Marius (1986, pp. 66-67) "Abelard believed that the gift of reason came from God and that it was intended by God to draw reasoning humankind back to Himself, that reason was what Aristotle said it was – the faculty of the mind to begin by doubts and then proceed, a step at a time, with what could be known, to build a system with all the parts in perfect harmony with one another, the harmony itself open to the human mind". Bernard, on the other hand, Marius continues, "wanted rapture – an aesthetic emotion – that lifted the soul from love of self to the complete love of God, a love that filled the soul with bliss, conveying an intuitive certainty far richer than intellectual understanding, a direct experience that was a foretaste of heaven itself".

I suppose the wish for rapture may be more than most moderns will acknowledge, but that some still with Bernard, and still in Marius' words, think "that to ask searching questions of the faith (is) to court eternal danger. It is also always a precarious thing to state that an age thought in this way or another. Renaissance ideas can be demonstrated in some Medieval thinkers, and quite a number of Renaissance people longed back to Medieval religious order. But it was during the Renaissance that distrust became general in the great scholastic system with logic permeating creation and demonstrating that reason could always be applied to human problems. The methodology of the system had demanded that the old authorities were always cited. Now it was discovered, and printed books made it widely known, that the authorities were often incorrectly quoted. Even that in some instances the old authorities had never existed. Philosophically this is pityingly primitive. But it was what happened.

Luther and Protestantism thus did not invent distrust in reason. But the political functions of the different theological denominations drew demarcation lines through Europe. Slowly human reason found its way back into the new world view of modern science, the new system. But fundamentally the distrust had been established, and now it is surfacing anew. For do we not now again experience distrust in reason also in the secular area? Is not the popular reactionary behaviour towards science and technology distrust in reason? If some technological applications are stupid or even morally wrong, it is the old mistake to fault reason itself. In the medical area we see flat and widespread superstition successfully exploited by the human vultures of the herb industry and healing and "therapeutic" branches.

## THE SPEAKING ONE IS LANGUAGE

This subheading has been taken from Kierkegaard (1923, IV, p. 352). He answers the possible objection to Genesis that not knowing the difference between good and evil, Adam would essentially be unable to understand the debarment. The imperfection of the Biblical account, Kierkegaard says, is resolved with the understanding that the speaking one is language, that it is Adam himself who speaks. In a more modern wording the term of language having arisen de novo was coined in the biologists' discussion of the genetic and neurological elements of the claim that man became human only once, and evolutionarily understood quite suddenly. The concept of primal, universal, or inate grammar is that the necessity of the rules followed in transformations must take them back to the mode of conversion of thought into extended speech. Unfortunately there is some confusion with elements from the methodological transformational grammar (which by the Chomskyan definition deals with the rules necessarily followed in the generation of one syntactic structure from another, i.e. the transformation). Chomsky (1976 a, p. 29) underlined himself that by the necessity of the rules followed in the transformations, he meant biological necessity. Details of my own research and clinical findings as well as references to the results of other workers may be found in my publications (Hansen, 1978; 1979; 1980; 1981 a; 1981 b; 1985; 1986 a; 1986 b; Hansen, Nerup & Holbek, 1986; 1987).

When we couple the grammatical logic with semantics we get what Jaynes (1976) calls the "holding power" of the words. And in this connexion there may be reason to heed Richard Jung's (1987) warning of the psychological inevitability of ontological commitment to the reality of a particular metaphor when employed. This would mean for example mistaking the latest "scientific" formulation for ontological evidence. But it would certainly also mean that we can forget to examine if the legacies of formulae from philosophy and religion really provide ontological evidence. It is even possible to get the idea of supplying a computer programme (Lefebvre, 1980; 1986) with ethical cognition.

So we see a leading systems theoretician (Ericson, 1987) not only find a conflict between personal values and group values both quantitatively increase and qualitatively exacerbate a moral dilemma in the modern organizational world. He also reaches the judgement that by the very impersonality of its mechanisms, cybernetic communication will cause some of our basic societal values to tend themselves towards hypocrisy. The technology shapes our values as much as our values shape.

The computerized decisions are absolutely methodologically rational in the sense that they are unbending executions of human grammatical logic. But for most computer users it was someone else who applied the symbolic logic of the Boolean functions to the programme, i.e. who decided which ethical value should correspond to the "1" and which to the "0". The distance is continuously increasing between that person who originally decided the answer the computer would give, and the person who will eventually act upon it. Even the legally responsible state official will more and more often have lost contact with the conditions of the actual decision. When in some instances for example the press has exposed an indisputable injustice, even high officials have been known to blame the computer.

The meanings we attach to symbols can be manipulated, in the human brain as in the computer, by the rules of syntax, and we may, again in the human brain as in the computer, get linguistic confusion of deductive logic. We have the infallible (in the biologically healthy human brain) language faculty of Noam Chomsky mapping experience into grammar. But the new behavioural problems we have introduced are ethical, as were the old ones. Ethics came into the world with language. Evolution has linked all other elements in human behaviour with language, as language caused man's dominance of the earth. Without that dominance there would be no moral choices to make; with it they had to be made (Hansen, 1989). And we can certainly both misunderstand right and wrong, good and evil – and lie about them.

REMEDIES?

Misunderstanding we can undoubtedly do something about by study. If efficiency does mean something different in southern and northern Europe because Aristotle and Thomas Aquinas called God the first efficient cause while the marriage between Protestantism and industrial revolution made mechanical productivity a virtue, then the metaphorical language of two old systems must be penetrated and elucidated before the computer is given the information to be applied in calculation of the justness of a deal. If terms as gain and loss really carry remembrance of two different ways to eternal life – the one essentially excluding the other – then we do have, as Churchman said, a lot of explaining to do in terms of our values.

If that may help to provide remedies may not be certain, but we got a new illustration with the fracture of Communist unity. Those countries which hastily left that unity show so many similarities to those that left the Universal Church. As a matter of fact I concluded this paper just when the missiles hit Israel after a protracted demonstration of different meanings given to all the central terms of human systems.

REFERENCES

Aquinatis, S. Thomæ Doctoris Angelici **Opera Omnia** (Sixteen Volumes). Iussu impensaque Leonis XIII P.M. edita ex Typographia Polyglotta, Rome 1882–1948.

Chomsky, Noam (1976 a) **Reflections on Language**. Pantheon Books, New York.

Chomsky, Noam (1976 b) On the nature of language. **Ann. N.Y. Acad. Sci.** 280, 46–57.

Churchman, C. West (1988) "Discoveries in an Exploration into Systems Thinking" in **Yearbook of the International Society for the Systems Sciences,** Vol. XXXI (ed.

William J. Reckmeyer). New York: ISSS, pp. 39–44.

Ericson, Richard F. (1987) "System-Induced Hypocrisies: Our Quintessential Moral Dilemma" in **Yearbook of the International Society for Systems Research,** Vol XXX (ed. John A. Dillon, Jr.). Louisville: ISGSR, pp. 77–81.

Hansen, Otto (1978) **The Genes of Universal Grammar.** Lund: University of Lund (ISBN 91-970270-0-6).

Hansen, Otto (1979) "Human Language as a Biological Behaviour Determinant" in **Improving the Human Condition** (ed. Richard F. Ericson). Berlin, Heidelberg, New York: Springer-Verlag, pp. 961–970.

Hansen, Otto (1980) **Speech and Language.** Lund: University of Lund (ISBN 91-970270-2-2).

Hansen, Otto (1981 a) Are the genes of universal grammar more than structural? **Hereditas** 95, 213–218.

Hansen, Otto (1981 b) "Biological Study of Language as a Key in General Systems Research" in **General Systems Research and Design** (ed. William J. Reckmeyer). Louisville: SGSR, pp. 547–555.

Hansen, Otto (1985) **Biological Linguistics.** Copenhagen: Akademisk Forlag (ISBN 87-500-2573-2).

Hansen, Otto (1986 a) Sociobiology and biological linguistics. **Essays in Human Sociobiology** 2, 129–139.

Hansen, Otto (1986 b) "Sign Language of the Deaf" in **Signs of Life** (ed. Bernard T. Vervoort) Amsterdam: IGL, pp. 57–61.

Hansen, Otto (1989) "Is Speaking Man an Ethical Creature?" in **Operational Research** (eds. M.C. Jackson, P. Keys & S.A. Cropper). New York: Plenum, pp. 259–264.

Hansen, Otto, Nerup, Jørn & Holbek, Bertha (1986) A common genetic origin of specific dyslexia and insulin-dependent diabetes mellitus? **Hereditas** 105, 165–167.

Hansen, Otto, Nerup, Jørn & Holbek, Bertha (1987) Further indication of a possible common genetic origin of specific dyslexia and insulin-dependent diabetes mellitus. **Hereditas** 107, 257–258.

Humboldt, Wilhelm von (1820) Ueber das vergleichende Sprachstudium. **Abh. der Akad.** (Berlin) as printed in **Die sprachphilosophischen Werke Wilhelm's von Humboldt** (ed. H. Steinthal). Berlin 1884: Ferd. Dümmlers Verlagsbuchhandlung.

Jaynes, Julian (1976) The evolution of language in the late pleistocene. **Ann. N.Y. Acad. Sci.** 280, 312–325.

Jung, Richard (1987) "A Quarternion of Metaphors for the Hermeneutics of Life" in **Yearbook of the International Society for General Systems Research,** Vol XXX (ed. John A Dillon, Jr.). Louisville: ISGSR, pp. 25–31.

Kierkegaard, Søren (1923) **Samlede Værker,** Vol. IV (eds. A.B. Drachmann, J.L. Heiberg & H.O. Lange). Copenhagen: Gyldendal.

Lefebvre, Vladimir A. (1980) An algebraic model of ethical cognition. **J. math. Psychol.** 22, 83–120.

Lefebvre, Vladimir A. (1986) "Modelling of Quantum-Mechanical Phenomena with the Help of the Algebraic Model of Ethical Cogniton" in **Yearbook of the Society for General Systems Research,** Vol. XXIX (ed. Rammohan K. Ragade). Louisville: SGSR, pp. 63–68.

Marius, Richard (1986) **Thomas More. A Biography.** London: Fount Paperbacks.

Prigogine, Ilya & Stengers, Isabelle (1985) **Order Out of Chaos. Man's New Dialogue with Nature.** London: Fontana Paperbacks.

# A UNIFIED SYSTEMS HYPOTHESIS

Derek K Hitchins

RMCS Shrivenham

## U.S.H. System Images

The Unified Systems Hypothesis (USH) seeks to establish Principles which apply to all systems. In so doing, it builds upon General Systems Theory and the work of many systems thinkers. This paper is a much-reduced presentation of the principal ideas behind the USH, which is best approached as a set of mutually-consistent models or images of Open Systems and, particularly of their interconnections. USH operates, not from the system viewpoint, so much as from the viewpoint of the interactions between systems

## A General System View

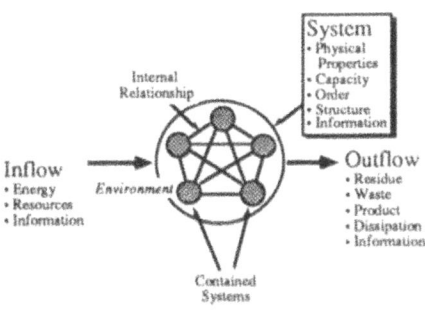

The first image is of a system receiving inflows, passing outflows and containing related and intraconnected systems. The inflows generally comprise energy, matter and information. The outflows are similar in substance but attract different titles. The system exhibits physical properties, it has order, structure or hierarchy, and it has capacity, intrinsic or explicit, to store/process energy, matter and/or information. Environment pervades and impinges upon the system and its contained systems. Evidently, this system image is of an Open System, connected to other systems not shown.

## Systems Hierarchy

The second image presents a three-level systems hierarchy in which a "System-in-Focus", that in which an observer has immediate interest, both contains systems (subsystems) and is itself contained in a Containing System along with other Sibling Systems. These siblings are related / interconnected to the System-in-Focus; its contained systems are intraconnected. Environment pervades the Containing System, but need not be homogeneous. Environment exists within the System-in-Focus, but need not be identical with that outside in the Containing System. Boundaries, shown as hard edges, may in fact be soft and fuzzy.

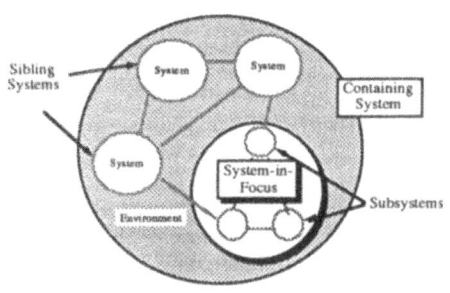

## Interacting Systems

The third image combines the first two into a networked set of contained systems with mutual interflows, such that the outflows from some form the inflows to others. One system's residue becomes another's resource; one system's dissipation becomes another's energy source. Information is, unlike energy and material, exchanged without significant loss to the supplier. The interacting systems exist within a container which also receives, dissipates and exchanges, so providing hierarchical consistency.

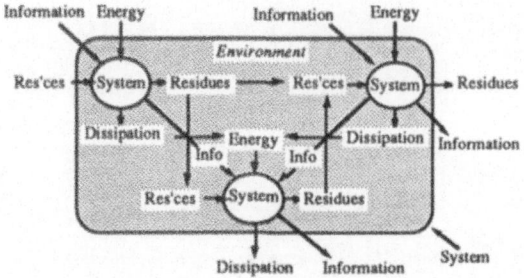

## Simultaneous Multiple Containment

The fourth image presents a different thought; that a system may be simultaneously contained within more than one container, as a bus-driver is simultaneously within a transportation system, a family system and a social system with his passengers. The potential complexity engendered by this image is staggering; if each system at each level of hierarchy can be simultaneously in a variety of containers then the resulting n-dimensional weave could be beyond untangling.

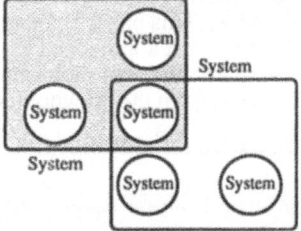

## Cohesion and Dispersion

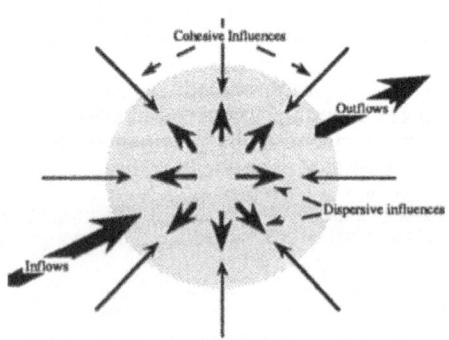

For a system to continue as an aggregation, it follows that there must be some cohesive influence attracting the contained systems one to another. That the system does not collapse to a point suggests that there must be counteracting influences tending to disperse the contained systems. Cohesive and dispersive influences must balance for a system to persist. Such balance could be static or dynamic (oscillatory) Since systems wax and wane, it must be possible for the balance to be changed in either or both directions. The fifth image presents system inflows and outflows as the mediators of change in this weakening or strengthening of binding influences.

## U.S.H. DEFINITIONS OF "SYSTEM", "ENVIRONMENT" AND EQUILIBRIUM

### System

Within a Unified Systems Hypothesis, the definition of "system" is of particular interest, since there have been many definitions. Most commentaries agree that there are concepts both of parts, and of relationships between those parts, in the notion of system. I would contend that it is the orderliness of the systems concept which is appealing, in that it reveals pattern in complexity or from obscurity. The following definition, used as a basis within the Unified Systems Hypothesis, is hopefully sufficiently vague to capture all kinds of systems, yet sufficiently explicit to be useful:

> *A system is a collection of interrelated entities such that both the collection and the interrelationships together reduce local entropy.*

In this definition, the relationships receive a degree of prominence equal to that of the entities, because the pattern or network of relationships reduces uncertainty just as much as the collecting of entities. The definition covers all kinds of systems, human activity, man made, natural, etc., and is compatible with open as well as closed classifications. This is not to suggest a relational structural approach; Angyal (1941) suggested that "systems cannot be deduced from relations, while the deduction of relations from systems still remains a possibility". Since systems could be related in many ways, a particular pattern of relationships carries information, reduces uncertainty—the definition seeks parity for structure with entity, but not precedence.

## Environment

Environment is a strange concept to define, yet it is an essential feature from the most abstract of system levels down to the air we breathe and the situations in which we live. I therefore propose a seemingly new definition, designed as with "system" to be both vague, yet precise:

> *Environment is that which mediates the interchanges between systems. Total environment is the sum of all such mediations*

Consider any two systems. Identify the exchanges between them. Identify that which mediates the interchanges; that is environment. For example, that which mediates the interchange between economic systems is money, barter and trade—we often speak of a "favourable trading environment." Consider a suburban dormitory system and a City business. That which mediates the interchange of people is the commuting facilities—we often refer to the travelling environment. Plants and animals exchange $CO_2$ and $O_2$ using the atmosphere and the biosphere as a mediator. In physics, forces are mediated by the exchange of particles. Heat being conducted along a metal rod is mediated by conduction electrons. Living, walking in the town and country, environment is that which mediates the multitude of interchanges between us and the surrounding features.

We humans tend to organize our environment into transport systems, communication systems, infrastructure systems and so on. This presents no problems within USH, since it is merely a hierarchy shift. It is, however, convenient to retain the notion of environment as mediating interchange between systems—it is a useful model.

## Equilibrium

As with environment, so the notion of equilibrium has been disturbed by systems thinking. Koehler (1938) held the view that equilibrium was essentially associated with a low state of energy, as for a marble running to the lowest level in a saucer, while for many organisms what was frequently referred to as equilibrium corresponded instead to a heightened energy state. A candle flame reaches its stable operating length from the wick when it is burning brightly, not at some minimum energy condition.

Evidently there can arise a static or dynamic balance between inflows to, and outflows from, an Open System such that it reaches a stationary or stable condition. Stability in Open Systems, then, is generally associated with *high* energy; the universal test for equilibrium cannot be one of minimum energy. Instead, I propose the following definition for all systems:

> *Interacting systems can be said to be in equilibrium when their environment is stable, statically or dynamically*

The USH definition is descriptive—if the environment is stable, then it may be deduced that interacting systems are in equilibrium. In USH, stable environment is the litmus test of equilibrium.

## U.S.H. PRINCIPLES

We are now in a position to identify some simple systems principles which are induced from observation, accepting Popper's (1968) admonition on the limited value of induction, but nonetheless presenting the principles in Popper's (1972) spirit of openness as the basis for progress.

## Interacting Systems

Le Chatelier's Principle is a general principle of interacting forces in physics and chemistry:

"If a system is in stable equilibrium and one of the conditions is changed then the equilibrium will change in such a way as to restore the original condition"

← Pulley →

In the diagram, the three forces at the left are in equilibrium. At the right, a fourth force is introduced and the original three readjust to a new point of equilibrium for all four. The example is of forces in a single plane, but the concept is seen so often in everyday life that a wider interpretation seems eminently reasonable.

The Principle of Interacting Systems flows simply from the images and definitions above, and is as follows:

*"If a set of interacting systems is in equilibrium and either a new system is introduced to the set, or any of the set or their interconnections undergoes change, then, in so far as they are able, the other members of the set will rearrange themselves so as to oppose the change"*

The principle is unexceptional for physical or chemical systems, to the point that it may seem axiomatic; they were expounded by the French chemist Le Chatelier (1850—1936) in 1888 in that context. There seems to have been no general statement concerning the applicability of the Principle to *all* systems, however. The contention of the Unified Systems Hypothesis is that the principle applies equally to economic, political, ecological, biological, stellar, particle or any other *interacting system*.

The principle does not indicate the *manner* of movement. There is certainly nothing in the principle to suggest that movement should be linear. For particular systems and their interactions, movement could be slow, fast, even explosive, as suggested by Catastrophe Theory and Chaos Theory. Both these theories would seem to interface with the Principle of Interacting Systems.

## System Cohesion

This principle derives simply from the fifth image above, and may seem axiomatic, particularly for physical systems:

*A system's form is maintained by a balance, static or dynamic, between cohesive and dispersive influences.*

Since the USH is intended to apply to all systems, this principle must apply not only to such physical systems, but also—for example—to social systems such as families or ethnic groups.

## Connected Variety

The Principle of Connected Variety is concerned with stability[1] of interacting systems. The third image above showed a small set of three interacting systems. As the number of interacting systems increases, and as their mutual interconnections increase both in number and in the variety of energy, matter and information exchanged, they develop a closer and more cross-coupled weave in which it is increasingly likely that system outflows will match other system inflows, leading to a stable environment. These considerations lead to the Principle of

---

[1] Stability is not always a desirable state. A set of stable interacting systems may be resistant to change. While such resistance may be admirable in the biosphere, it may be less so in, say, business or politics, where controlled change may be the objective.

Connected Variety:

*Interacting systems stability increases with variety, and with*
*the degree of connectivity of that variety within the environment*

The Principle emerges because increasing connections between interacting systems increases the potential for feedback, which can be positive or negative. If negative, stability is increased. If positive, then stability will also be reached but after a period of change—see Principle of Preferred Patterns below. Connection is not sufficient for feedback; the right substance or information must be fed back too; increasing variety increases the prospect of *effective* feedback. Evidently, there are shades of W. Ross Ashby's (1956) Law of Requisite Variety in this principle, but it is not intended as a cybernetic statement. Instead, the image evoked by the principle is one of Complementary Systems, sets of Open Systems whose outflows and inflows are mutually satisfying. The balance between floral and faunal $CO_2$ and $O_2$ exchanges was mentioned in the discussion of environment above, and is an ideal example of Complementary Systems; the balance depends upon variety and connectivity, and is evidenced by a stable environment.

## Limited Variety

The Principle of Limited Variety is stated as follows:

*Variety in Interacting Systems is limited by the available*
*space and the minimum degree of differentiation*

The principle is axiomatic once "space" and "minimum differentiation" have been established. To explain, consider a guitar string. It can vibrate in a variety of modes limited by the need for nodes at bridge and stop. This maximum set of modes is the available space; the minimum differentiation is set by the need for each mode to comprise waves in integer half wavelengths only.

## Preferred Patterns

As the weave of interactions between systems becomes more complex, it is increasingly likely that feedback loops will be set up, some perhaps existing through many successive systems and exchanges. The occurrence of positive feedback loops is to be expected, and leads to the Principle of Preferred Patterns:

*The probability that interacting systems will adopt locally-*
*stable configurations increases both with the variety of systems*
*and with their connectivity.*

Locally-stable, interacting systems abound. Cities, computer giants, international conglomerates, thunderclouds and tornadoes, molecular microclusters, ecological niches, bureaucracies—all are instances of positive feedback, or mutual causality as Maruyama (1968) described it, leading to stable configurations. The general expectation of positive feedback is that it will produce some form of regenerative runaway. That need not be the case when such positive feedback exists within a web of essentially-negative feedback loops. Instead, multiple points of stability arise.

## Cyclic Progression

The last of the USH principles addresses a phenomenon which we all recognize, that systems do not last for ever. Civilizations may be considered as systems and as H. G. Wells (1922) noted, they come and go:

> • Neolithic Civilization • Sumeria • Egypt, Babylon and Assyria • The Primitive Aryans • The Early Jews • The Greeks • Alexandria • The Romans • Carthage • China • The Barbarians • The Byzantine and Sassanid Empires • The Arab Nations • The Mongols • The Americans • The Industrial Revolution • And so on up to the present.

Such thoughts lead directly to the Principle of Cyclic Progression, expressed as follows:

*Interconnected systems driven by an external energy source will tend*
*to a cyclic progression in which system variety is generated, dominance*
*emerges to suppress the variety, the dominant mode decays or collapses,*
*and survivors emerge to regenerate variety.*

The principle does not imply that the *same* systems emerge. Clearly with civilizations, that is not so. Emerging systems may occupy the same "space" however, whatever that term implies in particular situations. Variety is generated in the space by influx from surroundings, or by mutation of systems (Maruyama, 1968), or both. A simple mechanical analogy might be that of plucking a guitar string off-centre, so as to create a wealth of harmonics. Gradually the overtones subside, leaving the dominant fundamental which decays in its turn. If the finger is moved along the fretboard while the harmonics are present, any may be picked out. If only the fundamental is left, moving along the fretboard will suppress the vibration. Response to change is better where the variety exists.

## U.S.H. PRINCIPLES AS A SET

### The U.S.H. Principles as One

Each of the six principles has been presented independently. It is evident, however, that they address complementary aspects of interacting systems:

- The *Principle of Interacting Systems* addresses the tendency to equilibrium
- The *Principle of Cohesion* addresses the changing form of an interacting system and limits to growth
- The *Principle of Connected Variety* addresses the bases of stability between interacting systems
- The *Principle of Limited Variety* addresses the limits to differentiation in interacting systems
- The *Principle of Preferred Patterns* addresses the emergence of dominance
- The *Principle of Cyclic Progression* examines life cycle.

The principles are best viewed in the context of system lifecycle. Interacting systems exhibit stability as a result of their (limited) connected variety, while the emergence of dominant systems does not *necessarily* reduce either the variety or the connectivity. Where dominance does result in, or is associated with, a reduction in connected variety, then decay and/or collapse will follow because of the reduction in stability and / or the reduction in cohesion associated with excessive growth and ponderousness, together with the concomitant reduction in ability to respond to change. Dominance can reduce variety by reducing the cohesion of lesser interacting systems so that they lose viability, or by effectively isolating such systems from the interactions, or both.

### Complementary Systems—A Systems Engineering Method

Systems engineering is concerned with optimization. The concept of cost-effectiveness, often at the heart of systems engineering projects, is one of optimization. The USH images above present a problem in this respect. If systems exist in containers like Babushka Russian Dolls and if systems are interconnected and intraconnected, how can any one system be optimized in its own right without disturbing the similar optimization of siblings, contained and containing systems? Since the foundation of systems engineering methods and procedures, steeped in the original Operations Research (see Hitch (1955)), is fundamentally aimed towards optimization, this is a serious question.

Is an optimized car one which makes most profit for its manufacturer, goes faster, handles best, uses least fuel, causes least pollution, sells best, absorbs least natural resources, takes least energy in production, provides most work for suppliers in a depressed area, etc? Any attempt to answer will show that optimization can only be local. Does this matter? The question says it all—the motor car is a classic example of local optimization, causing widespread pollution, absorbing resources, providing great pleasure and satisfaction, and connecting individuals and groups within society so as to improve societal stability and cohesion. The car designer does not concern himself with much of this when "optimizing" his design, and the accumulated effects of many "local optimizations" can be either good or bad, according to situation and viewpoint.

```
                    DESIGN GUIDELINES
    •   Establish requirements by reference to Containing
        System(s)
    •   Identify perturbed (Sibling) Systems and interactions
    •   Design system to complement Sibling Systems
    •   Partition system to promote internal variety, avoid
        dominance
    •   Intraconnect that variety to promote stability, mutual
        reward
    •   Enhance cohesives, diminish dispersives
    •   Interconnect that variety to promote external
        stability
    •   Adjust / Establish Complementary Systems to
        neutralize unwanted    perturbations.
    •   Interconnect to promote mutual reward
```

Perhaps a better alternative exists, as seen from the USH perspective. A new interacting system perturbs the fabric of existing interactions in many ways when it is introduced. It is possible, using the USH Principles , to develop a simple, new and effective approach to systems engineering, as shown in the panel above.

## Addressing Issues

The USH Images and Principles also permit a different approach to Issues. The concept is concerned with symptoms in the first instance. Issue-symptoms are evidence of disharmony or imbalance between interacting systems; these systems may thus be identified from their symptoms. Such systems are *implicit*—see panel—because they may not correspond to organizational systems in the Issue Domain.

```
                    Guidelines
•  Identify Issue and Issue Domain

•  Capture Issue Symptoms

•  Identify Disturbed Implicit Systems from Symptoms

•  Cluster to Identify Implicit Issue-Containing Systems

•  Identify Containing Systems Issues and Environment

•  Seek Resolution at Containing System level

If Containing Systems Issues can be resolved, then original Issue
                    is resolved too.
```

Many systems may emerge from this process, generating a high degree of complexity in their mutual interaction. The implicit systems are then clustered, using computer assistance if necessary, to reveal their Containing Systems. The number of Containing Systems and hence their interaction complexity is greatly reduced by clustering.

By examining the aggregated Issues, Containing Systems and Environment it is simpler to understand the Issue and may be simpler to resolve it too. Resolving at Containing System level automatically resolves at the original Issue level because of the aggregating procedure used.

## CONCLUSION

The Unified Systems Hypothesis brings together views and concepts from a wide variety of systems thinkers, old and new, and presents a set of system images, definitions and principles which are intended to provide a common basis for the perception, understanding, analysis, design and creation of all systems. This is a bold aim and it is difficult to prove—or disprove—many of the contentions presented. But then, it is a hypothesis and not a theory. The USH will have value if it provides an evolving basis for all systems practitioners to work together, soft with hard, open with closed, so that we may jointly improve our practices.

## REFERENCES

Angyal, A., (1941) "A Logic of Systems", Foundations for a Science of Personality, Harvard University Press, 1941, pp. 243—61

Ashby, W. Ross (1956) "Introduction to Cybernetics", Chap 11, Wiley 1956, pp. 202—18

Hitch, C., (1955) "An Appreciation of Systems Analysis", The RAND Corporation, pp. 699, 8—18,55.1—25

Maruyama, Magorah (1968) "Mutual Causality in General Systems", Positive Feedback, John H Milsum (Ed), 1968, Pergammon

Popper, K., (1968) "The Logic of Scientific Discovery", Hutchinson

Popper, K., (1972) "Conjectures and Refutations: the Growth of Scientific Knowledge", Routledge and Kegan Paul

Sachs, W.M., (1976) "Toward Formal Foundations of Teleological Systems Science", General Systems, xxi (1976), pp. 145-54

Von Bertalanffy, Ludwig, (1950) "The Theory of Open Systems in Physics and Biology", Science, Vol III, 1950, pp. 23—9

Wells, H.G. (1922) "A Short History of the World", Penguin Books, Harmondsworth, UK

APPLICATIONS OF METHODOLOGY (BOTH HARD AND SOFT)

APPLICATIONS OF METHODOLOGY (BOTH HARD AND SOFT): INTRODUCTION

We have expectations nowadays at any conference on systems thinking. There should be innovative developments on the theoretical front. Recent evidence shows that at last systems thinking has worked out credible intellectual foundations, as underlined in the section 'Problem Structuring and Critical Systems Thinking'. Another expectation, however, is for clear accounts of how theoretical developments lead to ever more "revolutionary" intervention. There is a link between the two expectations. This is particularly the case for systems thinking because it is traditionally practice driven where other disciplines are not. Papers in this section therefore must be read in the light of both expectations.

The immediate focus of this section concerns applications of methodology (both hard and soft). We can break this down into the following: methodology, method and the relationship between them; techniques of methodology and method; and application areas.

Exploration of the relationship between methodology and method is essential. Explaining how and why particular methodology or method lead to characteristic outcome is of equal concern. We should look forward to such analysis in this volume. The type of methodology and method we expect to read about includes hard and soft, but also embraces emancipatory approaches, and new innovative approaches yet to be understood according to these three categories. Starting with hard we could encounter system dynamics, general systems theory, systems engineering, hierarchy theory, systems analysis, and others. With soft we may investigate strategic assumption surfacing and testing, interactive planning, soft systems methodology, and others. Emancipatory as a category will be poorly represented because the necessary developmental work has only recently started in earnest. We await innovative approaches of all sorts with anticipation. Papers in this section dealing with methodology must be scrutinised critically to find out whether utility is dealt with. Discovering the strengths and the weaknesses of methodology and method in theory and practice is crucial to the medium and longer term viability of systems thinking.

For other purposes it is legitimate to look inward at the techniques and rules of individual approaches. Many questions can be raised: 'Has measurement been adequately understood and developed according to the logic and laws of the scales of measurement and in the face of the logic of the approach?'; 'Would automation of the principles complement or contradict those principles?'; 'Which current and new tools for methodology and method are appropriate and why?'; 'Are features of approaches properly or fully understood, eg Weltanschauung in root definitions in soft systems methodology, recursion in viable system modelling, or dialectical discussion in strategic assumption surfacing and testing?' Introspection, although wholly insular, is a necessary way of harnessing the full potential of individual approaches.

Another expectation participants have at modern day systems conferences is that application of methodology and method will be discussed and that the areas of application will be diverse. Heterogeneity is a "reality" for systems thinkers and poses an exacting challenge to would be systems problem solvers. Historic areas of systems thinking such as biology, physiology, and neurology, still play a role. Socioeconomic methods like urban modelling are relevant. But contemporary systems intervention increasingly deals with management information systems, software design, and broader organisational "problems" like quality management. This volume fulfils all expectations by showing wide use and novel applications of systems methodology and method.

The three areas outlined above - exploring the relationship between approaches, looking inside them, and considering application areas - are useful as individual study areas. They can and should be brought together by persistent questioning about which approach should be used when and why? This links methodology to problem context', ie to application area. By asking critical questions we will latch the current section to 'Problem Structuring and Critical Systems Thinking'. Critical systems thinking in one sense is the synergy that results from the bringing together of these study areas. It is innovative in dealing with theoretical and methodological difficulties that dog systems thinking. No longer, with critical systems thinking, do we look for the best approach. We more modestly seek to reveal the most relevant approach by critical reflection on issues of concern. Please note, choice is not made according to dominant thinking. We can argue about relevance using information arising from reflection on perceived problems and on the strengths and weaknesses of each methodology or method. We should expect that papers dealing with application of methodology or method will square up to these requirements.

# SYSTEMS THINKING AND INVESTIGATION OF

# DIGESTIVE SECRETION

Michael Yu. Chernyshov

Irkutsk Computing Centre
USSR Academy of Sciences
Irkutsk, 664033, U S S R

## INTRODUCTION

As a rule, a model of a system takes account of a few most obvious externally expressed parameters considered to be of paramount significance in the aspect of external operation of this system. This approach does not allow to consider such a system internally, as a homeostatic unit, and, consequently, to obtain reliable estimates of its stability, flexibility of its operation, its survivability, to make forecasts of its operation. When one investigates such a large-scale interconnected and multilevel system as human organism, such account of intra- and inter-system functions (considered in time) is of principal importance in order to understand and model it. Secondly, interconnected systems of high degree of complexity are generally investigated in part, and functions of a set of parts can hardly be understood systematically. Thirdly, an investigator must take into account that structures and organs of organisms are functionally specific. But direct decomposition into structural blocks, which seem to be obviously responsible for certain functions, only seems obvious: organs co-operate in systems intended for definite functions. Decomposition into co-operating functional systems is a problem, which necessitates the combined (functional + structural) approach. Finally, on some stage it appears obvious that operational logic must be taken into account. This logic may be violated or changed by external effects (with respect to an organism) and by internal or external controls. To work with such logic of functions in time is a serious problem, so there generally appears the temptation to simplify the approach by (i) reducing the number of substantial parameters, (ii) simplifying the operational logic, etc. This not only makes the resulting model hardly ever valuable at all, but also often leads the researcher to a sort of "unification tendency", when finally no difference can be found in operation between models of, e.g., a living system and a power plant.

For over than 15 years we investigated the digestive glandular functional system of mammals and men, principles of its intra- and intersystem operation and coordinated control.

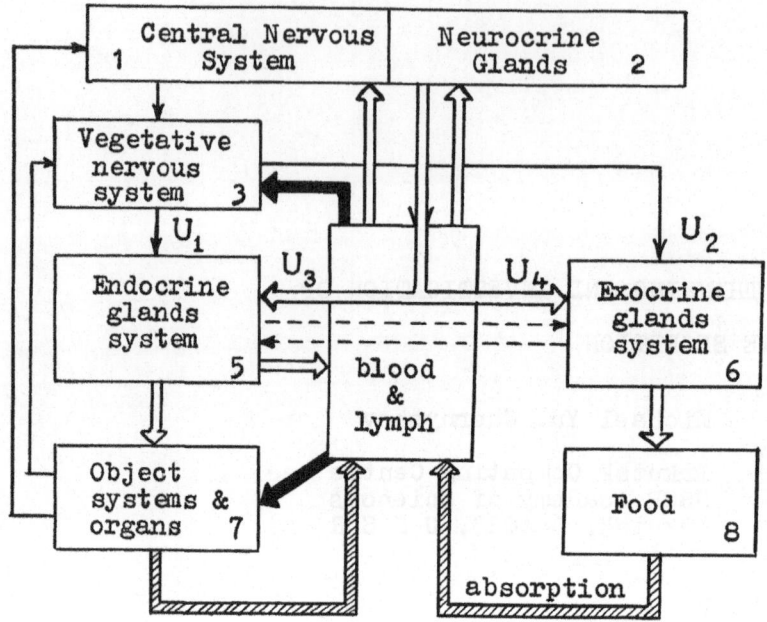

Fig. 1. A model of the organism neuro-humoral
control system of exo- and endosecretion.
(→ - reflex control; ⇒ - secretion of
products; ⇒ - humoral feedbacks; --→ -
functional interconnections; ⟹ - humoral control.)

    The principal model constructed: (a) is hierarchically
organized, the levels being included successively, the components being interconnected n:n;   (b) has controllers specific of a certain system (nervous, humoral (endocrine, exo-
crine)), which are of two types (central, periferal), two-
channel feedbacks being provided; (c) represents a functionally open system with a few types of interconnected "actuators"
(nervous, humoral);   (d) represents a structurally-functional-
ly two-fold system (there are compensatory and (or) adaptation
providing components (functional subsystems), functions (controlling systems and actuators), agents (information transmit-
ters, actuators), which together form a hierarchical two-fold
network of two-fold subsystems)(see Fig.1); (e) is characterized by internal coordination and efficient resolution of specific internal contradictions in functions that govern the
whole process of functional activity.

    All the above properties elucidated for living organisms
provide  principal qualities and advantages (to compare with
technical systems), which ensure high reliability , stability
of operation, survivability of living systems.  The two-fold
structural (and, note, also functional) organization is obvious: pairs of eyes, legs, nostrils, kidneys, pancreatic ducts,
etc.  For example, a pair of eyes provides binocular stereos-
copic vision (space-volume estimation abilities), and when
motion of the body is added, the so called stereokinetic mechanism works and provides the distance estimation capability.

So, there is the need to continue investigation of principles of intrasystem operation for interconnected systems
(such as living systems), wherein control to actuators and
subsystems is applied as a coordinated system of direct and
feedback signals.  We are sure that possible results of such
investigations will be very useful both for understanding
some functions of human organism and for refinement of technical systems .

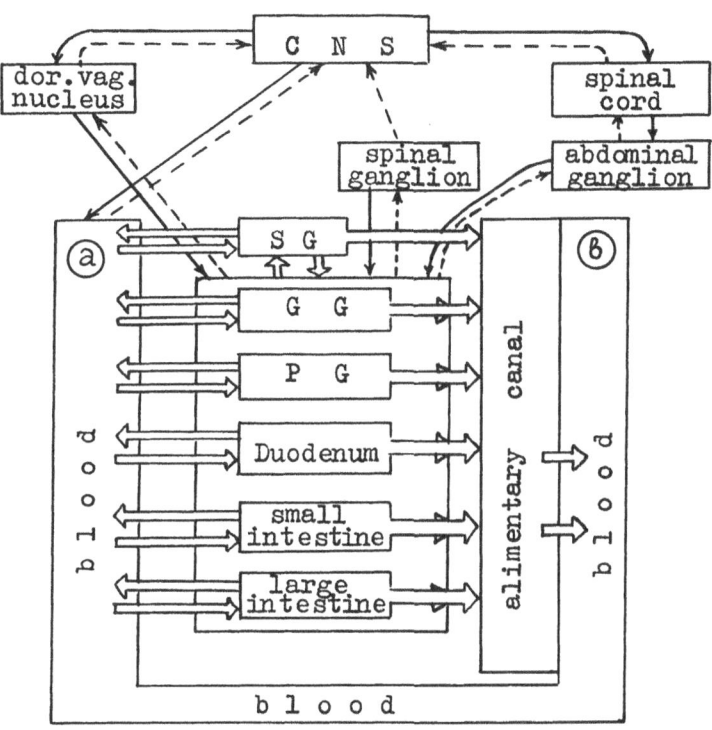

Fig. 2. A simplified model of reflex control
of EESS  functions. (--→- informational interconnections; ⟹- exchange
of controlling agents in EESS; ——→-
reflex control) ( SS - salivary
glands; GG - gastric glands; PG -
pancreas; CNS - central nervous system)

SOME APPROACHES TO INVESTIGATIONS

In the above aspect we considered the human digestive
secretory system, while trying to find out mutually controlled
pairs of "actuators" in it.  Fig.1  shows a model of the neuro-humoral control system (NHCS) including nervous (3), humoral (4) controlling channels, which coordinate and regulate
functions of two effector controlling systems (5, 6).  This

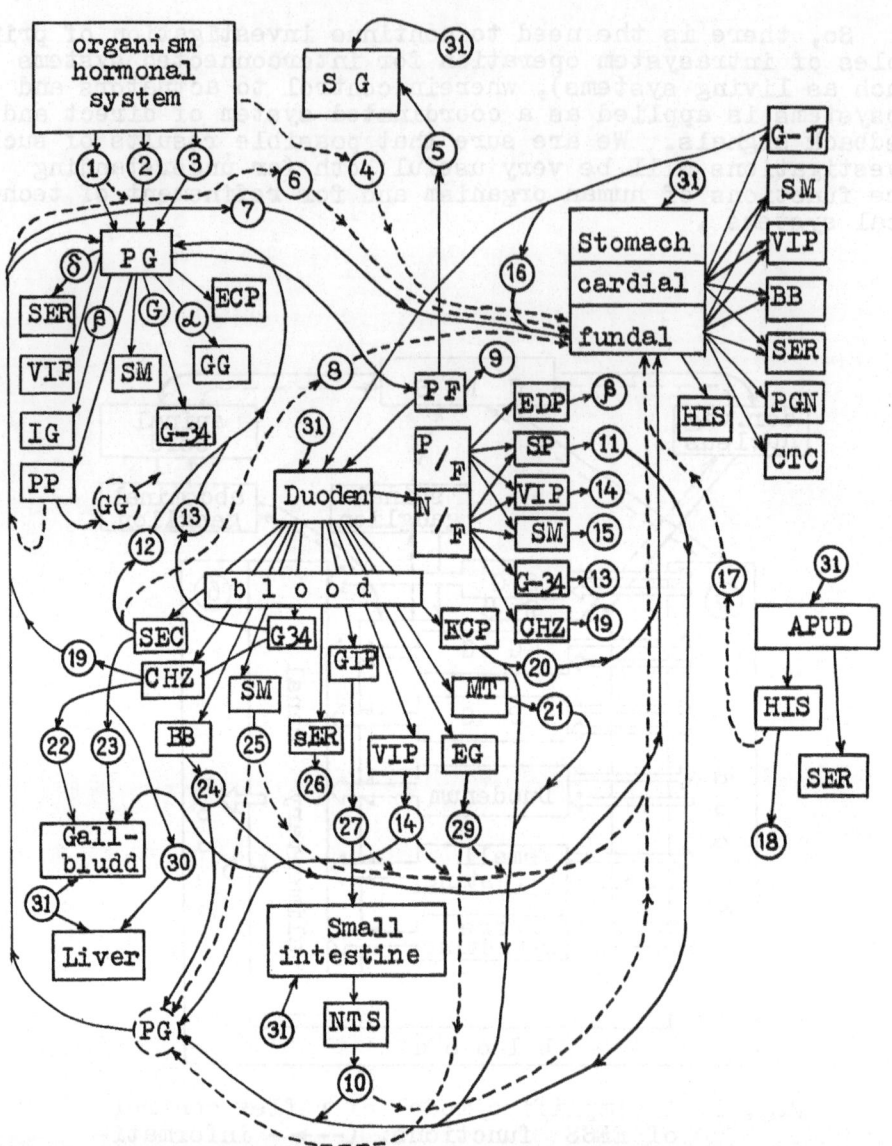

Fig. 3. A simplified model of hormonal control in EESS.
(→ - stimulation; --→ - inhibition)

model represents a network of hierarchically included functional sublevels (submodels), which are sequentially initiated in the process of operation under the supervision of CNS and in accordance with some logic of operation, for example :
(1&2)+{3+5+((3+(1&2))∨(4+(1&2)))}∨{4+5+((3+(1&2))∨(4+(1&2)))}
∨{3+6+((3+(1&2))∨(4+(1&2)))}∨{3+5+(3+6+((1&2)∨3∨4)∨4+6+((1&
2)∨3∨4))∨ etc. (here: "+" indicates the sequence of switching on system components; & and ∨ - logic signs). Now, to solve some practical control problem, it is necessary to expel some "forbidden" (not forming feedbacks) sequences of transitions from the network, and after that, at the subordinate level

(subsystems 5, 6), to expel the circuits forbidden for some secrets. The rest of the circuits will now form a network of "controllers", "actuators", control links and feedbacks, etc., which operates on the principle "condition-event" (i.e., transition - switching on a functional system - transition or resulting operation). Here, the parameter of time of switching plays the principal role in automated processes.

Figs. 2,3 demonstrate two main functional components of the NHCS model. The system of reflex control of secretion (its model is shown in Fig.2) is the main subsystem of quick-response control of secretion in the integrated Endo- and Exo-crine Secretion System (EESS). It is represented by submodels (1 and 3).

Fig.3 demonstrates a model of hormonal control of EESS functions (submodels 5, 4 and 6). The corresponding system is responsible for long-term adaptations of the whole system (NHCS, or the gastro-intestinal system, or even organism on the whole) to external perturbations violating homeostasis. It also participates in some slow-response reactions.

The total network for maintaining some optimal dynamic balance of secretory parameters by NHCS is composed of combinations of submodels 2, 3, 4, 5, 6 and the submodel of 1 (see Fig. 1) responsible for evolutionally provided effects and norms.

Consequently, for systems of the abovementioned type, it is insufficient to use only the traditional apparatus of differential equations. The requirements of adequate description of such systems in dynamics necessitate the account of the operational logic possible for such systems, which is made on, e.g., the basis of the semantic nets formalism. Employment of AI software to implement the apparatus of operational logic has substantially enriched the technique of investigations.

In our case, the description of variations of system parameters and states in time with respect to a set of chosen parameters is given in terms of states diagram. Operation (functional state) of the considered system in each node of the diagram is described by a system of ordinary first-order differential equations of the form

$$\dot{\vec{z}}_i = F_i(\vec{g}_1,\ldots,\vec{g}_k) + \widetilde{F}_i(\vec{g}_1,\ldots,\vec{g}_k) + G_i(\vec{g}_1,\ldots,\vec{g}_k,\vec{p}), \quad i=\overline{1,n}$$

where $\vec{g}_j$ is a generalized parameter : $g \to z,x,..,\lambda,t,u(t)$; Here, $z_i \in H_i \subseteq R^{n_i}$, $H_i$ - domains in $R^{n_i}$; $n=n_1+n_2+\ldots+$ $+n_m$; $\vec{z} = (z_1^T,\ldots,z_M^T)^T$ ; $x_i \in Q_i \subseteq K^{n_i}$ ; etc. Real functions

$F_i : T \times H_i \to R^{n_i}$ define an isolated subsystem, which in the vector form writes :

$$z = F(z,x,\ldots,\lambda,t,u(t)) ,$$

where each parameter, e.g., $x$ , is considered as a point in the n-space $R^n$ with the norm $\|x\| = |x^1| + \ldots + |x^n|$ and

$$x = \begin{pmatrix} x^1 \\ \cdots \\ x^n \end{pmatrix} \quad ; \quad F = \begin{pmatrix} F^1(z^1, \ldots, z^n, \ldots) \\ \cdots \cdots \cdots \cdots \cdots \\ F^n(z^1, \ldots, z^n, \ldots) \end{pmatrix}$$

Real functions $F_i : T \times H \to R^{ni}$ characterize the effect of other external subsystems on an isolated one (on account of cross connections in the system model). The parameter $p \in \mathcal{P}$ in the vector function $G$ characterizes perturbations in the model initiated by external or internal effects.

The parameters estimated with respect to time are: $m(t)_j$-mass of a sort of juice; $\rho_1$ - concentration of $NaHCO_3$; $\rho_2$ - concentration of proteins; $\varphi$ - electric controlling potential; etc. Time derivatives of these parameters are taken into account.

The description of sufficiently complete model of operation necessitated the account of digestion, absorption and deposition of nutrients. The problem of decomposition-aggregation was one of principal problems.

Each elementary functional subsystem was subdivided (on the second stage) into (i) a subsystem of quick response and (ii) a subsystem of long-term adaptation. The corresponding mathematical apparatus was provided.

CONCLUSION

To work with our states diagrams, we elaborated a software tool (the system PHYSICIAN , which employs the principles of System Dynamics) that allows: (1) to formulate a problem; (2) to construct a computer model of the system (taking account of more than 25 parameters); (3) to solve the system of up to 800 ordinary differential equations for each node of the states diagram; (4) by variating system parameters, functions, processes iteratively, to select modes of operation close to those obtained in natural long-term experiments on men and dogs.

We also investigated rhythms of adaptation, secretory phases, with switching on/off control mechanisms of the model, aspects of control of secretory functions and parameters. We have an approach to the method of cortico-visceral correction of exosecretion. Noteworthy, externally introduced disturbances of homeostasis in all our experiments activated the cortex reflex control mechanism and were entailed by intensive adaptive secretion in the first two hours, when (Chernyshov, 1990)

$$\sum_{2h} m_s \cong 0.65 \sum_{6h} m_s$$

REFERENCES

Chernyshov, M., 1990, System of homeostatic control mechanisms of main digestive glands, in: "Homeostatics of Living, Technical, Social and Ecological Systems," Y. Gorsky, ed., Nauka Publ., Novosibirsk, Ch. 4.5.

# VARIABLE RECEPTIVE FIELD SYSTEM TRANSFORMATION AND ITS APPLICATIONS TO VISUAL FIELD

J.C. QUEVEDO-LOSADA, J.A. MUÑOZ-BLANCO, AND
O. BOLIVAR-TOLEDO

*Departamento de Informática y Sistemas*
*Universidad de Las Palmas de Gran Canaria*
*Las Palmas. Canary Islands. Spain*

## GENERAL CONCEPTS

The sensory systems are defined to specify their input and output spaces and the relational structure which links them. As it has been pointed out by Moreno-Diaz [MORENO-84], and from the methodology point of view, it is essential in the beginning, in principle, a set of selections with regard to both the nature of the input and output spaces, and the type of language used to describe the relational structure between both.

We focus on the study of transformations that acting on the input space or data field and by means of operations which can vary from the analiticals to the algorithms, generates a new output space.

The Completness concepts of a transformation with regard to receptive fields and to the function were introduced by Candela [CANDELA 87] in 1987. According to him and from the analytical point of view, the complete description (that according to heuristic and the experiment could be "truncated" for a system of practical artificial vision), needs "a priori" the preservation of the number of freedom degrees or the number of independent properties, N, of the visual environment. This constancy allowed the establishment of a conservation principle which is determined by the computed functionals and by the receptive fields, that in addition reflects a situation of duality, in the sense that the completness requires to increase the number of them if the other decreases.

The concept of data field is obtained as the inmediate generalization of the representation of images in one, two or more dimensions. In this way a unidimentional data field of length N and resolution R is an ordered set of N places i, such as a number $I_i$ (real or complex) with resolution R can be assigned to each place.

The places i, must be understood as "file pages" whereas R must be understood as the number maximun of binary holes necessary for saving the possible $I_i$ data.

Given a data field D(N) of resolution R, we can define a set of L independent partitions of the addresses i {i = 1, ..., N}. Let's consider a subclass of partitions of the addresses i, where each address is taken, at least once.

A theorem recently proposed by Bolivar [BOLIVAR-89], said that given a data field of N addresses and a partition of L colums such as M = N/L is integer, then the computation of M functional coefficients, which are linearly independent and different from zero in each partition, provides a complete description of the data field.

In the design of a system of visual recognition, this complete (or truncated) algebraic-analytical descriptions are not better or worse than others which are conventional, since they do not affect the perceptive structure but the sensorial structure.


## APPLICATIONS TO VISUAL RECOGNITION

In this context, we use transforms that combine partitions and functionals and we design a classification system in base of these complete algebraic-analytical descriptions which can be easily truncated for purposes of visual recognitions. The efficacy of the classification system will depend, in principle, on factors, such as the complexity of data field, the larger or smaller extension of receptive fields, the number of descriptors in each of them and the resolution. Some previous works show that when the resolution is relatively small and the complexity of the data field is also small, a high number of receptive fields and descriptors causes an efficiency decrease in the classification, due to the inherent noise and to the increase of redundancies which are not controlled in the original data field.

The applications carried out are based on a reduced system of inference [MUÑOZ-87] which has the nature of an Expert System with the following components:

a) A referential data base, that contains a set of descriptive phrases of reference and a label is assigned to each of them.

b) A device or inference machine that starting from some rules, accepts an unknown descriptive phrase and after the interaction with referential data base, produces a decision or diagnostic.


Using the µVAX II, monitor and acquisition board FG-100 of Imaging Tec., we have acquired a set of 90 images. These consist of 10 different aquisitions from each of the 9 different shapes.

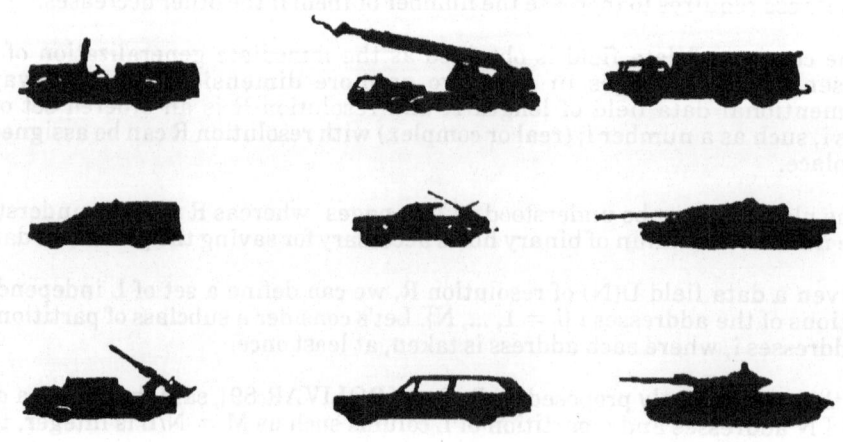

We apply a segmentation process to these images by making a space of monodimentional measure where there are only two clusters. Finally, we achieved a process to normalize and make them invariant against traslations, rotations and homotecias.

For the generation of the descriptive phrases that must label each class containing the forms, both patterns and unknown, for posterior recognition, we started from value N (freedom degrees or resolution). To obtain the values of L (partitions number) and d (freedom degrees of the receptive fields), the unique restriction must be imposed by the following expression:

$$L = [(N - d) / \delta] + 1$$

being $\delta$ the interpartitions shift.

This means that if we fix freedom degrees and the shift, we obtain the necessary number of partitions.

As an illustrative example, we have selected the values N = 128, d = 32 and $\delta$ = 16, both power of 2, which greatly simplifies the computational cost to implement the transformation. The seven partitions generated are enough for covering all the data field and by construction are linearly independent. According to the previous theorem, completness requires a number of functionals for the partitions of M = N/L in average. In this case, as this number is not integer the problem has been solved by choosing 18 functionals for the partitions 1,2,3,5,6,7 and 20 functionals for the partition 4. We present below a diagram of this partitions.

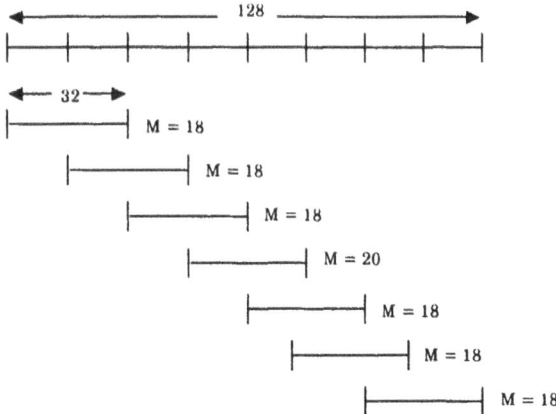

We have used moments as functionals and according to Alt [ALT-62], the best results for purposes of classification are given by low order moments. For this reason, we have chosen the moments $M_{0,3}$, $M_{0,4}$, $M_{1,2}$, $M_{1,3}$ in addition to the moments $M_{0,6}$, $M_{0,8}$, $M_{0,10}$, $M_{1,4}$, $M_{1,6}$, $M_{1,8}$, $M_{1,10}$, arbitrarily chosen.

The algebraic-analytical transformation represented by the M matrix is obtained by the application of each functional vector to each partition.

$$
\begin{bmatrix}
F11 & \cdots & F1d \\
F21 & \cdots & F2d \\
\vdots & \ddots & \vdots \\
Fm1 & \cdots & Fmd
\end{bmatrix}
\times
\begin{bmatrix}
P11 & \cdots & P1l \\
P21 & \cdots & P2d \\
\vdots & \ddots & \vdots \\
Pn1 & \cdots & PNL
\end{bmatrix}
= N
$$

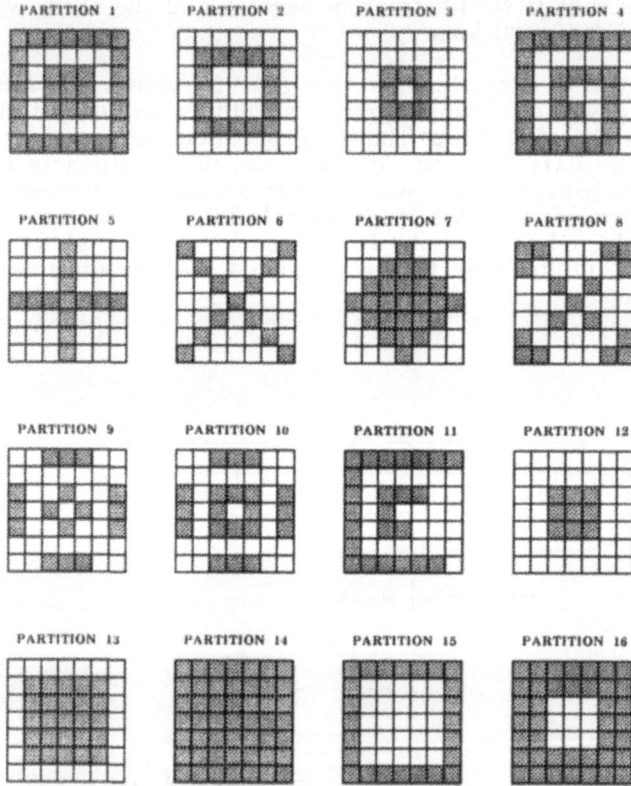

Next we present each one of the receptive fields considered for the generation of the descriptive phrases.

At last, we proceed to implement the classification process. To achieve this we have used a modified version of the Euclidean distance.

Given an unknown descriptive phrase, D'(i) and a set of descriptive phrases corresponding to each referential data field class, D(i,j), the modified Euclidean distance is defined as:

$$[\text{Distance (j)}]^2 = \Sigma\, [(D'(i) - D(i,j))/10^{E(j)}]^2$$

where

$$E(j) = \min(\exp.(D'(i)), \exp.(D(i,j)))\ (i = 1, ..., n)$$

being N the dimension of each descriptive phrase.

Once computed every distance between unknown descriptive phrase and each one of the descriptive phrases which conform the referential data base, the inference system associates the unknown pattern to that whose distance between them be lowest.

EXPERIMENTAL RESULTS AND CONCLUSIONS

In our original referential data base we consider 63 images as pattern and 27 images as unknown. The next table shows the failure number produced in the classification for each partition and moment.

# MOMENTS

|  | | $M_{0,3}$ | $M_{0,4}$ | $M_{0,6}$ | $M_{0,8}$ | $M_{0,10}$ | $M_{1,2}$ | $M_{1,3}$ | $M_{1,4}$ | $M_{1,6}$ | $M_{1,8}$ | $M_{1,10}$ |
|---|---|---|---|---|---|---|---|---|---|---|---|---|
| P | 1 | 1 | 3 | 1 | 2 | 5 | 0 | 1 | 2 | 1 | 3 | 1 |
|   | 2 | 0 | 2 | 1 | 1 | 1 | 1 | 1 | 1 | 1 | 1 | 1 |
| A | 3 | 1 | 0 | 0 | 0 | 0 | 0 | 1 | 0 | 0 | 0 | 0 |
|   | 4 | 2 | 1 | 2 | 4 | 2 | 0 | 3 | 1 | 3 | 2 | 3 |
| R | 5 | 0 | 0 | 0 | 0 | 0 | 0 | 0 | 0 | 0 | 0 | 0 |
| T | 6 | 2 | 2 | 2 | 1 | 1 | 2 | 2 | 2 | 2 | 1 | 1 |
|   | 7 | 1 | 1 | 1 | 2 | 2 | 1 | 1 | 1 | 1 | 2 | 2 |
| I | 8 | 1 | 1 | 0 | 3 | 0 | 0 | 0 | 0 | 0 | 0 | 0 |
| T | 9 | 1 | 4 | 1 | 2 | 3 | 1 | 2 | 2 | 1 | 3 | 1 |
|   | 10 | 1 | 4 | 1 | 2 | 3 | 0 | 2 | 2 | 1 | 3 | 1 |
| I | 11 | 0 | 0 | 1 | 1 | 0 | 0 | 0 | 0 | 0 | 0 | 0 |
| O | 12 | 0 | 0 | 0 | 0 | 0 | 0 | 1 | 0 | 0 | 0 | 0 |
|   | 13 | 0 | 2 | 1 | 1 | 1 | 1 | 1 | 1 | 1 | 1 | 1 |
| N | 14 | 0 | 1 | 0 | 0 | 0 | 1 | 0 | 0 | 0 | 0 | 0 |
| S | 15 | 1 | 3 | 2 | 4 | 5 | 0 | 3 | 6 | 1 | 5 | 1 |
|   | 16 | 0 | 1 | 0 | 0 | 0 | 1 | 0 | 0 | 0 | 0 | 0 |

We have selected a receptive field configuration set and a functionals set and we have realized a comparative study of the classification system efficiency as regards the different partitions and functionals previosly chosen.

From this study we remark on the following conclusions:

1) There is a set of configurations that given better results than others for purposes of classifying. It is emphasized the relevance of receptive fields against the classical functional. This can be checked by observing that the variability of partitions versus the functionals is smaller than the functionals versus the partitions.

2) Furthermore, an increase in the number of descriptors causes a decrease in the efficiency of the classification system. This is possibly due to the inherent noise and to the increase of redundancies which are not controlled in the original data field. We conclude that the important fact is not the number of descriptors but where and how they are selected.

## ACKNOWLEDGEMENTS

The authors wish to thank Prof. Roberto Moreno Díaz for his valuable suggestions.

## REFERENCES

[ALT-62]      ALT, F.L.: Digital pattern recognition by moments. J. - ACM9, 240-258 (1.962).

[BOLIVAR-89]  Bolivar Toledo, O.: "Hacia una Teoría de las Transformaciones en Campos de Datos Receptivos y Campos de Datos. Implicaciones en Teoría y Proceso de imágenes". Tesis Doctoral. Universidad de Las Palmas de Gran Canaria. 1.989.

[CANDELA-87]  Candela Solá, S.: "Transformaciones de Campo Receptivo Variable en Proceso de Imágenes y Visión Artificial". Tesis Doctoral. Universidad de Las Palmas de Gran Canaria. 1.987.

[MORENO-84]   Moreno-Díaz, R.; Mira Mira, J.: Un marco teórico para interpretar la función Neuronal a altos niveles. "Biocibernética". Ed. Siglo XXI. Madrid 1.984.

[MUÑOZ-87]    Muñoz Blanco, J.A.: "Jerarquización de Estructuras de nivel bajo y medio para Reconocimiento Visual. Aplicaciones a texturas de Formas". Tesis Doctoral. Universidad de Las Palmas de Gran Canaria. 1.987.

# An Experimental Approach to The Description of Contour Segments using the Fourier-Bessel Transform

J. Cabrera , A. Falcón , F.M. Hernández , J. Méndez

Departamento de Informática y Sistemas
Universidad de Las Palmas de Gran Canaria
Apartado de Correos 550, Las Palmas
SPAIN

## Introduction : The Fourier-Bessel Transform

The characterization techniques of forms can be usually classified grosso modo between global and local techniques. The complexity of the images to be analized imposes strong limitations to the freedom of choice about the model of description to be used. So the global techniques are used in enviroments where the objects to be classified are simple in nature and do not overlap. On the other side, the employement of local techniques is required as the starting point for the description of more complex scenes, linking this local characterization to some process of structural inference. When some measures, obtained from the raw data, are used in the description we find a set of techniques which can be grouped under the term of transformed representations. They all have in common the ability to reduce the amount of data to be analized without degrading substantially the semantics of the image.

The Fourier-Bessel transform is one among these representations obtained as a functional expansion, that is capable of generating a metric characterization space. In a broad sense, the most common expansions are those defined over orthonormal kernels so that a function $f(\omega)$, representing a form, can be expressed as :

$$f(\omega) = \sum_{j=1}^{N} F_j \Psi_j(\omega)$$

where $\Psi_1, \Psi_2, ... \Psi_N$ form a complete set of orthonormal kernels, and the $F_j$'s are the descriptors or the coordinates of the form in the representation space. These descriptors are determined by:

$$F_j = \int_{\Omega} f(\omega) \Psi_j d\omega$$

This sort of functional expansions have the characteristic property that the cuadratic error or distance between two original forms is related to the distance in the representation space.

$$d^2(f,g) = \int_{\Omega} [f(\omega) - g(\omega)]^2 d\omega = \sum_{j=1}^{N} |F_j - G_j|^2$$

When characterizing objects, different types of orthogonal expansions have been employed using both the area or the contour as the domain in which the form of the object is defined. Based on some expansions, it is possible to obtain descriptions which are invariant under translations, rotations and scale changes. To achieve the invariance under rotations it is convenient to treat the expansion of bidimensional forms factorizing the kernels in angular and radial parts:

$$F_{nm} = \int_{\Omega} \Gamma_{nm}(r) \Theta_n(\varphi) f(r,\varphi) dA$$

*Systems Thinking in Europe*, Edited by M.C. Jackson *et al.*
Plenum Press, New York, 1991

The radial part of the factorized kernel admits different representations. In (Falcón and Méndez, 1984), the use of radial kernels with a smooth evolution (undulatory) is recomended as it increases the precision of angular detection. Because the amount of information contained in both the function $f(r, \varphi)$, which describes the form, and its Fourier spectrum is the same, different global methods have been proposed based on functional expansions of the spectrum as it is the case of the Fourier-Bessel descriptors developed in (Méndez,1983, Méndez and Falcón, 1984). The characterization of a form from its spectrum has the interesting property of limiting the amount of information to be considered by selectively filtering the spectrum, without a significant lost in the identity of the form.

Let $F(\omega, \phi)$ be the Fourier transform of the form $f(r, \varphi)$, expressed in polar coordinates. Decomposing the spectrum in angular armonics, we obtain:

$$F(\omega,\phi) = \int_{\Omega} e^{-i\vec{\omega}\cdot\vec{r}} f(r,\varphi)\, dA = \sum_{n=-\infty}^{\infty} C_n(\omega) e^{in\phi}$$

where $C_n(\omega)$ is the function :

$$C_n(\omega) = (-i)^n \int J_n(\omega r) e^{-in\varphi} f(r,\varphi)\, dA$$

if $J_n$ is the Bessel function of the first kind of integer order. If now, in order to simplify the representation, the spectrum is limited to a circle of radius $\omega_0$, the $C_n(\omega)$ kernels can be developed as a Fourier-Bessel orthogonal serie :

$$C_n(\omega) = \sum_{m=1}^{\infty} \frac{2(-i)^n}{J_{n+1}(j_{nm})} F_{nm} J_n(j_{n,m}\frac{\omega}{\omega_0})$$

where $j_{nm}$ is the $m^{th}$ zero of $J_n$. The complete development of the limited spectrum results in :

$$F(\omega,\phi) = \sum_{n=-\infty}^{\infty} \sum_{m=1}^{\infty} \frac{2(-i)^n}{J_{n+1}(j_{nm})} F_{nm} J_n(j_{n,m}\frac{\omega}{\omega_0}) e^{in\phi}$$

The $F_{nm}$ coefficients are calculated by:

$$F_{nm} = j_{nm} \int \frac{J_n(\omega_0 r)}{j_{nm}^2 - (\omega_0 r)^2} e^{-in\varphi} f(r,\varphi)\, dA$$

The $F_{nm}$ descriptors are invariant under translations if they are obtained from the centroid of the form, and the modulus of these complex descriptors are invariant under rotations of the objects. Finally, if for every form, the radius, $R$, which circumscribes the form is determined from the centroid, it is possible to achieve the invariance under scale changes normalizing all the descriptors to a common radius $R_0$ by :

$$F_{nm} = \left(\frac{R_0}{R}\right)^2 j_{nm} \int \frac{J_n(\omega_0 \frac{R_0}{R} r)}{j_{nm}^2 - (\omega_0 \frac{R_0}{R} r)^2} e^{-in\varphi} f(r,\varphi)\, dA$$

As it has been stated at the introduction, our main objective has been to study the descriptional capability of this global characterization technique when applied locally on contour segments. The interest of this study comes from the good properties that the Fourier-Bessel transform has shown as a technique for global description (Méndez and Falcón 1984, Hernández 1987). With this aim, the function $f(r, \varphi)$ representing a contour primitive, can be expressed aproximately as a Dirac's delta, $f(r, \varphi) = \delta(r-g(\varphi))$, if $g(\varphi)$ is the distance from the centroid to every contour point as a function of the angle $\varphi$. If this definition of contour is accepted, the normalized Fourier-Bessel descriptors can be calculated by :

$$F_{nm} = \left(\frac{R_0}{R}\right) j_{nm} \int \frac{J_n(\omega_0 \frac{R_0}{R} g(\varphi))}{j_{nm}^2 - (\omega_0 \frac{R_0}{R} g(\varphi))^2} e^{-in\varphi} g(\varphi)\, d\varphi$$

## The Rule-Based Segmentation Machine

As in almost every structural method, special atention must be placed on the segmentation process as the goodness of the description will greatly depend on the stability of the segmentation. We have developed a contour rule-based segmentation machine starting from the calculus of the discrete curvature function of the contour. The segments are initially produced using the zero-crossing points of this function. The discrete curvature measure of the $i^{th}$ contour point is defined by:

$$K_i = \frac{(x_i - x_{i-M})(y_{i+M} - y_i) - (x_{i+M} - x_i)(y_i - y_{i-M})}{\left[(x_{i+M} - x_{i-M})^2 + (y_{i+M} - y_{i-M})^2\right]^{1/2}}$$

Fisically, $K_i$ represents the heigth of the triangle defined by the $i$-$M$, $i$ and $i+M$ contour points. In order to avoid the noise produced by the evaluation of the curvature from adjacent points, this is evaluated every $M$ points, beeing $M$ a constant. However, in this way small but significant details (i.e. small contour segments) with a great descriptional power may be lost. Our experience in this sense is that it is advisable to use a moderate value for $M$ and convolve the resulting curvature function with a gaussian of $\sigma$ width. This approach respects the small segments while smoothing the curvature function. The optimum values of $M$ and $\sigma$ depend on the typical contour length and morphology. The relevance of this last point can be appreciate in Fig.1 where the three forms used are shown with the resulting segmentation. In some papers, the optimum value of $M$ is determined as an increasing function of contour length; however, while simple, this is not adecuate in general. To overcome this limitation, our rule-based machine outputs a final segmentation in two steps. Firstly, a non-optimum segmentation is produced based on $M$ and $\sigma$ values determined from the contour length. From this segmentation, characteristic measures of the contour like the mean segment curvature and length are extracted. Secondly, from this contour depending parameters new values for $M$ and $\sigma$ are obtained and the machine proceeds with a second segmentation acting recursively over a previous segmentation until this is stable. In both stages, the segmentation is performed by the application of a serie of processes (refered as filters) based on heuristic rules. The exact implementation of these filters are explained with greater detail in (Cabrera et al.,1991).

## The Class Definition Problem

An important premise of structural methods for scene analysis is that the description must be based in a not too high number of primitive classes. In our approach, the classes are defined by grouping the contour segments resulting from the segmentation stage accordingly to a similarity criterion. The continuous nature of this problem, normally there will not be well defined and distinct typologies of segments, and the unavoidable noise introduced during the segmentation pose hard requirements to the clustering algorithm to be employed in this labelling process. In this application, we have tested six well known hierarchical clustering algorithms studied in (Jain et al.,1986), and the Ward's algorithm with a minimum cuadratic error criterion as it is presented in (Anderberg, 1973). Within this application the Ward's algorithm proved the best, followed by the complete linkage algorithm. So the definition of primitive classes was accomplished by means of Ward's algorithm. The number of final classes was determined studying the evolution of the cuadratic error function as the algorithm proceeded clustering the segments. Further details about the whole labelling process can be found in (Cabrera et al.,1991).

## The Classifier

The implemented classifier has been a hierarchical tree classifier. This type of classifier has some advantages over the so called one-shot classifiers when applied in a problem as the one we are studying. These advantages can be resumed saying that, although a tree classifier suffers from some problems (i.e. overlapping and error acumulation), it reflects the structure of the problem what is valuable in order to know the quality of the intended description. What constitutes the main distinction among different tree classifiers is the criterion used to select the descriptors that define the decision function associated with every non-terminal node of the tree. In order to explain the algorithm we have used, let A be the indicative of a non-terminal node containing N classes which must be separated in at least two groups, every one associated with a child node of A, using certain combination of descriptors. The branch selection is performed according to a minimun-distance criterion. So, if $c_s$ is the number of patterns contained in node c and $\Sigma_l$ is the dispersion matrix of node l, then the distance $D(c,l)$ from pattern c to the centroid of node l is calculated from:

$$D(c,l) = (\vec{x}_c - \vec{x}_l)^t \Sigma_l^{-1} (\vec{x}_c - \vec{x}_l)$$

The first part of the algorithm is as follows:

- Create N temporary child nodes of A and assign to each node one of the classes contained in A;
- Beginning of the algorithm;
- For every node, s, (s:1...N) {
  For every pattern, c, included in node s, (c:1...$c_s$) {
    For every node l, (l:1...N) {

$$q = \min_l \{D(c,l)\};$$

    }
    $e_{qs}++$;
  }
}

The NxN matrix, $e_{ij}$, is the confusion matrix where the element $e_{ij}$ indicates the number of patterns that belonging to node j have been classified as members of node i. Clearly, only the main diagonal elements are the correct classifications. The addition of all the elements of the $i^{th}$ column minus the diagonal element, $w_j = \Sigma e_{ij} - e_{jj}$, is the number of patterns from node j incorrectly classified in other nodes; and $g_i = \Sigma e_{ij} - e_{ii}$ is the total of patterns that have been incorrectly classified as members of node i. Then the algorithm uses this information to fusion the two child nodes which have been worse discriminated as indicated by the confusion matrix. This constitutes the second part of this algorithm:

- If (($\Sigma w_i = 0$) OR (N=2)) return N; (EXIT).
- else {

$$p = \max_i \{\frac{\omega_i}{c_i}\};$$

$$q = \max_i \{e_{ip}\};$$

  fusion p and q;
  N = N - 1;
}
-Return to the beginning of the algorithm;

When the process is finished N represents the resulting number of child nodes of A for the current combination of descriptors being tested. The discriminant power or success of this combination is measured by H, which is based on the Shannon entropy:

$$H = \frac{\sum_{i=0}^{N} n_i H_i}{\sum_{i=0}^{N} n_i}$$

$$n_i = \omega_i + g_i - e_{ii}$$

$$H_i = -y_i \log_2(y_i) - (1-y_i)\log_2(1-y_i)$$

$$y_i = 0.5 + 0.5 \frac{e_{ii}}{n_i}$$

where $n_i$ is the total of patterns related to the $i^{th}$ child node of A, and $H_i$ is the entropy of this child node. This process is repeated for every possible combination of descriptors not yet used in the branch of the tree that contains node A. The number of descriptors used to design the decision function associated to every non-terminal node determines the dimension of the subspace onto which the set of the node classes is proyected. The dimension of this subspace is a parameter that can change between nodes, although in the experiments we have kept it fixed and equal to 2. The minimum entropy criterion is employed to select the most discriminant combination of descriptors at every non-terminal node. However, it is possible to find several combinations with zero entropy. In these cases, it is selected the combination that proyects the child nodes as far away as possible from each other, weighting the distances between two nodes by the number of patterns in each node. If $c_i$ and $c_j$ are the number of patterns in child nodes i and j respectively, this idea can be formulated by:

$$D^2 = \frac{\sum_{i=1}^{N-1} \sum_{j=i+1}^{N} (c_i + c_j) D_{ij}^2}{\binom{N}{2}}$$

$$D_{ij}^2 = (\vec{x}_i - \vec{x}_j)^t \left(\frac{\Sigma_i + \Sigma_j}{2}\right)^{-1} (\vec{x}_i - \vec{x}_j)$$

As it was noted at the beginning of this section, the overlap of the classes is one of the most serious drawbacks of a tree classifier. In overcoming this problem we have selected the approach presented in (Wang and Suen,1987). In our implementation, every terminal node with non zero entropy is intended to be divided into two new terminal nodes: one containing the correctly classified patterns and the other all the misclassified ones. The original terminal node is expanded treating it as a non-terminal node containing two classes. The expansion only takes place if it reduces the entropy of this terminal node.

It is known that a tree classifier does not utilize the total amount of information avalaible in the training data. In (Shlien,1990) it is proposed the generation of several trees from a common training set using different entropy measures as a form of generalizing the structure of a decision tree and achieving higher accuracies. Then the distinct evidences produced by each tree are combined using the Barnett's formulation of the Dempster-Shafer model. However, in our experiments the application of the criteria presented in (Shlien,1990) resulted always in the same tree; this is the reason why , while keeping the methodology, in our approach new trees are generated using the best combination of descriptors never used before to resolve the same set of classes at a non-terminal node.

## Experiments and Results

Two experiments were developed to test the description method proposed. In both cases, the original description space was generated using 28 descriptors selected by the heuristic rule presented in (Méndez, 1983). Following this criterion, the selected descriptors were those with $j_{nm} \leq \omega_0 R_0$. The values of these parameters ($\omega_0 = 0.5, R_0 = 30.0$) employed within this study were chosen to produce a moderate number of descriptors. The segments employed with the first experiment were sinthetic smooth corners divided in ten classes according to the aperture of the corner ($10^\circ, 20^\circ, 40^\circ, 60^\circ, 80^\circ, 90^\circ, 120^\circ, 140^\circ, 165^\circ, 180^\circ$). Fifty-four patterns were included for each class: 18 different orientations ($10^\circ, 30^\circ, ..., 350^\circ$) at three scales (magnification factors: 0.6,1.0,1.4). The mean number of pixels per segment was about 61. Three trees were generated using the described set as a learning set all of which gave a 0% resubstitution error rate. Then three test sets were classified with these trees. The first test set contained the same patterns of the learning set but at different orientations. In the second set there were 45 patterns per class: 3 orientations in 15 sizes (0.3,0.4,...,1.7). Note that the learning set included only three of these sizes. Finally, the third set contained 30 patterns per class. The difference was now that the length of the segments was reduced by putting away pixels from only one segment ending (3,8,13,18,23) or from both (6,16,26,36,46). The numbers in parenthesis indicate the resulting length reductions.

The first and second sets were classified using only one tree at a time producing a 0% error rate with the first set, and only one misclassification at the smallest size with the second set. The results with the third set, obtained using only the first tree, are summarized in Table.I. The maximum recognition score by scales was 10. Table.II illustrates the results obtained with this test set when different trees are employed and their evidences are combined using the Dempster-Shafer model. In this table, the first two columns are the numbers of rejected patterns when the largest belief function is not greater than 0.6. The W.R.(wrong rejections) column contains the number of patterns that would be correctly classified according to a maximum belief criterion. The R.R.(right rejections) column contains all the potential errors if the last mentioned criterion would have been employed. In the last column appear the number of misclassified patterns.

Table. I

| % | 0.6 | 1.0 | 1.4 | Tot | % | 0.6 | 1.0 | 1.4 | Tot |
|---|-----|-----|-----|-----|----|-----|-----|-----|-----|
| 3 | 10 | 9 | 9 | 28 | 6 | 10 | 10 | 10 | 30 |
| 8 | 8 | 9 | 9 | 26 | 16 | 10 | 10 | 10 | 30 |
| 13 | 7 | 7 | 7 | 21 | 26 | 9 | 9 | 9 | 27 |
| 18 | 4 | 5 | 5 | 14 | 36 | 9 | 9 | 9 | 27 |
| 23 | 3 | 3 | 4 | 10 | 46 | 9 | 8 | 8 | 25 |

|       | Table. II |      |        |
|-------|-----------|------|--------|
| Trees | W.R.      | R.R. | Errors |
| 1     | 0         | 0    | 62     |
| 2     | 22        | 35   | 26     |
| 3     | 4         | 33   | 35     |

|       | Table. III |      |        |
|-------|------------|------|--------|
| Trees | W.R.       | R.R. | Errors |
| 1     | 11         | 2    | 33     |
| 2     | 15         | 13   | 26     |
| 3     | 2          | 2    | 37     |

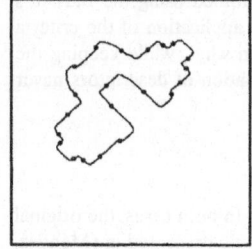 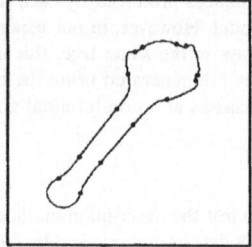 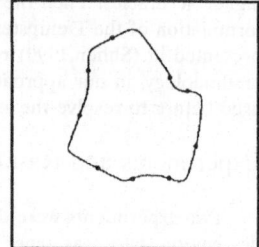

Fig 1. The forms used in the second experiment displayed at the same scale. The small circles indicate the segmentation points.

The only set employed in the second experiment contained 1056 segments produced by the contour segmentation of 24 prototypes (8 orientations in three sizes:0.8,1.0,1.2) of each of the forms showed in Figure.1. These segments had been previously grouped into 8 classes using Ward's algorithm. Employing this set as the learning set, up to three trees were created, and the resubstitution results are presented in Table.III as function of the number of trees used.

## Conclusions

The method described in this paper has proved in the experiments to possess a capability of description when applied on contour segments which is equivalent to that shown as a global description method. The invariance properties of this description are robust against the noise caused by space quantization and moderate loss of points. Also the orthogonal nature and the capability of description of this method favour the use of a classification schema based on the combination of the evidences produced by distinct trees.

## References

Anderberg, M.R., 1973, Cluster Analysis for Applications, Academic Press, New York.

Cabrera J., Falcón A., Hernández F.M., Méndez J., 1991, A Systematic Method for Exploring Contour Segments Descriptions, Proc. EUROCAST´91, Springer Lecture Notes in Computer Science, (To be published).

Falcón A., Méndez J., 1984, Localización y Orientación de Piezas en Sistemas visuales de Reconocimiento, Actas II Simposium Nacional IFAC de Automática en la industria, pp.209-212, Zaragoza.

Hernández F.M., 1987, Reconocimiento de Formas Tridimensionales mediante Caracterización Global de Vistas, Doctoral Thesis, Univ. Politécnica de Canarias. Las Palmas de Gran Canaria.

Jain N., Indrayan A., Goel L.R., 1986, Monte Carlo Comparison of six Hierarchical Clustering Methods on Random Data, Pattern Recog., 19:95.

Méndez J., 1983, Contribuciones a la Teoría de Caracterización de Formas para el Reconocimiento, Doctoral Thesis, Univ. Politécnica de Las Palmas, Las Palmas de Gran Canaria.

Méndez J., Falcón A., 1984, On the Orthogonal Expansion of Images for Low Level Recognition. Fourier-Bessel Representation, Proc. First Conf. Art. Int. Applic., pp. 167-169, IEEE Press, Denver, Col.

Shlien S., 1990, Multiple Binary Decision Tree Classifiers, Pattern Recog., 23:757.

Wang Q.R., Suen C.Y., 1987, Large Tree Classifier with Heuristic Search and Global Training, IEEE Trans. on Patttern Anal. Mach. Intell., 9:91.

SELF-ORGANIZATION IN RURAL ECONOMIC SYSTEM

Zhou Shipeng

Research Department 5
No.1 Beiyuan Da Yuan
An Ding Men Wai
Beijing 100012, CHINA

INTRODUCTION

The dual economic structure that modern industry and traditional agriculture exist simultaneously is the typical economic structure for most developing countries. The modernization of economy for them, especially China, so far as its main part is concerned, is essentially the modernization of the economy of rural areas in which the population is up to 80 per cent of the total population. Its aim is to make the rural areas transform from the closed system of traditional natural economy based on manual operations into the open system of commodity economy based on the mode of modern machine production.

How do we realize this aim? Lewis proposed the theory of dual economy in 1954, as shown by Gersovitz (1983), but more than thirty years practice shows that his theory can not be corroborated in any case in China. Just when the developing countries were puzzled by the traditional dual economic structure in the process of development, most socialist countries were difficult to advance or to retreat due to the dual system separating city from countryside in the process of economic reform, after carrying out the peasant family output-related system of contracted responsibilities the country enterprises bursted into activity in China, the great extremely intricate developing country, impacting irresistibly the closed system of the rural natural and semi-natural economy, which has been sunk in sleep for several thousand years and involving increasingly and extensively the rural economy in powerful current of the commodity economy. However, what role do the country enterprises play in the process of industrialization and modernization? And how can the rural economy develop coordinately? They are just the research purpose of this paper.

Because studying the rural dual economic structure formed between the country enterprises and the traditional agriculture is not only the key of studying the above mentioned problems but also the basis of studying the dual economic structure, this paper makes the model of the rural dual economic structure by the method of combining the Synergetics (Haken, 1977) with the System Dynamics (Forrester, 1968) as shown by Zhou (1988), the method makes full use of respective advantages of the two subjects and can give both the qualitative direction and the quantitative analysis. From the study of structure evolution, we obtained the two critical values of population growth and found out the conditions under which the system forms a stable and orderly self-organization structure. On the basis of it, we approached the subject on the developmental strategies of rural economy, especially on the problems of country enterprises development and rural labor transfer.

## THE MODEL OF THE RURAL DUAL ECONOMIC STRUCTURE

In order to study quantitatively and qualitatively the problems that face the development of rural economy and approach the developmental strategies of rural economy, we take the method of combining the synergetics with the system dynamics. According to the method, first we make the model of the rural dual economic structure by the system dynamics as the basic model of quantitative analysis, the causal-loop diagram of the model is illustrated in figure 1, the detail about the model as shown by Zhou (1987); Second we derive the synergetic development equation from the basic model as shown by Zhou (1989); Third we analyse the equations with the synergetics so that we obtain the qulitative result to direct the quantitative analysis and refrain from the blindfold policy test of the system dynamics. This paper gives mainly the part of qulitative study about the model. Assume the agriculture to be department No.1, and the country enterprises to be department No.2, we derive the synergetic development equations as follows:

Let $K_i$ and $L_i$ denote the capital input and labor input of department No.$i(i=1,2)$ respectively, and then the output $Q_i$ can be formulated by Cobb-Douglas production function

$$Q_i = A_i e^{a_i t} K_i^{b_i} L_i^{1-b_i} \qquad (i=1,2) \qquad (2.1)$$

where $A_i$ is a constant greater than zero, $e^{a_i t}$ is the output change caused by technological progress, $a_i$ is called the rate of technological progress, and $b_i$ is the elasticity of output to capital. Then assume that $C_i$ is the coefficient of the net product value of department No.i, and $y_i = C_i Q_i / L_i$ denote the per capital national income of department No.i, thus

$$y_i = C_i Q_i / L_i = C_i A_i e^{a_i t} (K_i / L_i)^{b_i} \qquad (2.2)$$

If we differentiate the above formula with respect to the factor of time, then we have

$$\dot{y}_i / y_i = a_i + b_i (\dot{K}_i / K_i - \dot{L}_i / L_i) \qquad (2.3)$$

Assume that $k_i = \dot{K}_i / K_i$ is the growth rate of capital, and $n_i = \dot{L}_i / L_i$ is the growth rate of labor of department No.i, then the above formula can be written as

$$\dot{y}_i = (a_i + b_i k_i - b_i n_i) y_i \qquad (i=1,2) \qquad (2.4)$$

Notice that when $a_i + b_i k_i > b_i n_i$, the per capital national income will be increased up to infinity. Because of the restriction of natural resources and other factors, it can not be realized, so we have to consider a very important factor in the economic development, namely saturation effect. Because the saturation effect becomes more and more remarkable with the increase of $y_i$, $a_i + b_i k_i - b_i n_i$ can be replaced by $a_i + b_i k_i - b_i n_i - h_i y_i$, and (2.4) finally becomes

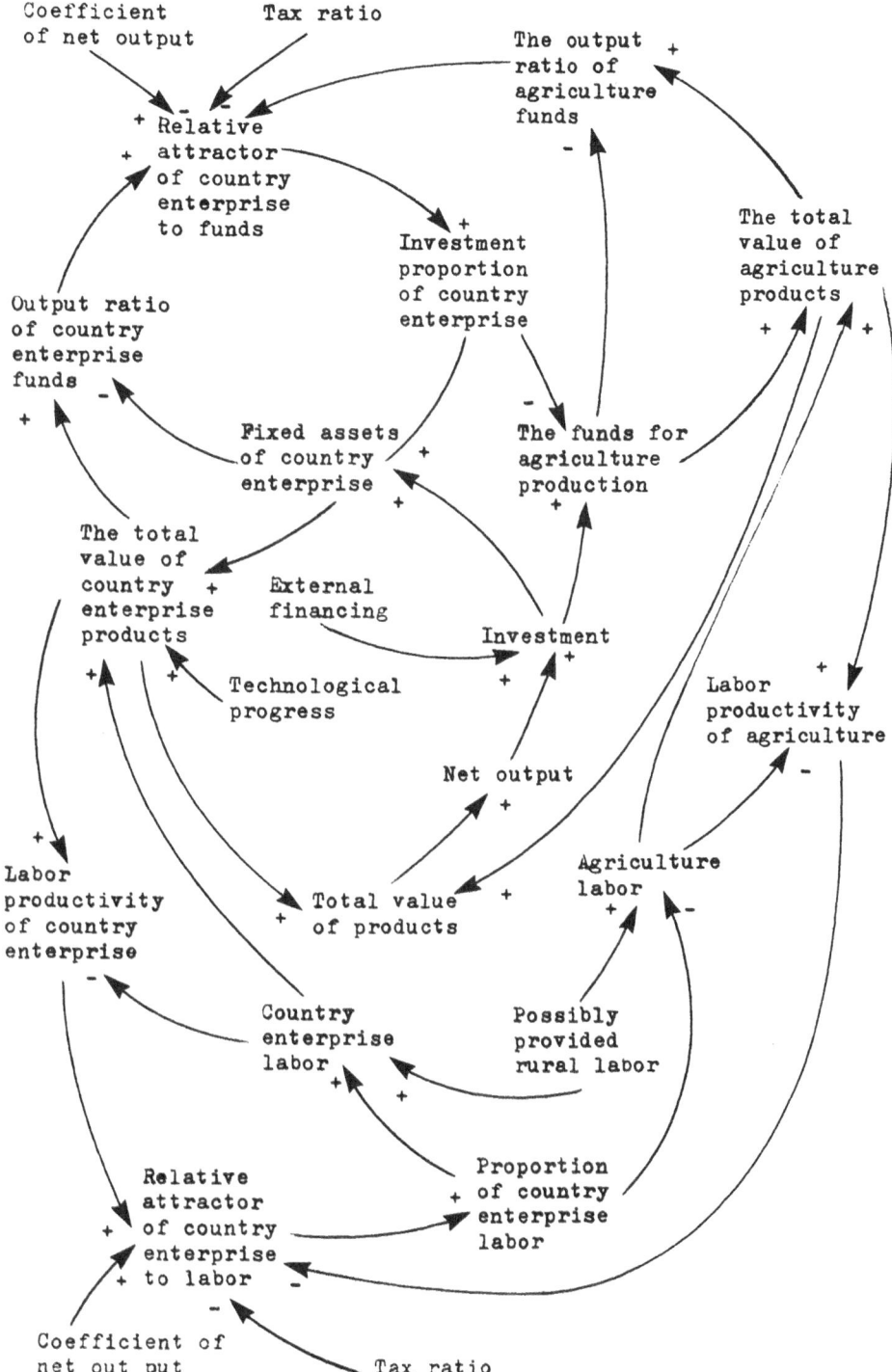

Figure 1. The causal-loop diagram of the model of the
rural dual economic structure system.

239

$$\dot{y}_i = (a_1 + b_1 k_i - b_1 n_i)y_i - h_i y_i^2 \qquad (i=1,2) \qquad (2.5)$$

where $h_i$ is the saturation coefficient.

Let L be the total rural labor that can be provided by the rural areas and n be its growth rate, and let the proportion of the rural labor to the population is a constant. Assume $u=L_2/L$, that is the proportion of the labor of the country enterprises to the total labor, and r to be the transfer rate of the labor of the traditional agriculture to the country enterprises. Then we can get

$$n = n_1(1-u) + n_2 u = \dot{L}/L \qquad (2.6)$$
$$r = \dot{L}_2/L_1 = n_2 u/(1-u) \qquad (2.7)$$

## THE STUDY OF STRUCTURE EVOLUTION AND THE POLICY ANALYSIS

Now we will qualitatively analyse the rural economic system according to the above model with the synergetics, study the conditions, characteristics and law that the system forms a stable and orderly structure, then make policies according to them and control the system.

### The Closed System of The Traditional Agriculture

When $u=0$, we can obtain $L_2=0$. This shows that the country enterprises has not emerged yet, the rural economic system is not dual. It is a closed system of natural and semi-natural economy which only comprises the traditional agriculture. Now the above model can be transformed into

$$\dot{y}_1 = (a_1 + b_1 k_1 - b_1 n_1)y_1 - h_1 y_1^2 \qquad (3.1)$$
$$n_1 = n = \dot{L}/L \qquad (3.2)$$

The most important variable that measures the level of the economic development of the developing countries is per capital national income, we choose it as the order parameter. The formula (3.1) is the equation of order parameter, its steady state equation has two steady state solutions:

$$y_{11} = (a_1 + b_1 k_1 - b_1 n_1)/h_1, \; y_{12} = 0$$

Assume $p_1 = k_1 + a_1/b_1$. According to the analysis of the linear stability, we know:

1) When $n_1 < p_1$, $y_{11} = (a_1 + b_1 k_1 - b_1 n_1)/h_1$ is stable, but $y_{12}=0$ is unstable. It shows that when the growth rate of the rural labor is less than the critical value $P_1$, the rural economic system is in a low-level stable and orderly state of development.

2) When $n_1 > p_1$, $y_{11}$ becomes negative. Since per capital national income can not be negative, this solution has no significance in economy, and the equation has only a disorderly solution $y_1=0$ and it is stable. This shows that when the growth rate of the rural labor is greater than the critical value $P_1$, no matter how small the tiny disturbance is, the system will transit from the ordered to the disordered state.

Conclusion 1. The condition under which the closed system of the rural natural and semi-natural economy develops stably and orderly at a low level is that the growth rate of population (the value of n) is less than the critical value $P_1$; When $n=P_1$, the rural economy will fall into the trap of low-level equilibrium; When $n > P_1$, the rural economic system will transit from the ordered to the disordered state.

In most underdeveloped countries, so far as the labor change is concerned, the output elasticity is quite stable, so assuming the value of $b_1$ to be constant is reasonable. And because the per capital national income of the agriculture is rather low in comparison with the modern industry, on one hand, the attraction for funds is limited; And on the other hand, the ability of investment is very weak, and the value of $k_1$ is comparatively small. Therefore, it is important to speed up the rate of technological progress (the value of $a_1$) or to reduce the growth rate of population. If possible, we can simultaneously deal with these two problems by taking some measures to increase the value of $y_1$, so that we can remarkably change the poor and underdeveloped state. But when the growth rate of population exceeds the critical value $P_1$, and the modern industry is unequal to the mission, namely absorbing rural redundant labor, given by Lewis, the only way that impel the rural economy to develop stably and orderly is to transfer the rural redundant labor to non-agriculture departments, it is the necessity that make the country enterprises come into being in China. The country enterprises realize the transfer by the way of leaving land without leaving countryside, and undertake the task that modern industry does not do. This increasingly shows that it can become not only a bridge that link the dual economic structure but also a channel that opens the dual system.

## The Open System of The Rural Dual Economic Structure

When $u \neq 0$, the rural economic system is the open system of the rural economic structure formed between the traditional agriculture and the country enterprises. From the steady state equations of the order parameter equations (2.5), we obtain four groups of steady state solutions:

$$y_{11} = 0, \quad y_{21} = (a_2 + b_2k_2 - b_2n_2)/h_2 \tag{3.3}$$
$$y_{12} = (a_1 + b_1k_1 - b_1n_1)/h_1, \quad y_{22} = (a_2 + b_2k_2 - b_2n_2)/h_2 \tag{3.4}$$
$$y_{13} = (a_1 + b_1k_1 - b_1n_1)/h_1, \quad y_{23} = 0 \tag{3.5}$$
$$y_{14} = 0, \quad y_{24} = 0 \tag{3.6}$$

Assume $P_2=k_2+a_2/b_2$, according to the analysis of the linear stability, we know:

1) When $n_1 > P_1$ and $n_2 < P_2$, only the first group of solutions (3.3) is stable. It shows that when the transfer rate of the labor of the traditional agriculture (the value of r) is comparatively small and the growth rate of the labor of the traditional agriculture (the value of $n_1$) can not be reduced below the critical value $P_1$, the traditional agriculture will still be in the disordered state.

2) When $n_1 < P_1$ and $n_2 < P_2$, only the second group (3.4) is stable. From (2.6) and (2.7) we can get $n < P_1(1-u)+P_2u$ and

$$n/(1-u) - p_1 < r < p_2u/(1-u) \tag{3.7}$$

It shows that when the growth rate of population is less than the critical value $P_1(1-u)+P_2u$, if we control properly the transfer rate of labor of the traditional agriculture to the country enterprises (as shown in formula (3.7)), the rural economic system will show a self-organization state of stable and orderly development.

3) When $n_1 < p_1$ and $n_2 > p_2$, only the third group (3.5) is stable. From formula (2.7), we can get $r > p_2u/(1-u)$. This indicates that when the transfer rate of the labor is too high and exceeds the critical value $p_2u/(1-u)$, the subsystem of the country enterprises will transit from the ordered to the disordered state.

4) When $n_1 > p_1$ and $n_2 > p_2$, only the fourth group (3.6) is stable. From formula (2.6) and (2.7), we know that when the growth rate n of population exceeds the critical value $p_1(1-u)+p_2u$, no matter what value the transfer rate r of the labor takes, the rural economic system will transit to the disordered state.

<u>Conclusion 2</u>. The conditions under which the open system of the rural dual economic structure forms a stable and orderly self-organization structure are that the growth rate n of population and the transfer rate r of the labor of the traditional agriculture to the country enterprises satisfy respectively

$$n < p_1(1-u) + p_2u$$
$$n/(1-u) - p_1 < r < p_2u/(1-u)$$

Therefore the <u>developmental strategies</u> of the rural economic system are: The first, transfering the rural redundant labor from the traditional agriculture to the non-agriculture departments, namely the country enterprises, and properly controlling its transfer rate. The second, opening to the outside and introducing technologies and funds, speeding up the rate of technological progress, reducing the human fertility. The third, devoting major efforts to developing the rural education. Developing education is a long-term plan of raising the scientific and technological level, reducing human fertility and realizing the modernization of the rural economy. The fourth, progressively carrying out the transfer of the country enterprises to modern industry.

REFERENCES

Forrester, J. W., 1968, "Principles of Systems," MIT Press, Cambridge.
Gersovitz, M., 1983, "Selected Economic Writings of W.Arthur Lewis," New York University, New York.
Haken, H., 1977, "Synergetics," Springer-Verlag, Berlin Heidelberg.
Zhou Shipeng, 1987, The Developmental Strategy Study of Country Enterprises of Dong Ting Lake Region, "The Technical Report on the Model of the Unified Plan for the Harnessing and Exploiting of Dong Ting Lake Region," Hunan Advisory Center on Science and Technology, Changsha.
Zhou Shipeng, 1988, "The Application of System Dynamics and Non-equilibrium System Theory to the Unified Plan for the Harnessing and Exploiting of Dong Ting Lake Region," The Master Dissertation, Changsha Institute of Technology, Changsha.
Zhou Shipeng, 1989, The Establishment of the Synergetic Development Equations of Social Economic System And Stability Analysis, Systems Engineering, 7:21.

# A SYSTEM DYNAMICS APPROACH TO ASSESSING

# THE IMPACT OF MANAGEMENT INFORMATION SYSTEMS

S. M. Henderson, A. W. Gavine, E. F. Wolstenholme[§], K. Watts[‡]

Royal Armament Research and Development Establishment
Sevenoaks
United Kingdom

## INTRODUCTION

This presentation briefly describes some work performed on appraising the System Dynamics method in assessing the operational benefits of management information systems. It includes a discussion on the knowledge elicitation and capture aspects of systems investigation with particular emphasis on the role that the System Dynamics methodology can play in this vital but largely unexplored area.

The applications selected for study in this work are military command, control and information systems (CIS). CIS are information systems designed to assist the military commander in collecting and assessing information on a rapidly changing and hostile environment, co-ordinating the resources at his disposal, planning his future actions and executing his instructions.

## BACKGROUND

This presentation results from several years collaborative effort involving RARDE and Bradford University investigating the application of System Dynamics to problems of military assessment generally, and the command and control process in particular.

The high level of investment required in developing and implementing a typical CIS has stimulated a search for an objective approach toward the assessment of its likely benefits. Benefit assessment, the evaluation of the benefits to an organisation resulting from the acquisition of computer-based information systems, is consequently assuming a higher profile

---

[§] Based at the Management Centre, Bradford University, United Kingdom.
[‡] Formerly of Bradford University, but currently based at Bradford City Council.

*Systems Thinking in Europe*, Edited by M.C. Jackson *et al.*
Plenum Press, New York, 1991

in the process of procurement of such systems.  Experience with
large scale CIS procurement has highlighted the inadequacies of
past approaches to this problem.

## The System Dynamics Method

System Dynamics is a method for explaining a given system's
behaviour in terms of its structure and policies and for
investigating changes to structure, policies, or both, which
will lead to an improvement in system behaviour.  It is
particularly appropriate to modelling feedback and the complex
structures arising from the interaction of physical and
information flows.  The technique has been increasing in
popularity since its initial development some 20 years ago and
has been applied in many diverse areas of policy evaluation
(Coyle 1978).  It is also proving popular in many other parts of
the defence science community (Wolstenholme & Al-Alusi 1987,
Wolstenholme 1988) and in policy analysis generally
(Wolstenholme 1989).  It is supported by a variety of software
and graphical tools and has the attraction of producing readily
understood models with relatively little effort.  It is not the
purpose of this presentation to supply a tutorial on the basic
method - many texts exist which can describe the origins,
development and application of the method (Forrester 1968, Pidd
1988, Roberts 1983, Wolstenholme 1990).

## The Role of System Dynamics in System Investigation

Unlike many traditional hard modelling environments, the aim of
a System Dynamics study extends beyond providing merely a
quantitative description of a system and a simulation of its
behaviour within the rigid, technique-dependent constraints
which characterise some traditionally founded hard modelling
environments.  System Dynamics is also a qualitative tool which
encourages the participation of all the relevant actors in a
holistic and educative debate and embodies many of the concepts
associated with soft systems thinking (Wolstenholme 1990).  As
such, it is one of an emergent generation of problem structuring
methods concerned with the management of complexity, the
handling of uncertainty and the resolution of conflict.  The
soft systems approach (Checkland 1981) and similar methods
(Rosenhead 1989) are reviewed elsewhere.

SD attempts to define the scope of the investigation, to
identify real-world symptoms and possible causes and to capture
them in the form of an influence diagram.  These diagrams form a
structural representation of the relationships and dependencies
between the entities within the problem space and focus on the
behavioural dynamics by means of an explicit realisation of the
feedback processes involved.

The thrust of the study was aimed at providing an holistic
appreciation of the CIS on the host system in terms of high
level performance measures relating to system achievement.  It
was concerned with assessing the need for information as a
function of the way in which the information will be used.  An

244

appraisal of the capability of System Dynamics in assessing CIS effectiveness is described elsewhere (Gavine 1990).

## The Application

One application comprised a battlefield logistics information system and is described elsewhere (Watts 1990). The application described in more detail in this presentation consisted of a hypothetical Battlefield Information System (BIS) which might, if deployed, assist battlefield commanders and their staffs by improving the present command and control process thereby giving commanders and staff more time for decision making based on accurate, relevant and timely information.

## THE APPROACH

Throughout the study, a three-stage methodology was employed (the full methodology is described by Wolstenholme et al 1990). The BIS study began by attempting to conceptualise and structure the analysis of the CIS problem area through discussions with the client. The model construction and its subsequent development provided a means of structuring the debate and stimulated insight by study of the model's output. By employing System Dynamics as a data capture and knowledge structuring aid, the case study highlighted some of the soft systems aspects of the method. The use of influence diagrams as a communications medium was particularly significant, instigating discussion on modes of functional representation. The construction of a baseline model, was based on these representations and subsequently enhanced by the superimposition of the CIS. This model is described more fully elsewhere (Henderson 1990).

## KNOWLEDGE CAPTURE ISSUES

The individual habitually seeks to understand the world in terms of mental concepts and images. His decisions, actions and behaviour all derive from this constantly evolving viewpoint. The System Dynamics model is a tool to support this process:- it provides a concrete statement of assumptions; it can provide a documented history of the growth of the model; it facilitates communication about the model stimulating the eduction of a consensus view; and it can be manipulated via experimentation facilitating assessment and impelling novelty of approach.

## SD as a Communication Medium

The first stage in an SD investigation is for the analyst to acquire knowledge about the the situation under study. One source of knowledge is documentation relating to the situation (e.g. empirical data; operational, technical and policy documents; specifications; proposals etc). Another, and perhaps more useful source of knowledge (in so far as its relevance can be considered more 'situational') comes from those people who have an active interest in the situation (in an encounter, referred to as clients). They can be managers; owners, participants and identifiers of the problem; system actors;

system specifiers etc. In order to facilitate knowledge acquisition through discussion with these people, it is necessary for the analyst to have three things: (i) a means of *eliciting* relevant knowledge; (ii) a mechanism for *representing* and capturing the elicited knowledge; and (iii) a way of *checking* that this representation is what the client actually meant.

A prerequisite to successful knowledge elicitation is client interest. In an SD study, interest is often instigated through the presentation of an initial 'catalytic' model. This can be totally unrelated to the problem area, or can be constructed from knowledge gained through reading relevant documentation[1]. Catalytic models serve both to illustrate the techniques employed by the analyst, and also to stimulate a response from the client.

It is important that any concepts captured by the analyst are recorded in a form which the client can understand, and importantly, can participate in. To quote from Garrod and Anderson (1978):

'...on most occasions when we speak or listen we do so within the... framework of dialogue, where a major goal of both parties is to achieve mutual intelligibility. To communicate effectively, speaker and listener must co-ordinate their respective use and interpretation of the language, within the context of that particular exchange. They need to establish that they share the same overall conception of what is being discussed and agree upon how each utterance should be interpreted with respect to this conception.'

System Dynamics represents knowledge through 'Influence Diagrams'. Put simply, an influence diagram is a pictorial representation of the structural interdependencies which exist between the components within a system. Its main attractions are (i) that it is very simple (its nomenclature consists of four different symbols) and (ii) that it facilitates easy transition to the latter parts of an SD study (e.g. quantitative model construction). Constructive Simplicity (Ward 1989) is incorporated into an influence diagram by restricting the number of variables, limiting the amount of uncertainty incorporated (or ignoring it all together), assuming a well defined objective function or set of decision criteria, and by aggregating variables with certain shared characteristics.

Influence diagrams have proved highly successful as a medium for communication with clients. Client interest is held from the start of an interaction, thus information is made more freely available. Influence diagrams do not inhibit the ability to express ideas through a barrage of technicalities, and importantly, the client is able to see directly the contribution

---

[1]     Interestingly, Morrison (Morrison 1987) has found that knowledge elicitation is more successful where the analyst is, or pretends to be, a complete novice, thus forcing a more thorough explanation in simple terms which might otherwise be assumed.

they have made to the structural representation. This fosters
interest and encourages implementation. Military officers
involved throughout this study have been very enthusiastic
towards the techniques which have been employed. In general,
they felt that other techniques failed to capture the 'Big
Picture', whereas influence diagrams enabled a high level view
of the system to be represented, and, more importantly,
understood. Military officers were not only receptive of the
techniques, they were proactive. On several occasions officers
started producing their own influence diagrams, having first
contributed ideas to an influence diagram on a white-board, and
seen how easily their ideas were incorporated.

## The Role of SD in Achieving Consensus

Influence diagrams represent knowledge *structurally*. Thus, an
influence diagram, captured through interaction with a client,
not only represents that client's perceptions as to the *area* of
the situation which they consider to be important; it also
represents their *understanding* of how the elements within that
perception influence each other. This is valuable, in that it
allows conflicting views about the situation (arising out of
both structure and operation) to be highlighted.

An important aspect of analysing any situation is to recognise
that the people within it each have their own perspectives.
Checkland (in Rosenhead 1989) exemplifies perspective using
conflicting descriptions of political activists as 'Terrorists'
and 'Freedom Fighters'. It is also important to recognise that
an individual's perspective has a boundary which does not
necessarily coincide with the organisational boundaries within
which he/she operates. For example, if asked what elements in a
battlefield scenario contribute to target acquisition rate, a
gunner in a tank might reply: "window size, field of view,
terrain, use of thermal imaging" etc (all low-level, front line
concerns). However, a Brigade commander might reply: "forward
recce and long-range radar" (an aggregated description of a
front line concern which subsumes the concerns of the gunner,
and also a rear line strategic concern). In actual practice,
even people with the *same* responsibilities in an organisation
who view a common concern will give different answers as to
which elements they believe are important.

By capturing several different viewpoints about a system in the
form of influence diagrams, it is possible to identify
conflicting views. Content and structure can be compared - for
example, a structural element in one diagram might have two
influences upon it. In another diagram it might have five. It
is important to identify where there is a genuine clash of
views, and where there is an inappropriate level of resolution
represented (which can be resolved through aggregation).

The recognition and highlighting of conflicting views is a
valuable and integral part of an SD investigation. Conflicts
can be taken back to clients, and misunderstanding about the
system demonstrated using the technique as a communicative

medium. By getting people to *understand* how they actually operate within the situation, it is possible for each participant to make compromises in their views, thereby enabling a *consensus view* of the situation to be achieved. Once this has been arrived at, it provides a firm, agreed foundation for further analysis.

## SD as an Aid to Conceptual Thinking

Influence diagrams are powerful tools for exploring the structure and operations within a situation. Analysis of the interdependencies within a situation facilitates the recognition of *functional boundaries*, and thus the identification of *component systems*. It is then possible to examine how the component system works internally (through identification of controlling negative feedback loops, resource transformation processes, capacities and decision processes). The interaction between component systems can also be studied, and this is particularly useful for identifying problems caused by localised decisions and actions. Often, action taken in one part of the system produces unexpected, and indeed, unwanted results in another part (this is explored in greater detail by Wolstenholme 1990). The *holistic perspective* employed by System Dynamics enables this type of problem to be detected.

## Further Exploration with SD

Whilst the latter stages of an SD study draw techniques from the Hard Systems paradigm, their use is still geared towards increasing understanding. A number of rigourous techniques can be applied to the influence diagram to check that it is 'structurally' and 'definitionally' coherent (Coyle 1988). It can then be quantified (note that this does not necessarily require precise measurement) and translated into a computerised model which facilitates behavioural simulation. The output from a System Dynamics simulation is expressed in terms of 'broad behavioural dynamics' - in other words, SD is more concerned with trends and behaviour than it is with precise mathematical prediction. The simulation can be used to generate insight and deeper understanding of how the situation behaves. It can be used to direct further exploration, and it provides an excellent basis for further discussions with clients. It can also encompass 'what if' experimentation, and can facilitate the cross-over into the hard world of optimisation and mathematical analysis.

## CONCLUSIONS

The study has concentrated on System Dynamics as a means of assessing the benefits arising from any improvement of operational effectiveness in the target application area. However, the BIS case study has also shown the benefits of System Dynamics as a qualitative, participative method, well suited to the ill-structured problems and ephemeral issues associated with the feasibility or pre-feasibility phases of system investigation.

248

REFERENCES

Checkland P. B., "Systems Thinking, Systems Practice", Wiley, 1981.

Coyle R. G., "Management System Dynamics", Wiley-Interscience, 1988.

Forrester J. W., "Principles of Systems", Wright-Allen Press, 1968.

Garrod S.C. and Anderson, A., "Saying what you mean in dialogue: A study in conceptual and semantic co-ordination", Cognition, no.27, 1978

Gavine A. W., Wolstenholme E. F., "An Appraisal of System Dynamics in Assessing the Impact of Computer Information Systems", International System Dynamics Conference, Boston, Mass., 1990.

Henderson S. M., Wolstenholme E. F., "The Application of a Dynamic Methodology to Assess the Benefit of a Battlefield Information System.", International System Dynamics Conference, Boston, Mass., 1990.

Morrison A., "Session chairman's remarks at the Third International Expert Systems Conference", London. 1987.

Pidd M., "Computer Simulation in Management Science", John Wiley, 1988.

Roberts N. et al, "Introduction to Computer Simulation – A System Dynamics Modelling Approach", Addison-Wesley, 1983.

Rosenhead J., "Rational Analysis for a Problematic World", John Wiley, 1989.

Ward S., "Arguments for Constructively Simple Models.", Journal of the Operational Research Society, Vol. 40, No. 2, 1989.

Watts K., Wolstenholme E. F., "The Application of a Dynamic Methodology to Assess the Benefit of a Logistic Information System in Defence.", International System Dynamics Conference, Boston, Mass., 1990.

Wolstenholme E. F., "Defence Operational Analysis Using System Dynamics", European Journal of Operational Research, Vol 34, No. 1, pp16-18, February 1988.

Wolstenholme E. F., "An Overview of Systems Dynamics", Transactions of the Institute of Management and Control, Vol. 11, No. 4, pp171-179, November 1989.

Wolstenholme E. F., "System Enquiry – A System Dynamics Approach", Wiley 1990.

Wolstenholme E. F., Al-Alusi S. A., "System Dynamics and Heuristic Optimisation in Defence", System Dynamics Review, Vol. 3, Pt. 2, 1987,

Wolstenholme E. F., Gavine A. W., Watts K., Henderson S. M., "The Development of a Dynamic Methodology for the Assessment of Computerised Information Systems.", International System Dynamics Conference, Boston, Mass., 1990.

# STELLA MODELLING PROCESS FOR A MANPOWER STRATEGY

Matthew Byrne
Andrew Davis

Business Consultants (ICA/41)
Shell International Petroleum Company
Shell Centre, London

## INTRODUCTION

This paper takes the reader through a consultancy process that was recently used by Shell Business Services. Shell Business Services is a part of the Shell International Petroleum Company, providing an internal consultancy in the field of decision support for Shell companies. A particular project will be discussed from the point at which contact was made with the client right through to the final deliverable. The discussion will focus more on the consulting process rather than a detailed explaination of the model that was built. In the first section we will discuss the client's organisation, it's concerns and the process by which these were distilled into a useful STELLA model. The second section explores the issues of model complexity and puts forward a matrix to help consultants classify the type of project they are involved in and what the main thrust of their consulting process should focus on. Finally, the third section takes a look at the way in which projects are presented to the client in terms of a deliverable. It contrasts previous approaches that we have used with a new approach that was used with this client.

## WHAT IS STELLA?

Stella is a software package used for modelling business thinking. It is particularly suited to modelling issues in a dynamic way and is ideal for coping with systems that contain feedback.

## ENTRY TO THE CLIENT

The client was a group of managers with responsibility for managing 1400 computing staff with varied skill categories, in different geographical regions and across different sectors of the business. They have responsibilities for determining overall numbers of staff at each level in the company, the calibre of those staff, the blend of short-term and medium-term contractors to permanent staff, recruitment and career progression, to mention but a few. As can be imagined the impact of strategic decisions in this arena can have very significant long term effects on businesses of Shell companies. The clients were interested in gaining a better understanding of this complex system and exploring the long term implications of specific strategies. Until recently, the

*Systems Thinking in Europe*, Edited by M.C. Jackson *et al.*
Plenum Press, New York, 1991

only tool available to these managers to aid them in their planning has been a linear and fairly static view of the short to medium term. This existing spreadsheet tool was felt not to offer the kind of long-term dynamic view which they needed.

The clients had already been targetted by a piece of earlier marketing. This had involved building a small "throwaway" model to demonstrate the capabilities of system dynamics and STELLA. It had been built around a specific problem in the personnel field and was presented to the management team to interest them in the applicability of such an approach to their wider strategic concerns. Thus, the client was already familiar with our work and to a certain extent familiar with the system dynamics approach.


INITIAL MEETING

Our first meeting was with the whole of the management team consisting of about a dozen people. The original "throwaway" model was demonstrated live and it was explained how this work could be expanded to address more of their initial concerns. However, as we have often found to be the case, time, and the process of building the first model had changed the particular problems that were foremost in their minds. The main management concern was to meet future target levels of staff at the lower technical levels whilst still bringing sufficient people through the system to staff the management positions of the future. It was decided, therefore, that the best way forward would be to take one of their business sectors and use it as a guinea-pig to test the usefulness of a new model. This new model would address "softer" and more wide-ranging issues than before.


STRUCTURING MEETING

The first meeting to structure a new model took place at the client's site. It was an informal meeting in which members of the management team would wander in and out of the discussion when they had the time to attend. On the face of it, this would seem a strange way to run a meeting, but from a practical point of view it was the only way to get all of the stakeholders together in a single session, if not all of them always at the same time! At the close of the day we managed to bring all of the team together and agree a series of causal loop diagrams that should form the logic of the model. It is interesting to note that while we had complete agreement amongst the management team as to the important issues and their relationships, when we showed the model logic to other personnel professionals, they brought out other issues which they felt to be of equal importance. This emphasises the fact that any model built depends crucially on the management team that was involved in the construction. It is therefore vital to have all the major players agreeing on the logic, or the model's credibility will be immediately lost.

A very interesting problem arose from this meeting which should be of general interest to modellers. We found that there was a difference between the model which had been explained to us by the management as being the theory of how the system operated and the actual system that operated in practise. We were faced with a set of logics and guidelines which the management believed they used, albeit liberally, but in practice they palpably did not. With the first stage of modelling we were able to show them the effects of managing the system under both sets of logics and demonstrate that if they were to use the "theoretical" logic in reality, then there would be serious problems in the future. Thus, the modelling process had already unearthed an interesting difference in perceptions of the system.

We believe that this is a major benefit of the "Modelling as Learning" approach (David Lane, 1988). The client is forced to take a clear and consistent view of their system and build a real understanding of the underlying logics. This is one of the areas where a traditional modelling approach might have failed as the client is encouraged to "leave it to the consultant". In this case, the consultant could have ended up presenting a model which gave obviously inconsistent results leading to the client rejecting the model altogether because the only client understanding of the model would have been at the level of the results it was giving rather than the logic it was using. Using the "Modelling as Learning" approach, we were able to explain the reasons for the results in terms of the causal loop diagrams that they themselves had built. This caused them to re-examine those logics and agree that they were not consistent with what they were doing in practise. They were learning!

WHAT VALUE COMPLEXITY ?

There is always a play off to be made between the complexity of the model and how understandable it is. Clients tend to want to model reality with every last detail. Unfortunately, the value of such a model ( which indeed it would no longer be) would be lost in a morass of detail and complexity. The real value of STELLA models is that they can quickly communicate a set of ideas and interactions and allow the user to check the consistency or sanity of those ideas. As soon as models start to contain life size helpings of complexity, the client often loses the thread of what is actually contained in the model (not to mention the consultant!). The only person who will really be able to run and benefit from the model will be the person who built it. This surely defeats many of the reasons for using such a user friendly tool and destroys the chance for any real learning to take place.

However, if models are going to be anything but trivial they must contain some complexity. The consultant's role then becomes important in managing the interface between the client and the model. Building in enough complexity to satisfy the customers need for a representative model while balancing this with ensuring the model does not become a "black box " in the client's eyes is a critical part of the consultancy process. We have experienced both extremes of this dilemma. Having built one of the most complex STELLA models and found that no-one really understands it or uses it, has proved to be an expensive and largely pointless exercise. On the other hand, going to clients with trivial "noddy" models that everyone can quickly understand, and having them ripped apart for being of no value at all is even more soul-destroying. So clearly complexity is of value, but so is balance.

In this particular study, the client wanted issues included which required complex pieces of STELLA modelling. Though the modelling from a system dynamics point of view proved to be successful, it is our opinion that the extra complexity led to a reduction in the client's understanding of the functioning of the model. Although the model was theoretically more "correct" with the complexity included, it seems clear that the client gained no real benefit from the extra work being done. It merely led to the model being a "black box" in their eyes. Yet, whether they would have accepted a model which had not included this extra level of complexity is unclear.

Complexity is thus a double-edged sword with significant implications for the consultancy process. Adding increasing amounts of complexity into a model should be approached with care and an awareness that it might be leading the consultant into a series of different management problems. A summary of our experiences in this area is contained in Fig.1, which we have found to be a useful model for classifying projects and making the consultant aware of the different process issues associated with different levels of complexity.

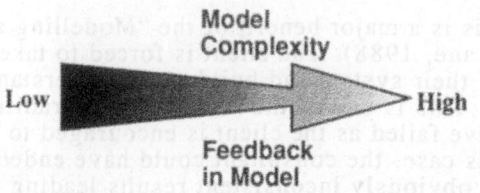

| Client Understanding of Model Results | Intuitive | Counter-Intuitive | Non-Intuitive |
|---|---|---|---|
| Model Behaviour | Transparent | Explicable | Non-Explicable |
| Deliverable | Understanding | Learning | Answer |
| Consultant's Focus | Ensure Usefulness | Ensure Participation | Ensure Trust / Belief |

Fig.1. The Value and Problems of Model Complexity

As explaination of this matrix, we will discuss a couple of cases.

The small "throwaway" model that was built for the management team in order to sell the system dynamics concept would have fallen somewhere in the intuitive column. The model had almost no feedback and was very simple with only a small number of variables. When we ran examples through the model, the clients were able to predict the result with ease due to the model's linearity and simplicity, thus their understanding of the results was intuitive and to them the model was transparent. For these sorts of models the deliverable is a reinforcement of the client's understanding of the system and the modelling technique. As the consultant, you should be concerned with ensuring that the model is performing some useful function. The danger with over-simplified models is that they are useless.

The first significant model that was built for this study would fall into the middle column. The model had a significant amount of feedback and was reasonably complex. When we ran examples through this model, the clients were surprised by the results that emerged. They had predicted that a variable would move in one direction and the model showed it moving in the opposite direction, the results were counter-intuitive. This is often due to the amount of feedback in the model. However, the results were explicable because we could refer to the causal loop logics that were developed with the client and trace the reason for results being counter-intuitive. This process of going back to the logics, explaining and then refining generates client learning about the system. Yet, we were only in the position to generate this learning because the client had built the logics and been actively participating in the whole process.

By the time we came to build the final the model, the clients had decided that they needed many more variables in the model which further increased it's complexity and took us into the final column. Now we had a very large model that was giving us results that were non-explicable due to the large amount of complex logic. The results were not only counter-intuitive they had become completely non-intuitive and all we could do was communicate an answer to the clients that they either believed or didn't. By this stage we had to ensure that they trusted us enough to trust the model. Thus, the management of the client had shifted from getting them to believe the model to much more focus on getting them to trust us and our ability.

Our feeling is that the importance of STELLA as a learning tool demands that the client has a grasp of the model. Once that has been lost or significantly reduced then the value of the process diminishes significantly. Thus, it is important that a balance be struck between complexity and value. Too little complexity and the model will be of no value as the client won't buy it, too much and the client won't understand it and the resultant value from the running of the model will be lost. It is all very well asking a large number of "what if ?" questions and seeing graphs generated in front of your eyes, but this is not the real learning experience. That comes when the user sees the results of a run and then asks "why am I getting this behaviour?". This circular process of running the model, comparing the results with the real world situation, going back to the model logic, analysing and UNDERSTANDING why this particular behaviour pattern has been obtained and then proceeding, is where the real learning is generated. This "Learning Wheel" is shown below. If any of the six spokes in the wheel are not present then the wheel is likely to buckle or collapse and destroy the chance of generating any real learning.

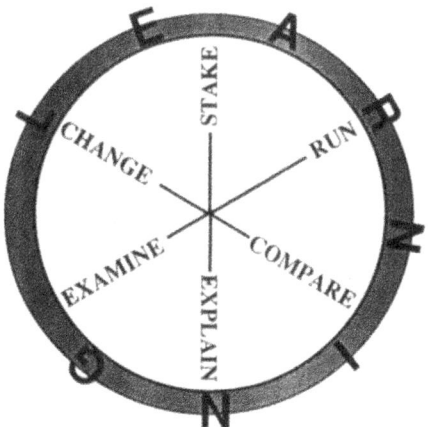

Fig. 2  The Learning Wheel

Expanding briefly on these six stages:

1.    The first step should be to get the client to "put a stake in the ground". By this we mean that they should commit themselves before any run of the model is made to what they think the results will be. This has two functions. It stops the inevitable comment after they have seen the

results of "Well, that's obvious..". Indeed with 20/20 hindsight it often is. It also gets them to really think before making a run rather than just treating it like a game of space invaders.

2.    After discussing what the results might look like, make the run.

3.    Then compare the actual results with those that were expected. Is there any difference?

4.    If there is a difference, then go back to the causal loop logic and explain why these results were obtained.

5.    If the logic is found to be wanting then go back and examine the assumptions that went into it.

6.    Change the model to fit the new logic described in phase 5.

This process continues until the team are satisfied with the model.

If the model has been built without any real participation of the client, then the consultant will find it very difficult to execute any of these phases except the RUN element. It then becomes almost impossible to generate any significant learning.


DELIVERABLES

The real aim of any management or business study should be to initiate tangible improvements in the business. This implies that the decision maker or power broker in the system must understand, believe and trust the results of the study. There is no better way of ensuring this than making the problem owner the problem solver. The consultants role is then to manage the thinking and problem solving processes rather than the actual problem solving itself.

In previous system dynamics projects it has always been deemed necessary to produce a report of the project. This has been the result of existing culture; the feeling that a report demonstrates by it's sheer size the importance of the work that has been carried out. Yet, the very nature of a report suggests that the client needs to learn about what has been done. If the project has been carried out successfully, then this reason is redundant. The client should understand what has been done, because he/she did most of it. The participative nature of the process should be aimed at involving the client in the work. That is where the learning and understanding is really generated, not in a report to be filed in a cupboard, never to surface again.

Indeed the whole reason for carrying out a system dynamics project according to the Modelling as Learning principle, is to ensure that the results from the study initiate action and real business benefit. The problem is, what is the most effective way of ramming home the issues and learning that are generated by a study?

The workshop environment represents an interactive and dynamic learning experience rather than the one-way, static report method. As such, the workshop is much closer to the philosophy of Modelling as Learning. The approach is also much better suited to the capabilities of STELLA. One of the real plus points of STELLA is that it is so easy to make changes to a model, run it and then watch that particular scenario unroll before your eyes. The process of then comparing the results of a run with perceptions of the real world, using the underlying logic in the model to explain why it is happening generates discussion, new ideas and learning. Not only that, it's exciting !! The package is an interactive learning tool. As such, we feel that the only way to make the project and model really come alive is to allow it to do just that. Make it come alive in a live environment, run a workshop!

Running a workshop is however no bed of roses for the consultant. It is far easier to sit down in a nice cosy office environment and write a safe report of the study. No one to ask silly awkward questions, no-one trying to do things to your model that you never intended in fact no-one really challenging you and your work to prove their worth. Also from the client's point of view it involves a committment of often scarce time and energy to something which won't give them the answer. Trying to sell the concept of learning rather than solving, understanding rather than mastery and generating more questions and ideas rather than giving a solution is difficult.

So clearly there is a problem here. How can we get the decision makers to commit the necessary time and energy to a study? If they don't make that committment and choose instead to delegate the responsibility, then how do we get them to buy into what has been done ? In other words, how do we make sure that our work is the catalyst for real change in the business ?

The approach we took with the clients in Aberdeen was to offer them belt and braces. The safe and instantly recognisable report, a presentation and the more challenging danger of a workshop.

With a model that is addressing strategic issues, it is very important to have the decision makers understanding and using the model. However, in this case the management had delegated much of the modelling process to their staff due to the scarceness of their own time.

Therefore, for the work to make any impact it was important that the managers first understood the underlying logic in the model. The presentation was used for this purpose. We decided that this presentation would have much more impact if their staff were to explain the model rather than ourselves. As one of the ideas behind Modelling as Learning is that the process generates learning and understanding, we thought it only appropriate that the client's team should be presenting. At first this may seem like us taking the easy ride. Being paid all that money and not even making our own presentation. However, it was far more of a test of our work and the success we had achieved than giving the presentation ourselves. If the clients stood up and made a mess of things then it would be a bad reflection on our work. We were putting our reputation in their hands! Yet it worked well. Not perfectly but the effect was as important as the content. It meant that the management team felt their department was still the problem owner. They hadn't handed over the problem to anybody else.

Having now read the report and listened to an explaination of the model logic, the management team were prepared for a workshop. The model was used live with an overhead projection screen so that the outputs from the model could be seen clearly by all. The first session of the workshop was spent developing two scenarios that the management felt were possible futures for their system. The scenarios were described in terms of the model variables; how many people they would need in this future, what balance they wanted between permanent and contract staff, how many people they would recruit each year and at what level. Once there was general agreement on the details of each scenario, we discussed what sort of management problems each scenario might develop and whether the system would be able to cope. Having detailed what we thought would happen under each scenario, we then ran the cases through the model and watched the results unfold before our eyes. As the graphs were produced an instant discussion would develop to explain why we were getting that type of behaviour and why it was different from our expectations. The STELLA model was acting as both a catalyst and a focus for those discussions. Several runs were made over a period of two hours as the management explored different options and discussed possible policies. Each time we followed the "learning wheel" round, and each time some new aspect of the system was discovered and debated. The workshop had succeeded. Managers left the session saying they had moved further forward in this one session than they had in the whole of the last year. Where there had been arguments over what the implications of a strategy might be, they were now able to run that strategy through a model of consistent logic that they had agreed upon. The process turned argument into discussion and understanding.

Putting together a package of a workshop, presentation and report certainly seems to have worked. The clients can get much more of a "feel" for their problem by interacting with the model directly rather than reading about the benefits that the consultants gained from it. And by running the model themselves they are a lot more likely to understand and believe the results (not to mention criticise them!).

## CONCLUSION

The modelling process that was used with these clients proved to be successful. Managers are focussing their attention on the key issues and are actively exploring policy options with the aim of developing a robust strategy for the future. We believe that much of the success of this project was due to the consultancy process rather than the specific modelling work itself which is why much of this paper has concerned itself with a discussion of these issues. We would be very interested to hear from other STELLA users as to how they have approached the problem of adding real business value.

## REFERENCES

Lane, D.C., 1989, Modelling as Learning, Proceedings of the European Simulation Congress

# AN URBANISTIC PROJECT BASED ON GENERAL SYSTEMS THEORY

Antonio Caselles and Josep-Lluis Uso

Departament de Matematica Aplicada i Astronomia
Universitat de Valencia
46100 Burjassot. Valencia. Spain

Rafael Carretero, Vicent Llorens, Victoria Solaz

Conselleria de Obres Publiques i Transports
Generalitat Valenciana
Av. Blasco Ibañez, 5. 46010 Valencia. Spain

ABSTRACT

This paper intends to be an outline of the Urbanistic Development
Planning of the Valencian Community in Spain to be performed during the
years 1991 to 1993. The following are considered as objectives of the
Planning:
 - Quality of Life improvement,              - protection of Nature,
 - rational management of natural resources,    - rational use of land;
all based on social demand. The territory and the mechanism of intervention
is considered as a system whose elements can be grouped as:
 - uses of land,            - infrastructure, equipment, services,
 - objectives,              - intervention and management.
The Planning proposes itself to be: integral, strategic, prospective,
programmed, controllable and participative. A simulation model will be
built on this base. The Quality of Life is defined as a linear function of
the "state indicators" (analphabetism, delinquency, hope of life,
morbility, etc.), and it is obtained from real data using Multivariate
Analysis. Parallelly, Quality of Life is obtained dynamically from the
"Level of Life", that is calculated as a linear function of "level
indicators" such as park surface, energy consumption, number of beds in
hospitals, etc. The relation between Quality of Life and Level of Life is a
non linear regression function obtained from data of the past.

## INTRODUCTION

Most spanish Territorial Planning works studied by the authors are
collections of sectorial studies, by themes and space, not well related
with the strategic objectives defined by law, that are:
(1) rational use of land, and (2) responsible management of natural
resources. These objectives have as a consequence another two:
(3) Quality of Life improvement, and (4) ambience protection.
From these objectives it can be deduced that any territorial action must be

*Systems Thinking in Europe*, Edited by M.C. Jackson *et al.*
Plenum Press, New York, 1991

addressed towards the benefit of the population resident on the territory base of the Planning. So, the key for optimisation of interventive actions have to be the study of a function of social wellbeing or quality of life. Connections between quality of life and interventive actions have to constitute a mathematical model that describes the behaviour of the most important parts of the socio-economical system, for example, infrastructure, production and distribution, but not only "what" and "how" but also "where", emphasizing the spatial problem, specific factor of Territorial Planning.

A construction of this kind requires to be based on some axioms or assumptions, that are obviously restrictions of reality, criticable, but the only way to handle an information that have to be elaborated, synthetic, measurable, and representable, in order to be able to work with it. These axioms and assumptions are the following.

1. Social Wellbeing is the wellbeing of all the society considered. That means that discriminations of any sense are not permitted.

2. Social Wellbeing is defined as the degree of satisfaction of social necessities. Social necessities are defined as individual human necessities plus group human necessities. In the following we understand for "state" the degree of satisfaction of a social necessity. We will call State of Wellbeing at a linear function of the states corresponding to all social necessities considered in the society object of study.

3. Material conditions or elements are useful in the measure they satisfy social necessities. The value of a material condition or element will be in relation to its usefulness. In the following we understand by "level" the amount of a condition or element that can satisfy a social necessity. We will call Level of Wellbeing at a growing function of the levels that considers not only the amount of goods and services but also its respective conditions of distribution or accessibility for the population.

MEASUREMENT OF SOCIAL WELLBEING: STATE AND LEVEL

When human necessities are studied, it is frequent to have the view distorted by economicism and confuse tools with targets. Use of variables such as happiness, satisfaction, realization, etc., results at less problematic in practice, because of its subjectivity and difficulty of measurement. Thus, it is preferable to use direct indicators over which there exist space-time statistical series. These indicators permit us to calculate a function that we have called Level of Wellbeing.

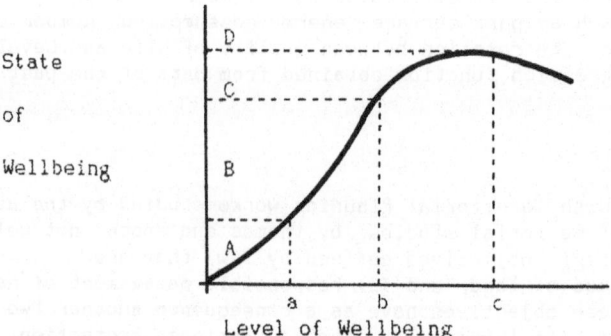

Fig. 1. Relationship between State of Wellbeing and Level of Wellbeing.

In order to obtain dynamically the State of Wellbeing from the Level of Wellbeing, we intend to obtain by regression adjustment over data of the past a function that is supposed to be as indicated in Figure 1. On this curve, we will have to determine several significative points in order to make classes of territories taking into account its actual Level of Wellbeing. For example, from point O to A we have a society that may be qualified as "poor"; from point A to B the society may be said "deficient"; from point B to C the qualification could be "sufficient"; and from C to D "spendthrift". Low increases in "level" in sections A and B produce high increases in "state". This increase is moderate in section C and insignificant, null and possibly negative in section D, because of degradation of resources.

The interest of this classification of regions or societies consists on demonstrate that not all level increase produce the same state increase and that the difference depends on the level of departure, or initial data.

OPERATING PROCESS

The team that prepares the Urbanistic Project proposes itself the following tasks and order of performance.

1. To observe the variation of the different state indicators for every society previously defined in the Valencian Community (cities, counties, provinces, etc.).

2. To set up the relationships existing between these indicators using Multivariate Analysis. Thus, synthesis variables expressed as linear functions of the indicators will be created.

3. To obtain non linear regression functions (Caselles and Uso, 1990) that explain the behaviour of some variables such as natality, mortality, emigration and immigration, as depending of the internal and external State of Wellbeing in the society considered.

4. To study the spatial distribution of selected variables with techniques of Dynamical Analysis (neighbourhood analysis, spatial concentration/dispersion, Gini index, functional specialization, etc.)

5. To build and validate a mathematical simulation model based on differential equations that permit us to obtain the evolution of the state indicators of Wellbeing corresponding to several possible interventive policies in the system..

6. To prepare sets of interventive actions, compatible with existing restrictions and limitations (financial resources, natural resources, minimal risks, etc.), transform them into values for the input variables of the model, and simulate them.

7. The set of tested interventive actions that leads the system to a higher value of the State of Wellbeing in the period considered, will be considered as the recommendable one.

POPULATIONAL MODEL PROPOSED FOR THE VALENCIAN COMMUNITY

The model proposed here is a submodel of the global model to be considered for the Valencian Community. It will be the first submodel to develop and perform, because of the priorly exposed importance of population and its wellbeing with relation to territorial planning.

We want to remark now the following characteristics of this submodel.
  - It is a deductive model. Deductive models are the only possible when a great number of variables interact.
  - Natality, mortality, emigration and immigration depend on the State of Wellbeing calculated for the area considered, and the State of Wellbeing estimated for the interacting external world.
  - These same variables and others are calculated using non linear multiple regression functions.
  - The variable State of Wellbeing is calculated dynamically using a regression function from the variable Level of Wellbeing.
  - The variable Level of Wellbeing is calculated from level indicators by means of a tree of linear functions determined from data of the past by Multivariate Analysis.
  - The variable State of Wellbeing may be calculated in a given moment by means of a tree of linear functions obtained from real data by Multivariate Analysis. Figure 2 shows an hypothetical way to calculate of this manner the State of Wellbeing and its interaction with population movements.
  - The time unit will be the year. The model pretends to be able to simulate the evolution of population at medium and long term.
  - The works are now in the conceptualization phase, creation of databases, and obtainment of concrete equations. The areas for parametrization and validation of the model have been defined already.

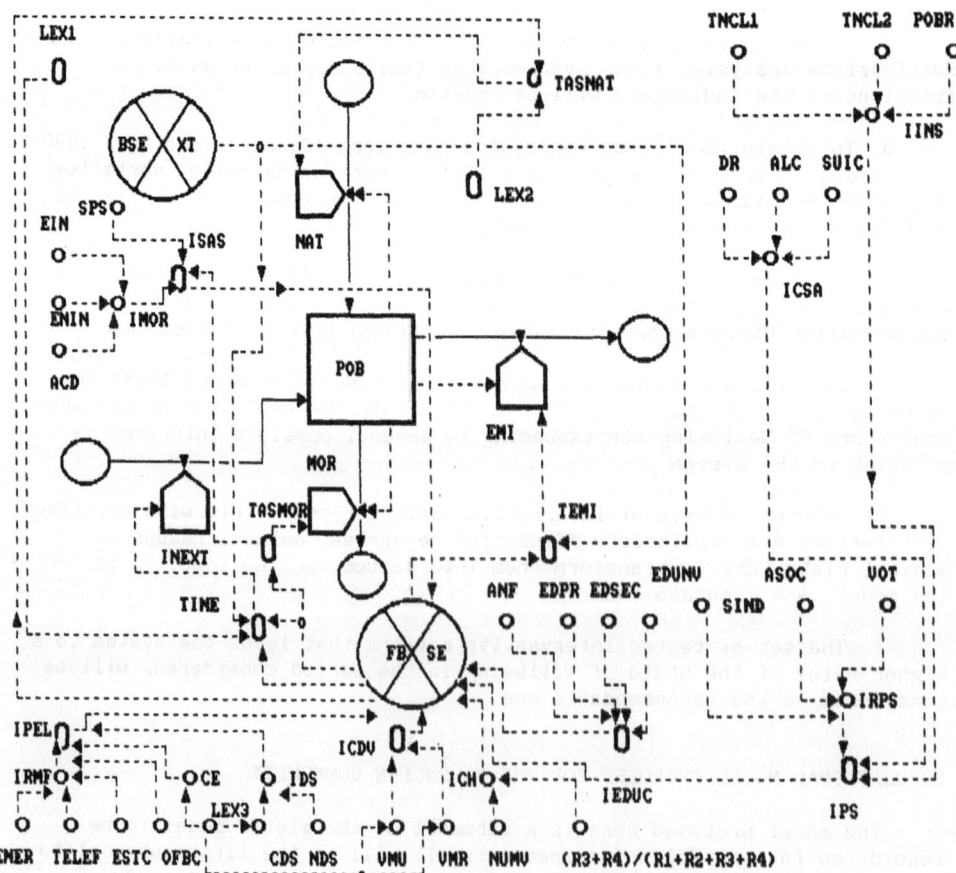

Fig. 2. Flow diagram of the population and quality of life submodel.

## Methodology

The model building methodology followed by the team may be found in Caselles (1988) and Caselles et al. (1990). Synthetically, the principal steps are: 1. formulation of objectives; 2. identification of the objects or elements to consider with relation to the objectives; 3. giving a form (a name and a unit of measure) to the elements identified; 4. identification of the causal dependencies among the elements; 5. giving a form (equation, table, rule, etc.) at the causal relationships identified; 6. building an ad hoc computer program (we use for this target the programme generator SIGEM (Caselles, 1988)); 7. Verification of results in the parametrization area; 8. Validation by contrast of the predictions given by the model against real data in another area; and 9. use of the model to attain the objectives proposed. It is normal to go back and reconsider prior steps a lot of times from any step.

In order to find synthesis variables with the so called state indicators and level indicators, we will group similar variables and apply the Principal Components technique.

In order to find an accurate non linear multiple regression function, we use the programme REGRESSUS (Caselles and Uso, 1990).

## Identification of the elements

According to the objectives already exposed, the following elements where identified by the team using Brainstorming techniques.
ACD: accidents index, number of accidents / population.
ALC: alcoholism index, number of alcoholic people / population.
ANF: analphabetism index, number of analphabets / population.
ASOC: associationism index, number of associates / population.
BSEXT: Social Wellbeing in the near exterior universe.
CDS: number of workers without an employment / number of workers.
CE: number of labour contracts for less than 3 years / number of contracts.
DR: number of drugs-dependants / population.
EDPR: number of basic students and graduates / population.
ADSEC: number of secondary students and graduates / population.
EDNUV: number of university students and graduates / population.
EIN: number of infectious diseases / number of cases.
EMI: emigration rate.
ENER: domestic electricity consumption (Mw.h).
ENIN: number of non infectious diseases / number of cases.
ENESP: number of students in special studies / total students.
ESTC: number of commercial enterprises / population.
FBSE: State of Wellbeing.
ICDV: life conditions synthetic indicator.
ICH: persons/space synthetic indicator.
ICSA: atypical social conducts synthetic indicator.
IDE: educational synthetic indicator.
IDS: social integration synthetic indicator.
IMOR: morbility synthetic indicator.
INCL1: number of catalan speaking people / population.
INCL2: number of castillan speaking people /population.
INEXT: immigration rate.
IFEL: economic-laboural position synthetic indicator.
IPRS: social membership synthetic indicator.
IPS: social integration synthetic indicator.
IRMF: average familiar rent.
ISAS: health and social assistance synthetic indicator.
LEX1: restrictions to emigration multiplier.
LEX2: natality prizes multiplier.

LEX3: social intervention multiplier.
MOR: rate of mortality.
NAT: rate of natality.
NDS: number of non employed people / active population.
NUMV: number of persons by housing.
OFBC: number of bank offices / population.
POB: population.
POBEXT: number of possible immigrants estimated.
POBR: number of cases of social assistance / population.
RI: number of high density lodgings / total lodgings
SIND: number of affiliates to syndicates / population.
SPS: number of psychic diseases / total diseases.
SUIC: number of suicides / population.
TASMOR: coefficient of mortality.
TASNAT: coefficient of natality.
TELEF: number of telephone lines / population.
TEMI: coefficient of emigration.
TINE: coefficient of immigration.
VMR: land taxes / number of contributors.
VV: urban taxes / number of contributors.
VOT: number of voters in elections / total possible voters.

## Causal relationships between the elements

The result of the process of relationships identification and giving them a form is the diagram of Figure 2 and the following list of equations.

1. $POB(t+\Delta t) = POB(t)+\Delta t * [NAT(t)+INEXT(t)-MOR(t)-EMI(t)]$
2. $NAT = POB * TASNAT$      3. $INEXT = POBEXT * TINE$
4. $MOR = POB * TASMOR$     5. $EMI = POB * TEMI$
6. $TASNAT = LEX2 * f_1(FBSE)$     7. $TINE = f_2(FBSE, BSEXT) * LEX1$
8. $TASMOR = f_3(FBSE)$     9. $IPEL = \alpha_1 * IRMF + \alpha_2 * IDS + \alpha_3 * CE * LEX3$
10. $FBSE = \beta_1 * IPEL + \beta_2 * ICDV + \beta_3 * ISAS + \beta_4 * IEDUC + \beta_5 * IPS$
11. $ICDV = \gamma_1 * ICH + \gamma_2 * VMU * LEX3 + \gamma_3 * VMR * LEX3$
12. $ISAS = \delta_1 * TASMOR + \delta_2 * SPS + \delta_3 * IMOR$     13. $IPS = \eta_1 * IPRS + \eta_2 * IINS + \eta_3 * ICSA$
14. $IEDUC = \lambda_1 * ANF + \lambda_2 * EDPR + \lambda_3 * EDSEC + \lambda4 * EDUNV$
15. $IRMF = .0014 * ESTC * ENER + .60 * TELEF + 84.69 * OFBC + .61 * EXP(.1 * ESTC) + 226.87$
16. $IDS = \mu_1 * NDS + \mu_2 * CDS * LEX3$     17. $ICN = \varepsilon_1 * RI + \varepsilon_2 * NUMV$
18. $IMOR = \nu_1 * EIN + \nu_2 * ENIN + \nu_3 * ACD$     19. $ICSA = \tau_1 * DR + \tau_2 * ALC + \tau_3 * SUIC$
20. $IINS = \sigma_1 * INCL1 + \sigma_2 * INCL2 + \sigma_3 * POBR$     21. $IPRS = \rho_1 * SIN + \rho_2 * ASOC + \rho_3 * VOT$

It can be observed that most parameters and regression functions are not yet determined. That is because of databases are still in building process.

REFERENCES

Caselles, A., 1988, SIGEM: A Realistic Models Generator Expert System, in: "Cybernetics and Systems'88", R. Trappl, ed., Kluwer A.P., Dordrecht.
Caselles, A., and Uso, J.L. 1990, REGRESSUS: A Finder of Functional Relationships in a System, in: "Advances in Support Systems Research", G.E. Lasker and R.R. Hough ed., International Institute for Advanced Studies in Systems Research and Cybernetics, Windsor (Canada).
Caselles, A., Nebot, V., and Uso, J.L. 1990, Modelling Mathematically Ecological Problems: Flowers Polinization and Fruit Dispersion, in: "Proceedings of the 12th International Congress on Cybernetics", International Association for Cybernetics. Namur (Belgium).

# ANALYSIS AND DESIGN OF SOCIO–ECONOMIC SYSTEMS

J. Korn, F. Huss, J.D. Cumbers

Middlesex Polytechnic
Bounds Green Road
London N11 2NQ, U.K.

## INTRODUCTION

The operation of systems can give rise to financial, environmental and social risks and to problems of a human and technical nature. Risks may be reduced and problems alleviated if the consequences of human and technical activities can be foreseen. Apart from inspired guesses, prediction requires a vehicle that can be manipulated and which enables the predictor to draw inferences. Such a vehicle, whether it be the entrails of an animal, or the insight into human nature possessed by the oracle of Delphi, is called here an inference machine. In the present context, general systems theories and other approaches using mathematical methods as inference machines, were developed (Bertalanffy, 1956; Klir, 1969; Chestnut, 1966; Churchman, et al., 1957). A feature of these methods is that they use 'variables' as the central theoretical construct which restricts their application to quantifiable aspects of a part of reality. Problem–solving, or diagnostic–oriented hard and soft system methodologies have also evolved (Jenkins, 1969; Checkland, 1981; anon; 1982). Such methodologies do not lead to the construction of an inference machine, they are directed towards alleviating problems.

Language, although commonly used in a predictive sense, has many shortcomings when employed as a formal model (Korn, 1989; Korn et al., 1990). Linguistic modelling, however, involves aspects of linguistic analysis (Halliday, 1982; Filmore, 1968) for unravelling linguistic complexities and then turning them into 'basic constituents' (Korn et al., 1989, 1990) which are the elements used for the construction of semantic diagrams. Such diagrams, through the application of propositional logic, can be translated into a Prolog program thus completing the construction of an inference machine. Aspects of linguistic modelling can also be applied to the design of systems and products (Korn, 1988). Design may be regarded as a mental process that deals with the construction of schemes for systems and plans for products. Thus, it refers to events which are to take place in the future.

The objective of this paper is to outline an application of linguistic modelling and design to a situation with intense human activity. The paper assumes a familiarity with the technique of linguistic modelling and, because of shortage of space, is only concerned with a demonstration of semantic diagrams and some use of logic.

## SOCIO–ECONOMIC SYSTEMS

Whether a person acts on his own or as a part of a group, all human activities

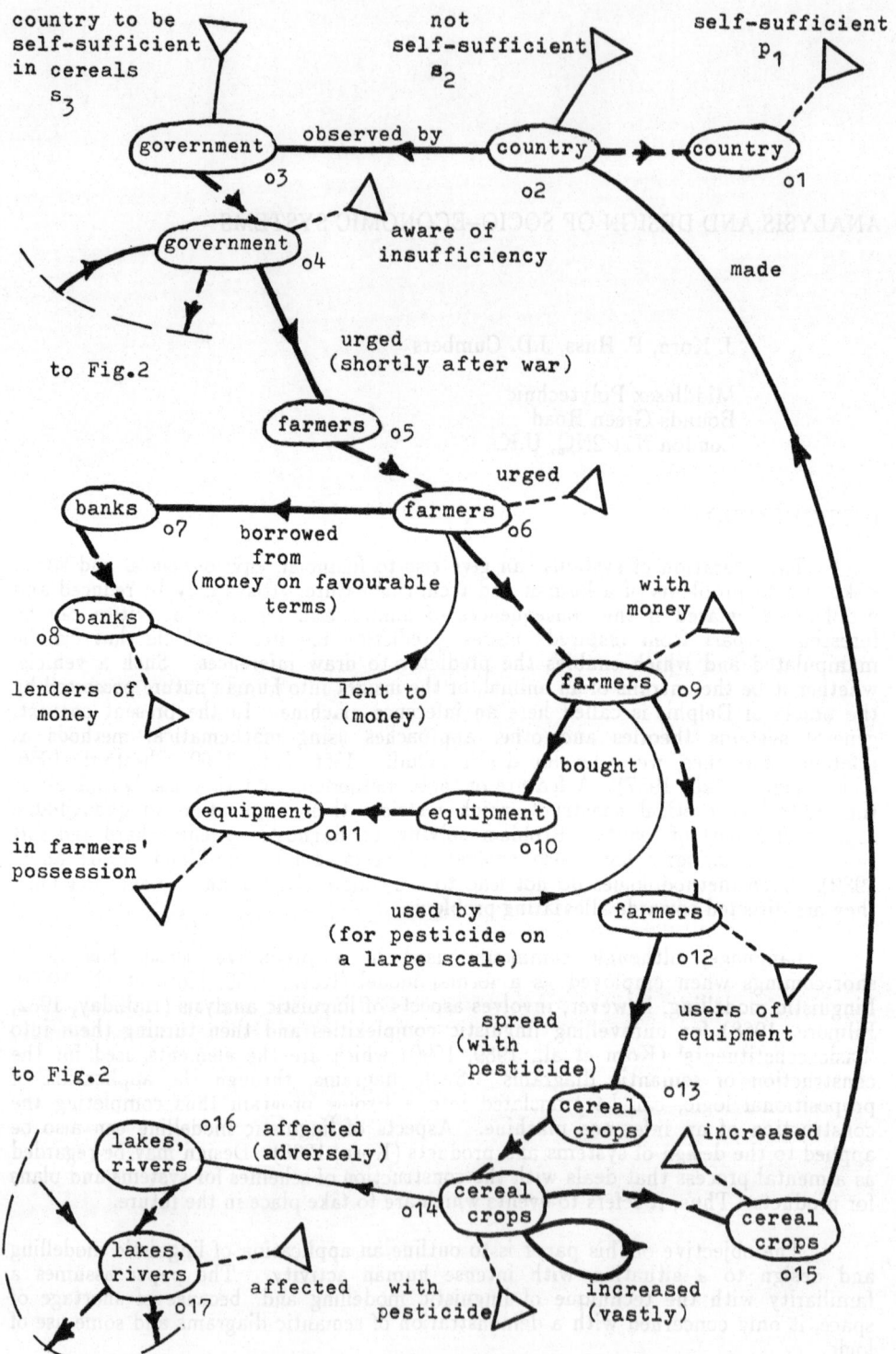

Fig.1    Semantic diagram of a socio-economic system

are directed towards the attainment of some new state of affairs which is embodied in properties (Korn, 1988). The achievement of a state requires a cause which is the immediate creator of a change of state and is generated by the concerted action of a group of entities referred to as a 'system' (Korn, 1989). The goal of the system is the envisaged state.

When people are gathered into interacting groups performing specific functions in the course of trying to attain a goal, we speak of a social system. However, since people need products for their existence created by manufacturing systems, money, or other means, is usually circulated. Thus, it is more appropriate to use the term 'socio–economic' system. When a number of socio–economic systems operate side by side without a common goal being set, changes of state, or events, appear to take place as a result of chance.

The intention now is to show how linguistic modelling can be used to construct an inference machine of a socio–economic system. Consider the following report: 'Shortly after the war, the government urged the farmers to increase their cereal crops in order to make the country self–sufficient. Farmers borrowed money on favourable terms from the banks, then bought equipment needed for large scale use of pesticide which they spread on the cereal crop leading to its vast increase. It subsequently turned out that pesticides adversely affected the country's lakes and rivers'. Based on this narrative, a semantic diagram is constructed as shown in Fig. 1. there are two outcomes: 'the country is self–sufficient' and 'lakes and rivers are affected'. Using the notation: o – objects (numbered), s – inherent properties (triangles connected to objects by a solid line), r – interactions (solid lines connecting two objects) with two–place verbs, q – interactions (solid line forming a loop over an object) with one–place verbs, the following causal chains are identified:

1) o1,o2,o15,o14,o13,o12,o9,o6,o5,o4,o3  2) o8,o7,o6
3) o11,o10,o9  4) o17,o16,14

For each causal chain a set of relations of implications can be written, for example, for 1) with dot between the terms designating an 'and' function:

$$o_1^2 \supset p_1 \qquad\qquad p_9 \cdot p_{11} \cdot r_9^{11}(_1) \supset o_{12}^9$$

$$s_2 \cdot p_{15} \supset o_1^2 \qquad\qquad o_9^6 \supset p_9$$

$$o_{15}^{14} \supset p_{15} \qquad\qquad p_6 \cdot p_8 \cdot r_6^8 \supset o_9^6$$

$$p_{14} \cdot q_{14}^{14}(_1) \supset o_{15}^{14} \qquad\qquad o_6^5 \supset p_6 \qquad\qquad 1.$$

$$o_{14}^{13} \supset p_{14} \qquad\qquad p_4 \cdot r_5^4(_1) \supset o_6^5$$

$$p_{12} \cdot r_{13}^{12}(_1) \supset o_{14}^{13} \qquad\qquad o_4^3 \supset p_4$$

$$o_{12}^9 \supset p_{12} \qquad\qquad s_2 \cdot s_3 \cdot r_3^2 \supset o_4^3$$

in which, for instance, the sixth term reads: 'if the farmers are users of equipment and they spread the cereal crops with pesticide then the cereal crops become covered with pesticide'. Similarly for the other terms. In eq. 1 the numbers in brackets designate the adverbs that qualify interactions. Eq. 1 leads to:

$$p_8 \cdot p_{11} \cdot r_2^{15} \cdot q_{14}^{14}(_1) \cdot r_{13}^{12}(_1) \cdot r_9^{11}(_1) \cdot r_6^8 \cdot r_5^4(_1) \cdot s_2 \cdot s_3 \cdot r_3^2 \supset p_1 \qquad\qquad 2.$$

In eq. 2 the terms $p_8, p_{11}$ are due to interaction with other causal chains. Expressions similar to eq. 1 can be written for all causal chains leading, after the appropriate

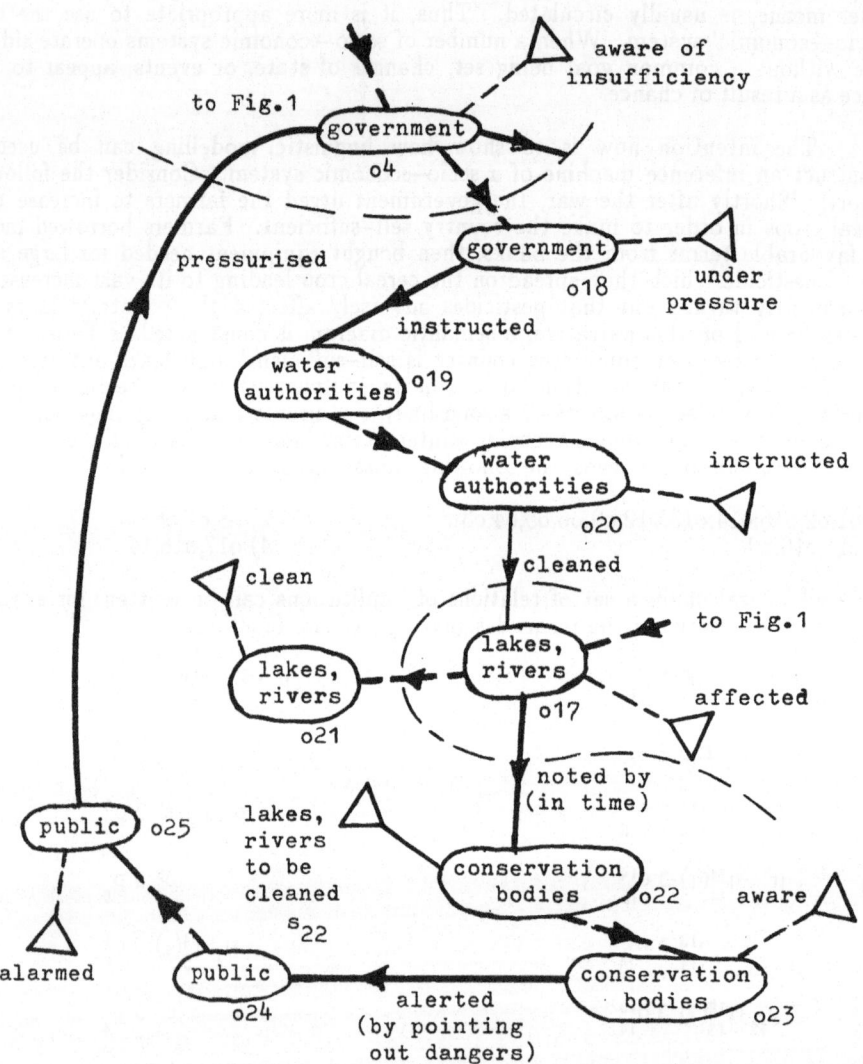

Fig.2   Semantic diagram of a socio-economic control system

manipulations, to

$$s_2 \cdot s_3 \ r^2_3 \cdot r^4_5(_1) \cdot r^6_7 \cdot r^8_6 \cdot r^9_{10} \cdot r^{11}_9(_1) \cdot r^{12}_{13} \cdot (_1) \cdot q^{14}_{14}(_1) \cdot r^{15}_2 \supset p_1 \qquad 3.$$

$$s_2 \cdot s_3 \ r^2_3 \cdot r^4_5(_1) \cdot r^6_7 \cdot r^8_6 \cdot r^9_{10} \cdot r^{11}_9(_1) \cdot r^{12}_{13} \cdot (_1) \cdot r^{14}_{16}(_1) \supset p_{17} \qquad 4.$$

Eq. 3,4 show the sequence of events which must all take place together with the existence of inherent properties, for the outcomes,'the country is self–sufficient' and 'lakes and rivers are affected', to be realised.

## SOCIO–ECONOMIC CONTROL SYSTEMS

Control systems operate so as to bring about a change of state in accordance with a goal; they have components connected in a specific topology and performing the function of: conversion (of energy or signal or influence), amplification (or information–energy or information–influence interface) and comparison of signal or information. (Korn, 1987,1989).

On the basis of the notion of 'change of state in accordance with a goal', these concepts are applicable to socio–economic systems although they may require a wider interpretation in terms of observable properties. Thus, in Fig. 1 it is the country which undergoes a change of state, cereal crops play the part of conversion, the farmers act as amplifier and the government has the multiple role of: generator of the goal, converter of information regarding the state of the country, the comparator of this information with that contained in the goal leading to a difference for prompting the farmers into action. The action of all the components involved in the operation of this socio–economic control system towards the successful production of an outcome, is expressed by eq. 3. However, there is an additional outcome, as given by eq. 4, 'lakes and rivers are adversely affected'. This seems to be considered undesirable and an additional control system is intentionally introduced to remedy: 'In time conservation bodies who wanted clean lakes and rivers, noted the adverse effects of pesticides and alerted the public. Under public pressure, the government instructed the water authorities to clean up the lakes and rivers'. The semantic diagram representing this description is shown in Fig. 2.

When Fig. 2 is superimposed on Fig. 1, we have the diagram of the complete situation. Analysing the system in Fig. 2 in the same way as for Fig. 1, leads to an expression which gives the conditions for the existence of outcome 'clean lakes and rivers', $p_{21}$,

$$s_2 \cdot s_3 \cdot r^2_3 \cdot r^4_5(_1) \cdot r^6_7 \cdot r^8_6 \cdot r^9_{10} \cdot r^{11}_{13}(_1) \cdot r^{12}_{16}(_1) . r^4_{22}(_1) \cdot s_{22} \cdot r^7_{24}(_1) \cdot r^{23}_4(_1) \cdot r^{25}_{19} \cdot r^{18}_{17} \cdot r^{20}_{21} \supset p_{21} \qquad 5.$$

In the present discussion only the functional aspect of system modelling has been considered; political, financial and other interferences have been ignored. The question of precise recognition of a particular functional component in terms of some properties, requires more investigation.

## AN APPLICATION OF SYSTEMS DESIGN

It has been possible to apply aspects of linguistic modelling to design of systems and products (Korn, 1988). An object may be described in terms of a set of stative statements each containing a property. A change of state takes place when any one of the properties is altered in some specified manner. This property is called dynamic, the rest is termed quiescent. In many cases empirical relations can be established between the dynamic and some of the quiescent properties. In such a case, the change of dynamic property prompts a change in the others. Such a change is to be considered and, in many cases, is to be avoided, or minimised so as to preserve the integrity of the object carrying the change.

A central point in system design is the specification of a postulated change of state, of users, suppliers, absorbers and structural effects which can all be known and represent the present situation. Systems design is then concerned with specifying the components and their topology acceptable to the situation, it is, thus, concerned with 'completing an incomplete situation' at some future time. However, in order to avoid changes other than those due to a dynamic property, additional considerations must be given to:

adjectives which describe properties assigned to components,
adverbs qualifying verbs which describe interactions,
additional systems which may be required to ensure that quiescent properties remain quiescent.

In the present case, further to Fig. 1, the carrier of change of state is the object 'country' which under the action of the socio–economic system, becomes 'self–sufficient' in cereal crops. Self–sufficient means that the country can satisfy all its need without imports, and to achieve the change, cereal crops are to be increased.

Such a projected change prompts changes of properties of constituents of the country. For example, the change will affect importers and transport, in particular shipping. Also diseases, insects and other agents which are harmful to cereal crops, and at present tolerated, are to be reduced so as to minimise losses. In order to achieve such reduction, pesticides are then spread over the crops. Pesticides, in liquid or powder form, eventually penetrate the soil and can infiltrate the lakes and rivers under the action of gravity and moisture in the ground.

This brief discussion has made some of the properties of the object 'country' explicit and has established a relation between the dynamic property and others. Having now explored at least some of the consequences of the postulated change of state, it may be possible to take action BEFORE the proposed change will be put into operation.

Further to Figs. 1 and 2, there are three courses of action:

1.      Adjectives describing properties of objects that can be used for the protection of lakes and rivers can be introduced, or invented.

2.      The adverbs of interactions between objects 11 − 9 and 12 − 13 can be specified to limit the supply of pesticide.

3.      The additional system for cleaning can be introduced from the start of, or before, the application of pesticide, rather than as the result of public pressure.

REFERENCES

Anon., 1982, "Systems Behaviour", The Open University Press, London.
Bertalanffy von, L., 1956, General systems theory, General Systems, v1.
Checkland, P. B., 1981, "Systems Thinking, Systems Practice", Wiley, Chichester.
Chestnut, H., 1966, "Systems Engineering Tools", Wiley, Chichester.
Churchman, C. W., Ackoff, R. L., and Arnoff, E. L., 1957, "Introduction to operation research", Wiley, Chichester.
Filmore, C., 1968, The case for case, in: "Universals in Linguistic Theory", E. Bach and R. T. Harms, ed., University of Texas, Holt, Reinhart and Winston, New York : 1.
Halliday, M. A. K., Fawcett, P., and Robins, R.,1982, "New developments in Systemic Linguistics", Batsford, London.
Jenkins, G. M., 1969, The systems approach, J. Sys. Eng., 1 : 1.
Klir, G. J., 1969, "An approach to General Systems Theory", Van Nostrand Reinhold, New York.
Korn, J., 1987, Modelling of devices as generalised system components, J. Franklin Inst., v324.

Korn, J., 1988, An alternative approach to design, ICED 88 Conference, Budapest, 23–25 August.

Korn, J., 1989, Systems and design as the basis of engineering knowledge, IEE Proceedings, v136, pt A, no2.

Korn, J., Huss, F., and Cumbers, J. D., 1989, Linguistic modelling of situations, in: "Systems Prospects", R. L. Flood, M. C. Jackson and P. Keys, eds., Plenum Press, New York.

Korn, J., Huss, F., and Cumbers, J. D., 1990, Development of a systems theory using natural language, Systems, Modelling, Control, 6, Conference, Technical University of Lodz, Zakopane, Poland, October 8–12.

# ARE ALL FAILURES SYSTEMS FAILURES?

Geoff Peters and Joyce Fortune

Centre for Technology Strategy, The Open University, Walton Hall
Milton Keynes MK7 6AA, UK

## Abstract

On a scale of "systemsness", approaches to the understanding of, and dealing with failures and disasters range from non-systemic techniques and methods to full blown systems methodologies. This paper outlines the area of failures and disasters, and various approaches and classifications that have been adopted. The authors suggest a further approach to classification which is concerned with the systemic complexity of failures and which can therefore be useful in the choice of approach.

## Background

The development of systems ideas in relation to the investigation of failures of various types has been traced back almost 20 years ( Peters 1990 ). Most recently it has been associated in some quarters with a failures specific system based method which in part derives from Checkland's "Formal system" model (1972) and hence Jenkins (1969) and Churchman (1971). A fuller explanation of this aspect of the Failures Method is given in Fortune (1991a).

Examples of the application of this failures specific method range across the relatively straightforward post hoc analysis of a particular accident, Powell (1987), comparative studies of sets of large scale disasters, Brearley (1990), to the investigation of "softer" situations in which the evidence and the extent of any failure are themselves matters of disagreement (Fortune 1991b).

The Failures Method had been initially formulated in such a way that if necessary it could subsequently lead to a full scale systemic method for action and change. In particular it was envisaged that following a thorough understanding gained from the application of this Failures Method, investigators would be able to easily select either a Soft Systems Method, Checkland (op cit) or a Hard Systems Method, Hughes(1984).

There are two classes of circumstances, however, in which the choice of an appropriate approach requires further elaboration. Firstly the individual or group conducting the investigation needs to decide which systems method to employ at the very beginning of the investigation. The distinguishing features of soft and hard systems methods, and the situations in which they are appropriately applied are by now familiar territory. See for example papers by Jackson(1984), Keys (1988).

However even when the more recent Failures Method has been included in such comparisons the guide-lines for practitioners have been either simplistic ["if in doubt use failures first" Waring(1989)] or overly ambiguous (Peters 1984).

The second decision which needs to be addressed at an early stage is whether it is necessary or appropriate to embark upon a "systems" method at all. In the area of safety engineering for example there are well established methods such as risk assessment. There is a temptation in this as in other aspects of applied systems work to ignore such achievements in other fields, and to view all situations as ones which demand a systems approach. A viewpoint which is close to that described by Buck (1956) " One is unable to think of anything, or any combination of things, which could not be regarded as a system. And of course a concept that is applied to everything is logically empty."

Consideration of both the question of methodology choice and selection of a systems method rather than a non-systemic alternative have led the authors to similar conclusions.

## Failures Hazards and Disasters

Much of the work which has led to the production of this paper has been concerned with the understanding of large scale disasters, accidents and catastrophes. Various definitions exist of these terms, some of which attempt to be objective whilst others acknowledge the judgmental nature of the attribution of such terms (see Horlick-Jones 1990). Even a relatively all embracing definition of a disaster as, "an event, or series of events, which seriously disrupts normal activities" (Cisin 1962), implies a personal view about that which is serious, and a debate about what activities are normal. Furthermore producing a simple method for distinguishing types of failures is not a straightforward matter. Single dimensions such as energy dissipated, or more commonly, financial cost and loss of life are available. The Bradford Disaster scale for example uses a logarithmic scale of fatalities in a similar manner to the geophysical Richter scale, (Keller 1989).

In every day language disasters are frequently divided according to whether the responsibility for their occurrence rests with God or Man. Acts of God or Natural disasters like earthquakes and blizzards can claim many thousands of lives, but are infrequent whereas "Man" made disasters like rail accidents or explosions more normally claim tens of lives, but are more common. Research in Canada (Wapner 1976) has indicated that people deploy a more subtle grouping of hazards into:

> Natural (e.g. Snowstorms, tornados)
>
> Quasi Natural (e.g. Air and water pollution)
>
> Social (e.g. Riots and epidemics)
>
> and Man-made (e.g. Industrial accidents).

Hazards and disasters are not synonymous. Hazardous events only become disasters if they are perceived by humans. Furthermore the impact of even natural disasters on loss of life and cost is partially dependent upon human action, such as settlement patterns, disaster preparedness and mitigation. Human intervention is at least a partial explanation of why the 6.9 Richter scale earthquake in the San Francisco Bay area in October 1989 resulted in one thousandths of the fatalities experienced as a result of a similar event in December 1988 in Armenia.

Figure 1 expresses this relationship between human perception and Natural and quasi natural hazards in terms of a systems model.

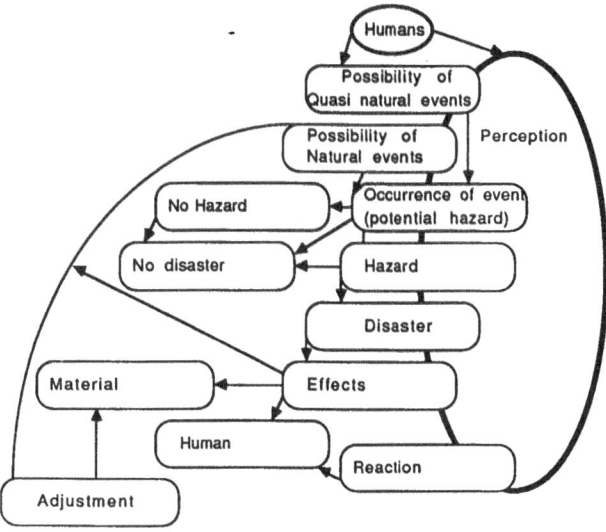

Fig 1  A systems model to illustrate the effect of environmental hazards and the possible adjustments to a disaster after Whittow (1980)

The failures with which the authors have been concerned are those which have been perceived as failures of human activity, and therefore a natural disaster might be considered as a failure if its impact could have been reduced, or if the emergency planning proved to be ineffective. One definition of disaster which goes some way towards acknowledging these facets states, "Disasters are the interface between an extreme physical event and a vulnerable human population." (Susman 1983)

## System Accidents and Systems Failures

Within the realm of human activity systems, and in particular high risk technologies, considerable attention has been focused on the design or engineering of systems which contain sophisticated safety features. In this regard much of the work on engineering reliability has understandably been developed in the context of military and nuclear technology. However Perrow(1984) has argued that these high risk technologies often display two features: interactive complexity and tight coupling.

Interactive complexity is as the name suggests an aspect of the system which means that failures and events are not independent, but related, and that the more complex the system the more chance there is that couplings of events will be unforeseeable, and unexpected. Attempts to subsequently engineer warnings or the like, in turn only increase the complexity still further thereby introducing new potential couplings of events.

Tight coupling refers to the speed of interaction, and the extent to which parts of the system can be isolated. Taken together, interactive complexity and tight coupling, will in Perrow's view inevitable lead to what he terms System accidents, or Normal accidents.

An earlier and more all embracing approach to the failures arising from a system refers to Systems Failures (Peters 1976).

Systems Failures rely on:

> (1) Human perception and identification as a failure, thereby acknowledging that one person's failure is another persons success.

and either

(2) Failure to meet system objectives attributed by those involved, such as designers and users.

or

(3) The production of outputs which are considered to be undesirable by those involved.

It has been this rather wide definition of significant systems failures, where significance is entirely in the eye of the beholder that has been the basis of the work referred to here.

## Systemic appraisal

### Causes and consequences

When faced with the analysis of failures there are a range of possible approaches available. Firstly there is the systems sphere, where in addition to the general systems methods for action and change like Checkland (op cit) there are failures specific systems methods. Van Gigch(1988) has proposed a systems failures methodology based upon the classification of failure modes into malfunctions of structure, behavior and technology. The authors and their colleagues have developed other systems failures methods related more closely to Checkland's (Peters 1990). In some circumstances, the use of individual systems techniques or concepts have apparently been sufficient e.g. Kirschman-Anderson(1980) sought to explain the Jonestown massacre in which 913 followers of the Rev Jim Jones religious cult the "People's Temple" died apparently after drinking a mixture of cyanide and soft drink. She found that systems concepts like open and closed systems, entropy and feedback provided an appropriate framework for analysis and explanation. Outside of the systems arena there are a range of approaches to accident investigation or failure analysis, which can be applied.

The authors have found that an initial classification of failures according to the extent to which their causes could be described as systemic was insufficient as a basis for deciding between a systems based approach and other schemes. However a two dimensional categorisation which considers the systemic aspects of both causes and consequences has proved illuminating.(See Figure 2 below)

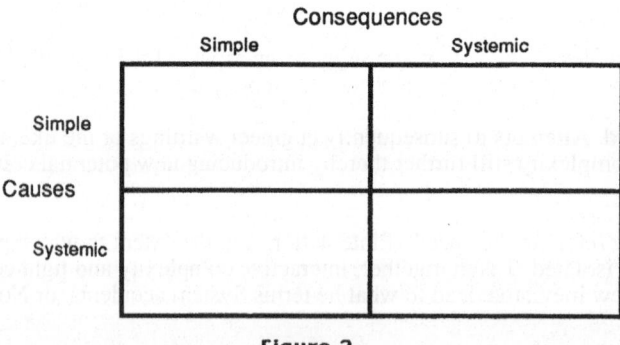

**Figure 2**

For example an individual road accident will normally have a relatively uncomplicated cause, and only limited consequences. However if it is on a particular road and/or at certain times the consequences can be more far reaching. Similarly a rail accident will normally have consequences which result in some disruption to the rail system, but may have a relatively simple cause. A series of rail accidents may however suggest a deeper level of investigation of the systemic features underlying the safety provision. Large scale public disasters such as Bhopal, Sveso and Three Mile Island usually have complex causes for the reasons outlined earlier. They certainly have dramatic consequences, on legislation, public pressure and the strategy of companies in related fields. See for example the study of the impact of the methyl isocyanate leak from the Union Carbide of India pesticide production plant in Bhopal on a large US chemical company, Bowman (1988).

## Appraisal methods

In order to ascertain the degree to which there are systemic aspects to the cause and consequences of a particular failure, it is necessary to conduct some initial appraisal of the situation. Rich Pictures (Checkland op cit) although deliberately avoiding a systems based structure never-the-less inevitably display both complexity and interaction. The authors have also found other diagrammatic techniques worthwhile, as well as interaction matrices and nets (Peters 1976), which give an almost quantifiable measure of complexity.

Within the current version of the Failures Method a thorough systems appraisal is achieved by means of a method for describing systems which includes the stages below.

*Stage 1 Awareness* Being aware of some activity; it is described, but not in systems terms *Stage 2 Commitment* Having reasons for staying with the study, e.g. to describe , understand, repair or maybe redesign. answering the question, "Why am I doing this?" *Stage 3 Systems Detection* Testing against a rigorous definition of system to reveal systems attributes *Stage 4 Separation* Separating some systems of interest; putting trial boundaries on them ; giving them titles. *Stage 5 Selection* Selecting one system and clarifying the nature of it. *Stage 6 Description* Describing in systems terms the components, subsystems environments, goals, flows, outputs, inputs, levels, states, connections and so on of the selected system in order to reach an adequate understanding of its processes, structure and behaviour.

## Conclusion and discussion

In simple schematic form, the approach being suggested by the authors is summarised in Fig 3

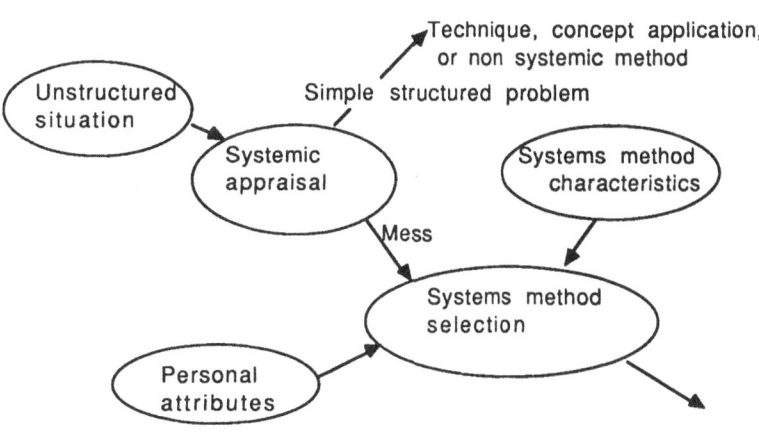

Figure 3

Within the framework of the investigation and understanding of systems failures, then an initial systemic appraisal of the situation ties in with Perrow's observations about the complexity of systems. Within the wider context of more general systems inquiry, there are also links for example with Churchman (op cit) and Peters (1979). The approach described here is however pragmatic, so although in general terms it may fit with earlier work on the design of inquiring systems, it does still leave some difficulties in relation to Checkland's Soft Systems Method. Specifically it is difficult to confidently embark on the use of Soft Systems Method as a result of a thorough appraisal in systems terms, when the first requirement of Soft Systems Method is that the situation is left unstructured. One possibility currently being explored is the deliberate incorporation in the Failures Method of rich pictures, and an early decision point about whether to switch to Soft Systems Method.

## References

Bowman E and Kunreuther H ,1988 Post-Bhopal Behaviour at a Chemical Company, J of Management Studies, 25:4

Brearley S A,1990 High Level Management and Disaster, paper presented to 2nd Disaster prevention and limitation conference. Bradford

Buck, R C,1956, On the logic of general behavior systems theory, in "Minnesota Studies in the Philosophy of Science vol 1," H Flegel and M Scriven (eds) U of Minnesota Press, Minneaplois USA

Checkland P B Towards a system-based methodology for real-world problem solving J Syst Eng ,3:2

Churchman C W, 1971, "The design of inquiring systems" Basic Books, New York

Cisin I H and Clark W B ,1962 The methodological challenge of disaster research in "Man and Society in Disaster," Baker G W and Chapman D W (eds) Basic Books New York

Fortune J, and Peters G ,1991a The Formal System Paradigm for studying failures' Technological and Strategic Management 3:1

Fortune J, Peters G and Rawlinson-Winder L ,1991b Science education in Primary schools, Open systems group report, Open University Milton Keynes

Horlick-Jones, T,1990 "Acts of God? An investigation into disasters", EPI Centre, London

Hughes J, Tait J ,1984 "The hard systems approach" Open University Press, Milton Keynes

Kirschman-Anderson E.,1980 Jamestown, Guyana - A systems autopsy "Proceedings of the 24th Annual meeting of the Society for General Systems Research 1980 SGSR Louisville USA

Jackson M C and Keys P ,1984 Towards a system of systems methodologies J. Operat. Res. Soc 35

Jenkins G M, 1969 The Systems Approach, J Sys.Eng. 1:1

Keller A Z ,1989 The Bradford disaster scale, paper presented to Disaster prevention and limitation conference. Bradford Sept 1989

Keys P ,1988 A methodology for methodology choice Syst. Res 5:1

Perrow C ,1984 "Normal Accidents" Basic Books, New York

Peters G and Fortune J,1990 A systemic method for the analysis of failure, OR90 conference paper

Peters G, 1979 On systems methodology in "Improving the Human Condition: Quality and Stability in Social Systems" Ericson R F, SGSR, Louisville USA

Peters G,1976 "Systems and Failures" Open University Press, Milton Keynes UK

Peters G,1984 "Comparing systems approaches" Open University Press, Milton Keynes UK

Powell D J A, 1987 Study of the Seer Green Railway accident ,1981 using a systems approach J of Systems Analysis 14 .

Susman P, O'Keefe P and Wisner B,1983 Global disaster a radical interpretation, in "Interpretation of Calamity" Hewitt K (ed),Allen and Unwin, Winchester, Mass

Van Gigch J P ,1988 Diagnosis and metamodeling of systems failures Systems Practice 1:1

Wapner S, Cohen, S B and Kaplan B, 1976 "Experiencing the Environment" Plenum Press, New York.

Waring A, 1989 "Systems methods for managers" Blackwell, Oxford

Whittow J ,1980 "Disasters:The Anatomy of environmental Hazards" Penguin, London.

# USE OF THE VSM TO ASSIST IN THE ESTABLISHMENT OF AN ENGINEERING CONSORTIUM

Graeme Britton

School of Mechanical and Production Engineering
Nanyang Technological Institute
Nanyang Avenue
Singapore 2263

## INTRODUCTION

This paper describes how the author assisted a group of managers establish an engineering consortium and how the VSM was used to provide critical insights and conceptual clarity. The paper is written to follow the chronological order of events so that the reader can better appreciate the difficulties faced by the author and the other participants.

## INITIATION OF AUTHOR'S INVOLVEMENT

In May 1988 Sid Goldsmith, the Managing Director and owner of a small engineering company, asked the author to assist him and several other companies establish an engineering consortium. The idea for the scheme was motivated by the difficult economic circumstances the companies faced and the fact that a consortium of small engineering companies was operating with considerable success in Australia.

I agreed to help and sent Sid a note on the different types of cooperative they could form (see Appendix 1). Subsequently I was invited to a meeting of the interested parties to present my ideas.

One morning soon after that meeting Sid rang me and said that he had a draft proposal for the consortium. He asked me to study it and to attend a meeting <u>that afternoon</u> to advise them on the proposal. I agreed and immediately started mapping the proposal onto the VSM using pre-prepared blank diagrams which I have made for this purpose.

## THE FIRST PROPOSAL

The first proposal was to establish the consortium as an incorporated society. The mapping onto the VSM is shown in Figure 1. The mapping was performed directly from the constitution and rules of the proposal. Given the short time most

*Systems Thinking in Europe*, Edited by M.C. Jackson *et al.*
Plenum Press, New York, 1991

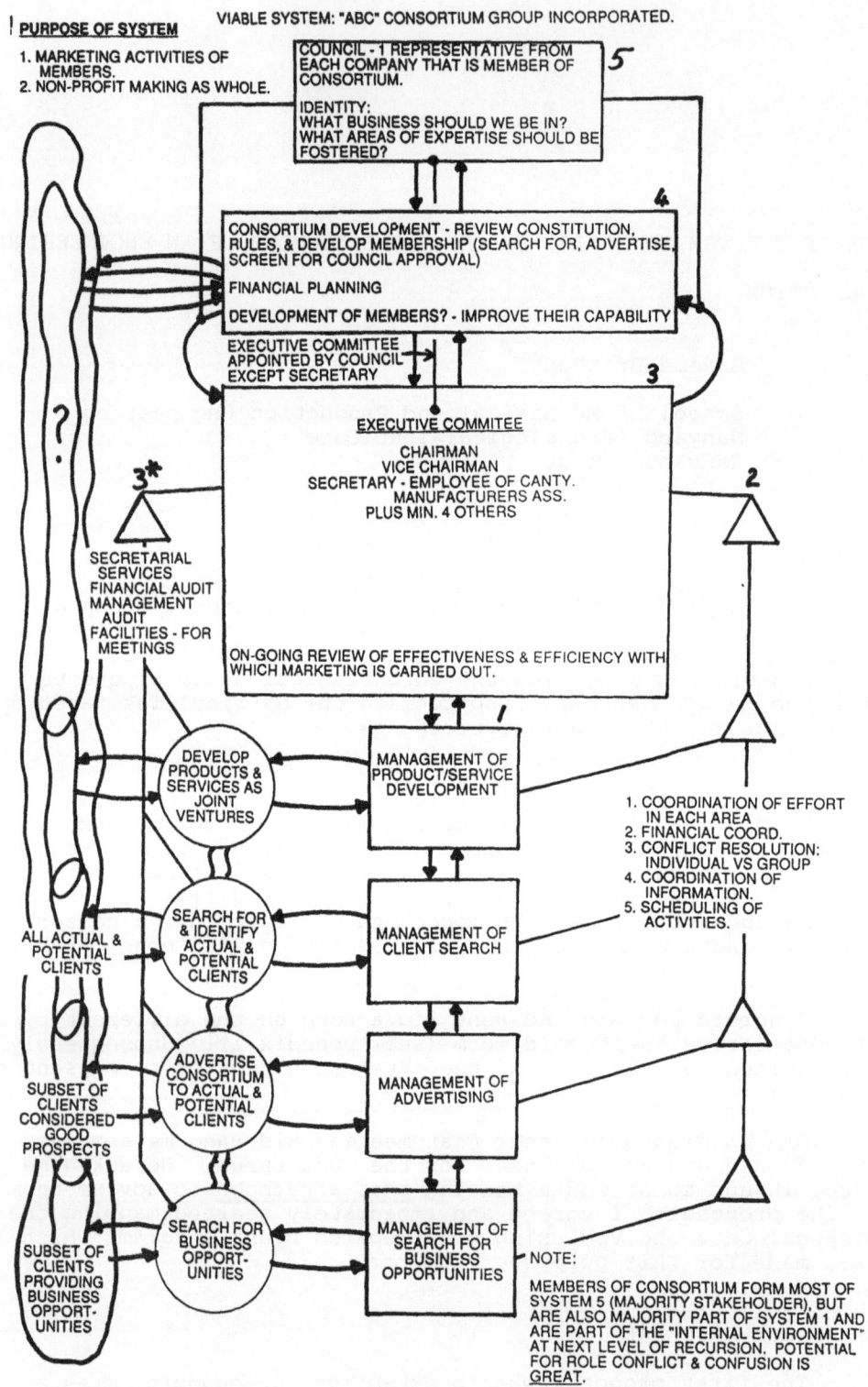

VIABLE SYSTEM: "ABC" CONSORTIUM GROUP INCORPORATED.

**PURPOSE OF SYSTEM**

1. MARKETING ACTIVITIES OF MEMBERS.
2. NON-PROFIT MAKING AS WHOLE.

COUNCIL - 1 REPRESENTATIVE FROM EACH COMPANY THAT IS MEMBER OF CONSORTIUM. **5**

IDENTITY:
WHAT BUSINESS SHOULD WE BE IN?
WHAT AREAS OF EXPERTISE SHOULD BE FOSTERED?

CONSORTIUM DEVELOPMENT - REVIEW CONSTITUTION, RULES, & DEVELOP MEMBERSHIP (SEARCH FOR, ADVERTISE, SCREEN FOR COUNCIL APPROVAL) **4**

FINANCIAL PLANNING

DEVELOPMENT OF MEMBERS? - IMPROVE THEIR CAPABILITY

EXECUTIVE COMMITTEE APPOINTED BY COUNCIL EXCEPT SECRETARY

**3**

EXECUTIVE COMMITEE

CHAIRMAN
VICE CHAIRMAN
SECRETARY - EMPLOYEE OF CANTY.
MANUFACTURERS ASS.
PLUS MIN. 4 OTHERS

**3\***
SECRETARIAL SERVICES
FINANCIAL AUDIT
MANAGEMENT AUDIT
FACILITIES - FOR MEETINGS

**2**

ON-GOING REVIEW OF EFFECTIVENESS & EFFICIENCY WITH WHICH MARKETING IS CARRIED OUT.

**1**

DEVELOP PRODUCTS & SERVICES AS JOINT VENTURES

MANAGEMENT OF PRODUCT/SERVICE DEVELOPMENT

1. COORDINATION OF EFFORT IN EACH AREA
2. FINANCIAL COORD.
3. CONFLICT RESOLUTION: INDIVIDUAL VS GROUP
4. COORDINATION OF INFORMATION.
5. SCHEDULING OF ACTIVITIES.

ALL ACTUAL & POTENTIAL CLIENTS

SEARCH FOR & IDENTIFY ACTUAL & POTENTIAL CLIENTS

MANAGEMENT OF CLIENT SEARCH

SUBSET OF CLIENTS CONSIDERED GOOD PROSPECTS

ADVERTISE CONSORTIUM TO ACTUAL & POTENTIAL CLIENTS

MANAGEMENT OF ADVERTISING

SUBSET OF CLIENTS PROVIDING BUSINESS OPPORT- UNITIES

SEARCH FOR BUSINESS OPPORT- UNITIES

MANAGEMENT OF SEARCH FOR BUSINESS OPPORTUNITIES

NOTE:

MEMBERS OF CONSORTIUM FORM MOST OF SYSTEM 5 (MAJORITY STAKEHOLDER), BUT ARE ALSO MAJORITY PART OF SYSTEM 1 AND ARE PART OF THE "INTERNAL ENVIRONMENT" AT NEXT LEVEL OF RECURSION. POTENTIAL FOR ROLE CONFLICT & CONFUSION IS GREAT.

**FIGURE 1.    REPRODUCTION OF VSM MAPPING**

of my detailed critique was written directly on to the draft constitution and cannot be reproduced here. However there were three major issues which were to significantly affect the final proposal:

(a)  Why establish the consortium as an incorporated society?

(b)  There was a lack of clarity as to the identity of the consortium. The purposes of the consortium were not clearly defined.

(c)  There appeared to be confusion between the roles in the consortium and those in the individual companies (see the note at the bottom of Figure 1).

It is worth noting two points. Firstly, the consortium itself was modelled at the same level of recursion as the individual companies because it was neither a higher nor lower level organisation compared with the participating companies.

Secondly, the major focus of modelling at this and subsequent stages was on Systems 3 & 5 as these can be legally defined under the Incorporated Societies and Companies Acts. In addition the major concerns of the participants were about issues relating to these systems. However emphasis was also placed on System 2 because some coordination issues needed to be formally resolved as well. No attempt was made to formalise the other systems (4 & $3^*$) because I wanted maximum flexibility in the operation of the consortium, and I thought this was best achieved through informal processes in the initial stages of development.

I attended the meeting that afternoon and presented my comments and questions. The reason given for establishing the consortium as an incorporated society was that the participants didn't want to pay a large sum of money for executive staff to manage the consortium. I pointed out that they, the Managing Directors of the companies, were in fact the consortium and that no other people need be involved, except perhaps some secretarial assistance. I also pointed out that there were limitations in incorporation because an incorporated society cannot make a profit and there were no tax advantages. Given these arguments the participants soon accepted a proposal to establish the consortium as a company.

Of greater concern to me at the time were the comments that kept surfacing relating to the mechanics of how the consortium would operate. Specific concerns were:

(a) What were the purposes of the consortium?

(b) How was entry and exit from the consortium to be controlled and how could any one company be prevented from gaining a controlling interest in the consortium.

(c) How was risk to be shared on projects undertaken by the consortium when not all members would be involved in a specific project?

(d) How was work obtained by the consortium to be allocated fairly among the members?

(e) What marketing information should be disseminated to the participating companies given that some of them were competitors?

Most of these concerns arose because of a lack of trust between the interested parties. Hence I considered my prime task at this stage to be the generation of trust among the participants. I spent some time discussing this issue and various means by which protection could be provided legally. I also stated that they had to design the consortium if it was to succeed.

My comments and stance produced some confusion. There was much animated discussion but nothing was resolved. A further two meetings were subsequently held in which these same issues were discussed thoroughly but with nothing being resolved. Towards the end of the second meeting I was sensing frustration among some of the participants who were eager to proceed. But I kept pointing out that the consortium would not work unless they trusted each other and built in protection at the very start. I was worried though that if discussion continued too long then the consortium might never be established.

Another meeting was scheduled with a lawyer present to offer advice on some of the legal points that had been raised. It was an amazing meeting in some ways. The participants very quickly came to agreement on the contentious issues and the lawyer stated that there were no legal obstacles except for one issue which is discussed later. The Chairman was given permission to instruct the lawyers to prepare the final proposal (described in the next section).

THE FINAL PROPOSAL

The major design features of the final proposal were:

(a) The purpose of the consortium is to be a marketing and business cooperative.

(b) Rules specify the conditions under which companies can gain shares in the consortium and when they must relinquish those shares. These rules also contain conditions preventing one company taking control of the consortium.

(c) Conditions were specified defining the financing and operation of the marketing activities of the consortium. Tax relief was to be obtained if possible.

(d) No conditions were placed on joint ventures as it was agreed that the risk for these must be carried by the participating companies and not by the consortium.

These features were met by three legal agreements:

(a) The Articles of Association defining the purpose and rules of the consortium under the Companies Act 1955 (NZ Act of Parliament). The consortium was established as a Limited Liability Company.

(b) A Participation Agreement that overrides the Articles of Association and regulates the conduct of shareholders. It also enables the participating companies to fund promotional activities on an annual basis as opposed to subscriptions for capital. The annual funding is tax deductible.

This Agreement was necessary for two reasons. Firstly and obviously to obtain a tax advantage not conferred by the Companies Act. Secondly, to regulate the conduct of the shareholders as required.

(c) A Service Agreement that is linked to the Participation Agreement. Shareholders can only hold shares if they agree to the Service Agreement. The Agreement specifies the marketing services to be provided by the consortium and how these are to be charged.

## PROBLEMS ASSOCIATED WITH COMMERCE ACT

One of the reasons that the companies wanted a consortium was to prevent cut-throat pricing which no-one would win. Unfortunately this is not allowed under the Commerce Act 1986 (NZ Act of Parliament). The Act expressly forbids companies formally or informally entering into agreements that would substantially lessen competition. This issue was so important that the Participation Agreement contained a clause about this:

""ABC Consortium" shall not be used as a means of reducing the competitiveness of bids by members for the same work or for any other practice giving rise to collusive pricing."

Regrettably the admirable aims of the Commerce Act in this case prevented the companies from reducing some of the turbulence in their environment by cooperating on bidding and by favouring consortium participants over non-participants.

## VIABILITY OF THE CONSORTIUM

How viable is the consortium? The time and care taken in building trust and providing legal protection paid off. Firstly, a large engineering company bought out two of the companies in the consortium, and could possibly have taken control of the consortium if the provisions in the Participation Agreement had not prevented this. What happened was the two companies resigned from the consortium. Then another company resigned for other reasons. Despite these withdrawals and a harsh economic environment the consortium is still surviving, but its future viability depends critically on the level of trust amongst its members. Unfortunately given the current environment of the consortium, this trust could easily be destroyed by just one error of judgment.

## APPENDIX 1: TYPES OF COOPERATIVES

There are 7 major types of cooperatives which different firms can form for mutual benefit: facility, personnel, purchasing, information, financial, business, and market.

Facility: a common facility cooperative is one in which several firms share a facility or an expensive plant or equipment e.g. a factory, a computer system, or an office. Costs can be apportioned according to the amount of use or to the relative size of the contributing firms. Alternatively the facility can be run as a profit or performance centre with the firms paying market prices for work but getting a dividend according to the amount of their investment in the facility.

Personnel: a personnel cooperative is one in which several firms share one or more people e.g. a salesperson or secretary. Costs can be apportioned according to the amount of use or to the relative size of the contributing firms. It can be very difficult to operate this type as a profit or performance centre and it will probably not be worth the effort.

Purchasing: a purchasing cooperative is one in which several firms jointly purchase materials and components in order to obtain quantity discounts or to reduce inventory holding costs. Costs and savings can be apportioned according to amount purchased. Alternatively the purchasing cooperative can be run as a profit or performance centre with the firms paying market prices for the materials but getting a dividend according to the amount of their investment in the cooperative.

Information: an information cooperative is one in which several firms share information e.g. market or technical information. The cost of obtaining the information can be apportioned according to the relative size of the firms or on a user pays basis. This could also be operated as a profit or performance centre.

Financial: a financial cooperative is one in which several firms jointly obtain equity or venture capital or share financial risks. The benefits or risks can be apportioned according to the relative size of the firms or to the amount of investment.

Business: a business cooperative is one in which several firms jointly undertake a project or an innovative venture or some kind of development. Costs and benefits can be apportioned according to the amount of investment and/or the amount of risk taken by each firm. This could be run as a profit or performance centre with each firm sub-contracting to the business cooperative on a competitive basis but getting a return according to amount of investment in the cooperative.

Market: a market cooperative is one in which several firms jointly market their products and services. The firms may produce or provide distinct products or services, which may be complementary, and distribute and sell these through a common network; and/or they may carry out joint advertising and market intelligence. Costs can be apportioned according to the relative size of the firms or to the amount of use. It is also possible to run the market cooperative as a profit or performance centre.

It is possible to form a cooperative which is a combination of the above types e.g. combining a business cooperative and a market cooperative.

HOW CAN WE USE A "HARD" METHOD IN A "SOFT" WAY?

LESSONS LEARNT FROM CASES USING "RATING CHART METHODS"

Ken'ichiro Senoh

Department of Systems and Information Management
Lancaster University
U.K.

INTRODUCTION

In considering the question 'how can we make an effective choice of systems method?" we generally assume that a method is to be used in only one way or in a few ways, which a user takes for granted. These usages are within the range of what the developer of the method prescribes. Usually a description of a method in a textbook gives us an idea of "how it is used", and users follow that description. If a user has problems with its usages, she/he would develop her/his own usage. However little attention on how to use them in real-world management seems to have been given so far. My questions here are; (a) How are methods actually used in various ways in real management situations?; (b) How can actual cases be generalized? (c) Why are methods used in various ways? Firstly, I relate my experience of how DA of the KT method was used in actual management; secondly, some lessons are drawn from the case; and finally some comments are made on "choice of methodology". (In this discussion the term "usage" means a way to use something either in principle or in practice.)

A: RATING CHART METHOD IN VARIOUS USAGES

In this paper I will describe my real experience in which a "hard" method was used in various ways, including "soft" ways. Let me take up the case of "Decision Analysis (DA)" part of the Kepner Tregoe (KT) method which has been developed in the USA for more than 30 years. It is said the popularity of KT is due to the effectiveness of KT in training managers for "rational thinking" and giving them expertise in problem solving. (Kepner & Tregoe, 1965, 1981). However KT is said to be effective for routine work and operation problems in which "the problem" is relatively clear (Brightman, 1980).

DA is a kind of mathematical method for the evaluation of alternatives for decision making. DA can be regarded as a kind of "rating strategy" which is a very popular decision making method/technique described in many current textbooks of management (Watson & Buede, 1987). Obviously it is a typical systematic decision making method using a rating chart and would normally be regarded as a "hard" method in systems study. I will examine this by describing and commenting on my actual experiences and interviews.

*Systems Thinking in Europe*, Edited by M.C. Jackson *et al.*
Plenum Press, New York, 1991

(CASE 1) KT METHOD WORKSHOP

The second largest photo film company in the world recognized as one of the most "excellent" manufacturing companies in Japan had KT training courses and workshops. They were conducted by internal instructors with the help of professional instructors from Kepner Tregoe (Japan). I joined a one-week workshop which was held for three real project teams. These included a marketing strategy team of which I was a member as the chief marketing planner at that time. Each team consisted of a senior manager, some managers and some staff. The workshop consisted of two parts; a workshop for providing training in the four KT processes with their ready-made drills; a second part devoted to real project work, making much use of KT learnt with the help and supervision of the instructors.

The task of our project team was (a) to review the market need for "professional photography for ceremonies such as weddings" based on a concern that demographic trends would threaten this market which required professional negative films and papers; (b) to build a strategy which would vitalize the market and to design a strategy for re-structuring the total market if this was thought necessary. This task was far from an everyday operational type of problem, and was of the strategic business and management issue type.

In the workshop, we were required to use the method as it was, or at least as it was supposed to be. We were assigned exercises which were fairly easy to "solve". We listed some alternatives and carried out three steps: the musts/the wants analysis; weighting items and valuing alternatives; potential risk analysis. Some questions occurred to us and we discussed them with the instructors. For example, how do we decide which are "the musts" and "the wants"?; why do we put alternatives that were apparently to be rejected since they failed to satisfy the criteria of "must"?; why should the decision process end when the alternatives have been evaluated and one of them is selected? Although we were not satisfied at all with the instructors' answers to these questions, the first part of the workshop was over. I shall mention those questions and answers in detail elsewhere (Senoh, forthcoming).

We were also supposed to use DA in the real project phase which followed the case study phase. However we never found a textbook-like situation where a process of rating alternatives took place automatically. We tried to follow the recipe and let members propose each evaluation. But interesting things happened. Instead of following the textbook process, we started to discuss why each member put weight on an item. For example, some stressed the importance of "traditional ceremonial occasions" while some argued to promote "casual photos for younger generations". We tried to know other members' reasons for selection and evaluation of alternatives and criteria. Sometime we went back to the reason for proposing alternatives. Sometime we considered what was actually required for the task, and the task itself was examined. The instructors were frustrated with this but the project members never felt it other than useful. We enjoyed the debate itself and learnt a lot. Actually it was a learning process. The DA chart triggered a debate. We used it as a device for generating debate. Some courses of action emerged during discussion and all of us knew why they did. Whether in spite of DA not being used in a proper way or because of it, we did not know, but the work of the project team was highly "successful" from our point of view.

Another chance of using DA in the successful but not "textbook" way came afterwards, which confirmed my view about the effective usage of DA.

(CASE 2) IN REVIEWING THE QUALITY OF A COLOUR REVERSAL FILM FOR PROFESSIONALS

After some months, I joined a two day planning meeting. The participants were the senior manager together with some managers and senior staff from planning, sales and technical support sections. The theme was an overall review of the current short-term marketing strategy for professional colour reversal films for commercial use.

Since members were from various sections, interpretations of the current market were different, and sometimes it was difficult for them to understand each other. One of the controversial issues was about the appreciation of the needs of professionals on colour quality of the reversal films. Although all those attending felt that the current quality, especially colour tone, should be improved, their views varied. The disagreement was not only on the quality aspect itself but also on the reasons they held these views. The discussion showed that the views of some participants did not mesh with each other. As the facilitator of the meeting I felt I had to organize a base for discussion. Then I came up with an idea; I intervened to suggest modelling DA.

I ask them what kind of specifications were necessary for this discussion? Which items should be "the musts" and why? What rating should be given to each item? Some sets of alternatives were discussed; the various types of pro-photographers were listed and rated according to the market size and the importance in the sense of influential effect on opinion; the various current films were listed and assessed.

The discussion was quite different from the "textbook" DA process. We focused on inquiring into appropriate definitions of meaning of "colour tone" and a better appreciation of market needs. DA itself did not work as it was supposed to do. But by trying to make a DA chart and filling it in we generated a productive debate. When I drew a matrix chart of DA on a white board, everybody recognized the point of the discussion. They tried to understand the real meaning of the terms others used, such as "richness of colour", "sizzle of the image". The appreciation of the market became much easier when they understood the different jargon used in different professional groups; we recognized why each member stressed a certain genre. Furthermore we discovered that each member had their own interpretation of the mission! One manager took a longer view than others. Another manager thought that, in the situation at that time, brand image should be preferred to profit, while another had the opposite view. We started to understand who stressed which aspect of the mission and why.

In this meeting we learnt much, not only about market needs, but also about our various interpretations of them; we were successful in gaining common understanding or interpretations and terminology. This became an asset of the team. In this sense the discussion was the highlight of the meeting although little was "decided". The participants were pleased and satisfied. The next meeting was again highly productive. Since we had a common ground, a course of action to be taken emerged easily.

As we have seen, DA of KT does not seem to be useful for managerial work, at least not if, ironically, we follow the directions of the KT instructors. Instead, we found some effective ways of using the method for managerial work, ways which are, however, different from the way the developers wish users to follow. After this experience I sometimes used DA as a device for productive debate rather than as a systematic decision making machine. In systems terms, in retrospect, I used DA, which is supposed to be a "hard" method, in a "soft" way.

B: LEARNING FROM THE REAL CASES

The cases suggest two points; First, it is difficult to evaluate a method without consideration of the particular ways in which it is used. Second, even if it is considered to be a "hard" method, if we use it in a "soft" way it may be effective in management situation, not only in operational situations. From these view points, I shall draw some lessons from the two cases, A and B.

B-1: TEXTBOOK USAGE AND ACTUAL USAGE, HARD USAGE AND SOFT USAGE

I find it is interesting to think about a distinction between "textbook usage" and "actual usage", and between "soft usage" and "hard usage".

In considering types of usage, various categories can be drawn  The distinction between "espoused theory" and "theory in use" is one of these distinctions (Argyris & Schon, 1974). However, in this paper, I would like to make a distinction between (a) textbook usage:  the founder/developer of a method assumes or implies in textbooks; and (b) actual usage:  a user uses the method in practice. These usages are not necessarily the same. Various actual usages are possible. The cases A and B can be regarded as the examples in which a "hard" method is used in a 'soft' way. This leads to the second distinction; "Hard" and "Soft" usage of a method.

The term "hard" and "soft" are not always used carefully.  The experience in the cases suggest that the concepts of "hard" and "soft" usage may be defined in this way;'

(1) Hard usage:  the way of using a method as a systematic and rigid procedure of producing a decision, one that is to be "controlled" when it is implemented.  This process may be done by designing a model which may be a map of the real world or a design which the designer wishes to exist in the world in the future, or by using charts as a mechanical way of producing a "decision".

(2) Soft usage:  the way of using a method to provide systemic, flexible guidelines to promote learning about a situation which the user wants to improve. This process uses a model (an intellectual construct) as a device to facilitate a debate aimed at inquiring into and interpreting a situation and encouraging the emergence of new meanings from it.

In short, the key point is how (systems) models are used in a method.  In hard usage models are used to support systematic decision making while in soft usage models are used  to facilitate an inquiry; a debate in which learning may occur.

Actual Usage

|  |  | Soft | Hard |
|---|---|---|---|
| S O F T | | "Normal" SSM | some SSM |
| H A R D | | Case A, B | "Normal" DA |

Textbook Usage

Fig. 1  The matrix of usage

Here it is useful to draw 2X2 matrix (Fig. 1.  The matrix of usage).  The row is of "textbook usage"; hard and soft.  The column is of "actual usage"; hard and soft.  The cases above show that DA was used to facilitate inquiring debate  are categorized as "hard" in textbook usage and a "soft" in actual usage, and so is placed in the lower left cell.  This matrix can be used to map on usages of other methodologies such as Soft Systems Methodology (SSM).  Since, according to the founding developer of SSM, it is to be used in "soft" way (Checkland, 1981. Checkland & Scholes, 1990), it is categorized in 'soft usage' in text book usage.  Also we find many examples in which SSM is used in various ways in the range of soft usage in action as well as some examples in which SSM is used in "hard" ways (Checkland & Scholes, 1990. Atkinson, 1987).  The latter cases are categorized as "hard" in actual usage.

B-2: EVOLUTION OF METHOD

The learning so far suggests two issues to us.  The first is about a relationship between textbook usages and actual usages.  This leads to a notion of "evolution of methods".  The second is about the relationship between soft usage and hard usage.

This leads to types of method usage. I will leave discussion of the latter to another paper (Senoh, forthcoming).

In many cases, users of a method take a textbook usage as given and use the method in that way as if it were the best in any situation. These users are not aware of how to make the most of the method by modifying it in a way suitable to his/her situation. When they fail to use it they usually blame the method itself without realizing the possibility of their own failure to adapt the method in an appropriate way. On the other hand some users do try to develop new ways of using the method. An effective usage in an actual situation may emerge and this may lead to a revision of the textbook usage.

For example, SSM has developed from "hard" to "soft" as the developers reflected on and learnt from "actual usage" of SSM. Action research made it possible. The interactive relationship between textbook usage and actual usage has promoted the development of SSM itself. Thus we may say that if once we establish a mutual cyclic learning process between textbook usage and actual usage, we could facilitate the evolution of the method, as if, metaphorically, the method is self-organizing itself.

Then we may ask; why DA of KT has not been developed as SSM has. The reasons are two fold. First, the developers' "Weltanschauung" or hidden image of "method" and "the uniqueness of management situations" are different from those suggested above. Actual usages are dependent on both. Let us see them briefly.

On the one hand if a user is interested in how to make the most of a method by developing new ways of using, she/he develops a system-to-use-the-method, a system which I have called "usage" here. SSM developers try to make SSM flexible to allow for unique situations. Their basic Weltanschauung is that problem situations are unique and a method used for improving the situation should be flexible enough to meet each situation.

"Those who develop SSM were very conscious of its status as mouldable methodology rather than rigid technique, and they wished to leave "how, exactly, to do it" as a strategic choice for the user to make." (Checkland & Scholes, 1990, p291).

On the other hand, if a user believes in a rational "logic of the situation" rather than the uniqueness of situations, she/he would try to develop not a usage but a method itself, a systematic process which produces a rational decision independent of who the user of the method is. The KT developers try to establish rational way of thinking which lead anyone to the correct solution to the problem.

Apparently, these different Weltanschauung shows the basic difference between "hard" and "soft" systems thinking (Checkland, 1981). Hence those who think any situation is unique try to make flexible use of a method so that it could support improvement in the particular situation. In most cases, this can be done by facilitating interpretive debate. Thus "soft" usage is developed to serve these users. On the other hand, those who think any situation has a common logic try to use a method rigidly so that it can support rational solution of "the problem". "Hard" usage is appropriate to these users.

Furthermore different images of method lead to different orientations of the evolution of method. "Logos" of method influence the direction of development of methods. "Soft" approaches tend to evolve to wider usage while "hard" techniques tends to narrow down the range of ways in which methods are used.

C: CONCLUSION

A choice of methodology might be better done by paying more attention to its practical usage. In some recent works a choice of methodology is considered as a "contingent" activity; that is to say, a method is chosen contingent upon users' perceptions of a problem situation (Jackson, 1987. Keys, 1988). This discussion seems to assume that a method is used in only one way or in a limited range of ways. However a method can be used in various ways. Thus without consideration of these usages it is not sufficient to assess the effectiveness of a method. To consider only the textbook procedure of a method is not enough. Actual or possible usages should also be considered. A choice of method is not only dependent on the images of the problem situation but is also dependent on how the user interprets the method and its usage.

Furthermore, we should not eliminate the possibility that we may find a new way of using a method during its actual use and so develop it. This may be an important learning process not only for the user but also for the developer. Indeed, at that time the "wall" between a user and a developer of a method gives way; a method (like democratic government!) can be of users, for users, and by users in a problem situation. This "liberation" of systems methods may be done when usage of a method is transferred from professional hands to the hands of users who actually know the problem situations.

## REFERENCES

Argyris, C. and Schon, D.A., 1974, "Theory in Practice: Increasing professional effectiveness", Jossey-Bass, San Francisco.

Atkinson, C. J., 1987, Towards a Plurality of Soft Systems Methodology, J. Appl. Sys. Anal., 16:19.

Brightman, H. J., 1980, "Problem Solving: A logical and creative approach", Georgia State University, Atlanta, Georgia.

Checkland, P. B., 1981, "Systems Thinking, Systems Practice". Wiley, Chichester.

Checkland, P. B., and Scholes, J., 1990, "Soft Systems Methodology in Action", Wiley Chichester.

Jackson, M. C., 1987, New Direction in Management Science, in: "New Directions in Management Science", M. C. Jackson and P. Keys, eds., Gower, Aldershot.

Kepner, C. H. and Tregoe, B. B., 1965, "The Rational Manager: A Systematic Approach Problem Solving and Decision Making", McGraw-Hill, N.Y.

Kepner, C. H. and Tregoe, B. B., 1981, "The New Rational Manager", Princeton Research Press, Princeton, N. J.

Keys, P., 1988, A Methodology for Methodology Choice, Sys. Res. 5:65.

Senoh, K., 1990, Information Generating and Editing Methodologies: SSM and the K J Methodology. J.Appl.Sys.Anal., 17:53.

Senoh, K., (forthcoming), Hard methods in soft usages: Methods for the interpretive Management. Working paper in Dept. of Systems. Lancaster Univ.

Watson, S.R. and Buede, D. M., 1987, "Decision Synthesis: The principles and Practice of Decision Analysis", Cambridge Uni.Press, Cambridge.

# UNTANGLING THE PERCEPTION WEB: A METHODOLOGY

André Dolbec and Georges Goulet

Université du Québec à Hull
Hull, Québec
Canada, J8X 3Y7

From a rational point of view, one is often astonished by the woolly and erratic behaviour which appears to take place in a given human system (Checkland, 1980). Paradoxically, however, when one talks to the people involved in such a system, they rarely consider their behavior as erratic and, when they do, they cannot tolerate such a situation. They see themselves as quite rational in consideration of their own personal goals and perceptions. Moreover, as an active member of an organisation, that can be seen as a purposive system or as Checkland (1980) puts it a "formal system which has an objective, a mission, a definition of a final desirable state or an ongoing purpose", the same person may encounter as many opinions or perceptions as there are people involved in the pursuit of the organisational aims. Furthermore, each of the protagonists often appears to be committed to the pursuit of his or her own personal aspirations. Indeed, when one decides to contribute to a collective endeavor of any kind, one does so because he or she considers that in pursuing the ends of the group, he or she will be pursuing his or her own aspirations or ambitions (Maslow, 1965). Thus, the ends of the group or organization can be somehow considered as a means used by that person to achieve his or her own ends.

Moreover, a group or organization is generally expected to contribute to a social mission in the service of the community. To that end, it generally has to produce some kind of an output in material or services which somehow acts as a referent in defining the measures of performance used in the monitoring system of the transformation which the organisation pursues. It also will plan different specific targets which will contribute to the production of the expected output. These targets may be considered as the group's or the organisation's means to achieve its output and accomplish it's mission. On the other hand, the organisation often selects its staff according to its competence to contribute to the pursuit of these targets which the employee sees as a direct means towards the fulfillment of his own ambitions and aspirations. Furthermore, each person involved in the organization may have his or her own perceptions of its targets and outputs. These different perceptions, in turn, bring about the emergence of sub-groups which often compete in order to force their own viewpoint upon the whole organization.

The least that can be said about the description that we have just presented is that it constitutes a complex network of differentiated personal and organisational goals or missions as well as that of multiple perceptions of each one of these. It is in this context that we are often called upon to solve a dysfunctional *how* problem that can be considered as the means to achieve the expected output of the system . Most of the time, we soon become aware that the people do not even agree on the *what* or the output that is expected from them. Through the years, it has become evident to us that most of the *woolliness* around the nature of the output , *the what*, of Human Systems is rooted in the varied perceptions and diverse commitments of the people involved, a situation which is often manifested through motivational problems.

It is our contention that in order to untangle the perception web that we have just described, the intervenor must heighten the awareness of the people involved concerning their own identity, the *who*. He must also assist them in increasing their awareness of their own aspirations as well as of the intended outputs and mission of the organisation or group in which they are called upon to take part: *the what for*. On this basis, such an intervention could be seen as that of taking up the challenge of transforming a woolly conglomerate of "I's" into a coherent and convergent community, a "we". This would allow the emergence of a synergistic community mobilized in a social human system enabling it to collectively move the rock , the *what*, which will bring about a fine landscape, the *what for* (Barnard, 1965).

The predominant assumption which serves as a guiding light in all our interventions is that any human experience can be considered as an emerging whole resulting from the energy channelled through each personal combination of the physical, affective and mental configuration of the human systems involved. It is also assumed that one of the most efficient way to make sense of any given situation involving human beings is through the use of systems theory as an epistemological instrument. As evidenced in the description of the problematic situation which was presented previously, many systemic principles can be taken into account in such an investigation.

First of all, let us consider the essential element of any human system: the human beings which bring it to life. Each of these human beings in themselves can be seen as purposive systems directed toward the achievement of an end (Bertalanffy, 1968; Baker, 1973; Scott, 1981). As living systems, they strive to insure their own survival (Maslow, 1954, 1968, 1971) while endeavouring in the pursuit of their own ideal as thinking systems (Darwin, 1909; Allport, 1955). They can also can be considered as a composite of sub-systems whose specific finalities contribute to the transformation process of the system in which they participate (Bertalanffy, 1967). They are also submitted to the systemic law of connectivity which states that each change which affects the whole human system entails a corresponding change in each of its sub-systems. It also states that each change in one or another of its sub-systems entails a corresponding change in the system as a whole (Goulet, 1986).

From the standpoint of its intrapersonal system, each human being perceives his or her environment in accordance to his or her own motor, cognitive and motor skills which provokes the emergence of diverse figures at the mercy of its affective states (Perls, 1969).

Since human beings are capable of deliberate action, they can create purposive systems which they decide to direct toward an ultimate mission in the service of the community, the *what for*. They then identify a secondary goal, the *what*, which can be seen as a means which will allow them to attain this ultimate end. In order to pursue this goal, they then can infer processes, the *how*, through which they will channel the energy mobilized through their commitment to the pursuit of the goal. As functional means to accomplish the *what for*, the *what* and the *how* must be submitted to a monitoring process that will ensure their compliance to the pursuit of the ultimate end. This ends/means trajectory, when situated in a spatial and temporal environment, can be seen as descriptive of any human activity system.

Although the human activity system concept is often used merely as an epistemological instrument without any real existence, the Gestalt Approach considers each human experience as a system which emerges at a given moment through a contact of a human system either with its inner self or with something or someone present in its environment. This contact is brought about by a deliberate mobilization of energy which is generated through a person's awareness of a new information. By its very nature, contact can only be real while it lasts in a given space and time. That is what Gestalt considers as experiencing in the *here and now*.

Each experience or contact can thus be considered as a *figure* which emerges from the background of many potential figures triggered by the initial information stimulus. Thus, at any given moment, each person involved in a given situation may experience a different *figure* emerging from his or her time or space background. This background is colored by past experiences, individual culture, personal ideals or wants, as well as by the present experiential situation in regards of the physical and social environment. This approach may be explained by the research on human perception reported by Kohler (1970) and other Gestalt theoreticians (Perls, 1951; Zinker, 1977; Polster, 1973) which states that one may perceive only through the succession of different *figures* emerging from a given *ground*.

In order to have a sufficiently adequate knowledge of a given situation, one must attempt to see it in its *whole* or *Gestalt* by considering as many *figures* as possible while keeping in mind that a multitude of *figures* could emerge from the *ground*. In order to keep comparative *figures* relevant to each other, they must be considered as emerging from a common *ground*. Thus, in order to make sense of the tangled perception web, this approach strives to make each protagonist of a given situation aware of his or her own *figure* and give each of them the opportunity to share it with his or her colleagues. This sharing process is expected to allow the emergence of a common *figure* which will bring about a synergistic mobilization of energy. In any organisation, to be functional, this common figure must be relevant either to the *how*, the *what* or the *what for* of the organisation itself.

Applied to the Gestalt concept of *figure/ground*, the systemic ideas, which state that a system can be considered as a network of systems and sub-systems hierarchically ordained and functional to the pursuit of a given direction, become a powerful tool in the understanding of the perceptual web. The information stimulus which initiates each *figure* can be either deficiency oriented or purposeful, which often generates internal conflicts in human systems. A person may be faced with the dilemma of following a deficiency oriented urge which is in contradiction with a deliberately chosen goal. Each polarity of this dilemma may, in turn, generate a process directed to its pursuit through the creation of a hierarchical network of sub-systems aimed at facilitating the pursuit.

Considering that any *figure* which emerges from its ground does so in a set period of time as well as in a given space setting, the Gestalt Approach suggests that one can only observe a given situation in its *here and now*. Indeed, the past can only be observed through its present consequences and the future through the observation of a prospective *here and now*, both realities, therefore, can only exist in the mind of whoever thinks about them and are not attainable through observation. One can only observe a present *figure* or a network of present *figures* as they emerge from the cultural, experiential, chronological and spatial background of its protagonists. This network of *figures* can be expressed thus: $HN^o = \sum HNi$ , where $HN^o$ represents the *here and now* of the organisation or the network of present *figures* and HNi represents each of the individual *here and now* or present *figures*. Thus one can say that the perceptual *here and now* figure of the organisation at a given time equals the sum of the individual *figures* of its members. This mere heap of *figures* may explain the wooly and erratic appearance of the organisational behaviour. It indeed lacks the necessary direction which will focus its energies into a synergistic quest. As such, it can rather be seen as multidirectional, thus unsystemic.

Considering that subjectivity can be seen as rooted in the very nature of any human being, which implies that no ideas, no concepts, no theories could ever be conceptualized without the imperfect mediation of it's generator's perceptions, feelings and sensations, one can express the individual present *figure* of each of the protagonists of the situation through the following formula:

$$HNi = fPi \ [(Who)(What \ for)(What)]^o \ x \ Pi \ [(Who)(What \ for)(What)]^i \ x \ Pi \ [(Who)(What \ for) \ (What)]^{oi}$$

In this formula, HNi represents the *here and now* of the *figure* present in each individual's awareness. It is considered to be a function of each individual perception of the *Who*, the *What for* and the *What* of the organisation [o] as well as his or her own perception of the *Who*, the *What for* and the *What* of the individual himself or herself [i] as well his or her perception of the other individuals [oi] involved in the situation.

If we take this into account, the main objective of any intervention is to create an experiment in which a common *ground* is provided to its participants from which each individual subjectivity will have the opportunity to emerge into a particular *figure*. By sharing these *figures*, the participants will be allowed to discover commonalities among their individual subjectivities and feel part of a somewhat synergistic human system. The intervention will then create diverse backgrounds in which the participants will become aware of certain constancies in their common subjectivities. They will learn to belong to a supporting community which allows for differences and endeavors to pursue common goals on which they have agreed. If we compare this experiment to a classical scientific experiment, the planned common *ground* acts as an independent variable and each *figure* as a dependent variable. This approach allows the intervention to lock into successive bounded pieces of work which constitute a workable method to identify and analyze the diverse subjectivities that exist in human systems. In order to make sure that all the data present can be treated, the intervention must constantly ascertain that the scope of the piece of work is limited enough to allow a rapid heightening of awareness around a circumscribed perception in each participant as well as a basic convergence among the participants through the sharing and discussion of the perceptions of which each of them have been made aware. In order to ensure that the scope of the intervention is realistic, the intervenor will often be required to downgrade or upgrade what he had planned initially in order to take into account the readiness of the protagonists.

## METHODOLOGY

In order to set up a well bounded experiment, the *Ends/Means Trajectory* is used as a framework which will facilitate successive convergences of the subjective figures present in the situation. Each phase of the trajectory will be used as a background out of which will emerge the diverse figures nourished by the culture and past experiences of the participants as well as by the *here and now* culture created by this theoretical framework. This situation is assumed to be a microcosm of any situation in which these same people would be called upon to work on as a task group in the organisation.

Phase 1: Establishing the Common Background

1.1 Assessing the needs. The needs assessment is used as an opportunity to demonstrate how the *Ends/Means Trajectory* can be used to make sense or plan a human activity system.

1.2 Learning about the Intervention Framework. T wo avenues of intervention are possible according to the readiness of the participants. In the first one, the intervenors introduce the main concepts which underlie the intervention as well as the *Ends/Means Trajectory* (figure 1) which is presented as a framework to the intervention as well as a planning instrument that can be used in a different organisation or growth situation. In the second instance, following a further needs assessment, the intervenors go on with the process which they have planned with the people without a formal presentation of the *Ends/Means* theoretical framework.

Phase 2: Analyzing the Perceptions of the *What for* Present in the Group.

2.1: Identifying the Individual Perceptions In this phase, each individual is asked to reflect on his own perceptions of the *What for* of different relevant systems (his personal *What for*, the organisation *What for*...).

**2.2: Sharing the Data.** In order to allow a more active participation of each member, the participants are invited to share their awarenesses in sub-goups which in turn will prepare a report on the data gathered which will be presented to the large group.

**2.3: Searching for Commonalities.** Using the data which is available to the large group, the participants are invited to identify the commonalities as well as the divergences present in the group.

## Phase 3: Identifying the *What.*

In this phase and the following one, the inquiry is conducted in the same manner as it was in the preceding phase. Its object is to identify the perceptions concerning *What* the organisation is to produce, either as a material product or as a service. This production goal must be considered as a means towards the attainment of the *What for* which should be used in the monitoring of the processes which will be considered as necessary to attain this production goal.

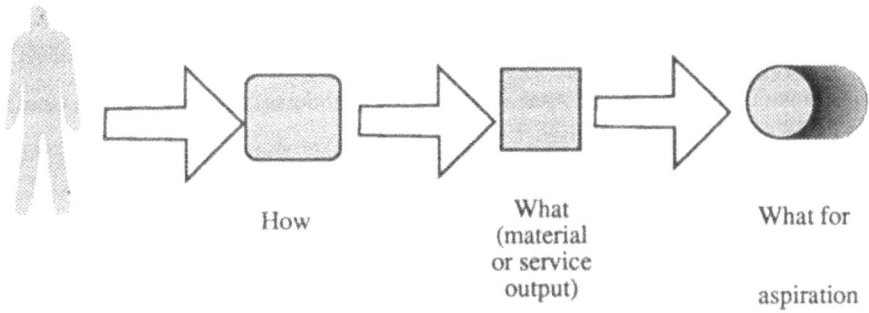

How               What            What for
(material
or service
output)

aspiration

Ends/Means Trajectory

Figure 1

## Phase 4: Identifying the *How*

The object of this phase is to identify the perceptions concerning the ways in which each specialized groups or sub-systems of the organisation contribute to the production of the *What* in order to ensure the pursuit of the *What for* which in this phase also is used in the monitoring of the activities. This *How* thus becomes the *What* of each of the sub-groups which in turn will be called upon to ask their protagonists how they contribute to the *sub-what*.

## The Identification of the *Who*

In the course of this process, each participant will heighten his or her awareness of his or her own identity and of *How* he or she responds to the other members of the group. He or she will also become aware of the relationship which exist between his or her identity and that of the other members present. He or she will also learn about the subjectivities that he or she shares with others as well as those that are in conflict with his or her peers. Thus, he or she will become more aware of the norms of inclusion or exclusion in the group. He will also learn how he can influence the group and how he is influenced by it. He or she will discover the processes of control and authority which are latent in its midst. If the process of community building is allowed to come to its term, he or she will be allowed to experience a degree of intimacy sufficient to respect his or her own independence. This degree of intimacy and respect for independence will allow a real contact among the members of the group. Thus he or she will have experienced a *We* stage in the group development instead of the *I-Thou* relationship which existed previously.

## CONCLUSION

In conclusion, we would like to share with you our astonishment of the power of personal thinking as an anchor for group sharing and planning. We would also like to express our heartfelt conviction that if people are given the opportunity to live an experience of true fraternizing in a nurturing context, they often discover that they have more in common that they previously thought and that, in some way, their differences can be worked out. This awareness can be seen as a stepping stone for a greater organisational synergy and productivity.

## REFERENCES

Allport, Gordon W. (1955). "Becoming, Basic Considerations for a Psychology of Personality". New Haven: Yale Univ. Press.

Baker, Frank (1973). "Organizational Systems: General Systems Approaches to Complex Organizations". Homewood, Ill.: Irwin-Dorsey.

Barnard, Chester (1965). "The Functions of the Executive". Cambridge Cambridge, Mass.: Harvard Univ.

Bertalanffy, L.V. (1967). "Robots, Men and Minds: Psychology in the Modern World".N.Y.: Braziller.

Checkland, Peter B. (1980). "Systems Thinking, Systems Practice". Chichester: Wiley.

Darwin, Charles (1909). "The Origin of Species". New York: P.F. Collier.

Goulet, Georges (1986). "Vers une théorie de l'éducation/apprentissage". International Review of Education, 32 (4): 439-457.

Kohler, Wolfgan (1970). "Gestalt Psychology: an Introduction as New Concepts in Modern Psychology", New York: New York.

Maslow, Abraham (1954). "Motivation and Personality". New York: Harper and Bros.

Maslow, Abraham (1968). "Toward a Psychology of Being". New York: Van Nostrand Reinhold Co.

Maslow, Abraham (1971). "The Farther Reaches of Human Nature". N.Y.: The Viking Press.

Perls, Frederick, Hefferline, R.F. and Goodman, P. (1951). "Gestalt Therapy", New York, Julian Press.

Perls, Frederick (1969). "Ego, Hunger and Aggression". New York: Vintage Books.

Polster, Erving and Polster, Miriam (1973). "Gestalt Therapy Integrated". New York: Brunner/Mazel.

Scott, W. Richard (1981). "Organizations, Rational, Natural, and Open Systems". Englewood Cliffs, N.J.: Prentice-Hall.

Zinker (1977). "Creative Process in Gestalt Therapy", New York: Brunner/Mazel.

# UNTANGLING THE PERCEPTION WEB: AN APPLICATION

Georges Goulet and André Dolbec

Université du Québec à Hull
Hull, Québec
Canada, J8X 3Y7

If one assumes that the whole question of defining what Checkland calls a relevant system is based upon whatever appears to be significant to people striving to solve a given organisational problem, it seems important to explore the perceptions of those who complain about the situation seen as problematic. The methodology presented her is based on the Gestalt notion of *figure/ground*, as explained in a previous article, has proven to be, in our own consulting work, quite a powerful instrument in working with people to discover whatever seems relevant in a problem-solving situation. Throughout our experience in working with human systems, we have often been faced with a multitude of different perceptions and ideas about the problematic situations that we have been called upon to facilitate. When people do not agree on the best way to accomplish a given task, they look for some kind of a point of reference in order to assess the situation. It may be their own opinion, that of colleagues or superiors, as well as the operations manual, etc.. In research, this process is called the triangulation technique (Cohen and Manion, 1985). In an organisation in which none of these points of reference is recognized enough authority to rally the opponents, consultants are generally called upon to facilitate the appreciative process (Vickers,1983) needed to untangle the perception web. As in most approaches, the suggested methodology does not pretend to find a definite solution to the situation perceived as problematic. It rather aims at helping the protagonists find a certain degree of commonality of opinion concerning the goals and means to be investigated.

In the following lines, we will present you an account of a problematic situation which we were called upon to work on as well as a description of the intervention that took place and the learning outcome that emerged.

We were invited to prepare and lead a session to be given to the Directors General of a large Research Agency which had been faced with drastic change in the last several years. Budgets had been slashed, staff had been significantly reduced, the top management had been replaced, a large restructuring operation had just taken place. Furthermore, the Government, its main funding source, had asked its top management to redirect its activities from science-centered to client-centered research. A working session was organized with the personel people in which we were briefed on the prevailing situation at that moment. We were told that the people involved considered the change to have swept through the organisation like a hurricane. It had indeed uprooted the very tradition of this respectable institution. Motivation was at its lowest and mistrust was overwhelming. Those who could be considered as authentically aware of the changes that had to be done were often considered as traitors to the orthodox mission of the organisation. Paradoxically, many perceived the necessary changes as

*Systems Thinking in Europe*, Edited by M.C. Jackson *et al*.
Plenum Press, New York, 1991

a means for organisational survival in an environment of prevailing scarcity of resources. On the other hand, others considered them as a threat to the very existence of the organisation as it had been known since the beginning of the Century.

The four vice-presidents' newly defined functions had been enlarged and required them to exercise authority over the newly created institutes which had regrouped different specific research centers whose Directors General had, for a long time, been sole masters on board . On the other hand, the Directors General, which had a direct link to the President, felt that they were caught between the Top management requirements for change and their employees' pressures to resist change as well as what they considered to be the meddling of the Head Office.

In order to broaden and triangulate the information which had been given to us by the Personnel Service, we asked to meet with a sample of the participants that we would have to train. Five were chosen: one Vice-President and four Directors General. These meetings reinforced our awareness concerning the motivation and trust problems as well as the insecurity and frustration that the people felt in regards to changes that they had considered as too rapid and poorly planned. We also discovered that there were serious divergences of opinion concerning the *What for* of the organisation as well as its *What*. Some considered that the organisation's ultimate end, its *What for*, was the development of knowledge, others that it had been founded to serve the Public. In the first case, the corresponding *What* could be considered to be fundamental or practical scientific discovery for the sake of knowledge, in the other, practical scientific discovery oriented towards increasing the well-being of the citizens.

Following this data gathering, a proposal was made to the Personnel Service who submitted it to the President. His main concern was to establish our credibility in the midst of a highly scientific culture which included a few Nobel Prize winners. In order to do so, he required that the session be centered more on content than on process. He had assumed that the more intellectual the session would be, the more the people would be interested. He therefore considered that the needs assessment that had been done before hand was sufficient. After a few discussions, a plan of the session was agreed upon. The plan was to be very detailed and include theoretical lectures on the framework of the intervention. The participants were to be provided with reading material which was to be read before the session. We were also told that a formal mission statement had been painfully established and that the participants should be asked to cover the reading related to its establishment. This mission statement was to be used as a basis for the discussions to be undertaken.

THE AGREED PLAN

Day 1:

Evening  -Dinner: Getting acquainted.
　　　　　　-Information and organisation Session.
　　　　　　　　1.1 Presentation of the objectives and goals of the session.
　　　　　　　　1.2 Short presentation by the consultants of the approach to be used during the
　　　　　　　　　　three days and introduction to the theme of cooperation.

Day 2:

9 a.m.　Presentation by the consultants:
　　　　　"Cooperation: an organized pursuit of a known and owned mission".
　　　　　　°thinking about human activity systems.
　　　　　　°the concept of mission.
　　　　　　°boundaries and subsystems.
　　　　　　°the whole system: a network of deliberate interrelationships between sub-systems.
11 a.m.　"The Perceptions of the Individuals and Those of Sub-Groups: Factors of Organisational Deliberation".

          1.1 Individually: Clarification of each one's perception of the mission and of the relative importance of its different components.

          1.2 In sub-groups: Sharing personal points of view in order to come to a sub-group position.

Noon:     Lunch Break

2 p.m.     Plenary: Reporting an discussing priorities.

3.30 p.m.  Presentation by the consultants of a methodology to build a rich picture of the organisational context: "The Context of the Organisation".

3.45 p.m.    Individual thinking time aimed at producing rich pictures of the organisational context.

4.30 p.m.  In randomly selected sub-groups: Sharing the individual rich pictures and building a construction of a group rich picture which will synthesize the individual presentations.

6.00.p.m.  Break

8.00 p.m.  Plenary session:     °Presentation of the sub-group rich pictures.
                           °Discussion.
10.00 p.m. Evening break.

Day 3
9 a.m.     Presentation by the consultants:
            "Management as a Self-Regulating Management Process".

10 a.m.    Individual thinking in order to determine the objectives that each of the Institutes should pursue as a means to accomplish the organisational mission
            (This reflection could be based on the discussions concerning the mission of the organisation and the relative importance of its different components which would have taken place on Day 2 as well as on the different official documents.

10.45 a.m.  In randomly selected sub-groups: Sharing the individual rich pictures and building a construction of a group rich picture which will synthesize the individual presentations.

2.00 p.m.  Plenary session: Reporting and discussing.

2.30 p.m.  In sub-groups corresponding to each Institute: answer the following questions:
            °Are the objectives suggested for your Institute by the different groups realistic?
            °Which ones should be modified, taken out or added?

4.00 p.m.  Plenary Session:  Reporting and closure of the session by the consultants and the President.

## THE INTERVENTION ITSELF

Although in its phase one, the methodology presented in the preceding article suggested an initial needs assessment at the beginning of the intervention, the plan of the intervention did not allow time for needs assessment. It was indeed considered by our client that the data gathered in preparing the plan was sufficient. Nevertheless as the President had travelled by bus to the place of the session with his staff, he became aware that his initial requirements had to be questioned. He therefore asked us to suspend the planned evening process and go on with a formal needs assessment.

We went back to the original methodology and asked the participants to share with us their expectations about the outcome and process of the session. In order to do so, each person was asked to share what he or she had in mind at the beginning of the evening session. The main message that was conveyed was that people did not want to be taught to neither did they want to talk about, they wanted to talk to each other according to an agenda that they would set for themselves. In the course of this exchange of ideas, people discovered that after hearing what their colleagues had to say, they already felt that they had to change their own statement and at the end of the evening, many expressed the desire to restate their original statement. They had already learnt about the fragility of their own perceptions as well as about the power of hearing others. In order to take their wishes into account, it was decided to scrap the whole pre-planned session and work with whatever would be *figure* at any given moment in the session. The substance of the evening discussion focussed around the questions of team building and common cause. Indeed they wondered if they could work together towards the pursuit of a shared goal and how they could go about learning how to share whatever they thought. These *figures* happened to coincide with what had been pre-planned for the second day. At the same time, they precluded the necessity of a theoretical presentation of the intervention framework as well as the opening conference that had been planned for the second day. We instead proceeded to phase two of the methodology analyzing the perceptions of the *What for* present in the group.

In order to do so, the participants where asked to read and reflect upon the mission statement that had already been put forward by the organisation. They were asked to clarify what it meant to them as well as their willingness to take ownership of it in whole or in part. They were then asked to share the data in randomly selected sub-goups of five or six people. Each group was to report on a group perception of the mission as well as on the disagreements that had arisen in the course of the discussion.

In the following plenary session, each group was asked to report on the commonalities. Each sub-group was then reconvened and was asked to prepare a new statement using the data gathered in the larger group. At this moment, a certain confusion emerged about the notion of mission. There was a discussion around the question of the nature of the statement that was being worked on. People wondered if it was the statement of a mission or the statement of a purpose or whatever... In the evening, each sub-group reported to the larger group and discussions took place around a possible statement that would suit each and every participant. At the end of the evening, people had agreed on the substance of the statement, and that a few vocabulary kinks had to be ironed out, nothing substantial as they said.

The following morning, the people that had been asked to iron out the vocabulary problems came back with a completely restated proposition. At the beginning of the session, people were asked if they were willing to live with such a statement and contribute to its fulfillment. It soon appeared that under the slight vocabulary differences, there laid profound preoccupations for the survival of some sub-systems of the organisation. The group was then faced with a choice of a statement which by clarifying the *What* or the *What for* of the organisation would possibly eliminate some complete sub-systems or parts of them. There other choice was to concoct some kind of an umbrella mission statement which would protect everyone from the threatening winds of change. Finally, accommodation was chosen and by the same token team building became *figure* instead of harsh rationalization which would be exclusive rather than inclusive. At the end of the morning, following very active discussions, they had agreed on a mission statement. Some participants wondered how this statement could be used as a rallying cry while others wondered if this agreement would be confined to the session itself and forgotten in the real work situation. Aware that team building had become *figure* during the discussions and also considering the importance of symbols as an instrument of cohesiveness in human groups, it was suggested that the mission statement be printed on a formal sheet of paper which would be considered as the Magna Carta of the organisation as well as a concrete outcome of the two day session. After lunch, in a solemn ceremony, each participant was asked to sign the statement which would be reprinted and distributed to all the employees as soon as possible.

As people considered that the mission statement included both the *What for* and the *What*, it was decides to go on to phase four of the methodology which consists of identifying the *How*. In doing so, the process reverted to the initial plan which consisted in dividing the large group into smaller sub-groups each of which corresponded to a Sector of the organisation. Following the same procedure, each individual was asked to answer the following questions: How can your Sector contribute to the mission and, in turn, how can your Institute contribute to the mission of your Sector, taking into account the mission of the organisation. Each Director General clarified for himself how his Institute should contribute to the mission of the organisation and to the mission of his Sector while each Vice-President did the same for his Sector. Each Sector representatives were regrouped together and shared their individual views and clarified what they expected from each other as well as the ways in which they could contribute to the mission of their Sector and, in turn, to the mission of the organisation.

Following the sectorial discussions, a plenary session was summoned in which each sub-group (sector and Institute) presented its own *How* and defined its boundaries. Afterwards, each member of the large group was invited to comment on what he or she had heard and discussions were undertaken to make accommodations around the boundaries that had been proposed. At the end of the discussion, people were willing to agree on each others boundaries as well as on each other prospective contribution to the organisation. Following the discussion, participants came to an agreement on what the next step should be in order to push forward the implementation of the mission statement.

According to Gestalt theory, any cycle of experience needs to have good closure if one wants to be able to undertake a new experience. Considering the session as a global experience, we decided to facilitate such a closure. In order to do so, we asked each participant to express whatever he or she had to say concerning both the process and the content of the session. It was underlined that if one wants to be able to let a new *figure* emerge, one has to make sure that once one has left the closing experience, nothing must be stuck in one's crop. The main ideas that were expressed on that occasion were that people felt that they had not completely attained their objectives but that they were very satisfied of the work that had been accomplished. They also expressed their uneasiness with the fact that it had taken two days of stumbling over words to agree on their perceptions about what they had to do together and about how each individual in each sub-groups could contribute to that pursuit. They felt good that they had experienced openness and collaboration from each other and that they had discovered a great deal of commonality in their values and aspirations. They also felt very grateful that the consultants had heeded to their request and had not talked to them but rather had been patient and skillful in facilitating smooth communication among them. They felt that the framework under which they had worked was very conducive of frank discussions with people that they hardly knew before the session.

As mentioned in the methodology reported in the preceding article, the identification of the *Who* of each participant as well of that of the group is an important part of untangling the perception web. In the course of the whole process, as each participant had to cope with different choice points concerning *what when, where, how* he or she wanted to say, he or she became more aware of his or her resistances and his or her limitations as well as potential as communicator and as a sharing member of a group.

Each participant also became aware of the potential for productive relationships which emerged as people witnessed what they were and shared their ideas. They also learnt about possible sources of conflict as well as sources of collaboration among peers. They became aware of the norms of inclusion and exclusion that can be risen in the context of their organisational culture. In their feedback, they also underlined that the community that had emerged during the session was respectful enough to allow each person to feel at ease in expressing his or her own ideas or to refuse to share them. The quality of I-Thou relationships that was experienced in many instances was also a great opportunity to discover very valuable collaborators.

# THE LEARNINGS

Once again, this intervention gave us the opportunity to experience the fact that the main quality of a plan is that its form can be put aside but that its substance must be completely appropriated so that the intervenors can use it as a guideline in focussing the *figures* that emerge as the process evolves.

We were also reminded of the importance of clarifying the contract with the real leader of the organisation rather than with impersonal services which negotiate contracts. The contract must set clear boundaries which in term define clear roles. It must clarify if the consultants are the leaders and planners of the session or if they are merely process experts that can be consultant if need be. We also learnt the major importance of clarifying the theoretical framework of the intervention in order to avoid confusion and misunderstanding. One must not sacrifice clarifying the background of the intervention in order to respond to momentary needs of the participants. Whatever the degree of sophistication of the participants may be, it appears imperative to agree on operational definitions of terms and not assume shared meanings on any of them.

# REFERENCES

Checkland, Peter B. (1980). "Systems Thinking, Systems Practice". Chichester: Wiley.

Cohen, Louis and Manion, Lawrence (1985). "Research Methods in Education". London: Croom Helm.

Vickers, Sir Geoffrey. (1983). "The Art of Judgement: A Study of Policy Making". London: Harper and Row

SYSTEMS PRACTICE: AN APPLICATION OF "SOFT" SYSTEMS THINKING

IN THE CONTEXT OF IS PLANNING FOR SMALL BUSINESS

Julie Travis

School of Information Systems
Curtin University of Technology
Bentley, Western Australia

INTRODUCTION

In order to facilitate learning and improve the situation of the small business, action research is being carried out with an independently owned pharmacy in Western Australia (as an example of small business) using the soft systems methodology (SSM) to help determine strategic directions for small business owners. This study is an exercise in rational intervention to support action as well as research. It is concerned about the needs of small business and how small business might be able to  better empower themselves by using a constructive framework for information systems planning.


WHY ISP IN THE CONTEXT OF THE SMALL BUSINESS?

Small business has been defined by The Small Business Development Corporation of Australia, (1990), as those businesses having ... "fewer than 20 employees, which are run by its owners, have a relatively small share of the market they operate in and do not form part of a larger organisation". This study has followed those guide-lines.

While small business comprises such a large percentage of industry in Australia - between 94 and 96%, there are some issues of concern which continue to thwart  the progress of small business owners.  In retrospect, the failure of small firms have always shown patterns of some inadequacy or inappropriateness (Back 1979).  Business and finances are often inadequate (Johnson, 1978), access to long term loans may be difficult, and government assistance may often be inappropriate (Back, op cit).  However, there are some primary issues of concern which this study is  addressing.

The first issue is the extremely high failure rate of small business. This is said to be approximately 80% however, according to The Small Business Development Corporation of Perth, Western Australia, it is hard to match such a percentage against such a general description such as "failure" without  qualifying the meaning of "failure".

It may be that the business in question must determine those definitions according to their own situation. In a survey undertaken by Galliers in 1986, information systems professionals were asked to define success factors and to evaluate the success of ISP from the viewpoint of their senior management. From this UK survey, 41% admitted that they were unsuccessful even though they had defined their own success measures

Although no distinction was made between small and large firms in the survey by Galliers (1986), the study served as a fair indication of the failure of direction or application of information systems planning.

The second issue is the lack of information represented in the literature distinguishing small business practice from large business practice. Furthermore, little, if any, of the literature distinguishes between large and small business on the subject of information systems planning. The existing literature describes mostly large systems either in America or the United Kingdom and mostly concerns itself with topics related to user satisfaction (Raymond 1985).

The third, and possibly one of the the most significant issues, is that managers often try to make important decisions about the business with inadequate or inaccurate information as demonstrated by a report of the Committee on Small Business (1971). Even though this report was carried out seventeen years ago it is still currently a topic of debate.

The fourth issue, is the lack of managerial experience or expertise. According to Taylor (1987), results from longitudinal studies carried out by Professor Alan Williams of the University of Newcastle in Australia, show that it is chiefly inexperience, incompetence or lack of managerial expertise which contribute to the causes of the demise of the small business. However, Taylor (1987) also states that considering the study was conducted on small business in the United States, the evidence available was unable to support the figures represented as an example of the Australian situation.

THE RESEARCH METHOD AND THE METHODOLOGY APPLIED

As the topic under investigation is a largely unexplored area having no literature specifically on Information Systems Planning for Small Business to draw upon, action research is considered to be the most appropriate.

The nature of action research is documented by Galliers (1984) who states that the major purposes are to improve efficiency and effectiveness of the organisation, to investigate IS failures and develop IS approaches. The method considers the impact of IT and IS where individuals, the organisation and society are concerned. Key features include applied research where there is an attempt to obtain practical results resulting in valued information to the relevant groups concerned. The groups would be those with whom the researcher is allied while simultaneously adding to the body of knowledge.

However, as subjectivity and ethical standards are concerns of the researcher applying this method and may be a threat to research findings in one way or another, it was decided that Seedhouses' Ethical Grid, as suggested by C J Atkinson (1989), would be a way to overcome this concern.

This grid consists of four layers: external considerations, consequences of possible actions, "oughts" and "core concerns or rationale", and is further divided into more specific ethical considerations. These considerations are matched against any phase of the application of SSM where major decisions are considered to be made (see Figure 1).

The approach to planning used in this study is adopted from the soft systems methodology: Checkland (1981), Wilson (1984) and Galliers (1988). Although these authors may vary their approach , the commonality lies in the two states of communication, ie the "existing" and "conceptual" worlds.

Within these spheres of thinking the methodology supports the chosen method of research in the facilitating of organisational change without being driven by theory. Moreover, the soft systems methodology (SSM) is recognised as being doubly systemic in that it is itself a methodological framework, which acts as a learning system and uses system models of human activities  Checkland (1985).

BACKGROUND OF THE SMALL BUSINESS IN BRIEF

The retail pharmacy has only been in operation under the present owners for a few years.  The business has seven full-time members of staff and three part-time.  The organisational structure places two owner managers at the top, a manageress/ stock controller beneath.  Three senior full-time staff answer to the mangeress and a full-time junior plus three other part-timers answer to the three full-time seniors.

While lines of authority are mapped out quite clearly, a factor identified as being of utmost importance to the all staff, is the ability for senior staff to think at all hierarchical levels of management so that the nature of the business changes from being reactive (in having to rush from one crisis situation to another) to being proactive.

THE APPLICATION OF SYSTEMS THINKING

The main aim of this research project is to facilitate learning in a small business using an adaptable methodological framework to support and encourage desirable and feasible change towards what the business recognises as being the desired outcomes. The conclusion of this research will document findings on how and where the methodological framework is successful or not for small business.  However, this paper serves to demonstrate how systems thinking brought about the activities for a framework that was constructed considering the environment and situation of a particular small business, and how it might further facilitate learning about this business.  The framework is illustrated as figure 1 below.

The methodologial framework may be applied to any organisation regardless of size, but the key activities are directed specifically at the small business situation.  The activities described below contain some empirical findings arising from the application of this study.as far as it has been completed.  The outcomes described below are to phase 3.  Even though the first iteration of phases four and five are complete, it is only the expected deliverables of the study that can be documented from phase 4 onwards.

1.Prepare for the study is a loose term, but  includes discussion about past activities as a continuous assessment for the application of the framework to the study.
Outcomes:  One outcome of this activity was formulated by examining historical data and formalising the business plan before starting an information systems planning study.  Overall objectives and statements were discussed at this point.  Another outcome was to determine the major focus of the study.  This was to identify and optimise information systems.  Among the expected long-term outcomes was: profitability, competitive advantage, a self-sufficient system with devolved responsibility, homoeostasis/synergism,

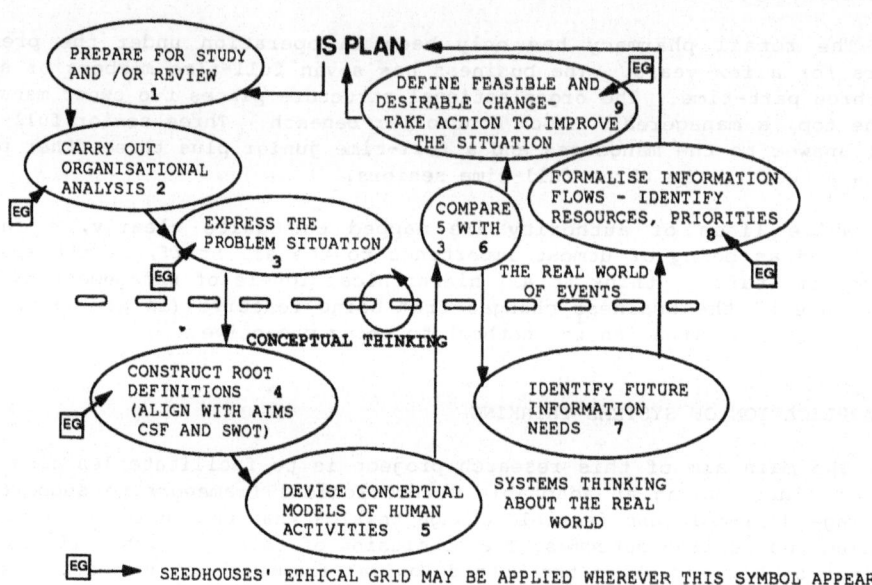

**IS PLAN**

PREPARE FOR STUDY AND /OR REVIEW 1

DEFINE FEASIBLE AND DESIRABLE CHANGE- TAKE ACTION TO IMPROVE THE SITUATION 9

CARRY OUT ORGANISATIONAL ANALYSIS 2

EXPRESS THE PROBLEM SITUATION 3

COMPARE 5 WITH 3 6

FORMALISE INFORMATION FLOWS - IDENTIFY RESOURCES, PRIORITIES 8

THE REAL WORLD OF EVENTS

CONCEPTUAL THINKING

CONSTRUCT ROOT DEFINITIONS 4 (ALIGN WITH AIMS CSF AND SWOT)

IDENTIFY FUTURE INFORMATION NEEDS 7

DEVISE CONCEPTUAL MODELS OF HUMAN ACTIVITIES 5

SYSTEMS THINKING ABOUT THE REAL WORLD

EG ➝ SEEDHOUSES' ETHICAL GRID MAY BE APPLIED WHEREVER THIS SYMBOL APPEARS

FIGURE 1.

THE SOFT SYSTEMS METHODOLOGICAL FRAMEWORK IN THE CONTEXT OF THE SMALL BUSINESS innovative staff activities, and more effective staff training and development programmes.

2.<u>Carrying out organisational analysis</u> is the activity where the small business owner would consider who are the stakeholders, who are the competitors and recognise the environmental situation.
<u>Outcomes</u>:A determination of opportunities and threats from strengths and weaknesses led to examine future growth areas. Major factors were identified as follows:
> Weakness-Restriction on building location
> Weakness-Restriction on decision making place and time
> Threat-Too many competitors - and too close
> Weakness-No adequate competitor information
> Threat-Government restrictions on conditions of practice
> Opportunity-Capacity for innovation and diversity
> Strength-Relatively small overheads

> Major <u>critical success factors</u> identified were to:
> -achieve good human resource management
> -recognise environmental needs
> -achieve more structured planning and learn better
>  managerial skills
> -achieve competitive advantage through profitability,
>  productivity and marketing strategy

3.<u>Expressing the problem situation</u> may be accomplished using various techniques. In this case a brainstorming technique, such as a problematique enabled the situation to be expressed with links from one situation to another.
<u>Outcomes</u>:The factors that were perceived to be most problematic were:

> -No planning framework
> -System currently financially unstable
> -Current system rules and policies need to be formalised
> -Staff lines of responsibility and leadership are not
>  clearly mapped
> -The stock system is too complex
> -Inadequate staff training and development

4-7. Root definitions are devised from CATWOEs and formulate the system objective from a holistic view of the organisation down to operational level sub-systems: Checkland (1981).
<u>Expected Deliverables</u>: Steps from 4 through to 7 allow the business to identify systems objectives and future models of the business systems. In this case the most useful outcome was considered to be the realisation of how the organisation is mapped systemically rather than departmentally.

8.<u>Formalising information flows</u> was considered to be a necessary step for small business. To document and analyse information flows within systems and inter-systems, tools such as the Maltese Cross as described in Wilson (1984) were discussed as being a possible tool for analysis and design.
<u>Expected Deliverables</u>: The outcomes for this step are expected to allow the small business to arrive, full circle from top-down planning at the strategic level to organising desired inputs, outputs and through-flows at a more practical level. It was understood by the organisation that this would be an important exercise involving all members of staff, and one which would also lay the foundation for the specifications for a computerised system should it be required.

**9.Define feasible and Desirable Change - Take Action To Improve The Situation**. Whilst identifying information flows, this activity attempts to balance that which is perceived to be (systemically) desirable with that which is (culturally) feasible.

**Expected Deliverables**: Recommendations for action were considered by key players to be the basis for the information systems plan, which may include a comprehensive set of considerations for IT acquisition matching identified information systems. While the aim of this activity is an agenda for action, all members of the business agreed that there must also be provision for review.

CONCLUSION

This study provides the small business with a means to identify valuable connections throughout the internal and external environments and to understand that small business must plan tactically and strategically as well as concentrate on the day-to-day issues.

Through this study the small business is expected to identify what success and failure means to them and to question their existing situation in the real world. It then permits the individual to pull together "real world thinking" and map this with "conceptual thinking" to actually create that which is visualised.

After examining the problem of how staff become more aware of the use of information as a resource at the appropriate levels of planning, there was one possible solution favoured by the majority of staff. This was that in order to become more competitive, the firm must be more aware of the external environment. To be more aware of the kind of information needed at tactical and strategic levels of planning, the possible solution was not to employ more managers but to actually carry out planning at all three levels This was considered by the business to have to involve better time management.It was deemed important to find the place and time away from daily activities so that the senior members of staff have time apart in which to plan.

There are a few limitations, in that this study does not take into account the growth process of the small business, nor at which point the business may fail, so this is clearly a limitation of action research. Conversely, it is not an objective of this study to seek out a small business that has considered itself to have failed, but rather to facilitate a surviving business into a state of awareness about how it might plan more effectively.

REFERENCES

| | | |
|---|---|---|
| Atkinson, CJ | (1989) | Ethics: A Lost Dimension in Soft Systems Practice - Journal of Applied Systems Analysis-Vol 16 pp 43-53 |
| Back, R D | (1978) | The Practising Accountant in Queensland as an Advisor to Small Firms: Working Paper - Dept of Commerce; James Cook University |
| Checkland, P | (1981) | Systems Thinking, Systems Practice - Wiley |
| Checkland, P | (1985) | Achieving "Desirable and Feasible" Change: an Application of Soft Systems Methodology-Journal of the Operational Research Society-Vol 36 No 9 pp 821 - 831 |

Delone, W H        (1988)    Determinants of Success for Computer Usage in
                             Small Business- MIS Quarterly - (March )  pp 51-
                             62

Galliers, R D      (1984)    In Search of a Paradigm for Information Systems
                             Research - Research Methods in Information
                             Systems. Proceedings of the IFIP WG 8.2
                             Colloquium-Enid Mumford et al ed-, Manchester
                             Business School, 1-3 September pp 281-297

Galliers, R D      (1986)    Strategic Planning: A Failure of Direction -
                             Business Computing and Communication-July/August,
                             pp 34-38

Galliers, R D      (1988)    Information   Technology   Management   for
                             Productivity and Strategic Advantage - IFIP TC8
                             Open Conference, Institute of System Science,
                             National University of Singapore - March 7-8

Johnson, P         (1978)    Policies Towards Small Firms: Time for Caution? -
                             Lloyds Bank Review-No 129 pp 1-11

Montazemi, Ali R   (1988)    Factors affecting information satisfaction in the
                             context of the small business environment - MIS
                             Quarterly - (June )  pp 239-256

Raymond, Louis     (1985)    Organisational characteristics and MIS success in
                             the Context of Small Business - MIS Quarterly
                             (March )

Report of the..    (1971)    Committee on Small Business  (Wiltshire Report) -
                             Canberra:AGPS

Small Business     (1989)    Development  Corporation:  The  Small  Business
                             Handbook, SBDC

Small Business     (1990)    in  Australia:  Challenges,  Problems  and
                             Opportunities:  Report  by  the  House  of
                             Representatives Standing Committee on Industry,
                             Science  and  Technology  -  January  1990  -
                             Parliament  of  the  Commonwealth  of  Australia  -
                             Canberra: AGPS

Taylor, L G        (1987)    Starting  and  Managing  a  Small  Business:An
                             Australian Guide, TAFE Educational Books

Wilson, Brian      (1984)    Systems Concepts, Methodologies and Applications
                             - Wiley

# SOFT SYSTEMS IN SOFTWARE DESIGN

Lars Mathiassen
Andreas Munk-Madsen*
Peter A. Nielsen
Jan Stage

Aalborg University, Frederik Bajers Vej 7, DK-9220 Aalborg Ø, Denmark
*Metodica, Nyvej 19, DK-1851 Frederiksberg C, Denmark

## INTRODUCTION

This paper explores the possibility of applying soft systems thinking as a basis for designing application software and it outlines a new method for software design (Mathiassen *et al.* 1991). The method is called 'Rapid Systems Modelling'. It supports systems developers and users in going from a problematic organisational situation to the design of a new and modified computer application for that situation.

Rapid Systems Modelling combines a set of widely appreciated principles and methods into one coherent framework. The approach taken to design emphasises learning as in Soft Systems Methodology (Checkland 1981). Rapid Systems Modelling combines this approach to learning with techniques and tools for modelling and experimenting with systems based on object-oriented thinking and the use of prototypes (Birtwistle *et al.* 1973, Jackson 1983, Coad and Yourdon 1990, Budde *et al.* 1984). The use of Rapid Systems Modelling is controlled through risk management (Boehm 1988, Boehm 1989) and a coherent design proposal is produced based on the idea of faking a rational design process (Parnas and Clements 1986).

Until now, Soft Systems Methodology has been used in various organisational settings and disciplines (Checkland 1981, Checkland and Scholes 1990). Attempts have also been made to adapt soft systems ideas to information systems development, cf. (Wilson 1984, Wood-Harper *et al.* 1985, Avison and Wood-Harper 1990, Stowell *et al.* 1990). All of these efforts are concerned with organisational change and the modelling involved is based on rigorous use of *human* activity systems. This paper reports from ongoing research where we attempt to adapt and supplement soft systems ideas to make them useful in a specific *technical* domain, i.e. the design of computer applications as an integral part of organisational change.

## EXPLORING SOFT SYSTEMS IDEAS

We start by briefly reviewing soft systems ideas in relation to traditional software design methods. On that basis we discuss possibilities for adapting soft systems ideas to the design of computer applications.

*Systems Thinking in Europe*, Edited by M.C. Jackson *et al.*
Plenum Press, New York, 1991

## Idea 1: Systems as Intellectual Constructs

The very idea of Soft Systems Methodology is that we may inquire into a problematic situation by means of the notion of system. Systems are intellectual constructs making explicit our subjective meanings attributed to reality and our visions about reality. Multiple perceptions are exploited to learn about and eventually improve a problematic situation.

In software engineering the term 'system' is seldom defined, but computer applications are taken to *be* systems. The term 'system' gets its semantics implicitly through a set of tools and techniques for specification of computer systems. Well-known examples are Structured Analysis/Structured Design (DeMarco 1979, Yourdon 1989), Jackson System Development (Jackson 1983), and Object-Oriented Analysis (Coad and Yourdon 1990). The missing or weak distinction within software engineering between the world of phenomena and the world of perceptions has practical consequences. Traditionally, software engineers conceive a computer application as restricted to the automatic execution of the corresponding program on a computer disregarding the perceptions and actions of users. Concerns are seperated and the computer system is thought of as something in itself. The designers' task is reduced to specification of a program meeting predefined and stable requirements (Floyd 1987).

The development of Rapid Systems Modelling is rooted in a tradition where systems consistently have been viewed as intellectual constructs dialectically related to the phenomena of computer applications. Thus, the exploitation of multiple viewpoints on the same computer application has been emphasized (Mathiassen 1981, Nygaard and Sørgaard 1987, Stage 1989, Nielsen 1990). In Rapid Systems Modelling we take the position that the idea of 'systems as intellectual constructs' is applicable not only in learning about human activity but also in designing computer applications.

## Idea 2: Learning Through Action and Reflection

Soft Systems Methodology is based on the idea that effective learning takes place as an interaction between real world activities and thinking about the real world in terms of systems. The problematic situation is experienced and expressed, different systems are defined and modelled, and these models are then in turn confronted with the real situation. In this way, models are used to structure and orchestrate a debate amongst actors in the situation with the purpose of learning about the problematic situation.

Conventional software development methods strongly emphasize description, specification and modelling of the system to-be. There is a number of widely accepted techniques for evaluating specifications, e.g. structured walkthroughs and reviews, cf. (Freedman and Weinberg 1982). But only a small number of techniques are provided to express situational characteristics in an informal and loosely structured way. Some of the rare examples are: event lists (Yourdon 1989) and lists of nouns and verbs (Jackson 1983). Generally, there is a growing appreciatiation of the idea of learning in software engineering, but there are still few frameworks that utilises the relationship between action and reflection in a systematic way. Instead, there seem to be two competing strategies: the specification approach relying strongly on reflection before action, and the prototype approach relying mainly on experiments (actions) without emphasizing systematic reflection (Mathiassen and Stage 1990).

Rapid Systems Modelling is based on the idea that software design requires learning and methods should thus support this by exploiting the relationship between action and reflection. One example of the application of this idea is the Spiral Model (Boehm 1988). Rapid Systems Modelling rejects the standpoint that prototypes and specifications represents two competing strategies. Instead, prototypes and specifications are seen as two complementary ways of expressing reflection in software design.

## Idea 3: Systems as Wholes

At the heart of soft systems thinking is the principle that whole entities exhibit emergent properties which are meaningful only when attributed to the whole, not to its parts. In this sense, Soft Systems Methodology utilizes holistic thinking. A conceptual distinction is made between what a system is (emergent properties) and what it does (constituent activities and relationships), and a practical distinction is made between defining the system and modelling its activities.

Methods for software development emphasize detailed and elaborate specification of systems. The methods distinguish between different levels of abstraction and different aspects, e.g. data flow and data definition. But overview and detail is provided without explicit conception of the system as a whole. A few methods suggest to define the purpose of the computer system, e.g. 'statement of purpose' in Yourdon's modern version of Structured Analysis/Structured Design (Yourdon 1989). Despite this, traditional software engineering methods support development of reductionistic models.

Rapid Systems Modelling supports designers in defining emergent properties of the systems explicitly *in addition* to modelling their contents. This is in accordance with the ideas behind Soft Systems Methodology. This position is further discussed in the following two sections.

## Idea 4: Defining Systems

One of the main activities of Soft Systems Methodology is the definition of systems by formulation of root definitions. A root definition is a precise description of the emergent properties of a system. It is suggested that a root definition should contain the CATWOE elements explicitly: Customers, Actors, Transformation, Weltanschauung, Owner, and Environment. These six elements are closely connected to the idea of human activity systems.

In Rapid Systems Modelling, systems are to be defined in a similar way. The exact form of a definition is yet to be found, but certain differences seem obvious. Firstly, when understanding computer applications 'transformation' is questionable as the key aspect of a system. The strong interactive nature of modern computer applications suggests metaphors like 'actor', 'agent', 'medium', and 'tool' each implying somewhat different systems concepts. Secondly, the notion of Weltanschauung plays a crucial role in Rapid Systems Modelling, but it needs to be specifically oriented towards the assumptions underlying a particular computer system and its relation to wider human activity systems. Thirdly, the other CATWOE elements have to be reconsidered and possibly supplemented by other aspects relevant to the technical domain of computer applications, e.g. interface facilities and technological platform.

## Idea 5: Modelling Systems

In Soft Systems Methodology, a conceptual model contains the minimal set of related (human) activities needed to carry out the transformation described in the corresponding root definition. A system is thought of as being adaptive. A set of monitoring and controlling activities are therefore included in each model. The conceptual model must be defensible against the root definition and vice versa.

In Rapid Systems Modelling each system is going to be modelled. The models are evaluated and compared with the purpose of eventually arriving at a design proposal. The flavour of the models in Soft Systems Methodology is inherited, but the models have to be different from conceptual models to support reflection on the technical domain. As a consequence, the method supports the use of two types of models: object-oriented specifications and prototypes.

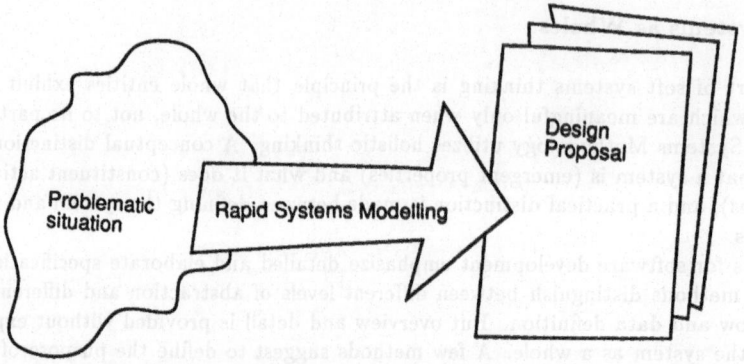

Figure 1. The overall transformation of Rapid Systems Modelling

The method also recommends to use different versions of models displaying different levels of detail. The purpose is to support organisational and technical learning and to facilitate choice among alternative systems.

## OUTLINE OF 'RAPID SYSTEMS MODELLING'

The software design method, Rapid Systems Modelling, combines and adapts already established ideas and methods about learning, modelling and management. The ideas and methods are combined and projected into the domain of designing computer applications for specific organisational settings. We are in the midst of trying our ideas in practice and in education. This, in turn, will reshape the proposed method and hopefully make it more useful. In the following we present a first version of the method based on our experience with each of the ideas and methods underlying Rapid Systems Modelling.

### Overall Transformation and Basic Activities

The area of concern is analysis and design of computer applications. We are interested in supporting systems developers and users in going from an unstructured organisational situation with an expressed need for improved application of computers to an agreed-upon proposal for a new or modified computer application. This overall transformation of Rapid Systems Modelling is illustrated in Figure 1.

Our approach to this transformation is shown in Figure 2. Rapid Systems Modelling consists of three strongly related activities. The method emphasizes learning about the problematic situation, technical possibilities in terms of computer systems, and the relationship to the organisational setting. The approach to learning is Soft Systems Methodology, but the specific techniques are adapted from software development. Learning is in our view a necessary and highly underrated activity in software design, see (Floyd 1987). Still, other activities

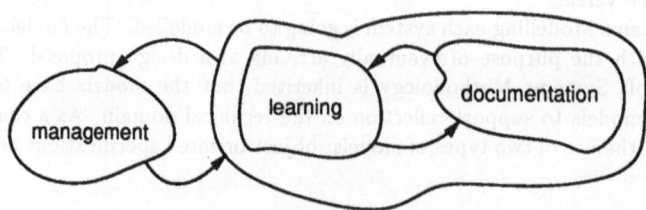

Figure 2. The main elements of Rapid Systems Modelling

are also important: proper documentation is crucial and effective management of resources is required.

## Modelling Based on Prototypes and Object-Orientation

Our approach to learning is illustrated in Figure 3. In applying Rapid Systems Modelling, several concrete learning processes are initiated. The initiation of a learning process is based on management considerations. Learning processes can be performed in parallel each requiring different amounts of resources and applying different types of modelling techniques.

Rapid Systems Modelling supports learning processes based on object-oriented specification. Object-oriented thinking supports designers in dealing with complexity by extracting in a condenced form the fundamental properties of a computer system. Techniques for defining systems and for modelling them as interacting sets of objects are provided. The specific outlook of well-formulated root definitions is still to be found. The object-oriented models are based on an integration of the abstract datatype approach of Object-Oriented Analysis (Coad and Yourdon 1990) and the communicating sequential processes approach of Jackson System Development (Jackson 1983). Our approach to object-oriented modelling utilizes the encapsulation and abstraction mechanisms and the focus on data relationships as suggested by Coad and Yourdon, and the application of the idea of simulation of the real world as suggested by Jackson.

Rapid Systems Modelling supports learning processes based on prototyping. Prototyping supports learning about the practical effect of specific design proposals. Rapid Systems Modelling provides techniques for defining systems and techniques for modelling these as computer-based prototypes. Also in this context, the specific outlook of well-formulated root definitions has yet to be found, though initial experiments have been performed (Bondgård *et al.* 1990). A significant aspect of this type of learning process is to ensure the systematic experimentation with use of protypes in realistic settings.

## Producing a Consistent Outcome

Proper documentation is crucial in software development. The outcome of Rapid Systems Modelling is a design proposal to be used as the basis for further development. The documentation is used as the basis for technical design and implementation, it is used to support division of labour between systems developers, and it plays a key role in assuring a satisfactory quality of the final computer application. The documentation activity of Rapid Systems Modelling evaluates and documents relevant insigths gained through the learning activities. The

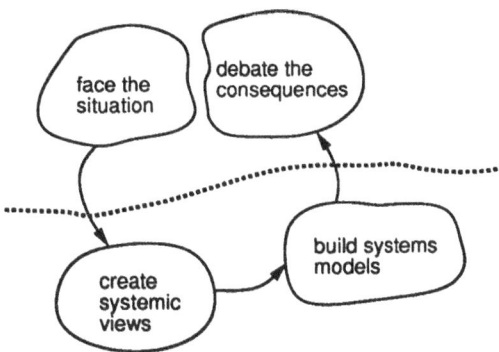

Figure 3. The learning activities of Rapid Systems Modelling

approach taken is inspired by Parnas and Clements (Parnas and Clements 1986) and principles for the resulting design document are being developed (Parnas 1972, Parnas *et al.* 1985, Parnas and Clements 1986, Stage 1989).

## Managing Risks

Learning in general as well as software development in particular are intrinsically open and experimental in nature, see for example (Floyd 1987). At the same time, software development is a resource demanding activity and effective management is required. Risk management, as proposed by Boehm (1988, 1989) offers an approach to management in software development that seems to handle this dilemma. Focus is on situational risks, i.e. uncertainties and complexities involved in deciding on relevant and useful actions. Techniques are provided for identification of risks (i.e. a need for learning) in the design situation, for assigning priorities to identified risks, and for practical planning of learning processes to resolve risks. A specific version of this approach is being developed that also involves monitoring and controlling the learning and documentation activities (Larsen *et al.* 1991).

## SUMMARY

The purpose of this paper has been to argue for the application of soft systems ideas in relation to software design. We have done this by discussing some of the fundamental aspects of soft systems thinking in relation to software design. The argument has been further substantiated by outlining how soft systems ideas can be supplemented and projected into this area of technical and organisational change.

The paper reports from an ongoing research programme of which the basic assumptions and ideas have been expressed at an overall level. Substantial questions and many details are yet to be investigated.

*The research behind this paper has been partially financed by the Danish Natural Science Research Counsil, Programme No. 11-8394.*

## REFERENCES

Avison, D. and A. T. Wood-Harper. *Multiview: An Exploration in Information Systems Development.* Blackwell Scientific Publications, Oxford, 1990.

Birtwistle, G. M., O.-J. Dahl, B. Myrhhaug, and K. Nygaard. *Simula BEGIN.* Studentlitteratur and Petrocelli/Charter, Lund and New York, 1973.

Boehm, B.W. A spiral model of software development and enhancement. *Computer*, May 1988.

Boehm, B. W. *Software Risk Management.* IEEE Computer Society Press, Washington, D. C., 1989.

Bondgård, P., E. Degn, and K. Vraagaard. Prototyping: Forståelse og forandring (Prototyping: Understanding and change. Master's thesis, Dept. Computer Science, Aalborg University, 1990.

Budde, R., K. Kuhlenkamp, L. Mathiassen and H. Züllighoven, editors. *Approaches to Prototyping*, Berlin, 1984. Springer-Verlag.

Checkland, P. B. and J. Scholes. *Soft Systems Methodology in Action.* Wiley, Chichester, 1990.

Checkland, P.B. *Systems Thinking, Systems Practice.* John Wiley and Sons, Chichester, 1981.

Coad, P. and E. Yourdon. *Object-Oriented Analysis.* Yourdon Press and Prentice-Hall, Englewood Cliffs, New Jersey, 1990.

DeMarco, T. *Structured Analysis and System Specification.* Yourdon Inc. & Prentice-Hall, Englewood Cliffs, NJ, 1979.

Floyd, C. Outline of a paradigm change in software engineering. In G. Bjerknes et al., editors, *Computers and Democracy*, pages 191–210, Aldershot, 1987. Avebury.

Freedman, D. P. and G. M. Weinberg. *Handbook of Walkthroughs, Inspections, and Technical Reviews.* Little, Brown and Company, Boston, 1982.

Jackson, M. *System Development.* Prentice-Hall, Englewood Cliffs, NJ, 1983.

Larsen, T., C. Millum, H. Solberg, and F. Tolstrup. *En risikoledelsesmodel til design af information systemer (A risk-based model for designing information systems)*. Master's thesis, Institute for Electronic Systems, Aalborg University, 1991.

Mathiassen, L. and J. Stage. Complexity and uncertainty in software design. In *Proceedings of the IEEE International Conference on Computer Systems and Software Engineering*, pages 482–489. IEEE Computer Society Press, Washington DC, 1990.

Mathiassen, L., A. Munk-Madsen, P. A. Nielsen, and J. Stage. Rapid systems modelling: The soul of a new method. Technical report, Dept. of Computer Science, Aalborg University, February 1991.

Mathiassen, L. *Systemudvikling og systemudviklingsmetode (System Development and System Development Method)*. Computer Science Department, Aarhus University, 1981. PB-136.

Nielsen, P. A. *Learning and Using Methodologies in Information Systems Analysis and Design*. PhD thesis, Dept. of Systems and Information Management, Lancaster University, July 1990.

Nygaard, K. and P. Sørgaard. The perspective concept in informatics. In G. Bjerknes et al., editors, *Computers and Democracy*, pages 371–393, Aldershot, 1987. Avebury.

Parnas, D. L. and P. C. Clements. A rational design process: How and why to fake it. *IEEE Trans. Software Eng.*, SE-12(2):251–257, February 1986.

Parnas, D. L., P.C. Clements, and D.M. Weiss. The modular structure of complex systems. *IEEE Trans. Software Eng.*, SE-11(3):259–266, March 1985.

Parnas, D. L. On the criteria to be used in decomposing systems into modules. *Comm. ACM*, 15(12):1053–1058, December 1972.

Stage, J. *Mellem tradition og nyskabelse. Analyse og design i systemudvikling (Between Tradition and Transcendence. Analysis and Design in System Development)*. Institute for Electronic Systems, Aalborg University, 1989.

Stowell, F. A., P. Holland, P. Muller, and R. Prior. Applications of SSM in information systems design: Some reflections. *Journal of Applied Systems Analysis*, 17:63–70, 1990.

Wilson, B. *Systems: Concepts, Methodologies, and Applications*. Wiley, Chichester, 1984.

Wood-Harper, A. T., L. Antill, and D. Avison. *Information Systems Definition: The Multiview Approach*. Blackwell Scientific Publications, Oxford, 1985.

Yourdon, E. *Modern Structured Analysis*. Prentice-Hall, 1989.

# A FIRST STEP TOWARDS THE AUTOMATION OF SSM?

Dr. F.A. Stowell and Mark Stansfield
School of Information Science
Science Faculty
Portsmouth Polytechnic
Portsmouth Tel. (0705) 843017; Fax (0705) 843334

## INTRODUCTION

A strength of SSM lies in the hermeneutic/phenomenological process of inquiry. The methodology is an operationalisation of Vicker's notion of appreciation which helps both practitioner and client(s) to gain an understanding of the perceived problem situation. The appreciative process applies equally to the practitioner as it does to the client(s) and seems to rely to a great extent upon the interaction that takes place between them. An important feature of SSM is that it offers a means of appreciating the problem without imposing the structure of some predetermined model. It would seem therefore, that the idea of automating SSM is dialectically opposed to the principles that underpin the methodology. However, an investigation has been undertaken to explore the possibilities of applying information technology to some aspects of the methodology. This paper will describe the research which has explored the possibility of incorporating the methodology within an expert system framework.

## CAN SSM BE AUTOMATED?

Our reaction to this proposal was, at first, negative but upon reflection we realised that this was to take the proposal too literally. The idea of attempting to program a methodology which is claimed to be phenomenological in concept (Checkland, 1981) was to take on an impossible task. However, the idea of providing a computer based leaning package to aid the student of SSM in his endeavours was less daunting. The computer as an aid to learning has many advantages. It never tires nor loses patience, it can provide a stimulating interaction and possesses novelty (at least for the first few attempts at using it). We decided to con-template the use of an expert system as a means of supplementing tutorials.

## EXPERT SYSTEMS

There have been many different ideas expressed about the nature of an expert system, but few find universal acceptance. We

*Systems Thinking in Europe*, Edited by M.C. Jackson *et al.*
Plenum Press, New York, 1991

would argue that an expert system should be viewed as a tool for use by a human to aid both decision-making and learning. The expert system should be seen as an aid but the responsibility for resultant action is that of the human and not the machine. Waterman (1986) argues that in order to carry out this function expert systems must have: (i) a knowledge base which contains a representation of the knowledge that is required, and deal with subject matter of realistic complexity that normally requires a large amount of human expertise. (ii) An inference mechanism which is the means by which this knowledge is handled, and through which it must be capable of explaining and justifying solutions and recommendations. (iii) An input/output interface which enables the user to supply facts and data, and enables the expert system to ask questions or supply advice and explanations.


## SHOULD IT BE DONE?

In many instances SSM is taught as one of a number of problem-solving methodologies and all within a tight allocation of time. Students who find the methodology of interest often use the tutorial time to discuss the concept of SSM and their own time to practice the ideas. It seemed to us that the addition of a computer aided learning package to the reference texts (e.g. Checkland, 1981, 1990 and the Open University) could be a useful tutorial aid.


## DEVELOPING THE EXPERT SYSTEM

One important question we needed to address before developing the expert system was to identify which aspects of SSM were to be chosen for incorporation into the expert system framework. Second, and an equally important problem, was to identify the limits of the knowledge bases which were to make up the expert system. The answers to these questions were to be found in an examination of the fundamental uses of expert systems, namely in providing advice and solving problems. Thus a potential area of SSM which might form the basis of an expert system application was in the teaching of the methodology. In adopting this aspect of SSM, the limits of the knowledge bases were to be determined by the seven stages which make up the methodology itself. Thus, the aspect of SSM to be adopted for incorporation into an expert system framework was to be a computer aided learning approach in developing the skills and knowledge required to become a competent user of SSM. The area of application was therefore identified, as was the boundary of the knowledge bases.

### Knowledge Elicitation

An area of fundamental importance to the design of an expert system was that of acquiring the knowledge. The process of knowledge acquisition is the process of acquiring the knowledge needed to power an expert system and structuring that knowledge into a usable form. The knowledge acquisition process used to form the basis for the design of the expert system was both structured and iterative. The main methods of knowledge elicitation that were used were prototyping and interviewing with the use of questionnaires and "talk throughs" for interesting and "live" examples in which SSM could be applied. Many inadequacies and problems were found to exist with these so called 'traditional' knowledge elicitation techniques, which were highlighted by this exercise. The difficulty, we would argue, in using SSM relates to the practitioners appreciation, consciously or unconsciously, of these underpinning philosophies. Checkland and Scholes make a similar point when discussing the way in which SSM is used and the results obtained by the various practitioners (Checkland and Scholes, 1990). But this has been discussed elsewhere and is not the subject of this paper, (see Stowell and West, 1987; 1988; 1989; 1990).

Building the Tutoring System

The approach adopted in the design and development of the tutoring system was a modular approach. This involved splitting it into sections, designing and developing a prototype for each section and after validation linking the separate sections together to form the overall expert system itself which was subsequently called "SSM AID". The tutor system was developed using a rule-based expert system shell. The shell fits well into the modular approach as a knowledge base could be designed and developed for each of the sections which made up the expert system and the separate knowledge bases linked together to allow a smooth interaction between the various sections which make up the overall tutoring system itself.

A number of important considerations had to be addressed in designing the second part of the tutor system, which dealt with the learning and teaching aspect of SSM. These were that it should, (i) be user friendly. (ii) easy to use and involve minimal training. (iii) be designed to account for different levels of userability. (iv) encompass all of the stages of SSM in some way, whilst preserving the essence of the methodology by not imposing tight constraints upon the user.

Overview of the Tutoring System

The tutoring system was designed to take the user through each of the stages of SSM, providing tutorial exercises at each stage. Each stage of SSM was divided into three levels: an overview of the stage, specific guidelines in carrying it out and a tutorial where the user would be able to compare their answers to that of an expert's. It was also considered important that the user should be guided through the stages of SSM using one specific example as well as more general examples. Each stage then comprised of examples to help the student with understanding and one of the examples was developed through the whole exercise.

The main direction of the stages 1 and 2 section was confined to providing the user with advice and guidelines in the compilation of rich pictures. There are no established techniques to guide the user of SSM in carrying out stage 1 of the methodology which involves a detailed exploration of the problem situation whilst trying not to impose a structure. The recognition that much of this stage of SSM is dependent on the judgement of the analyst through experience which cannot be gained through the use of a computer-based tutor system.

The overview section of the rich picture stage provides the user with a basic background and introduction to the "finding out" stage and its relevance within SSM. The guidelines section provides the user with advice and guidelines for drawing rich pictures. This is followed by the tutorial exercise which is based on presenting the user with a problem situation and a series of rich pictures. Initially these were drawn using a software graphics package and shown on the screen. The user is asked to choose which rich picture they think best depicts the problem situation. Following this they are given an expert's view of the rich pictures. The final part of the tutorial centres around the user being presented with another problem situation scenario and asked to draw a rich picture of their own relating to the problem situation. The student is then given the opportunity to compare their rich picture with that of an expert's.

A fundamental problem exists in the reproduction of hand drawn rich pictures to graphically drawn pictures using software packages. It is our view that while the structure and the process may be covered, the climate is largely lost (A similar point about this aspect of representaion is made by Anderton, 1990). A rich picture produced using a graphics software package is largely 'clinical' and 'clean' preventing the viewer from gaining a 'feel' for the problem situation. However, this problem is cuurently being addressed through the use of a scanner which has made possible the display of hand drawn rich pictures onto the visual display screen.

Stage 3 of SSM is centred around providing the student with advice and guidelines relating to the formulation of root definitions. An important point which is highlighted is the two main types, issue-based and primary task systems. The 'CATWOE' test forms the basis of the tutorial exercise in which the student is asked to carry out the test on three root definitions, which include both abstract and "real world" examples. The user is then able to compare their ideas with that of an expert's which is shown to them after their attempt.

We considered the possibility of carrying out a word check to identify key words that should be included in an answer to the 'CATWOE' test, however, some of the answers to the 'CATWOE' test may be implied or not stated in the root definition. More importantly the "Weltanschauung", or world view is dependent upon the personal view of an individual. We decided therefore to reject the idea of a word check as in practice this was a difficult stage to develop since each expert had their own view regarding "Weltanschauung". We warn the student that the expert's answers are only meant as a guide and should not be taken as definitive answers.

The part of the tutoring system aimed at providing advice for stage 4 of SSM centres around the building of conceptual models. Stage 4 of SSM is concerned with modelling the activities required to achieve the transformation process described in the root definition, and is in no sense a description of any part of the "real world".

The guidelines section at this stage concentrates upon providing the student with advice on building conceptual models and the use of the 'formal system' model. The 'formal system' model forms the basis of the stage 4 tutorial. This is achieved through the redrawing of an expert's conceptual model, gained from the knowledge acquisition stage. This is achieved by using a

322

graphics software package which was then saved as 'pix' files and called up in the knowledge base and displayed on the visual display unit.

The student is offered a choice of three conceptual models, which include both abstract and "real world" examples, and asked to fill in their answers to the 'formal system' model. Upon completion of this stage the student is able to compare their answers with that of an expert's view. Again the expert's answer is only meant as a guide.

The part of the tutoring system concerned with stages 5, 6 and 7 of SSM concentrates on advising the user on thinking about and then bringing the relevant system back into the "real world". The purpose of stage 5 in SSM is to decide if the conceptual model does offer useful ideas about the system relevant to the perceived problem situation, and if so what steps could be taken to change the situation so as to improve it. This section of the tutor system is concerned with first, providing the user with guidelines and different ways of carrying out the comparison stage, and, second a tutorial exercise to help the user to gain some practice in developing an agenda. The guidance is provided through the use of a tabular display on the visual display unit.

The advice provided to the user in carrying out stage 6 of SSM centres around the fact that the agenda should be used to structure a debate and any changes should be both systemically desirable and culturally feasible. The final part suggests useful references that we consider to be important for the student to read.

Although the tutoring system covers all seven stages of SSM, the stages that seem to work best are stages 2, 3, 4 and 5, where clear guidelines are provided in carrying out these stages. Stages 1, 6 and 7 do not seem to integrate well into an expert system framework since no clear guidelines and methods exist to carry out these parts of the methodology, relying in some part on the experience and judgement of the user: This difficulty, perhaps, highlights the deficiency of computer-based expertise.

EVALUATION OF SSM AID

The final version of SSM AID offers more than we originally conceived in as much as it does guide the students through the methodology as a whole. The feedback that we have had from students is encouraging. The tests to date have been limited to a small number of students and also to students who have already completed the SSM part of the MSc course. Consequently, it is not possible at this stage of the project to report accurately if the developed package does help with the teaching phase of SSM. Our observations, therefore, should be taken within this context.

The advantages of using the package seem to be; (i) The facility to work privately through several graded examples of each stage of the methodology. (ii) Feedback is provided for each example attempted. (iii) The tutorial examples include a case-study to enable a student to work through one iteration of SSM. (iv) An explanation of each stage of the methodology is provided as the student progresses through the tutorial.

We see an important feature of the computer-aided tutorial package as allowing a student to work at a series of exercises at their own pace. Each exercise is graded in terms of difficulty thus providing the facility for a student to select an exercise according to their needs. There is also a case-study which allows a student to work through one complete iteration of SSM with feedback comments provided through the package.

Attempts to make the Rich Picture and Conceptual models as "life-like" as possible have met with only partial success so far. There is still a hint of computer graphics about the results which, in our view, detract from the essence of SSM as a subjective methodology. We believe that the models used within the tutorial package should seek to preserve the "cultural" implications of the problem situation rather than a technological interpretation. The way that an individual wishes to represent a problem or conceptual model is, in our view, an aspect of the methodology that should be preserved. The limitations that a graphics package may impose upon stages 3 and 4 is one undesirable feature that we feel a computer-based approach may impose. However, we expect to overcome in this difficulty in the near future.

Our purpose in producing this package was provide an additional aid to conventional tutorials rather than as a replacement for tutor/student interface. We see the package as one means of supplementing, in a dynamic fashion, the material available on SSM. However, a potential danger that a computer-based tutorial package may impose is that the way in which the tutorial is presented may convince the user that SSM can be learnt in a mechanical fashion. One danger of the approach is that the student may be constrained into thinking that once the machine has been defeated, by answering the questions correctly, then the task is also completed. We have attempted to overcome this by stressing within the tutorial sessions themselves that the "answers" provided are expert views of the situation and should be used for guidance and illustration rather than a model answers. Our decision to adopt this posture does of course provide those critics of SSM with the opportunity to suggest that here is an example of the equivocal nature of the approach. However, we feel that the alternative would be to create a package which would be more suited to teaching a technique than SSM. We consequently rejected the idea of infering a set of "right" answers.

The package will take the average student approximately 4 hours to work through but it is not necessary to work through the whole tutorial in one sitting. The package is designed so that the students can use if for short sessions if that is the preferred mode of working.


CONCLUSIONS

The production of SSM-AID has broken the spell, for us at least, of employing computer assistance with SSM. We feel that the exercise shows the potential for information technology to be used with SSM and without jeopardising the essence of the methodology. Results so far are encouraging and it is intended to produce a more robust version for wider application. The production of a computer-based tutorial package as an aid to the teaching of SSM has, we believe, much to offer to the student of SSM.

# BIBLIOGRAPHY

ATKINSON, C.J. (1986), Towards a Plurality of Soft Systems Methodo-
    logy, Journal of Applied Systems Analysis, vol 13. pp19-31.
ANDERTON, R    (1990), The need for the Formal Development of the
    VSM, in "The Viable System Model", (Ed. R. Espejo and R.
    Harnden) Wiley: Chichester, pp39-50.
BELL, J and  HARDEMAN, R.J (1989), The Third Role, in "The Natural-
    istic Knowledge Engineer, Knowledge Elicitation, Principles
    Techniques and Applications. (Ed. D.DIAPER). Ellis Harwood
    Ltd. Chichester, pp49-85.
BRYANT, N. (1988), "Managing Expert Systems", Wiley, Chichester.
CHECKLAND, P.B. (1989), "Systems Thinking, Systems Practice, Wiley,
    Chichester.
CHECKLAND, P.B. (1989), Soft Systems Methodology, in "Rational
    Analysis for a Problematic World" (Ed. J. Rosenhead) Wiley,
    Chichester pp71-100).
HART, A. (1986), "Knowledge Acquisition for Expert Systems", Kogan
    Page
JACKSON, P. (1986), "Introduction to Expert Systems", Addison-
    Wesley: Edinburgh.
OPEN UNIVERSITY T301 BLOCK IV., The Soft Systems Approach, "Soft
    Systems Analysis: An Introductory Guide".
STOWELL, F.A and WEST, D. (1990), The Contribution of Systems Ideas
    during the Process of Knowledge Elicitation, in "Systems
    Prospects. The Next Ten Years of Systems Research", (Eds: R.L
    FLOOD, M.C JACKSON AND P.KEYS), Plenham Press: New York.
STOWELL, F.A and WEST, D. 1988b, Expert systems and the Systems
    Epistemomolgy, in "Cybernetics and Systems; pt2",
    (Ed.R.TRAPPL) Kluwer Academic Publishers: Dordrecht,941.
STOWELL, F.A and WEST, D. (1989), Expert systems: Ramifications for
    the Knowledge Engineer, Systems Analysis, Modelling andSimula-
    tion,Vol 6, No9, East Germany,  pp673-678.
STOWELL, F.A. (1991), Towards Client-Led Development of Information
    Systems, Journal of Information Systems, (In press).
VICKERS, G. (1983),"The Art of Judgement:A study of Policy Making"-
    Harper and Row: London.
WATERMAN, D.A. (1986),"A Guide to Expert Systems", Addison Wesley
    Edinburgh.
WELLBANK, M. (1983),"A Review of Knowledge Acquisition Techniques
    for Expert Systems, Ipswich: Martlesham Consultancy Services.
WEST, D. (1990), Knowledge Elicitation as an Inquiring System:
    Towards a Knowledge Elicitation Methodology, The 34th Annual
    Conference of the International Society for Systems Science,
    8th - 13th July, Portland Oregan, U.S.A.

# THE NEED FOR TOOL SUPPORT FOR SOFT SYSTEMS

D. E. Avison* and Paul Golder+

*Department of Accounting and Management Sciences
Southampton University, SO9 5NH, UK

+ Department of Computer Science
Aston University, Birmingham, B4 7ET, UK

## INTRODUCTION

The arguments for the soft systems approach in developing information systems are well known and documented (see, for example, Checkland, 1981 and Checkland & Scholes, 1990). The proponents of this approach argue that a better understanding of complex problem situations is more likely to result using this approach than with the more simplistic structured or data orientated approaches commonly in use. But we are left with the following dilemma: most information systems development methods are reductionist but pragmatic; the soft systems approach, on the other hand, takes account of complexity but is difficult to use. Another reason for the adoption of conventional methods lies in the many support tools which make developing information systems using conventional methods easier. There are few tools supporting the soft systems approach.

## SUPPORT TOOLS IN CONVENTIONAL METHODOLOGIES

Many techniques are common to a number of more conventional information systems methodologies (Avison & Fitzgerald, 1988). These techniques include entity modelling, data flow diagramming, normalisation, entity life histories, and process logic techniques such as decision trees, decision tables, structured English and action diagramming. The tools available are usually linked to these techniques (see, for example, Rock-Evans, 1989) and serve a number of purposes:

### Drawing aids

Most of these techniques are graphic and the tools aid their drawing. Standard sets of symbols can be easily and quickly included into the main diagrams used. Some tools will relate to one methodology and only have the symbols standard to that approach, whilst others will allow the user to choose which symbol set is required. Because they make the implementation of changes very easy, being designed to take the drudgery out of revising documents, they are in effect the diagrammatic equivalents of word processors.

*Systems Thinking in Europe*, Edited by M.C. Jackson *et al.*
Plenum Press, New York, 1991

## Validation

They can contribute to the accuracy and consistency of diagrams. The user can, for example, cross-check that levels of data flow diagrams are accurate and that terminology is consistent. Many tools are used for validation purposes, so that, for example, 'illegal' links in a drawing, such as a data flow connecting two data stores are not permitted.

## Cross referencing

Some tools link different part of a methodology, so that an entity created in an entity-relationship diagram is linked to an entry in the data dictionary and an entity document description created automatically.

## Speed

The use of tools may speed up the development of an application and hence reduce the applications backlog. It is frequently possible to generate source code and default screen forms directly from the data dictionary.

## Project management and standardisation

Some tools provide support for project management, either integral to the tool itself or in the form of compatible 'add-ons'. They can also be used to ensure that certain documentation standards are adhered to. In particular, the standards of a particular methodology can be indirectly ensured through the restrictions of the support tool. Standard documentation can be automatically generated as a by-product of using the tool.

## Simplifying methodologies

Tools originally had very simple objectives of providing automated support to some previously manual tasks such as documentation, but it is becoming clear that they are having a more fundamental effect. They are beginning to simplify methodologies, making the various activities more consistent, and in some instances shortening the process itself (Macdonald, 1987). Apart from speeding up manual processes, some steps have become redundant and fewer manual iterations and cross-checks are required.


## THE NEED FOR TOOL SUPPORT IN SOFT SYSTEMS

As we have seen, support tools can help information systems developers as they facilitate the drawing process related to some of the analysis and design techniques, validate the drawings, cross reference the techniques, speed up the development process, enforce standards, help project management and control and simplify the use of methodologies. Indeed, people adopt a particular methodology very often because of the quality of tool support.

Given that the soft systems approach has much to offer the world of information systems, the lack of tools supporting the approach reduces its impact and its adoption. On the other hand, we do not want to use tools if they make the approach simplistic. In soft systems we are trying to map realistic levels of complexity and the area is necessarily more difficult for analysts to work with. There is a tendency to move to harder systems methods for convenience, but whereas hard systems tools are there to enforce formalisms of the approach, soft tools could be used to reinforce the lack of formalism of the approach.

## PROBLEMS IN DEVELOPING TOOLS SUPPORTING SOFT SYSTEMS

There are many features of the soft systems approach which inhibit tool support. In this section we draw attention to some of those problem areas which distinguish the soft systems approach from structured methods.

### Informality

The strength of the soft systems approach to the analysis of organisational problems is that it has little formal structure. There is a view that system development tools are easiest to create and to use in a formal environment. In this context, tools appear as the means to automate a methodology. They are frequently used in a pre-specified phase of system development and in a well defined way to a well defined structure in order to transform this into a 'deliverable' ready for the next specified stage in the process. This way of perceiving a tool is significantly different from the more traditional view of a tool or a set of tools. The craftsman's tool box contains a range of tools, some 'high-tech', some rather primitive. The use of these is limited only by the imagination of the craftsman and the properties of the materials being worked.

### Fuzziness

In many analyses, especially in the early stages, some important concepts are ill defined. We may wish to record their relevance, and may need to discuss their relationships with other concepts, without wishing to define them at an early stage (if ever). This necessary fuzziness, which may extend to vagueness as to the existence of the object under discussion, is in strong contrast with the requirements of a tool builder who would want the nature of an object to be well defined so that its interaction with the tools can be specified.

### Richness

Soft systems address the wider environment of the information system and as such refer to a much richer range of objects and concepts than do structured methods. The objective of creating a simple structure in most methodologies has resulted in the management of a very small number of different types of object. The richness of soft systems poses a problem for the designer of support tools as an excessively complex range of facilities may seem to be necessary in order to enable the analyst to manage a very rich set of object types.

## Variety of views

In real organisations, it will be common to find that there are a variety of views not only about subjective issues, such as policy and objectives, but also about more concrete issues, such as responsibilities and the nature of customers. For example, with part-time work and sub-contracting, 'concrete' issues such as 'who is an employee?' become vague. That different individuals will have different views is not some inconsistency to be ironed out in the analysis process, but a proper reaction to the different responsibilities that individuals in organisations have. It is clearly very difficult to develop tools which will manage such conflicting views

## Uncertainty

The world of conventional structured methodologies is a deterministic one where the structure of things is certain. However in soft systems we recognise the variability and uncertainty of structures. Uncertainty takes many forms. In some cases it is a lack of knowledge about issues which one may expect to clear over time. In many cases, however, it is intrinsic to real organisations. The head of a department might change at any time and this may have implications for the behaviour and structure of the department. The environment is subject to all kinds of random forces and any adequate model of an organisation should have some way of coping with this. Tools which have ways of acknowledging uncertainty are difficult to design.

## Organisational issues

Organisations are a complex mix of hierarchical structures and networks of relationships. An opinion about an issue is different in kind if it is held by the managing director or by a clerical assistant. The politics of organisations are a major problem for most analysts (Armenakis et al, 1990). A typical example of such a problem occurs when some facts about the organisation cannot be acknowledged and cannot be published so that there is often a hidden agenda at meetings between the analyst and senior members of the organisation. This need to acknowledge organisational structure and organisational politics is also a hindrance to tool support for soft systems.

## Conflict

In many situations there are nodes of conflict, that is, areas where the participants have different views and find that those of others are an impediment to meeting their own objectives. To acknowledge conflict means more than recording the conflicting positions. It requires that all views, the costs of continued conflict and its resolution (if feasible) are recorded and included in the analysis.

## Chaos

The theory of chaos (Gleick, 1987) as applied to this domain, denies the explicit assumption of most structured methods that complexity will diminish as one lowers the level of analysis. In practice, the complexity of any one department is no less than the complexity of the organisation, and the relationships within a small group in that department may be equally complex as those of the department as a whole. Just as in the familiar Mandelbrot set, the analyst finds that as he/she focuses in on a department the apparent simple structure breaks down into a rich pattern of complexity. Chaotic behaviour has been ascribed to business markets, and we may well expect that small disturbances on the organisation or the people in it will have dramatic and unpredictable consequences.

## Multi-media

A rich soft systems document will contain many different types of information. To give a few examples, it could include tape records of peoples' opinions, photographs of the site from the public's point of view, or video film of the operations in part of the plant. All of these elements need to be made accessible so that they can be easily brought into the discussion. This is well beyond the scope of conventional tools which have tended to reduce richness and are not well equipped to handle the explosion of material which a soft systems analyst would require.

## CONCLUSION

We have identified the constructive contribution that systems development tools could make to the application of soft systems techniques. We have also reviewed the characteristics of the soft systems approach which make the development of support tools very difficult. In doing this we believe that we have established the main requirements for a tool-set specifically aimed at facilitating the application of soft systems methods and thereby encouraging their use and in consequence, we believe, to improve the specification and subsequent implementation of information systems. In a subsequent paper (Avison & Golder, 1991), we discuss prototypes of tools supporting soft systems.

## BIBLIOGRAPHY

Armenakis, A. A, Mossholder K. W., Harris S. G. (1990) Diagnostic Bias in Organisational Consultation, *Omega*, 18(6) pp 563-572.

Avison, D. E & Fitzgerald, G (1988) *Information Systems Development: Methodologies, Techniques and Tools*, Blackwell Scientific, Oxford.

Avison, D E & Golder, P (1991) *Tools Supporting Soft Systems*, this volume.

Checkland, P. B (1981) *Systems Thinking, Systems Practice*, Wiley, Chichester.

Checkland, P. B. & Scholes, J. (1990) *Soft Systems Methodology in Action*, John Wiley, Chichester.

Gleick, J. (1987), *Chaos: Making a New Science*, Viking Press, New York.

Macdonald, I. G (1987) *Automating Information Engineering*, unpublished paper, James Martin Associates, London.

Rock-Evans, R (1989) *CASE Analyst Workbenches: A Detailed Product Evaluation*, Ovum, London.

TOOLS SUPPORTING SOFT SYSTEMS

D. E. Avison* and Paul Golder+

*Department of Accounting and Management Sciences
Southampton University, SO9 5NH, UK

+ Department of Computer Science
Aston University, Birmingham, B4 7ET, UK

EMERGING TECHNOLOGIES FOR SOFT TOOL SUPPORT

An earlier paper (Avison and Golder, 1991) argues that there is a need for tool support for soft systems. It discusses the problems associated with the attempt to build such tools. However, there are a number of developments which would suggest that this is now plausible, for example:

## Graphics

The development of powerful colour graphics workstations with well designed user interfaces must be the most promising development in terms of soft systems tools. The current state of the graphic workstation technology is well illustrated in a special issue of the IBM Systems Journal(1990). The major advantage of informal graphical models over text and formal models (like the entity-relationship diagram) is that is is possible to make statements which communicate the intention of the user without formal definition. Thus the size and position of an icon may be used to imply importance and its relationship with other objects. Colour can be used to convey warning messages, to provide an indication of neutrality or to associate classes of objects which are spacially dispersed. Thus the graphic workstation will help us to model directly some of the important features of a soft systems model, such as fuzzyness in the nature of objects and their relationships.

## Object oriented

The object oriented technology which makes a contribution to the infrastructure related to user interfaces also offers a very promising approach to the building of soft systems tools and interfaces. The basic approach of the graphic user interface, by which the user can perform a large number of different actions in any order, already starts to meet some of our requirements. The analyst is now freed from the hierarchy of menus and the proscribed routes to the creation and manipulation of objects. Furthermore, the underlying object oriented model which portrays objects as independent entities exchanging messages and reacting to each other in an asynchronous way, is much closer to the soft systems approach than traditional procedural applications.

## Multi-media systems

The soft systems document is a complex multi-media document and the user should not have to read it linearly but be able to browse through it at random. Although we are some way from the complete technology, recent advances in multi-media technology, for example the compact laser disc, have given us a single document incorporating most features at a reasonable cost.

## Interface design

The development of interface technology in the management of diagrams has already provided such facilities as panning, zooming and the use of transparent or opaque layers. Together with the mathematical modelling tools which enable us to rotate a model in hyperspace, these facilities have enriched the number of ways that a model can be viewed. This gives us the ability to deal with organisational aspects, which in turn enables the representation of a multiplicity of views and thereby represent conflict.

## Fuzzy logic and reasoning

Some of the technologies discussed tend to be very deterministic. Thus, an object may inherit characteristics from a parent or even distinct characteristics from more than one parent. Even so, it will still be difficult to model some aspects of uncertainty and fuzzyness. Recent work in applying fuzzy logic and fuzzy reasoning to control processes (Kandel, 1985) suggests that we may well be able to incorporate this technology into the specification of structures. This should mean that not only may fuzzyness be implied by, for example, colour on a screen, but that this fuzzyness may be a direct representation of the fuzzyness modelled in the system. Apart from the use of fuzzy reasoning, a number of authors have drawn on other newer and more established approaches to the issue of fuzziness and uncertainty in systems. Mobasheri et al (1989), for example, draws on the theory of chaos, whilst Morrissey (1990) uses information theory to cope with such problems.

## Psychological mapping tools

The development of tools for the structuring of problems has potential for contributing to the creation of soft systems tools. Thus, such tools as repertory grid analysis and multi-dimensional scaling provide the ability to explore and map complex ideas and feelings. Other soft techniques for exploring conflict situations such as hypergames (Bennett et al 1981) have been used by systems consultants for some years.

## TOOLS FOR SOFT SYSTEMS

There are a number of areas where tools supporting soft systems work may be helpful. The drawing of rich pictures, the creation of root definitions and Checkland's conceptual models come immediately to mind. We will look at the rich picture as one area for tool support.

The rich picture represents a subjective and objective perception of the problem situation in diagrammatic and pictorial form, showing the structures of the processes and their relationship to each other. It can be used to identify problem themes, conflicts, an absence of communication lines, shortages of supply, and so on. Through debate within the organisation, it is possible to identify relevant systems which may relieve problem themes.

Typically, a rich picture is constructed first by putting the name of the organisation that is the concern of the analyst into a large 'bubble', perhaps at the centre of the page. Other symbols are sketched to represent the people and things that inter-relate within and outside that organisation. Arrows are included to show these relationships. Other important aspects of the human activity system can be incorporated. Crossed-swords indicate conflict and the 'think' bubbles indicate the worries of the major characters. All these can be standard symbols included in a rich picture tool.

Differences of opinion can be exposed, and sometimes resolved, by pointing at the picture and trying to get it changed so that it more accurately reflects people's perceptions of the organisation and their role in it. A strength of an automated tool is to make these adaptations easily. Another possibility is to enable the drawing of different rich pictures, dependent on the different views of the same situation, and then using a merge facility to attempt to produce one rich picture from these various interpretations. Further, rich pictures could be decomposed so that one top level rich picture is decomposed into several second level rich pictures.

Another possibility of a tool would be for the user to be able to double click on an item within the rich picture to get further details of that aspect, to 'zoom in'. Thus, there could be narrative description of the formal role of an actor or some informal views of a conflict or problem. All these are feasible with present technology.

Figure 1 shows a rich picture of a distance learning unit of a UK polytechnic taken from Avison & Wood-Harper (1990). It was drawn using a prototype tool for rich picture development being constructed at Aston University. The various symbols used are part of the symbol set provided by the tool. Figure 2 shows how the tool may be developed so that lower level rich pictures or other rich pictures can be called in or more detail provided, such as narrative, annotated forms, graphs, photographs or video pictures.

CONCLUSION

The need and technology exists to provide tool support for soft systems, and this paper has described one particular direction that tool support might take. We would, however, like to raise two issues:

## The danger of being reductionist

Tools may give a hard appearance to soft issues, encouraging excessive formality and structure and therefore being reductionist. This is obviously undesirable, as it leads

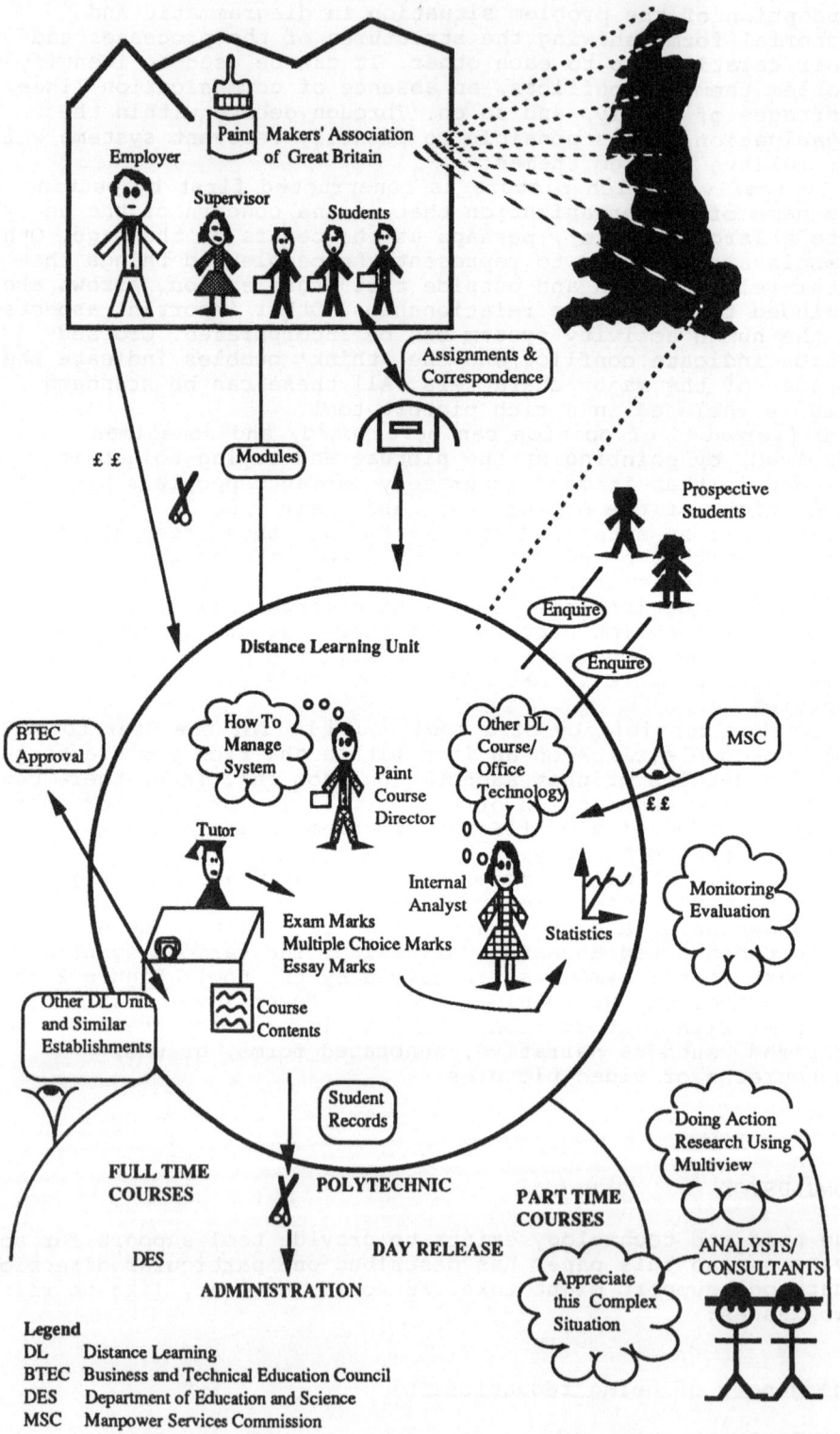

Figure 1. Rich picture of polytechnic distance learning unit

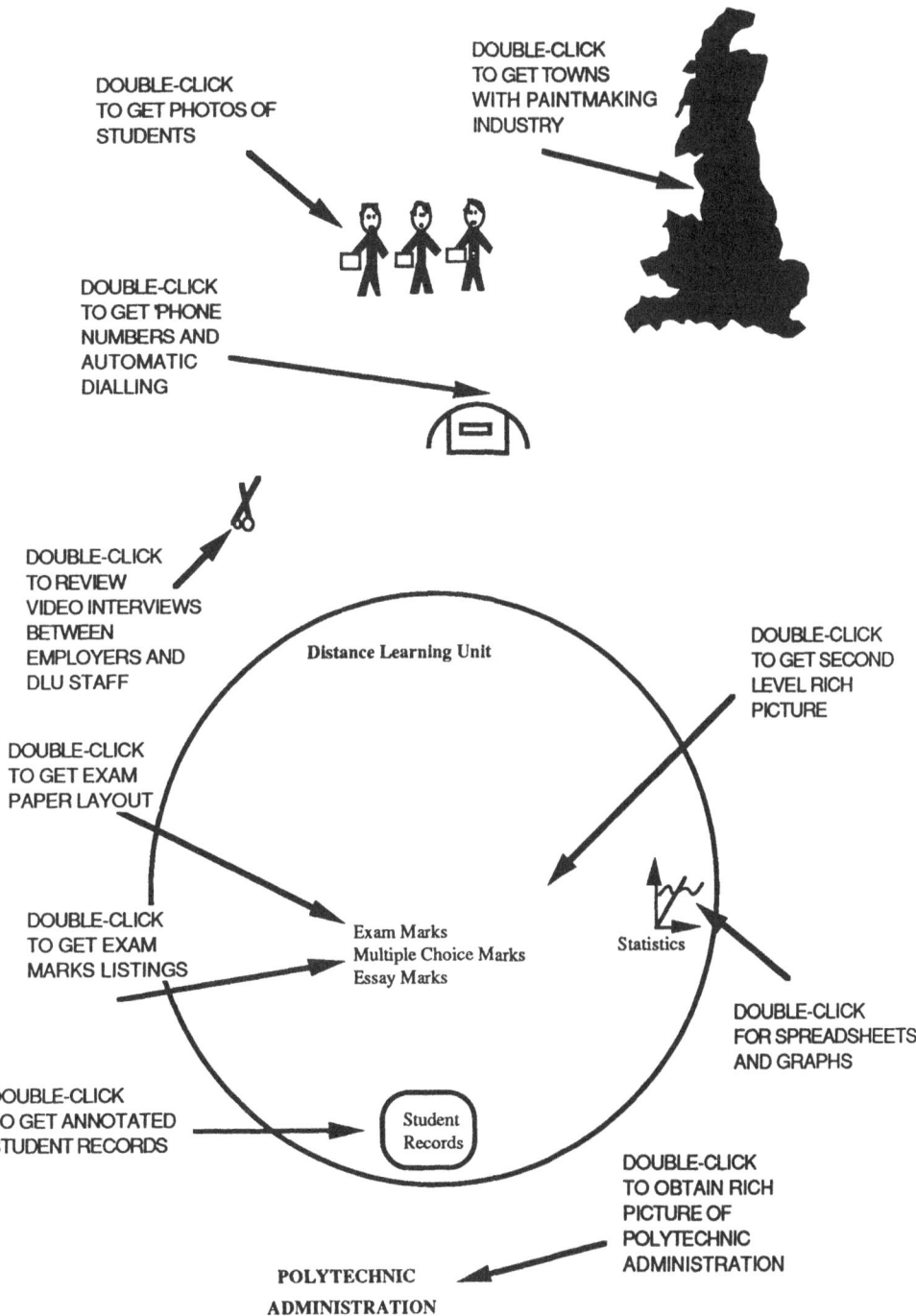

Figure 2. Typical requests to rich picture support tool

the user of the tool in directions totally against the ideals
of soft systems.

The danger of over-complexity

Suitable tools may make it possible to create very complex
and rich soft systems models and it may be that some of the
power of the soft systems approach would be lost in the mass of
detail.

These issues should be reviewed when tools sets supporting
soft systems become more widely available.

BIBLIOGRAPHY

Avison, D. E & Golder, P. A (1991) *The need for tools
supporting soft systems methods*, this volume.
Avison, D. E & Wood-Harper. A. T (1990) *Multiview: An
exploration in information systems development*, Blackwell
Scientific, Oxford.
Bennett, P., Huxham, C. & Dando, M (1981) Shipping in crisis: A
trial run for live applications of the hypergame approach,
*Omega*, 9(6) 579-594
*IBM Systems Journal* (1990) (whole volume), 29, 3, 1990.
Kandel, A. (1985) *Fuzzy mathematical techniques with
applications*, Addison Wesley, Reading Mass, 1985.
Mobasheri, F., Orren, L.H. & Sioshansi, F. P (1989) Senario
Planning at Southern California Edison, *Interfaces*, 19, 5,
1989 pp 31-44
Morrissey, J.M (1990) Imprecise Information and Uncertainty in
Information Systems, *ACM Transactions on Information
Systems*, 8, 2 April 1990, pp 159-180.

PROBLEM STRUCTURING AND CRITICAL SYSTEMS THINKING

PROBLEM STRUCTURING AND OPTICAL SYSTEMS TUNING

PROBLEM STRUCTURING AND

CRITICAL SYSTEMS THINKING : INTRODUCTION


The question of how to structure problems has been an issue in management science since its inception. In general the systems community has produced two possible answers to this question. The first, a hard systems response, advocates looking at the world in systems terms, identifying systems, sub-systems and supra-systems, and examining the relationships between them. It then becomes possible to build a systems model which represents the real-world. The second, a soft systems response, eschews looking at the world in systems terms because it fears this might distort the problem situation and lead to an early foreclosing of possible ways forward. It advocates instead the collecting of as much relevant information as possible about the problem situation and the construction, implicitly or explicitly, verbally or pictorially, of a rich picture or rich pictures of 'reality'. The rich picture then becomes a resource against which can be set the design of a desirable future, various notional 'ideal-type' representations of systems, or alternative world-views, depending upon which soft approach is being followed. The soft systems response abandons the notion that it is possible to find some structure in the real-world which can be used as a basis for building systems models.

The papers which follow may be read as commenting on the correctness or otherwise of the 'hard' and 'soft' approaches to problem structuring and as offering refinements to the broad division in method just outlined. They deal with issues such as the way methodologies cope with the involvement of humans in the change process, how different observers themselves perceive problems in radically different ways, with the role assigned to the analyst by different methodologies, with organising debate between involved actors, and with choosing appropriate methodologies for the problem-context at hand.

The choice of an appropriate methodology is obviously a matter of overwhelming importance to anyone concerned about problem structuring, because different systems methodologies go about problem structuring in very different ways. This is our link to the second type of paper contained in this section of the book - the papers on critical systems thinking. Critical systems thinking is a new development in management science but it is already attracting considerable interest and generating much enthusiasm. Critical systems thinking sets out to make use of all the various methodologies developed in the systems movement but does so in a way which takes advantage of the strengths of each of them. It aims to specify when particular approaches are the most suitable for use. These points will become clearer if we now attempt to summarise the philosophy of critical systems thinking.

Essentially critical systems thinking embraces five commitments. The first of these is critical awareness. This comes from closely examining the assumptions and values entering into actually existing systems designs or any proposals for a systems design. It also concerns understanding the strengths and weaknesses and the theoretical underpinnings of available systems methods, techniques and methodologies. The second is social awareness, which involves recognising that there are organizational and societal pressures which lead to certain systems theories and methodologies being popular for guiding interventions at particular times. Social awareness should also make users of systems methodologies contemplate the consequences of use of the approaches they employ. Third is a dedication to human emancipation, seeking to achieve for all individuals the maximum development of their potential. This is to be achieved by raising the quality of work and life in the organizations and societies in which they participate. The fourth commitment is to the complementary and informed use of systems methodologies in practice. The final commitment, which gives bones to the third and fourth, is to the complementary and informed development of all varieties of systems approach. Different strands of the systems movement express different rationalities stemming from alternative theoretical positions. These alternative positions must be respected, and the different theoretical underpinnings, and the methodologies to which they give rise, developed in partnership. This can be achieved by relating different systems epistemologies to the three fundamental human interests unearthed by Habermas – the technical, the practical and the emancipatory interest.

The papers on critical systems thinking in this section will be seen each to contribute to one or more of these five commitments. The discussions which ensue focus on, among other things, pluralism, systems boundaries and who or what is marginalised by them, power, appropriate social theory to guide emancipatory methodologies, and how and whether emancipatory approaches can be successfully employed. The practical side of critical systems thinking, involving the use of alternative consultancy methods, is also considered.

If critical systems thinking is successful in its endeavour to provide appropriate systems methodologies for the rich array of problem-contexts around, and to provide guidance on the question of 'Which methodology when?' then those concerned with problem structuring obviously stand to benefit. The papers on problem structuring similarly provide an input to critical systems thinking by commenting on and enriching available approaches. The editors hope that you, the readers, find this section of the book as interesting as we do.

# A GENERIC MODEL FOR SCENARIO ANALYSIS AND MODELLING

Simon Peck

Faculty of Information and Engineering Systems
Leeds Polytechnic, Beckett Park, Leeds LS6 3QS

## INTRODUCTION

The concept of a generic model showing the basic features of an arbitrary system and its environment is not new. Indeed its simplest form, the familiar diagram of a 'system' inside its 'environment' is present throughout much of the systems writing. However, rather than making the point that all systems exist within a context / environment, such a simple generic model has little further use. More powerful generic models include those of Petit (1976), Checkland (1981a), Beer (1981, 1985) and Finlay (1985). However in most cases, the generic models are constructed within a particular methodology rather than coming before them. A case can be made for a generic model which demonstrates more structure than the system / environment model but which is, to a great extent, free of a particular methodology. Such a model could then be used during the initial stages of a study, before the style of the investigation is chosen, or, alternatively, allow access to a variety of approaches simultaneously. Of the features which could be included in such a model, the following are thought to be desirable:

## Hierarchy

The model should show the hierarchical nature of systems and facilitate change of emphasis both to supra- and sub-systems. This is of course true of the simple system / environment model.

## Access to Specific Modelling Techniques

The generic model should allow access to specific modelling techniques such as influence diagrams, process diagrams and conceptual models as necessary during the progress of an investigation.

## Role of analyst

A major criticism of the system / environment view is that it is based on a Newtonian point of view i.e. the viewer of the system is considered to be on a higher plane adopting a purely observational role. Such a viewpoint is inappropriate for many organisational tasks and, indeed, is totally inconsistent with the viewpoint of most soft methodologies such as SSM (Checkland, 1981b; Wilson, 1984) and cognitive mapping (Eden et al., 1983).

## THE MODEL

The generic model proposed here consists of five elements together with interactions between (some) pairs of them. Each element, whilst not necessarily physically, or even organisationally distinct represents a particular aspect of the scenario and, consequently has a particular language. The elements are as follows:

## System Body

The System body forms part of the System as would be defined in the system / environment model, and focuses on those aspects which can be described in terms of the concept of state. Typically this will include operational elements of the system, including resources, materials and primary information. In all but the simplest cases there will, however, be no simple boundary between the system body and the next element which we discuss.

## Autonomous Control

Also part of the 'system proper' the Autonomous control element represents the aspects of control at the level immediately below that of the Analyst. Whereas the Body can be thought of as consisting of states, the autonomous control will consist of rules and procedures operating on the System body. In all but the simplest cases, the autonomous control will not be a physically distinct element, but rather is an aspect of the system distinct from its raw dynamics.

## Analyst

This represents the viewpoint and thinking of the viewer of the system. The term analyst does not imply that this represents just the analyst or modeller but, rather, that it includes the level at which the decision making / design which is of interest is to be made. The language for this element is typically in terms of objectives. The analyst element is purposely placed in an ambiguous position between the 'total system' and the 'total environment'; thus allowing for the two alternative views that the analyst is external or internal to the system.

## Environment

As an in the familiar system / environment model, the environment is based on the idea of those aspects which, whilst not of primary interest, need to be considered because

344

of their impact upon the system. In this five element model, however, we include in the environment the non-managerial aspects and consider it to be made up of parameters.

## Decision Environment

The final element of the generic model also forms part of the total environment and, particularly, the environment of the analyst. It is referred to here as the decision environment and is considered to consist of policies.

Figure 1 shows the relative positions of the five elements via a diagrammatic representation. Common boundaries between elements imply some communication between them; this is further discussed in the next section.

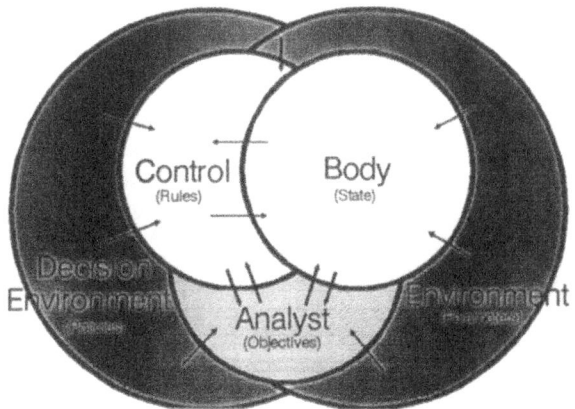

Figure 1 - The five element model

COMMUNICATION

Central to the five element model is the nature of the communication between the elements. Communication is defined between (not all) pairs of elements and, just as each element had its own natural language, similarly, each communication has a particular aspect. Table 1 shows the communications.

At this stage, we are considering the direction of communication in terms of causality. Of course a communication from the environment to the system may correspond to a physical flow of material from the system to

### Table 1 - Communication routes and names

| To From | Analyst | Control | Body |
|---|---|---|---|
| Dec. env. | Policies | Policies | |
| Environment | Parameters | Parameters | Parameters |
| Analyst | | Decisions | Decisions |
| Control | Observations | | Actions |
| Body | Observations | Measurements | |

the environment; we will return to this in a later section.
Notice also that no communication into the environment or
decision environment is included since they are considered as
passive elements in the model. A need to  consider these
elements in greater detail will imply an inappropriate
definition of the boundaries has been taken.


HIERARCHY

As was mentioned previously, a desirable feature of a
generic model is its ability to support a hierarchical view
of systems and to facilitate changes of emphasis during the
course of study. It is intended that the five element model
has this feature; Table 2 illustrates the relationship
between the elements at different levels of system view.

Of course, the diagram shows the relation between
elements when the level of view is changed by a whole level.
In practice such complete steps are unlikely and so, for
example, a change to a slightly smaller system would give a
new environment made up of parts of the original environment
together with parts of the original system body.


### Table 2 - Relation between elements of a system and those of its sub- and supra- systems

| Suprasystem | Initial system | Subsystem |
|---|---|---|
| Analyst | Dec. env. | |
| Control | Analyst | Dec. env. |
| | Environment | |
| Body | Control | Analyst |
| | | Environment |
| | Body | Control |
| | | Body |

MODELLING

As suggested in the introduction, a desirable feature of a generic model is for it to lead naturally into a range of modelling techniques which can be used to explore the system in greater detail. This, indeed, is a property of the five element model; we shall describe, in particular, how it relates to influence diagrams and to process oriented diagrams.

## Influence diagrams

The communication view taken above, based on the notion of causality feeds naturally into influence diagrams; in fact the generic model itself can be thought of as the 'level 0' influence diagram. Taking the analyst as external to the system leads us into traditional 'hard' influence diagrams (Bodily, 1985), whereas the inclusion of the analyst as within the system gives something more akin to a cognitive map (Eden et al., 1983) showing the analyst's perception of the situation.

## Process oriented diagrams

If instead of the causal view, communication between the elements is viewed as output / input (i.e. the process of communication is the accent rather than its logical implications) then the generic model approaches a 'level 0' process diagram. If the emphasis is on information flows then this corresponds to a top level dataflow (or context) diagram (Gane and Sarson, 1979; Curtis, 1989).

## Other modelling techniques

It appears possible to relate Checkland's conceptual models to the generic model; indeed Checkland has suggested using the elements of operational system, monitoring and control system, and planning system as an aid to developing conceptual models (Checkland, 1988). Correspondences with further types of model are still to be investigated.

DISCUSSION

The five element generic model has been constructed using many of the familiar ideas from systems thinking and has some similarity with other such models. Its advantage lies, however, in the intermediate level of its complexity which is simple enough to be used in addressing an area initially, yet which interfaces to a variety of modelling techniques which can be used during detailed analysis and exploration. Currently, the model has been used, largely, as a teaching vehicle for introducing undergraduate and post-graduate students of information systems to systems concepts at an early stage of their courses. It is hoped to use the model within a range of industrial based projects and to explore its potential as a tool for use by non-systems specialists.

REFERENCES

Beer, S., 1981, The Brain of the Firm (2nd Edition), Wiley, Chichester.

Beer, S., 1985, Diagnosing the System for Organisations, Wiley, Chichester.

Bodily, S., 1985, Modern Decision Making, McGraw-Hill, New York.

Checkland, P.B., 1981a, Towards a Systems-based methodology for real world problem solving, in Systems Behaviour, Open Systems Group (Eds.), Harper and Row, London.

Checkland, P.B., 1981b, Systems Thinking, Systems Practice, Wiley, Chichester.

Curtis, G., 1989, Business Information Systems: Analysis, Design and Practice, Addison-Wesley, Wokingham.

Eden, C., Jones, S., Sims, D., 1983, Messing about in Problems, Pergamon Press, Oxford.

Finlay, P., 1985, Mathematical modelling in Business Decision Making, Croom Helm.

Gane, C., Sarson, T., 1979, Structured Systems Analysis: tools and techniques, Prentice-Hall, Eaglewood Cliffs.

Petit, T.A., 1976, 'A Behavioral theory of Management', Academy of Management Journal, December.

Wilson, B., 1984, Systems: Concepts, Methodologies and Applications, Wiley, Chichester.

A PROBLEM STRUCTURING METHOD

Antonio Caselles

Departament de Matematica Aplicada i Astronomia
Universitat de Valencia
46100 Burjassot. Valencia. Spain

ABSTRACT

Given a formal definition of problem and a formal definition of system, the equivalence between both concepts is studied. Considering a problem as a 3-tuple $\langle D, R, P \rangle$, where D is the set of possible data, R is the set of possible results, and P the set of conditions of the problem, classes of problems are constructed as combinations of types of data, types of results and types of conditions. For example, data can be either literal or numerical, either with uncertainty or not; conditions can be determined by rules, tables, equations, it may have uncertainty, etc. As a case of application it is outlined how some of the most common problems (knowledge representation, search, reasoning and planning, etc.) can be adapted to the formal definitions given for problem and system and can be structured by the method proposed. This method is used in the realistic models generator SIGEM (Caselles, 1988) (a General Systems Theory based programme generator).

INTRODUCTION

We are going to treat about what we understand for general problems and for systemic problems, about the equivalence between both concepts and about a methodology ad hoc for systemic problems. Our target is to propose a method to structure problems in order to facilitate its solving process.

General Problems

With respect to the concepts of problem and solving a problem, we refer to the ideas of Veloso (1984) and Haeberer et al. (1989). For these authors a problem over a theoretical universe U is a 3-tuple $\langle D, R, P \rangle$, where D and R are subsets of U and $P \subseteq D \times R$. D represents all possible data, R is the set of all possible results, and P is the set of all pares datum-result that are permissible, namely the problem condition. A problem is said to be viable when $(\forall d)(d \in D \to (\exists r)(r \in R \land (d, r) \in P))$, in other words, when exists a permissible result for any possible datum.

So, we have three sets of objects to classify: Data, Results, and Conditions of the problem, in order to build classes of problems and to assign a method for solving each one. The classes of problems we are going to present here are those are used in the programme generator SIGEM (Caselles, 1988).

## Systemic problems

Given a universe U, we define a system (Caselles, 1988) as a pair $\langle S, D \rangle$ where $S \subseteq U$, and $D \subseteq S \times S$. We call structuring set at D and structural mapping of a system at $g: A \to P(S)$, where $A = \{a \: / \: a \in S \wedge (x, a) \in D\}$ and $(\forall \: a \in A) \: g(a) = \{x \: / \: (x, a) \in D\}$

We consider the concept of "variable" as synonymous of undetermined object that can be represented either by a name or a symbol. We call value of a variable at a determined object that can be represented by that variable, and range of a variable at the set of its possible values. We call formal system at a system $\langle S, D \rangle$ where S is a set of variables.

It can be observed that this definition of system is analogous to "set of interrelated elements", being D who defines the connections among the elements of the system, that is, its structure. The mapping "g" indicates which are the elements influencing a given one. We call input variable of a formal system at a variable x that has no influencers, that is to say: $(x \in S) \wedge (\forall \: (a, b)) \: ((a, b) \in D \to x \neq b)$, and output variable at a variable that is not an input variable. We call behaviour of an output variable at a correspondence between the cartesian product of the ranges of the influencing variables and the range of the influencee one, that is to say, let $y \in S$, $x \in P(S)$, $x = \{x_1, \ldots, x_n\}$, $n \in N$, $x = g(y)$; let Ry be the range of y, and $Rx_1, \ldots, Rx_n$ be the ranges of $x_1, \ldots, x_n$; the behaviour of y will be defined by $By \subseteq (Rx_1 \times \ldots \times Rx_n \times Ry)$. We call a loop at a subset A of D whose elements determine a closed chain in S, that is to say $(A \subseteq D) \wedge (A = \{(a, a)\} \vee (A = \{(a, x_1), (x_1, x_2), \ldots, (x_n, a)\}$ $\wedge (n \in N) \wedge (n \geq 1))$, and a hierarchical system at a system having no loops.

We consider systemic problem as synonymous of formal system. We say that a systemic problem is solvable when it is hierarchical and the behaviour of its output variables is defined by mappings. In this case it will be possible to write either $y = f(x)$, or $f: (Rx_1 \times Rx_2 \times \ldots \times Rx_n) \to Ry$. In other words, we can solve a problem when given the data we are able to find the corresponding results, that is to say, when given $d \in (Ri_1 \times Ri_2 \times \ldots \times Ri_n)$, being $Ri_i$ the ranges of the input variables, we can find an $r \in (Ro_1 \times Ro_2 \times \ldots \times Ro_m)$, being $Ro_i$ the ranges of the output variables, and that is only possible when the system has no loops and the behaviours of the output variables are mappings.

## Equivalence between general problem and systemic problem

Let $\langle D, R, P \rangle$ be a general problem, and $\langle S, D' \rangle$ be a systemic problem. We say that both problems are equivalent when:
(a) the set of data of the first coincides with the cartesian product of the ranges of the input variables of the second; that is to say, $D = Ri_1 \times Ri_2 \times \ldots \times Ri_n$;
(b) the set of results of the first coincides with the cartesian product of the ranges of the output variables of the second; that is to say, $R = Ro_1 \times Ro_2 \times \ldots \times Ro_m$;
(c) the set of acceptable pairs datum-result in the first, can be obtained as a consequence of the behaviours of the output variables in the second.

## CLASSES OF GENERAL PROBLEMS

If a general problem is a 3-tuple composed by Data, Results, and Conditions, then it will be necessary to combine types of data, types of results and types on conditions in order to build classes of problems. Given that any classification is discretional, its validity is related with its operativity. The classification proposed in the following has demonstrated its operativity during three years of experience with the programme generator SIGEM (Caselles, 1988).

## Types of data

Data may be classified combining two criteria.

Criterion 1: either qualitative or quantitative data. Qualitative data are literal data, that is to say, strings of characters under computers point of view. Quantitative data are numerical data, that is to say, either reals or integers.

Criterion 2: data either uncertain or not. Uncertain data may be concreted of different manners, depending on the type of uncertainty considered (probabilistic, possibilistic, etc.). The programme generator SIGEM considers, by the moment, only probabilistic uncertainties. So, uncertain data may enter either as pairs $(m, s)$, that is to say (mean, standard deviation), or in form of table of frequencies.

Combining both criteria, we have:

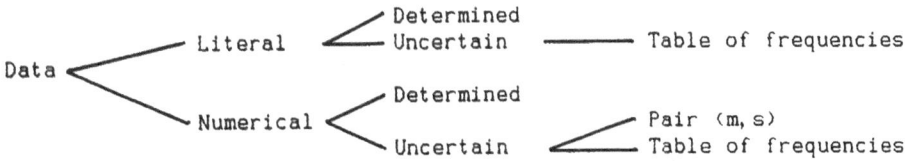

Several types of data, even all of them, may coexist in a same systemic problem.

## Types of Results

Types of results coincide exactly with types of data. However, when any data or condition with uncertainty exists, automatically all results are considered as uncertain. In this case, determined results will appear as pairs $(m, s)$ with $s=0$, or as tables of frequencies with an only value. Programmes generated by SIGEM can give interval estimations of uncertain results and tests of normality for numerical distributions of results.

## Types of Conditions

Conditions of a problem, in other words, what permits us to determine the pairs datum-result acceptable, are in practice the rules and/or functional relationships between the variables considered that make possible, given the necessary data, to determine the corresponding results. Such relationships may be classified analogously to data and results.

Criterion 1: functions may be either literal or numerical. Literal functions may consist on tables or on collections of rules. Tables permit us to assign to every literal or set of independent literals one dependent literal by means of a consulting operation. Rules permit us to perform the same operation by application of Logic. Numerical functions may be given by means of tables, individual equations or sets of equations and/or logical rules.

Criterion 2: functions may be either deterministic or with uncertainty. A function with uncertainty is a correspondence such that when for some given values of the independent variables there exist more than one possible value of the dependent variable, and an additional arbitrary rule is used in order to choose only one of them. This additional rule is a random selection in the case of probabilistic uncertainties.

Combining both criteria we have the following possibilities.

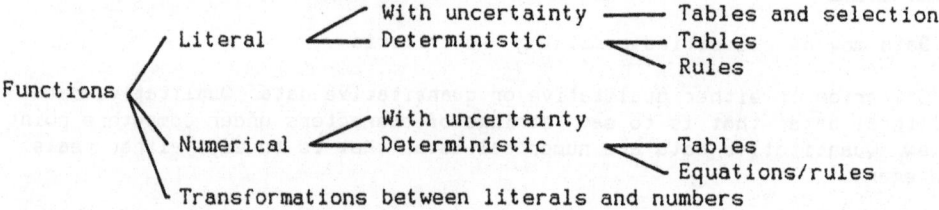

Several types of functions, even all of them, may coexist in a same systemic problem.

## Classes of Problems

Independently of data, results and conditions, it is possible to classify problems with relation to time into static and dynamic problems. We define a 1.dimension variable as a totally ordered set of variables with the same range, and a n.dimension variable as a totally ordered set of (n-1).dimension variables with the same range (Caselles, 1988). We consider time as the physical sense of one dimension of a variable, that is called a time variable. A dynamic problem is for us a problem that includes at least one time variable. In other case it is static.

A static problem may be, for example, database design and consulting. The only difference between a static and a dynamic problem for the programme generator SIGEM is that the first is monoperiodic and the second is multiperiodic. In other words: SIGEM asks the user for that character of the system, and answering "dynamic" it puts a period generator in the programme. Successively, the design of the database and the design of the consulting method go across the identification of the variables to consider and the identification of the functional relationships between them, as in any case.

Combining types of data, results and/or conditions, and considering the character either static or dynamic of the system, we may form different classes of problems, among which we remark the following.
- Class H: static and deterministic problems.
- Class G: static problems, with uncertainty in data (and consequently in results), and deterministic functional relationships.
- Class F: static problems with uncertainty in functional relationships.
- Class E: static problems with uncertainty in data and in functional relationships.
- Class D: dynamic and deterministic problems.
- Class C: dynamic problems with uncertainty in data (and consequently in results).
- Class B: dynamic problems with uncertainty in functional relationships.
- Class A: dynamic problems with uncertainty in data and in functional relationships.

Three subclasses may be considered in each of the prior classes, taking into account whether literal variables, numerical variables, or both types of variables are used. Taking into account, if desired, that functional relationships are given by tables, by rules, by equations or by combinations of these types, the number of classes of problems would be multiplied by six. Less important aspects such as that data with uncertainty are given either by a pair (m, s) or by a table of frequencies, may be also considered for the classification. We would have a base for a taxonomy of general problems developing exhaustively the classification with the criteria exposed.

## APPLICATIONS

As examples of application, we are going to outline how some of the traditional problems treated in Artificial Intelligence may be classified and solved by the method proposed. Let us recall them briefly.
- Knowledge representation: concepts, rules, and frames.
- Search.      - Reasoning and planing.      - Learning.      - Design.

### Knowledge representation

A concept may be considered as a set of objects with some attributes in common that receives a name. For example: "portico". The common attributes of all porticos may be the following: three components of rectilinear form, two of them in vertical position, and the third one in horizontal position, over the first two, linked with them, and being the vertical two as separated as possible. Concept representation may be considered as a class H problem with literal variables. Functional relationships may be represented by tables or by rules. The word "portico" would be a value of the variable CONCEPT, which range would be the set: {portico, arc, beam, pillar, ... }. The attributes considered would be values of variables such as:
HORIZONTAL {none, one, two, ... },      VERTICAL {none, one, two, ... },
HOR-POSITION {over, below, close up, near, ... }, etc. Functional relationships would be given as tables relating each range of the variables considered with the range of the variable CONCEPT.

Frame representation may be considered as a problem of class D, C or B with literal and numerical variables, being the functional relationships represented by tables or by rules. For example, the operating frame of a restaurant would include variables such as:
CUSTOMER {comes in, seats down, calls for, asks for, ... },
WAITER {attends, asks for, confirms, waits on, ... },
KITCHEN {receives a command, prepares, cooks, delivers, ... }, etc.
Functional relationships would include uncertainty because what the waiter, for example, makes in a given instant, depends on what kitchen, customer and the same waiter have done in the prior instant, but his action may not be unique. Uncertainty may be broken with a table of frequencies of possible responses to each set of values of the influencing variables and with a random numbers function.

### Search

A non solvable problem because of its correspondences are not mappings may be converted into a solvable one by forcing all correspondences to be mappings and repeating the process of calculus until an objective be attained. So, a search problem may be either static or dynamic, normally with deterministic data, but any time with functional relationships with non probabilistic uncertainty. For example, the classical problem of the farmer who intends to cross a river in a little boat with a fox, a chicken and a bag of corn, may be considered as a search problem. It may be a problem of class B with literal variables, for us. The farmer has to find a manner to transport his properties at the right side of the river, one by one, without any damage, and with a minimum number of trips.

Let us consider the following six variables:
LSB: Left side status at the beginning of a period (go and return).
LSE: Left side status at the end of a period.

RSB and RSE: the same for the right side.
BLR: Boat status in the way left to right.
BRL: Boat status in the way right to left.
The range of the first four variables is:
{Ø, fox, chicken, corn, foxcorn, foxchicken, chickencorn, foxchickencorn}.
The range of the last two variables is: {Ø, fox, chicken, corn}.
The functional relationships to use for calculus are:
BLR = $f_1$(LSB); BRL = $f_2$(RSB); LSE = $f_3$(LSB,BLR); RSE = $f_4$(RSB,BLR);
and, LSB = LSE; RSB = RSE; in order to be able to begin a new period.
Functions $f_1$ and $f_2$ may be done by rules and functions $f_3$ and $f_4$ by tables as
we are going to see immediately. The search process includes a period
generator (for example, a FOR loop in BASIC), a selector of options in $f_1$ and
$f_2$, a memory variable for the number of periods of a run, and a rule that
ends the process when the minimum number of periods is attained. In this case
the number of options in $f_1$ and $f_2$ is very low and the search process may be
exhaustive. In other cases (missionaries and cannibals problem, 8-puzzle
problem, etc.), the search process may study only a sample or may be combined
with learning (saving the best prior finding).

The function BLR = $f_1$(LSB) may be constituted by the following rules,
where RND is a random number between 0 and 1.
IF LSB = "foxchickencorn" THEN BLR = "chicken"
IF LSB = "foxchicken" THEN IF RND<.5 THEN BLR = "fox" ELSE BLR = "chicken"
IF LSB = "chickencorn" THEN IF RND<.5 THEN BLR = "chicken" ELSE BLR = "corn"
IF LSB = "foxcorn" THEN IF RND<.5 THEN BLR = "fox" ELSE BLR = "corn"
IF LSB = "fox" THEN BLR = "fox"
IF LSB = "chicken" THEN BLR = "chicken"
IF LSB = "corn" THEN BLR = "corn"
IF LSB = "" THEN BLR = ""

The function BRL = $f_1$(RSB) may be constituted by the following rules:
IF RSB = "foxchickencorn" THEN BRL = "": MEMORY = NUMBER-OF-PERIODS: GOTO ??
IF RSB = "foxchicken" THEN IF RND<.5 THEN BRL = "fox" ELSE BRL = "chicken"
IF RSB = "chickencorn" THEN IF RND<.5 THEN BRL = "chicken" ELSE BRL = "corn"
IF RSB = "foxcorn" THEN BRL = ""
IF RSB = "fox" OR RSB = "corn" THEN PRINT "ERROR": GOTO ??
IF RSB = "chicken" THEN BRL = ""
IF RSB = "" THEN BRL = ""

Functions LSE = $f_3$(LSB,BRL) and RSE = $f_4$(RSB,BLR) may be given by the
following tables, where Fo means fox, Ch means chicken and Co means corn:

| LSB | BRL | LSE | RSB | BLR | RSE |
| --- | --- | --- | --- | --- | --- |
| FoChCo | Ø | FoChCo | FoChCo | Ø | FoChCo |
| FoCo | Ø | FoCo | FoCo | Ch | FoChCo |
| Fo | Ch, Co, Ø | FoCh, FoCo, Fo | Fo | Ch, Co, Ø | FoCh, FoCo, Fo |
| Ch | Fo, Co, Ø | FoCh, ChCo, Ch | Ch | Fo, Co, Ø | FoCh, ChCo, Ch |
| Co | Fo, Ch, Ø | FoCo, ChCo, Co | Co | Fo, Ch, Ø | FoCo, ChCo, Co |
| Ø | Ø | Ø | Ø | Ch | Ch |

REFERENCES

Caselles. A., 1988, SIGEM: A Realistic Models Generator Expert System, in:
"Cybernetics and Systems'88", R. Trappl ed., Kluwer A.P., Dordrecht.
Haeberer, A.M., Veloso, P.A. and Elustondo, P., 1989. Towards a Relational
Calculus for Software Construction, Pont. Univ. Catolica: Res. Rept.
MCC 19/89, Rio de Janeiro.
Veloso, P.A.S., 1984, Outlines of a Mathematical Theory of General Problems.
Philosophia Naturalis, 354-362.

# LEVELS OF PROBLEM STRUCTURING AND PROBLEM DEFINITIONS

Jon-Arild Johannessen

Bodø Graduate School of Business
Dept.Information Management
Pb.6003-8016 Mørkved
Norway

## Abstract

An epistemological hierarchy of system levels are discussed to uncover the premises and presuppositions that lie behind every observer observing a system. Theese assumptions etc. will lead the observer when observing a system.

In the paper there is a main distinction made between the observed-system level and the descriptive statements one make of the observed system. There is also made distinctions in the epistemological hierarchy between method-level, model-level, theory-level and meta theory level. The epistemological hierarchy is here seen as important to uncover the role to the observer in his relationship to the observed system.

Two observers observing a system are both operating within their own epistemological hierarchies, which have to be done explicit. The structuring in epistemological hierarchies is necessary to avoid shifts of logical types in the problem solving process, which can lead to problem messes not problem solvings.

In addition to the epistemological hierarchy which the observers operates within, there is a relationship between the observers which is important to conceptualize, becauce different types of relationship can lead to different results. The relatioships between the observers are classified as Third order cybernetics and a relationship-matrix is developed, to uncover the main types of relationships that can exist between two (or more) observers.

In the process of defining and solving problems there have been emphasis on problem solving, and this is incorporated in the concept of "anasynthesis". The focus on problemsolving opposite to problemdefining is in this paper seen as counterproductive, and a developing from the concept of anasynthesis to panasynthesis (where P stands for problemdefinition) is done.

## I. THIRD ORDER CYBERNETICS

The concept relata refers to properties of person A,B,C etc., that are parts of, but do not determine, the constitution of the relationship between A, B, C etc. The concept of relationship is an entity of its own. Relata relates to relationship as parts relate to wholes, where the sum of the parts is not the whole. A relationship is on a higher logical level, or higher logical type (Russel, B & Whitehead, A.N 1913) than relata. Relationship can be looked upon as third order cybernetics, as shown in fig.1

*Systems Thinking in Europe*, Edited by M.C. Jackson *et al.*
Plenum Press, New York, 1991

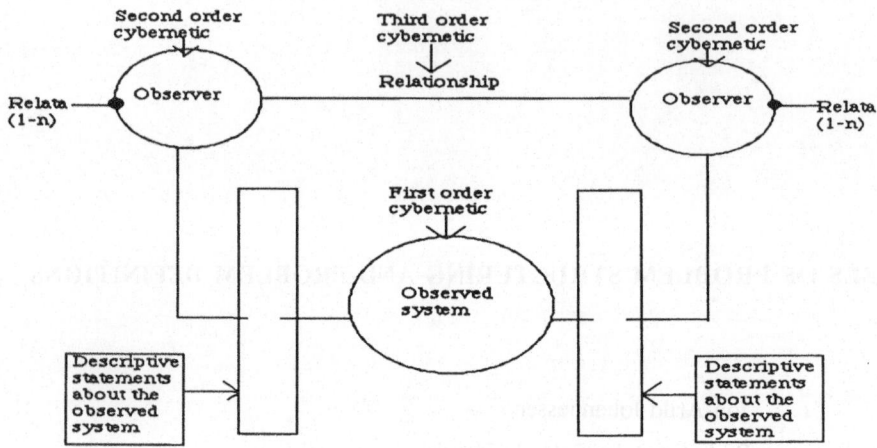

Fig.1 Relationship-Third order cybernetic

This view of relationship is contrasted with K.D.Mackenzie (1976 vol.1 and 2, 1979 pp.73-89) who uses relationship the way I use relata.
In the following I discuss the concepts which build up the relationship-matrix.
Relationships in a communicative process can basically have only two main forms (Bateson,G. 1979 p.151)
1. Symmetrical relationship : When actor A is doing something it is probable that actor B will do more of the same thing. Relationship can also be inverse, which means that increase in one variable is associated with decrease in another, (Chaffee,S.H & Berger,C.R. 1987 p.101) but this is, in my point of view, only an undergroup of symetrical relationship.
2. Complementary relationship : The behaviour of actor B is complementary to the behaviour of actor A.
Both of these relationships can have positive and negative feedback loops. When negative feedback loops are designed or enter the symmetrical or the complementary relationship, then self-correction will occur. Positive feedback in a relationship determines that the relation exceed a maximum and then collapse,or reach a climax and then decrease. We have to look for the pattern of circular loops which links a given occurrence, action etc.
Context is also included, because symmetrical and complementary relations can alternate in a given action sequence, but are digital in a given context-situation.
In social interaction there are two main forms of contexts that can exist, competitive and/ or cooperative.
The mechanisms that underlie coordination in a communicative network are : markets, clans, and hierarchies (Clark,P. & Staunton,N. 1989 p.170)
Market relationship are specified in formal contracts, clan relationship in trust and honor, hierarchical relationship in which the operation of certain systems is orchestrated by a focal firm which imposes detailed rules.
Both symmetrical and complementary relationship can be either healthy or unhealthy. In symmetrical relationship, health is linked to the concepts of mutual respect and trust, and poor health to the opposite. In complementary relationship, health is coupled with flexibility, and poor health with rigidity.
Both in symmetrical and complementary relationship, there will be a hierarchy of constraints on the relationship. It will be necessary to uncover this hierarchy of constraints to make the true relationship explicit. The communication structure in an organization can be described as

the invisible set of functional demands that organize the way in which organizational members interact (Minuchin,S 1974 p.71). These functional demands constitute the constraints of the relationship. The system of constraints is looked upon as the opposite to the seventeen demands in communicative competence (Johannessen, J-A. 1990 ).

The concept of calibration (Bateson,G 1979 p.269-279 & Bateson,G and Bateson,M.C. 1988 p.47-60 ) is extremely important because it has to do with change processes in relationships, and therefore change processes in the organizations too. The concept has to do with changes in the basic relation. Feedback and calibration alternate all the time in hierarchical ways. One has to do the dichotomy calibration and feedback explicit for oneselves on every recursivity level. The main point about the method of calibration is the absence of error-correction in the actual change process, and the use of many choices of action in a given action-sequence to gain change. The information used in calibration is of a higher logical type that which is used in feedback. In feedback we use information about errors in the actual situation. In calibration we use one or more classes of errors in a sequence of actions. Calibration can be coupled to the concept of "learning to learn", while feedback can be linked to the type of learning that is lower in the hierarchy of learning. (The hierarchy of learning is described by Bateson,G 1972 pp. 279-309)

The concepts discused so far are coupled in the relationship-matrix. By using the relationship-matrix, the relationship-part of a message can be analyzed. The matrix is a way of getting a systematic framework or classification-pattern, that will show how the relationship in a network effects the social interaction process, looked from a communicative point of view. The relationship-matrix will also be valuable in getting information that is important for strategic action.

In the relationship-matrix for simplicity we cultivate the two main relationship types, but "in reality relation between people and groups will neither be purely symmetrical or purely complementary, but that every such relationship contains element of the other type" (Bateson,G. 1979 p.70)

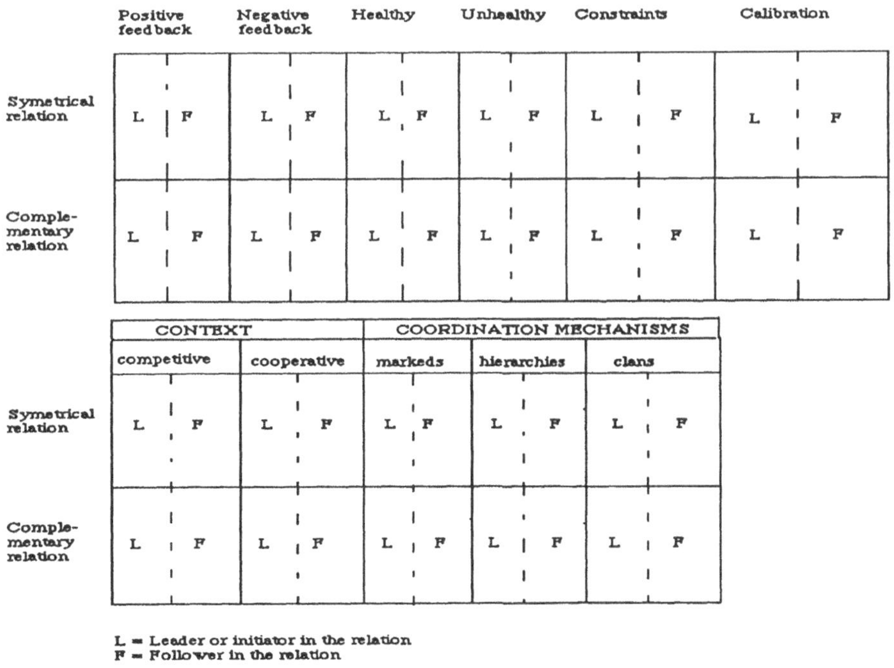

| | Positive feedback | Negative feedback | Healthy | Unhealthy | Constraints | Calibration |
|---|---|---|---|---|---|---|
| Symetrical relation | L   F | L   F | L   F | L   F | L   F | L   F |
| Complementary relation | L   F | L   F | L   F | L   F | L   F | L   F |

| | CONTEXT | | COORDINATION MECHANISMS | | |
|---|---|---|---|---|---|
| | competitive | cooperative | markeds | hierarchies | clans |
| Symetrical relation | L   F | L   F | L   F | L   F | L   F |
| Complementary relation | L   F | L   F | L   F | L   F | L   F |

L = Leader or initiator in the relation
F = Follower in the relation

Fig.2 The relationship-matrix

The relationship-matrix can be used to study the questions :
How do relationship processes change as organizations move from periods of growth and or stability into periods of decline ?
What is the relationship pattern of top management during periods of reorganization ?
Are there differences in relationship pattern in hierarchical and network organizations ?
Are there any couplings between management performance and relationship patterns ?

What patterns combine systems in a network ? To answer this question, we have to study the relationship that exists. A framework for the pattern that combines is shown in the relationship-matrix. Patterns between systems are never static, but can be understood as interacting alternating relationships. To understand a network of interacting elements, we have to focus on relationship, not the single element, and relationship is something more than the sum of the involved relata. The knowledge that is created in the relationship is qualitatively new. The description of one or more element does not tell us anything about the relationship, because it is of a higher logical level.

If linear thinking is the basis for evaluating communicative processes, then Faustian solutions will be the result, i.e. we quickly arrive at solutions, but these solutions will generate new problems of a higher complexity, and we end where we began. If, on the contrary, we take system theory and cybernetic thinking as the theoretical starting point, we will not so quickly arrive at solutions, because the problem definition phase is strongly focused, but probably we will not end with larger problem.

Every change process in an organization is based upon information and communication, which has social interaction or organizational assimilation as a catchment area. A communication model therefore must take as a supposition the complexity-level that exists in the organization, or the pattern of network that exists between interacting systems. If we oversimplify, it will lead us to the Faustian solution.

Every organization, and every network has its own interaction rules, and these rules are always understood by an observer in a certain epistemology. Therefore it is the pattern of redundant interactions in the communication-process which we have to focus upon.

A matrix for studying the pattern of relationship that connect systems in a network is shown in fig.3. In every cell in the matrix, all the factors named in the relationship-matrix shall be embedded.

| To<br>From | Element A | Element B | Element C | Element D | Element E |
|------------|-----------|-----------|-----------|-----------|-----------|
| Element A  |           |           |           |           |           |
| Element B  |           |           |           |           |           |
| Element C  |           |           |           |           |           |
| Element D  |           |           |           |           |           |
| Element E  |           |           |           |           |           |

Fig.3 Patterns of relationships

With the help of the relationship-matrix, the patterns of relationships and communicative competence ( Johannessen,J-A. 1990 pp.49-50) we have tools that can help us to :
1. explore the interactive levels of communication activity occurring in organizations. When using the relationship matrix and the pattern of relation matrix at different times, we can bring in the time and process dimension.

2. Communicative competence provides us with a concept for developing a communication policy in the organization. When the relationships are analyzed we can develop the communication policy further.

## II. THE EPISTEMOLOGICAL HIERARCHY

A and B observing a system have both immanent cognitive structures that can be understood in an epistemological hierarchy. From my point of view there will always be an important distinction between statements describing the object- system, and statements describing the conceptual frame which open up for describing the object-system, i.e. the epistemological hierarchy. The first type of statements describe the object level of science, while the second type describe the epistemological hierarchy of science, i.e. the meta-theory level (Weltanscauung), the theory level, the model level, the method level and the description language one uses to describe statements about the object-system. The two types of statements have to do with two different recursivity levels and must not be kept apart.

The observer has to be conscious of his own epistemological hierarchy, and as far as possible of that of the other observer. Communication between observer A and B, if not at the same epistemological level, will lead to confusions in logical types, and the communication process will eventually be blocked.

It is the correspondence rules in the description language that gives the theoretical terms, and thus the statements their meaning.

Every statements of epistemological hierarcical type will be immanent in the description language in statements that describe the object system. A distinction between theese two types of statements is therefore necesarry, to get rigor in problem solutions in the object-system. Problem solutions become more rigorous to the extent that the distinction between these two types of statements are observed.

The distinction between what one describes and the description is fundamental. The description can always be changed, never what one describes (in the moment one describes it).

When evaluating (falsification or verification), one always evaluate the description, not that what one describes, so there is not necessarily "pure empirical truths" that comes up, because one always evaluate one´s own descriptions, never the object-system per se. This insight of the distinction between a descriptive statement about the object-system and the object-system, makes it important to realize the problematical nature of the description, or better the transformation between the descriptive statement and that what is described. One way of doing it is to make explicit one´s own epistemological hierarchy, which in a subtle way lead the transformation between the descriptive statement and that which one shal describe.

The description of the object-system mirror to a large degree one´s own epistemological hierarchy and if this is forgotten there is a collapse of two domains, with Bateson between Creatura and Pleroma (Bateson,G 1972 & 1979).

In systems with high complexity, a high level of subjectivity is necessary, and the philosophy of science for such a complexity could be grounded in what could be called "Clarified Subjectivity". (Olaisen,J 1990. Johannessen,J-A. 1991)

## III. PANASYNTHESIS

Focusing upon problem solutions opposed to problem defining, taken what is said earlier in the paper, will be counterproductive. The use of the cybernetic concept anasynthesis, must therefore be replaced with the concept panasynthesis, where p- stands for problemdefining.

The field of system theory the way it is conceptualized in this paper is what G.M.Weinberger (Weinberger,G.M. 1975 p.18) name as organized complexity. Focusing upon problemdefining is a "new" way to look at this area upon, or with H.V.Foerster (Foerster,H.V. 1977 p.3) "we must learn how to see".

Complex systems are to be viewed as a whole and cannot be understood with classical analysis. The problemdefining phase is seeking from establishing an initial problemdefining (Chuchman,C.W. 1973 p. 209) to a definitive problemdefining. It is seeking for enough information to explain, understand and establish the problemdefinition (Pena,W. 1986 p.2). It is a coupling here to Checkland´s Soft-System-Methodology (Checkland,P. 1981 p.155).

What the system wants to gain with its actions must in the problemdefinition phase be emphasized, not why the system acts as it does. W.R.Ashby also make it clear that transformations is preoccupied with <u>what</u> happens, not <u>why</u> it happens (Ashby,W.R. 1961 p.11)

If the problemdefinition phase is taken to easily on the solutions we come up with could be those which created the new problems in the system, and more of the same will increase the future problems. This is excemplified by Koestler (Koestler,A. 1964) and Watzlawick (Watzlawick,P. 1980)

The starting from a wide point of view is also emphasized by Klir (Klir,G.J. 1985 p. 18). To little information in this phase can lead to a partial problemdefinition and a partial solution. On the other hand "the scope" must not be so wide that nothing comes out of it. (Pena,W.1986 p. 9). The transition from the problemdefinition phase to analysis is therefore critical.

M.Bunge (Bunge,M. 1967 p.200) says it in the following way; "the first operation, that of stating the problem, is often the most difficult of all,---. A good statement is half the solution".

# REFERENCES

Ashby,W.R.  "An Introduction to Cybernetics" 1961 Chapman & Hall LTD London (First Publ. 1956).
Bateson,G.  "Steps to an Ecology of Mind" 1972 Intex Book London.
Bateson,G.  "Mind and Nature.A Necessary Unity" 1979 (Swedish Translation 1988 Symposium Tryckeri Stockholm).
Bateson,G. & Bateson,M.C.  "Angels Fear" 1987 (Swedish Translation 1988 Symposium Tryckeri Stockholm).
Bunge,M.  "Scientific Research" Vol.1. 1967 Springer Verlag, Berlin.
Chaffe,S.H. & Berger,C.R.  "What Communication Scientists Do" pp.99-123 in
Berger,C.R. & Chaffe,S.H. (Eds.) "Handbook of Communication Science" 1987 Sage Publication,Inc New- York.
Checkland,P.  "Systems Thinking, Systems Practice" 1981 John Wiley, Chichester
Chuchman,C.W."The Systems Approach" (Swedish Translation 1973 Raben & Sjøgren Stockholm).
Clark,P & Stauton,N  "Innovation in Technology and Organization" 1989 Routledge, London.
Foerster,H.V. In "Future Research" 1977 Linstone,H.A. (Ed.) Addison Wesley Publishing Company, London.
Johannessen,J-A."The Holographic Organization- A Design Model" in Journal of Cybernetics and Systems, 22,1990 pp.41-55.
Johannessen,J-A."Clarified Subjectivity, as the Foundation for Information Management and Creative Management" 1991 Unpublished paper Bodø Graduate School of Business, Norway.
Klir,G.J.  "Architecture of System Problem Solving" 1985 Plenum Press, New York.
Koestler,A.  "The Act of Creation" 1964 Mc.Milliam, London.
Mackenzie,K.D."Where is Mr. Structure" pp.73-89 in Klippendorf,K. (Ed.) "Communication and Control in Society" 1979 Gordon and Beach Science Publisher, London.
Minuchin,S.  "Families and Family Therapy" 1974 Cambridge,M.A. Harvard University Press.
Olaissen,J.  "Pluralism or Positivistic Trivialism : Toward Criteria for a Clarified Subjectivity in Information Science" 1990 Paper presented ,ICIS 1990 des. Kopenhagen.
Pena,W.  "Problem Seeking" 1986 Excepts, University of Stockholm.
Russel,B. & Whitehead,A.N. "Principa Mathematica" 3.Vols. 2nd ed. Cambridge 1910-1913.
Wazlawick,P. "Change" (Norwegian Translation 1980 Gyldendal, Oslo.)
Weinberger,G.M."An Introduction to General System Thinking" 1975 John Wiley & Sons, London.

# A MULTIPLE CRITERIA APPROACH TO COMPLEX SITUATIONS

MARIA FRANCA NORESE

POLITECNICO DI TORINO, DIPARTIMENTO DI SISTEMI
DI PRODUZIONE ED ECONOMIA DELL'AZIENDA
Corso Duca degli Abruzzi, 24 - 10129 TORINO (ITALY)

## INTRODUCTION

The analyst's intervention in an organizational context, in relation to a problematic situation, implies an approach to learning through the interaction with the actors and the different logics of action and representation. In organizational contexts, such as the management of public services, where a lot of actors are involved from other organizations at different levels, with specific roles, functions and competences, limited formal interactions between these various actors may make the coordination of a collective action difficult; the evolutive and often conflicting context hinders actual forms of cooperation among the participant actors of these "macro organizations".

The analyst's intervention, inducing the participants' investment in the process of information gathering, evaluation, debate and presentation, helps to produce consensus, or at least an accepted integration of the different points of view. The multiplicity of representations, developed in this process or those that the actors of the Decision Process propose, become the principal object of the analyst's action. It must make different points of view explicit and then integrate them for the organizational action. In order to act on these representations the analyst needs schemes sufficiently elaborate to deal with complex situations and free enough from structural limits to become a vocabulary suitable for suggesting ideas, associating concepts, clarifying points of view and measuring compatibility and inconsistency.

Having always worked in the field of "Multicriteria Decision Aiding" as a research field and general approach to interventions, we have realized (cf Norese and Ostanello, 1988; 1989) that the *Multicriteria Approach* (cf De Montgolfier and Bertier, 1978; Moscarola, 1984) is a useful "representation tool" in the organization action. The main elements of this approach will be discussed in the paper, structures of multicriteria representation will be presented to show some examples of how these representations can support the individual and collective learning processes and to describe some ways to represent and evolve representations and to propose them in interactive contexts.

## LEARNING PROCESS AND ACTION ON THE REPRESENTATIONS

The actor who "moves" in the organization in order to understand the nature of the "boundaries" of his action (cf Crozier, 1977) activates learning processes "by which he modifies his knowledge about event contingencies, and adjusts and integrates behavior in terms of this contextual understanding" (cf Hunt and Sanders, 1986). His interventions in the system of action is in relationship with his a priori freedom and the specific courses of action that he must discover, create or choose and explain. Only with a real knowledge of the operating system can he discover his actual freedom of action, control it and if necessary, expand it and make the produced alternatives stand out, connecting them with the changing system, the old and new logics of action and the facts that generated the situation.

Among the models that explicitly consider learning, Piaget's (1974) Model of the development of Logico-Mathematical Structure has been used (for instance in Ramaprasad, 1987) to illustrate the potential role of cognitive processes as a basis for MIS and DSS design, to explain (cf Landry et al., 1983) the necessity of dynamic interactions between modelling and validating as a condition for learning and adaptation and (for instance in Courbon, 1984) to propose a support structure "allowing the manager to iterate smoothly through centrifugal loops of destabilization of representations and unstructuring of models, and through centipetal loops of restabilization of representations and structuring of usable models".

The *abstract representations* developed in a cognitive process may be presented very simply or by a very complex theory. They evolve and induce more refined observation and inference that, in their turn, develop the representation structure. These representations (or orienting schemes in Neissar, 1976) can become effective tools to acquire, organize, present and reorganize knowledge, at both an individual and a collective level. *Acting on the representations*, either developed ad hoc or that the involved actors propose, is the individual or collective process that always precedes the attribution of the representations to the reality (cf Ramaprasad, 1987) and then the implementation and the action on the object (cf Bourgine and Espinasse, 1987).

Two different kinds of activity characterize the action on the representations in an organizational context:

a) developing representations, acquiring and organizing knowledge of the problem, the decisional process, the involvement of actors, the state of information, ...

b) proposing representations in communication contexts where they can be explained, construed, validated, reinforced, made more precise or general, reconstructed (cf Courbon, 1984), refined and attributed to the reality (cf Ramaprasad, 1987) in a shared meaning.

Both the activities are related to the learning process, the first principally at a subjective level and the second in a more interactive meaning, and they are valid for both the individual and the collective or organization learning process. Developing representations and presenting them in an interaction context are radically different activities that should be supported by representation tools such as

- schemes suitable for directing the exploration which can easily be modified by the exploration results and

- structured representations of a language, that are frequently graphic or textual, adapted for stating global ideas and then fragmenting them and for declaring the meaning of every included element and controlling its ambiguity level.

Each support should be pertinent to the context in which it is used, judged pertinent by the actors who have to deal with the problem situation and in conformity with other representation structures supporting different phases of the same action (cf Landry et al., 1983). A global approach, common to each phase of the action on the representations, might guarantee specific pertinence and global consistency (validation activity); in order to become an actual support to the individual and collective action it should integrate *validation* activities to *conceptualization* and *communication* ones.

## MULTICRITERIA APPROACH

The analysis of twenty five Decision Aiding processes by the Multicriteria Approach (cf Norese and Ostanello, 1988; 1989) obliged us to distinguish between the analyst's activities at an individual level (mainly devoted to the modelling and model and method validation) and activities in interaction with the client, the information sources, the organization. A large variety of tools help the analyst in the individual work, few of them offer an actual support in collective contexts.

The *Multiple Criteria Approach* (MA), that *connects the possibility of using multicriteria methods with a specific way of acting on the representations,* is a valid tool. It can mainly be used both individually and collectively and proposes one only language for every phase of the intervention, a language that is rich, operational, immediate and accepted by all actors.

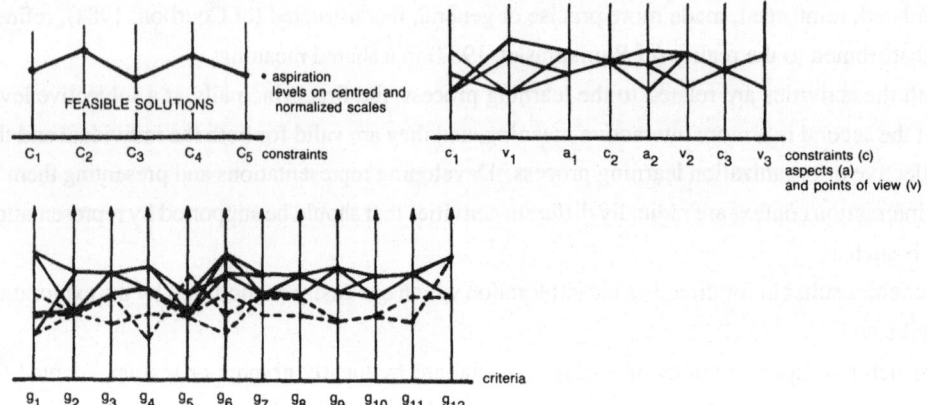

Fig. 1 From a feasible solution set (a) in a multiple constraint context to a multiple new logics (b) and then to a multiple criteria Reference System (c).

| Potential users | CLASSIFICATION Criteria | Profiles | Number of Objects for each profile | EVALUATION Criteria | Rankings |
|---|---|---|---|---|---|
| H | (1,3,5,6,7) | $P_1$ $P_2$ | (11) (11) | (1,2,3,5,6,7,9,10) | /18/12,23,31/24/1,19,26/21/3,5/ /12/18/23/31/1,24/26/19/21/3,5/ |
| S | (1,5,6,3) | $P_3$ | (6) | (1,2,3,5,6,7,8,10) | /8/7/20/25/14/29/ |
| M | (1,2,5,7) | $P_4$ | (8) | (1,2,3,6,7,9,10) | /8/15/28/22/13/10/17/11/ |
| BA | (7,8) | $P_5$ $P_6$ $P_7$ | (6) (14) (3) | (1,2,3,6,7,9,11,12) | /5/1/31/21/3/19/ /14/7/20/12/25,6/27,2/24/26/9/16/18/23/ /16/18/23/3/4/29/ |
| P | (1,4,7) | $P_8$ | (4) | (1,4,6,7,9,10,11) | /21/1/31/19/ |
| T/P | (1,7,8) | $P_9$ $P_{10}$ | (6) (9) | (1,2,3,5,6,7,8,9) | /1,31/5,19/21/3/ /20/25/14/12/2/27/23/26/18,24/ |

Fig. 2 Two phase multicriteria modelling.

The possibility of aggregating different points of view and problem dimensions, at the appropriate moment and by one of the multicriteria methods, induces a *non problematic disaggregation of the perceived problem* and a multidimensional and consistent inquiry system. The possibility of dealing with all kinds of data and choosing a multicriteria method consistent with the data state avoids a lot of descriptive, interpretative and communication constraints.

Different representations, coming from different proposers or reflecting multiple logics, may be taken into account completely (without neglecting any element) and developed in a proper context, by relevant criteria of the context itself. With the MA all the different attributes connected to the finalities and the value systems of the actors, may be dealt with and evolved with the aim of obtaining a family of criteria. The *internal consistency of a multicriteria representation*, guaranteed by tests of exhaustivity and non-redundancy (cf Roy, 1985; Roy and Bouyssou, 1987), is an aim for the modeler but it is also perceived by the users as a rational control of the global and local representations. The possibility of dealing rationally with situations neither easily nor globally modellizable becomes, like the "rational myth" in (Hatchuel and Molet, 1983), a new stimulus in organizational life which can create dialectical mechanisms in the collective learning process.

The analyst has to propose clear and operational representations in the dialogue with the client, in the inquiry that involves conflictual or non-communicant actors of the process, in the context of cooperative work where knowledge and expertise is shared, and in every other collective situation. Then he evolves these representations in a process of destructuring and restructuring, i) to validate and stabilize, ii) to induce incremental modifications and new interpretations and iii) to destablize them and introduce discontinuous shifts in interpretative schemes. By the MA he may move from the *action space* (action generation) to the *criteria space* (criteria generation) and vice versa without a substantial change of formalisms (see fig.1 from Norese, 1989). He may use the same word, *criteria*, in successive steps of the intervention process and then formalize descriptive, conceptive, judgemental and evaluative criteria; evolve them and use the multicriteria method suitable for specific modelling (see fig.2). He may *multiply representations* starting from a few conflicting alternatives (formalized as *multicriteria profiles*) and elaborate new strategic alternatives (see fig.3).

The Multicriteria Approach helps analysts and actors translate facts, proposals, points of view and preferences into formal models where mental, written and numerical data, from different sources, are organically synthesized and easily "linkable" to the proposing actors and context and process conditions (see for instance Norese, 1988). Every new element deduced in the inquiry, on the basis of a multidimensional scheme of interaction with the information sources and of interpretation of data, modifies this scheme. In this context the analyst's perception and representations can be modified through the different interactions, without substantial transformation of the model structure however.

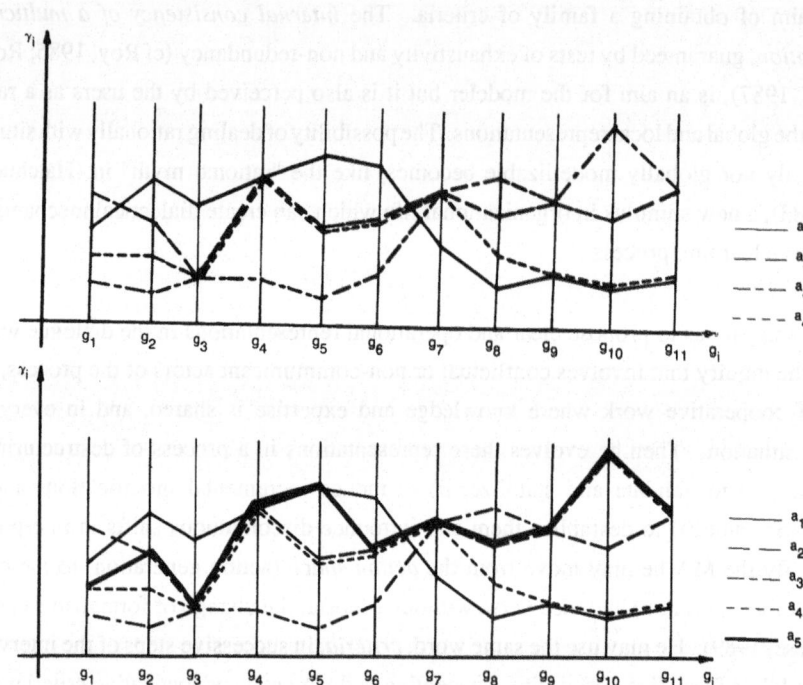

Fig. 3 Evaluation of multicriteria alternatives and elaboration of a new alternative.

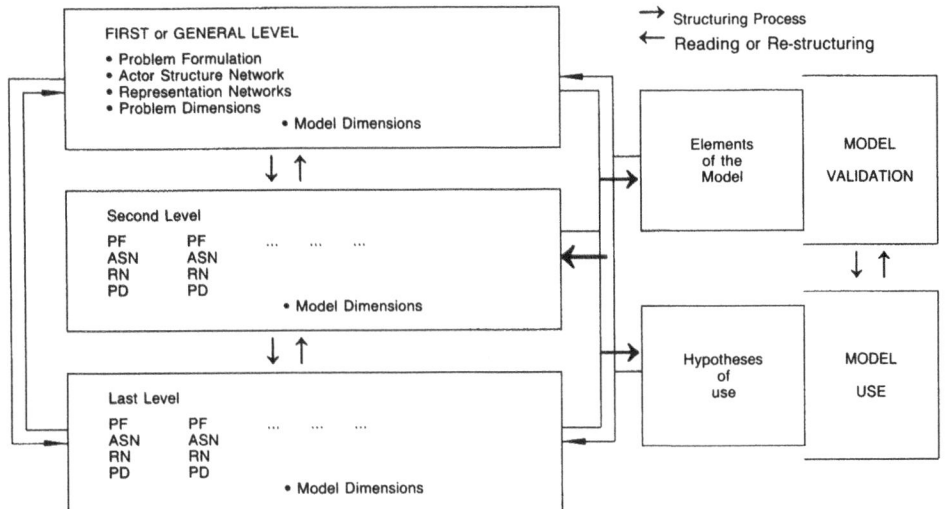

Fig. 4 Orienting scheme to problem and model structure.

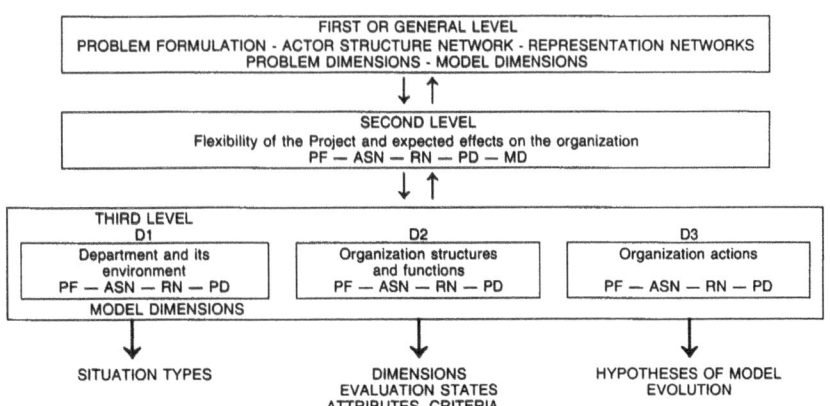

Fig. 5 Multiple Actor Representation Modelling Map, in a context of organizational change.

A *multidimensional orienting scheme* (see fig.4 and 5 from Mucci and Norese, 1990) may help the analyst to structure problem and model; it evolves with the structuring process, documents logic and context of every technical choice, controls the coherence of the process and the use of models which have arisen from this structured and systematic reading of interactions between representations.

## CONCLUDING REMARKS

The Multicriteria Approach supplies a common unambiguous vocabulary, rich enough to deal with every situation of data processing, which is rationally tested, easily updated and helpful in the communication contexts. In can help actors and analysts coordinate the global action when the dynamics of the process, in terms of information, role of the actors, preferences and constraints, impose a general approach supporting the action on the representations. It has appeared and appears valid in very different operational contexts and in relation with representation structures of several types. Connecting this approach operationally to the different levels of complexity of the organization and problematic situations and to the principal forms of communication that have to be activated may be a basis for a valid Group Decision Support System design.

## REFERENCES

Bourgine, P., Espinasse, B., 1987, Aide a la decision: une approche constructiviste, in: "Aide a la decision dans l'organisation", AFCET, Paris.

Courbon, J.C., 1984, Transparency of Data, information and models in Decision Support System, in: "Operational Research '84, J.P. Brans, ed., North-Holland, Amsterdam.

Crozier, M., and Friedberg, E., 1977, "L'acteur et le systeme", Editions du Seuil, Paris.

De Montgolfier, J., Bertier, P., 1978, "Approche multicritere des problemes de decision", Editions Hommes et Techniques, Paris.

Hatchuel, A., Molet, H., 1986, Rational modelling in understanding and aiding human decision-making: about two case studies, Eur. J. Opl. Res., 24:178.

Hunt, R.G., Sanders, G.L., 1986, Propaedeutics of Decision-making: Supporting Managerial Learning and Innovation, Decision Support Systems, 2:125.

Landry, M., Malouin, J.L., Oral, M., 1983, Model Validation in Operations Research, Eur. J. Opl. Res., 17:207.

Moscarola, J., 1984, Organizational Decision Process and ORASA Intervention, in: "Rethinking the process of Operational Research and Systems Analysis", R. Tomlinson and I. Kiss, eds., Pergamon, Oxford.

Mucci, L., Norese, M.F., 1990, A multidimensional model of the organizational answer to a change project, Presented to the IX-th International MCDM Conference, 5 - 8 August 1990, Fairfax.

Neisser, U., 1976, "Cognition and Reality. Principles and implications of cognitive psychology", Freeman, New York.

Norese, M.F., 1988, A Multidimensional Model by a Multiactor System, in: "Compromise, Negotiation and Group Decision", B.R. Munier and M.F. Shakun, eds., Reidel, Dordrecht.

Norese, M.F., 1989, Evolution from mono to multiple criteria in a process of interactive design, in: "Proceedings of the International Conference on Multiple Criteria Decision making: Applications in Industry and service", AIT, Bangkok.

Norese, M.F., Ostanello, A., 1988, Decision Aid process typologies and operational tools, in: "Ricerca Operativa e Intelligenza Artificiale", Proceedings of the AIRO'88 Meeting, Centro ricerche IBM, Pisa.

Norese, M.F., and Ostanello, A., 1989, Identification and Development of Alternatives: Introduction to the recognition of process typologies, in: "Improving decision making in organisations", Springer-Verlag, Berlin.

Piaget, J., 1974, "Recherches sur l'Abstraction Reflechissante", Paris, PUF.

Ramaprasad, A., 1987, Cognitive Process as a Basis for MIS and DSS Design, Management Sci., 33:139.

Roy, B., 1985, "Methodologie Multicritere d'Aide a la Decision", Economica, Paris.

Roy, B., Bouyssou, D., 1987, Famille de Criteres: Probleme de Coherence et de Dependance, Cahiers du LAMSADE, 37.

# THE ROLE OF THE ANALYST AND THE CHOICE OF THE METHODOLOGY

Peter Holland

Software development
Microtec Management Systems Ltd
Bath
England BA1 1UL

This paper examines the importance of an analyst considering his own role in a project when choosing a problem solving methodology. The author outlines his feeling that the effectiveness of a project can be improved if the analyst is more aware of his role and his responsibilities. Some of the roles the analyst may take are explored and attention is given to the effect these roles may have on the definition, progression and outcome of the project. Methodologies are examined with respect to the roles they may cast for the analyst. ( For the purposes of this paper the term 'methodology' includes what some may regard as approaches and techniques.)

A reason for the author following this line of thought is the sense, after some projects, that there was confusion, a dislocation between himself, the analyst, and the other people involved. One of the symptoms being a general dissatisfaction with the result of the work. The following is an example of one of these studies.

The analyst was a new member of an organisation. He was asked by a person, lets call Bert, to help him find a way to record the stock levels of some documents that he was finding difficult to manage. Looking into the situation further, the analyst found that the required literature was constantly changing, the 'order lead time' for new stock was about four weeks and it was often out of date when it arrived. The required numbers of each document were small and there were many of them. The literature was produced abroad by the organisations professional writers and illustrators on desk top publishing (DTP) equipment. It was delivered by courier service and was bulky and cumbersome to store. The accuracy of the information was very important and therefore only up-to-date documents were used.

The analyst put forward the idea that the writers and illustrators could return the document in electronic form using their existing computer equipment. The document would then be stored on a similar DTP system. This would be located in what had previously been the store room. Documents could be printed off as required. The occasional documents, requiring an extremely high quality of print and large numbers of copies, could be sent directly to the local printers using a modem.  This would eliminate the requirement for bulk storage, would pay for itself in the first year and reduce the stock control problem to maintaining up-to-date versions of each document on the DTP system.  One person required training in the use of this equipment.

All the people involved thought that it was viable and the finance was provided for the equipment.  The managers however could not agree on who should supply the person to operate it.  The old system of operation was maintained.  The result of the project was that a DTP system was sitting unused in the stock room, Bert still had the problem of recording stock levels and the analyst felt frustrated and annoyed that his solution was not operating.  The analyst began to see that although the problems he had identified were present and acknowledged by the people in the situation, he was the one who had invested his energy to drive the project.  When the time came for the managers to act they did not have the desire to do so.  After all they had not approached him for help.  If he had stuck to applying his ability to the problem of recording stock levels the result of his work may have been more satisfying.  This was all he had been asked to do by Bert.

This example shows how it is possible for an analyst to introduce many new factors into the problem situation.  His inclusion and presence in a project group is a part of the problem and the problem solving process.  It is obvious that the analyst did not focus his energy on the area that concerned Bert.  He worked outside the bounds of the stated problem.  Both parties are responsible for the maintenance of these boundaries.

In problem solving it is necessary for change to take place, to bring the conceptual solution into a real world form.  The change in the situation will involve an alteration of peoples views.  Preconceived ideas need to be released and new ones formed.  This is uncomfortable and stressful.  The group undergoing this process may not have the energy to invest in maintaining their boundaries with the analyst.  On the contrary, they may wish to pull him into other supportive roles.  This may deepen any misinterpretation that the analyst has of his role and the boundaries of his work.  The possible result could be that the original aims of the analysts involvement in the project may become lost and blurred.  The concepts of roles and boundaries may provide a means of perceiving the project/analyst relationship in a form that is easily maintained.  The effort of ensuring that this relationship remains focused on the primary task of the group may be eased.

There are many roles in project groups. eg. manager, leader, planner, reformer, mathematician, problem solver, accountant, expert, analyzer, advisor, seducer, preacher, mourner, thinker, organiser, theorist, practitioner, employer and employee.  These are only a fraction.  They can be identified from the different perspectives of problem solving, business and organisational management, group psychology, the social sciences and other disciplines.  Individuals will identify differently with each.  Their understanding and knowledge being dependant on their awareness of the role, their experience in the role and their experience with others in the role.

Roles have both functional and emotional components.  These will consist of conscious and subconscious parts.  They can be held by groups of people, organisations, societies, animals and inanimate objects. They can be regarded as being essential to the existence of a project and therefore are integral parts of the environment in which the project exists. A project can be viewed as a set of interacting roles, each one influencing the expression of the other; reinforcing, challenging, suppressing and supporting. The success and failure of the project will depend on the extent to which the roles associated with the objectives of the project (primary roles) are fulfilled. For the purposes of this paper primary roles will be identified with the general group objectives of problem solving and decision making.

Klein (1961) discusses decision making in relation to the four roles of 'Expert', 'Facilitator', 'Coordinator' and 'Morale builder'. The expert supplies information to the group. The facilitator requests information from the group. The coordinator puts forward suggestions to the group seeking agreement. The morale builder contributes to the cohesiveness of the group. The members of the group will interact in these four roles until a decision is made. The analyst will take part in one or more of these roles in a project.

The process of problem solving can be viewed in the following way. The problem is first perceived as a feeling, it is then defined, solutions are identified, a solution is chosen and then the solution is implemented. These are simultaneous activities which are all affected by a controlling process. This control process is driven by the motivation or desire of the problem solving group and determines whether the problem solving will continue. The problem perceiver is the problem owner. The problem definer requires expertise in the expression of problems. The solution identifier may have skills in conceptual thinking and creativity. The solution selector may use analytical techniques to test the feasibility of solutions. The solution implementor is the problem solver and requires belief in the solution.

In the case of Bert's stock management problem the analyst can be seen to have acted in any of these roles. Bert and the analyst could have agreed on a primary task, 'to investigate solutions to Bert's problem'. Using the roles described above they could have identified their primary roles. In the decision making Bert was a coordinator and facilitator. Both the analyst and Bert were experts and could contribute to their own morale. In the primary problem solving process the analyst was a problem definer and solution identifier. Bert was a problem perceiver, a problem definer and a solution selector. The role of solution implementor could remain undefined until Bert had chosen a solution. In this way the analyst would have been able to focus on Bert's problem, define the problem of stock level recording, design solutions and let Bert decide which one, if any, to use. He would have had clearly defined roles and boundaries in which to operate.

This is only one side of the relationship, the analysts contribution to the group. For a fuller picture it is necessary to investigate the groups contribution to the analyst. This is more difficult to identify. Here attention is given to the person's prior group experience. The theory is that an individual will join a group and regard it in the light of this previous experience. The desire to join the group will be due to the person wanting that which they have had, or wanted, in their membership of groups in the past.

The origins of a persons social behaviour and identity can be traced back to their experience in their family.  Here interactions in terms of a conscious primary task are less useful.  Klein (1961) outlines the interactions connected to self expression.  These constitute an expression of feeling.  A child will learn what to consider as valued self expression by the response it receives from the other family members.  The child will mould itself to conform so that it becomes more accepted.  It will learn how to behave in a socially acceptable way.  It will develop a sense of its own value and of its own worth.  A family will maintain its own set of values.

Any group can be regarded as an extension of this family group experience.  A group will have common values.  A newcomer will only be accepted if his values are similar to those of the group.  If they are different the group will challenge them.  It will be necessary for him to change and alter his values before becoming accepted.

It is important for the analyst to recognise the significance he places on his membership to the project group and therefore understand his motives for joining it.  He will then be able to distinguish more easily those values he holds which are commensurate to his involvement in the project and those which he maintains that are excessive to his task which involve his deeper membership of the group.  In the case of Bert's problem the analyst was new to the organisation.  It was important for him to feel needed.  He had experienced this feeling in the past when he had involved himself in solving problematic issues that existed in previous groups.  In this case he was allowed to abuse his role in this project to achieve this familiar position.  His ultimate confusion at the end of the project was not due to the failure of his analysis but to him associating the rejection of his solution to a sense of not being needed.  By satisfying, or just realising, this fundamental need the analyst would have been freer to operate on the primary task, his insecurity having been recognised.

The example used is small and relatively simple.  It does however illustrate that a useful understanding of a project situation can be gained through the identification of the primary roles associated with the task of the project and the individuals or groups responsible for fulfilling them.  In this paper these have been the primary roles of decision making and the primary roles of problem solving.  Further understanding can be achieved if the need that the group has of the analyst within the primary roles and the analyst's basic needs of the group are identified.  It is suggested that by maintaining a clear perception of these roles and relationships any confusion, lapse of responsibility or transgression of their boundaries can be identified and followed up more consciously.  This will free some of the effort required in the subconscious work of the group.  The end result of the project may therefore be improved.

The definition of the primary task of the group, ownership of the primary roles and the associated boundaries of responsibility are probably best viewed as a contract agreed by the group and the analyst.  This will be continually updated.  The choice of methodology should be made with strict reference to the part of this contract that the analyst is aiming to satisfy.  The methodology should be interpreted and implemented without breach of this contract.  The analyst and group can review the contract when necessary.  The end of the analysts involvement with the project will be signified by the fulfilment of this contract. If the group and the analyst have not breached the contract then they will both be satisfied within its terms.

A methodology can be viewed as a means of supporting a person while they fulfil a role or roles. Each methodology can be associated with an underlying approach. An approach incorporates and implies many different values. These will influence the way in which a person fulfils their role. The choice of methodology is viewed in terms of these values and those of the group and the analyst.

A choice implies that judgements must be made of the possible options, a decision is made based on these judgements and one option is then chosen (Hogarth R, 1987). Hogarth goes on to identify four areas of possible bias in judgement: the acquisition of the information from the environment and memory, the processing of this information, the manner of the response required and the outcome of a similar previous decision. Vickers (1965) describes what he calls the appreciative system, as being necessary and fundamental to any decision making process. He argues that two inseparable constituents of appreciation are 'reality' judgements ( judgments about the state of the system... both internally and in its external relations) and 'value' judgements (judgments about the significance of facts to the appreciator or to the body for whom the appreciation is made).

It is important that the chosen methodology incorporates values that are acceptable to the group. The analysts values will influence his role as an analyst and therefore his decision as to which methodology he will use. It has been shown that the analyst while joining the group will change his values. His choice of methodology should be made once this process is well under way. It can be stressful and uncomfortable. The author feels that in the past he has chosen a methodology to use before joining the group. He then used it to support him through this uncomfortable time. The groups sometimes did not accept him in the roles the chosen methodology cast. His work became dislocated from that of the group. The roles required by the methodology being unfulfilled leading to the analyst failing in his task.

The analyst grows with his experience in groups. This growth will depend on the group challenging the values he holds as an analyst. Groups in difficulty may not be able to do this effectively. The analyst is therefore devoid of this essential part of his growth process. It is beneficial for the analyst to be aware of his role and the groups response. In this way he will be more able to identify the weakest challenge, examine and consider it, rather than responding in an unaware and confused way. He is responsible for his own awareness. The development of his analyst skills are therefore closely bound to his personal development.

REFERENCES

Adams, J.L., 1974 "Conceptual Blockbusting," Penguin Books Ltd, Harmondsworth.

Avison, D.E., Fitzgerald, G., 1988 "Information Systems Development: Methodologies, Techniques and Tools," Blackwell Scientific Publications, Oxford.

Galinsky, M.D., Schaffer, J.B.P., 1974 "Models of Group Therapy & Sensitivity Training," Prentice-Hall inc, New Jersey.

Grainger, A.J., 1970 "The Bull Ring" Pergamon Press Ltd, London.

Hinshelwood, R.D., 1987 "What happens in groups," Free Association
        Books, London.

Hogarth, R., 1987 "Judgement and Choice" John Wiley & Sons Ltd, New
        York.

Klein, J., 1961 "Working with Groups," Hutchinson & Co (Publishers)
        Ltd, London.

Meredith Belbin, R., 1981 "Management teams: Why they succeed or
        fail," Heinemann Professional Publishing.

Vickers, G., 1965 "The Art of Judgement" Chapman & Hall, London.

# THE USE OF SOCIAL PARADIGMS IN THE ANALYSIS OF TEAM BEHAVIOUR DURING ORGANISATIONAL CHANGE

Roger Stewart
School of Information Systems
Kingston Polytechnic
Kingston upon Thames,Surrey KT1 2EE

## INTRODUCTION

The nature of the environment in which many organisations are currently operating can be characterised by rapidly changing markets and customer requirements, with increasing complexity in the technologies being utilised. This turbulent and continuously changing demand is reflected in the different strategies adopted by management to ensure the survival and growth of their companies. Waterman (1987) observed that, "Somehow there are organisations that effectively manage change, continuously adapting their bureaucracies, strategies, systems, products, and cultures to survive the shocks and to prosper from the forces that decimate their competion."

One of the current trends is to create a flatter, more flexible organisation structure. (For example Froonhof (1990), Kanter (1989), Peters (1987), and Bronner et al (1990).) In parallel there is a greater awareness of the dependance on the understanding of the social processes involved in the change from tall hierarchical structures to the flatter project oriented structures as discussed by Estibals (1990).

There are many perceived benefits from a flatter structure, however there are also possible counter productive effects. This paper takes the perspective that in order to understand the social processes involved in this type of organisational change, models and concepts from the discplines of Sociology, Social Psychology, and Organisational Psychology can be utilised to form paradigms that specifically address areas of less desirable emergent behaviour. Each paradigm describes the behaviour pattern and suggests possible causes that can be addressed with corrective action.

## PROCESS OF CHANGE

The creation of "flatter, flexible organisations" implies two changes. The first is flattening or reducing the number of levels of management and applying a different mode of strategic goal setting, authority levels and control of performance. Essentially these are structural and procedural changes, however the social nature of the central functions of the organisation will evolve from authoritarian control to strategic guidance and support. The second is the change from a role oriented organisation to a task based or federal organisation in order to increase the flexibility to be more responsive and adaptive to the rapidly changing demands of the environment. In both changes there is an element of "pattern breaking" to free the organisation from the structures, processes or functions that are no longer effective. (Barczak, et al. 1987).

The key element in the process of change is the creation of teams set up to provide a continuing function or service to the whole organisation, or a multi-disciplinary task based team with a high degree of autonomy set up for a specific project. (A team in this sense could be a small group or a larger department). It is this latter type of team, situated in the lower part of a new flatter pyramid, that this paper will consider in more detail. Descriptive social paradigms are suggested that could explain emergent counter-productive behaviour during three developmental or operational phases of a team. These phases should not be considered as discrete but rather points on a continuum.

### PHASE 1

Phase 1 covers the period from the restructuring of the organisation and the definition of the new team, to the early period of operation. This will involve setting the purpose, objectives, authority, resources, financial structure and nomination of the personnel involved. Almost certainly this will necessitate incorporating new members into the team from other areas of the organisation and new functions that were previously central responsiblilties. For example finance or personnel. The team will then begin to operate and if everything has been defined well and conducted with due consideration to the people involved, then this new 'social' system will perform as required. However there are a few pitfalls that can contribute to some emergent counter-productive behaviour of the team. One such pattern of behaviour can be described as follows.

### The Anomic Reaction Paradigm

Description   The team has been formed with the members coming from different areas of the organisation. There is a degree of uncertainty as to exactly what they are meant to be doing and the scope of their authority. This manifests itself in a continuous checking of work and decisions, both internally within the team and with other areas of the company. The level of communication of team members with previous formal and informal networks is high, showing a propensity to discuss previous activites. There is a reluctance to let drop past 'favoured' activities and a possible low commitment to, or understanding of, the new tasks. There is not yet a full identification with the new team, nor an understanding of how this new appointment may serve individual member's long term needs.

378

Possible causes   People have been moved around in organisations since organisations were formed, however the fluidity now required in task based organisations is such that as soon as they have settled into one team, they can be moved. This has the effect that as one set of norms and relationships are stabilised, then a new set has to be formed. Hence change is continuous. The new responsibilities, goals and authority may not be crystal clear and the relationships to other people and teams are yet to be firmly established. The old networks of peer groups and informal networks has been disrupted.

Theoretical basis   Durkheim (1952) has described his concept of Anomie as the feeling of aimlessness or purposelessness being provoked by rapid social change, whereby traditional norms and standards become undermined without being replaced by new ones. Anxiety and disorientation results from this situation. Merton (1957) modified this concept to refer to the strain put on an individual's behaviour when accepted norms conflict with social reality. Recently, Watson (1987) has used Anomie in the context of lack of discipline, principle or guiding norms. There is also the problem of satisfying the individual's personal needs, Maslow (1954). The case of Anomic Reaction described in this paper is where the old norms of the team the individual has left are not yet replaced by the norms still to be formed in the new team. It is also unclear to the individual how within the new team personal needs and aspirations will be fulfilled.

PHASE 2

This phase is where the team has developed as an entity and is performing normally. Members have been assimilated and old links to previous work areas have changed or decayed and new links established. The roles, responsibilities and authority of members have been clearly defined.

## The Team Cult Paradigm

Description   This paradigm examines the relationship of individuals to the team. Intra-team norms, work and social patterns have stabilised. A tightly knit society has developed with a 'clan' type of operation of close working ties and personal connections within the team and to other teams. A strong or charismatic leader may have arisen, with pressure on members to conform to the team. A team developed in this form can cause dissatisfaction and indeed some form of cognitive dissonance to some members. Other problems may surface as a result of the formation of a strong team. For example, 'deindividuation' may cause a lack of self-recogniton, 'groupthink' and 'group polarisation' may run counter to an individual member's strongly held attitudes and beliefs all leading to disassociation from the team or indeed 'reactance' or rebellion.

Possible causes   The group has developed a strong culture whereby the team cohesiveness is more important than individual members. The emergence of a strong leader is driving the team in directions that are counter to members wishes. Authority and responsibility levels are being disregarded or abused. Team members are not being consulted or are overidden in decision taking processes and there is no place for minority views. The social and personal activities of the team are given undue importance.

Theoretical basis  The paradigm addresses the areas of group formation and group dynamics drawing upon social psychology and organisational behaviour. The main concepts are 'clan groups' W. Ouchi (1981), 'cognitive dissonance' and 'deindividuation' Festinger et al (1957,1952), 'groupthink' I. Janis (1982), 'group polarisation' M. Isozaki (1984) and 'reactance' Brehm & Brehm (1981).

### PHASE 3

This phase is occurs when the team is well established and has been operating for a period of time successfully. There is strong cohesion in the team coupled with a high self-belief. It has been meeting its original objectives although not without what it perceives as constraints on current and future activities imposed from higher in the organisation. In the light of its success and the experience it has gained, the team believes that it knows best what should be done for itself and for the organisation. Also that it has the best expertise in the organisation to achieve this. There are inherent dangers of estrangement of the team in this situation.

### The Team Primacy Paradigm

Description  This paradigm examines the relationship of the team to the organisation. The team has developed a strong identity, self-belief and self-importance bordering upon arrogance. It is very clear on how to run its activities successfully, the resources it should have, how it should be structured and demands the authority to define its own boundaries of operation. It perceives as interference, hierararchical and lateral organisation attempts to control its activities by setting detailed objectives, goals and budgetary limits in areas where the team believes it knows best. The team feels that it does not have the power to follow its own strategies and is not being given the support or recognition warranted. This leads to some stormy external relationships. The belief of the team is that its activities are of prime importance to the success of the organisation, and hence a 'primacy' or pre-eminent status should be attributed. The result of not obtaining this status is an alienation of the team to the organisation. This team alienation is demonstrated by a resentment to the way in which the team is forced to operate and frustration at the way in which they are required to apply their skills and knowledge  without the discretion and autonomy they expect. The end product is an estrangement of the team to the organisation and from the senior management's perspective, counter-productive behaviour.

Possible causes  In the process of flattening the structure and the creation of task teams, too much emphasis is given to the importance of specific teams. Whilst this may be important in the motivation of the members, there is a danger of elitism. The devolving of tasks and functional responsibilty to teams without the corresponding devolvment of authority can cause.either a high degree of annoyance or projection of responsibility. Similarly where a semi-autonomous unit has been set up, the nature of the performance control mechanisms is important. A problem arises if feedforward control is not given predominance over feedback. A lack of explanation of where in the overall strategy this team's activities contribute, with associated defined importance, can cause a focussing on the team not balanced by the overall picture. This causes a sense of isolation and exploitation of the team.

Theoretical basis  In developing this paradigm the starting point was the concept of alienation, Marx (1963), whereby the division of labour alienates human beings from their work. This has some relevance in that division of labour is part of flattening the structure and creating task based teams, but does not totally relate to an internally well functioning team who are happy with work in the team but not with relationships to the rest of the organisation. The study by Blauner (Alienation and Freedom,1964) introduced the concept of dimensions of alienation and whilst this is a study of individuals and different technological settings, the dimensions of 'powerlesness', 'isolation' and 'self estrangement' can be applied to the relationship of a team to an organisation. The 'dependent participation' alienation of A.Touraine (1971) where specific skills and knowledge groupings of people are not given the discretion and autonomy they expect is very appropriate to a team based paradigm. With regard to the team self- perception, they can be regarded as a 'psychological' group, for example, "where the members interact with one another, are psychologically aware of one another, and perceive themselves as a group" (Schein,1988). This team psychological bonding coupled with a feeling of expertise, success and and yet a sense of alienation, combines to produce the behaviour of team primacy.

## CONCLUSIONS

The current trend of creating successful, flatter, more flexible, task based structures in organisations requires a better understanding of the social processes involved. This paper has attempted to demonstrate that by using wider fields of existing research it is possible to construct 'social' paradigms that are specifically applicable to the analysis of team behaviour in this trend.

## REFERENCES

Barczak, G., Smith, C., Wilemon, D., 1987, Managing Large-Scale Organisation Change, Organizational Dynamics, Summer, 1987.

Blauner, R., 1964, "Alienation and Freedom", University of Chicago Press, Chicago.

Brehm, S., Brehm, J. W., 1981, "Psychological reactance: A theory of freedom and control", Academic Press, New York.

Bronner, N., Moberg, S., 1990, Organizational Development of Vattenfall Engineering Group, "Management by Projects Vol. 1", Internet 90, Vienna.

Durkheim, E., 1952, "Suicide: A Study in Sociology", Routledge & Kegan Paul, London. First pub. 1897.

Estibals, D., 1990, Artificial Intelligence to Improve Project Management Flexibility, "Management by Projects Vol. 1", Internet 90, Vienna.

Festinger, L., 1957, "A theory of cognitive dissonance", Stanford University Press, Stanford.

Festinger, L., Pepitone, A., Newcombe, T.,1952, Some consequences of deindividuation in a group, Journal of Abnormal and Social Psychology., 47.

Froonhof, H., 1990, Total Quality Management in a Flat and Flexible Organizational Structure, "Management by Projects Vol. 1", Internet 90, Vienna.

Isozaki, M., 1984, The effect of discussion on polarization of judgements, Japanese Psychological Research, 26.

Janis, I. L., 1982, "Groupthink", 2nd ed., Houghton Miffin, Boston.

Kanter, R. M., 1989 The New Managerial Work, Harvard Business Review, Nov.

Marx, K., 1963, Alienated labour, T. B. Bottomore (ed.), in "Karl Marx: Early Writings", Penguin, Harmondsworth.

Maslow, A., 1954, "Motivation and Personality", Harper and Row, New York.

Merton, R. K. ,1957, "Social Theory and Social Structure", Free Press, Glencoe.

Ouchi, W. G., 1981, "Theory Z: How American Business Can Meet the Japanese Challenge", Addison-Wesley, Reading, Mass.

Peters, T., 1987, 'Thriving on Chaos", Alfred A. Knopf Inc., New York.

Schein, E. H. I988, "Organizational Psychology", 3rd Edition, Prentice-Hall, Englewood Cliffs.

Touraine, A., 1971, "The Post-Industrial Society", Random House, New York.

Waterman Jr. R. H., 1987, "How the Best Get and Keep the Competitive Edge, New York.

Watson, T. J., 1987, "Sociology, Work & Industry", 2nd ed., Routledge & Kegan Paul, London.

# A BRIEF INTRODUCTION TO COMPLEMENTARISM

Robert L. Flood
Department of Management Systems and Sciences
University of Hull
Hull, HU6 7RX, U.K.

## INTRODUCTION

All healthy disciplines continue to change. Critical systems thinking consolidates recent innovative changes into a whole new brand of systems science. Its main idea is openness and conciliation between people, and their knowledges and methods. This philosophy is called complementarism. The purpose of this paper is to provide a brief introduction to complementarism.

## BOHR'S AND APEL'S COMPLEMENTARISM

Complementarism is the proposition of a meta-science that respects human well-being. It is a meta-science that can coordinate other sciences (knowledges and methods) in an informed manner. It embraces emancipatory science among others. Modern complementarism does not propose to control any form of science with another dominant science. To the contrary, it is a meta-science that harnesses the worth of methods and knowledges according to their strengths and weaknesses. Complementarism therefore is not another dictator and must itself remain open and ready for change.

A comprehensive study on pre-contemporary meta-science can be found in Radnitzky (1970, p. 59 onwards). He undertakes a thorough and informative study of complementarism, recounting

*Systems Thinking in Europe*, Edited by M.C. Jackson *et al.*
Plenum Press, New York, 1991

Bohr's and Apel's ideas and leading up to Habermas' knowledge-constitutive interests.

Constructing a meta-science, Radnitzky argues, is no easy task. "There is no lazy way, nor short-cut to overcome the ethnocentricity of scientific subcultures expressed and reinforced by the special sublanguages." Its achievement hinges on developing an understanding of ideas about dialectic mediation, polarity and complementarity.

Dialectics plays a role in the development of totalities and their parts. Bohr's complementarity thesis is the classic model that inaugurated the application of the dialectic method. Complementarity does not intimate presupposition between theories of each other. Different viewpoints make us see different aspects of a theory. No single theory can help us catch all aspects of reality. A complete description is elliptic. Aspects of each theory may complement each other to give an ever fuller picture. Bohr, for example, argued that wave and particle theories in elementary particle phenomena complement each other in this way.

The impression of polarity of two theories, or of their base explanatory models, typically is due to claims of totalisation of either. Each model may claim that it enjoys universal application within a sector of reality of concern. But this causes destructive tension. Such tension between knowledge-systems or theories in polarity is a crisis. It is not merely a dialectical tension such as thesis-antithesis, but a logical contradiction.

Apel directs his complementarity thesis against all totalisation, considering knowledge-systems as complementary. We can employ many knowledge-systems to round and fill our picture of man. Apel argues that natural science and human science--in the science of man--and quasi-naturalistic and human approaches--in human science--are mediating each other. In each the developments of knowledge proceed in a continuous tacking between two approaches or levels. This is a move toward detotalisation.

While Apel explores these key ideas of complementarity, it is Habermas' (1971) argument for knowledge-constitutive interests that establishes a sound epistemological position.

# HABERMAS' COMPLEMENTARISM

## Introduction

Habermas' knowledge-constitutive interests sets out to assimilate seemingly disparate approaches through Kantian, Hegelian and Marxian poles in his thought. We will study relevant portions of Habermas' work so that we may grasp a better understanding of his ideas. There are two important contributions from Habermas' of interest to us.

First, that accounts of social practice may be thought of as labour and interaction. The production and reproduction of human lives occur through the transformation of nature with the aid of technical rules and procedures, and through communication of needs and interests under rule-governed institutions. This adds a 'social turn' to the 'linguistic turn' established in Hegel's work. Abstracting from either turn in isolation distorts and is mistaken. Social constraints and power relations always dominate dialogue.

The second contribution is Habermas' search for a foundation for critical social theory, while accepting an anti-foundational notion that there are no theory-neutral facts for inquiry. Habermas proposed that rationally motivated agreement among participants in argumentation is the only foundation. Although this ground is arguably unsettled, Habermas still rejects the extension of this uncertainty toward a sceptical abandoning of the search for justification and theoretical grounding. This can be sought after in ethical and cognitive stances (a modern scepticism Habermas says). These foundations can be established by analysis of perceptions of communication. Social theory, therefore, must be critical so that fundamental norms that guide theory may be reconstructed (communicative rationality, see Habermas, 1976). Accepting this, Habermas developed a foundation for critical social theory by examining the presuppositions of communication that reveal a rational dimension within the conversation itself, which can be reconstructed using Kant's transcendental mode of posing questions. A rational consensus will be achieved through free and equal discussions, within a framework of an ideal speech situation. This suggests truth, freedom and justice. This also assumes a process free from unnecessary domination in all its forms. The process must be emancipatory. In this Habermas offers the notion of legitimate authority, which poses a significant challenge to systems practice and practitioners (see Fairtlough, 1989).

Let us now look in some detail at knowledge-constitutive interests.

## Traditional and Critical Theories

When building a substantive critical theory based on Horkheimer's (1968) essay "Traditional and Critical Theories," Habermas set up the following proposition to refute: '... the only knowledge that can truly orient action is knowledge that frees itself from 'mere' human interests and is based on ideas, which states that knowledge has taken a theoretical attitude.'

According to Horkheimer, theory in a traditional sense is like 'looking-on' where we abandon ourselves to the events. In philosophical language, we abandon ourselves to contemplation of the cosmos. Theory can, therefore, enter the conduct of life and moulds life to its form. For instance we may assume: 'that is the way of nature,' 'I am of nature' and according to this position find theory reflected in the conduct of those who subject themselves to its discipline. If we contemplate the cosmos then our soul might be likened to motions of nature.

This is a conception that places life in theory. It inevitably leads to totalisation. It is a conception that has defined philosophy since its beginnings. To find such totalising positions characterised by strong intellectual forces will be no surprise. They are hard to escape without fundamental reconceptualisation.

Habermas made much use of contrasting Horkheimer's essay with Husserl's (1950) "The Crisis of European Sciences," these essays first appearing about the same time. Husserl's crisis was the degeneration of advanced disciplines from the status of true theory. This is a crisis of science where it has nothing to say to us, as if the information content of theories produces scientific culture. An alternate critical view is the production of scientific culture by formation among theorists of a thoughtful way and enlightened mode of life. This last position is the basis of knowledge-constitutive interests and complementarism.

## Traditional and Hermeneutic Positivism: Totalisation

Habermas made a further proposition for our consideration, that '... there is a real connection between the positivistic self-understanding of the sciences and traditional ontology.' With empirical-analytical sciences theories develop in a self-understanding, through dogmatic association with the natural interests of life. This science has the cosmological intention of describing the Universe theoretically in its law like order, just as it is.

With historical-hermeneutic sciences the concern is with transitory things and mere opinion. Yet it shares the tradition of noncritical theory though it has nothing to do with cosmology. Approaches to historical-hermeneutics are composed of a scientistic consciousness based on the model of science. We do not understand science as one form of knowledge, but identify knowledge with science. In this sense we are dealing with a totalising stance, one that introduces another form of positivism.

Cultural sciences do comprehend facts through understanding. They are concerned with discovering general laws. But these sciences share with the empirical-analytical sciences a methodological consciousness describing a structured reality within the horizon of the theoretical attitude. As Habermas said, historicism had become the positivism of the cultural and social sciences.

Positivism, we have argued, permeated the self-understanding of social sciences: whether they obey the methodological demands of empirical-analytical science or orient themselves to the pattern of normative-analytical science.

In essence, this idea of value-freedom promotes psychologically an unconditional commitment to theory and epistemologically the severance of knowledge from interest. It prevents consciousness of the interlocking of knowledge with interests from the life-world.


## Knowledge-Constitutive Interests: Complementarism

Three categories, proposed by Habermas, form the relationship between logical methodological rules and knowledge-constitutive interests. Three fundamental knowledge interests are presuppositions for the possibility of a differentiated constitution of meaning of possible objects of experience. These are nonreducible quasi-transcendental cognitive interests, to which we will now turn our attention.

Habermas' central idea is that orientation toward technical control, mutual understanding in the conduct of life and emancipation,

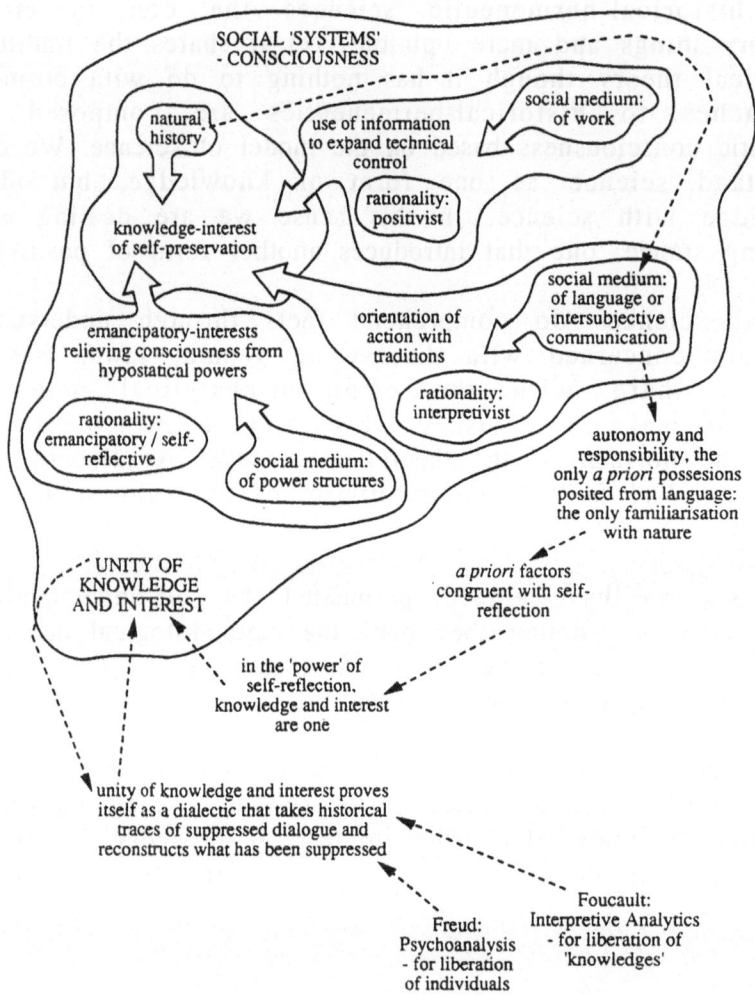

Figure 1.    A general thesis for knowledge-constitutive interests: complementarism

establishes viewpoints from which we may apprehend reality in any way. Respectively, the cognitive interests are technical, practical and critically self-reflective; exhibited in the paradigms empirical-analytical, historical-hermeneutic and emancipatory sciences. The first two represent the dichotomy between natural sciences vs. humanities, that complementarism struggles to avoid. The third category is Habermas' creation. His whole conception surpasses this short introduction (see Figure 1 from Flood, 1990).

## CONCLUSION

We have seen that complementarism, in particular Habermas' knowledge-constitutive interests, is a sophisticated response to the powerful forces of totalising thought. Each of Habermas' three categories are essential for critical systems thinking as it grows and makes even more relevant its ideas to practice. But systems methods based on them must be subject to critical reflection (see Flood, 1990; Flood and Jackson, 1991a, b; and Jackson 1991 for first attempts).

## REFERENCES

Fairtlough, G. (1990). Systems practice from the start: Some experiences in a biotechnology company. Systems Practice, 2(4), 397-412.

Flood, R.L. (1990). Liberating Systems Theory. Plenum, New York.

Flood, R.L., and Jackson, M.C. (1991a). Creative Problem Solving: Total Systems Intervention. Wiley, Chichester.

Flood, R.L., and Jackson, M.C. (1991b). Critical Systems Thinking: Directed Readings. Wiley, Chichester.

Habermas, J. (1971). Knowledge and Human Interests. Beacon Press, Boston.

Habermas, J. (1976). On systematically distorted communication. Inquiry, 13, 205-218.

Horkheimer, M. (1968). Critical Theory: Selected Essays. Herder and Herder, New York.

Husserl, E. (1950). *Die Krisis der europaischen Wissenschaften und die transzendentale Phanomenologie*, in <u>Gesammelte Werke</u>. Martinus Nijhoff, Hague.

Jackson, M.C. (1991). <u>Systems Methodology for the Management Sciences</u>. Plenum, New York.

Radnitzky, G. (1970). <u>Contemporary Schools of Metascience. Volume 2: Continental Schools of Metascience</u>. Scandinavian University Books, Goteborg.

# PARTICIPATORY RESEARCH:

# AN EMANCIPATORY METHODOLOGY FOR SYSTEMS PRACTICE

D.Schecter

Science and Technology Department
Pacific Bell
San Ramon, California, U.S.A.

Department of Management Systems and Sciences
University of Hull
Hull, U.K.

## INTRODUCTION

Over the last decade, a growing number of systems scholars
have called for the development of explicitly emancipatory
methods for systems practice (Flood, 1990, Flood & Jackson,
1991) - methods that deal with issues of power and social
justice, and are oriented toward serving the disadvantaged and
toward social change. With the exception of Ulrich's critical
systems heuristics (Ulrich, 1983), these methods have not
emerged yet. One reason the methods have been slow to emerge is
that so far there is no emancipatory systems methodology (Oliga,
1988). However, adult educators and community health care
workers have developed an emancipatory methodology that some
systems practitioners have adopted - participatory research.
This paper describes participatory research and outlines its
implications for systems practice.

## WHAT IS PARTICIPATORY RESEARCH?

Participatory research is not a research method - a
specific set of procedures. Rather, it is a methodology - a set
of principles about research methods. Participatory research
emerged out of experience with adult education and community
development projects in Third World countries, in situations
characterized by great inequalities of power. Practitioners
found that the dominant (positivist) approach to research was
inadequate in such conditions. More specifically, it was
inadequate for working towards social justice and the
empowerment of the people involved in the situation being
researched. The dominant form of research presented an
oversimplified form of social reality, was often oppressive in
nature, and did not provide easy links to subsequent action
(Hall, 1975).

Participatory research differs from most other methodologies in relation to the goals of research, who is served, where research questions come from, the roles of researchers and researched, the outputs of research, the relationship between research and action, and the relationship between research and education. This section describes the position of participatory research on each of these issues.

What are the goals of research? In the hard systems approach, as in positivist science in general, the goal of research is the production of generalizable knowledge, in order to enable better prediction and control. In the soft systems approach, the goals are mutual understanding and learning, leading to "desirable and feasible change" (Checkland, 1981). These are both important goals for research, but inadequate for emancipatory systems practice. It is also important to note that with all existing approaches there is frequently an additional, unstated goal - career advancement for the researchers.

The goal of participatory research is human emancipation. This means improvement of the lives of the people involved in the research situation, not only in the sense of greater material well-being, but also in the sense of greater control over their own destiny. Beyond this, the ultimate goal of participatory research is to create a more just society, where no groups are denied material well-being, human dignity, or self-determination.

Who should research serve? In the vast majority of cases, social research serves two groups: first, the research sponsors (generally large corporations, government, and the military); second, the researchers. A third group is left out - the participants in the research situation. In contrast, participatory research serves the community involved in the research. Specifically, in coercive situations, participatory research serves the have-nots, the oppressed.

In addition, it is not sufficient that the participants benefit from the results of research. They must benefit from the research process as well. These considerations are currently lacking from most systems research. However, this is precisely the orientation of the Community O.R. movement (Jackson, 1987; Rosenhead, 1987), and some Community O.R. practitioners have already adopted participatory research (Thunhurst, 1988).

How should research questions be chosen? Research programs don't just exist. They come from somewhere. Someone chooses what questions are worth researching and what questions are not. In participatory research, these questions originate with the needs of the community - not with what intrigues researchers most, or what is most likely to be published in professional journals.

What should be the outputs of research? Traditionally, the outputs of research are scholarly papers (for a small intellectual elite), policy studies (for powerful clients), and career advancement (for the researchers). Any output for the people whose reality is researched is incidental. As participatory research is primarily designed to serve the participants, its outputs must reflect this consideration. Tandon (1988) has described five outputs of participatory research: relegitimizing people's knowledge, refining people's capacities for critical analysis, appropriating knowledge from the dominant system for use by the participants, developing new knowledge that is relevant to the participants, and liberating the minds of the oppressed from ideology that is opposed to their interests. Tandon (1981) stresses the importance of empowerment of the participants as a primary output of participatory research.

Who does what in the research process? In most research approaches there is a strong separation in roles between researchers and "researched" - that is, those people whose reality is the object of research. The researchers study, the researched are studied; the researchers act, the researched are acted upon. Participatory researchers reject this dichotomy. In their view, all participants must be, as much as possible, researchers. All researchers must be, as much as possible, participants. But participation without control is not enough. The community being researched must have some control over the research process.

What relationship between research and action? In the dominant research approaches, there is a strong separation between research and action, so much so that many scholars object that action-oriented research is not research at all (for a discussion of this debate, see Conchelos & Kassam, 1981). Participatory researchers reject this dichotomy, arguing that this idea is one of the reasons that traditional research has been so hard to use in action, especially with the poor and oppressed.

What relationship between research and education? The practice of participatory research is intended to be a dialogue over time, not a one-time, one-way process of information gathering. Participatory researchers insist that research must be an educational experience for the participants. It shares this commitment with soft systems thinking (Checkland, 1981) and Action Science (Argyris, Putnam & Smith, 1985). However, unlike these two approaches, the concept of education and dialogue in participatory research is strongly connected to liberation and social justice.

## IMPLICATIONS FOR SYSTEMS PRACTICE

In this paper I am advocating that systems practitioners who want to work in coercive contexts should adopt the basic principles of participatory research. However, I do not imply that participatory research should replace existing systems approaches. Rather, it add to them. In the terms of Flood and Jackson (1990), I am advocating complementarism, not imperialism. While hard systems approaches are very appropriate for the technical interest and unitary contexts, and soft systems approaches for the practical interest and pluralist contexts, participatory research supplies the foundation for an approach that is appropriate to the emancipatory interest and coercive contexts.

What are the implications for emancipatory systems practice of adopting the principles of participatory research? Systems practitioners would be more likely to make clear commitments about who they serve. Questions that are vitally important to underserved populations, but so far have escaped the notice of scholars and professional consultants, would get onto research agendas for the first time. Resources might be devoted to training non-academics in how to do their own research. The expert status of scholars and consultants would be questioned much more vigorously than it is today. Systems practice would have to become more empowering, more egalitarian, and closer to social action.

Adoption of the participatory research methodology would also raise new issues for systems practitioners. First, the question of who to work for - the preferred clients of participatory research generally don't have the money to hire researchers. There is the problem of how to make research academically rigorous as well as practically relevant. Perhaps most importantly, there is the question of how to democratize the researcher/participant relationship. Although these are difficult questions, they are precisely the questions that must be faced by systems practitioners who wish to make their practice relevant to social justice.

## CONCLUSION - PARTICIPATORY RESEARCH AND EMANCIPATORY SYSTEMS PRACTICE

Systems practitioners interested in emancipatory practice in coercive contexts would do well to familiarize themselves with participatory research, adopt its basic principles, and learn from what is already a rich store of experience with emancipatory practice in social systems. Adopting the principles of participatory research will provide the foundation to develop effective methods for emancipatory systems practice. In the words of Bud Hall (1975),

*"We need not more highly trained and sophisticated researchers operating with ever more esoteric techniques, but whole neighborhoods, communities and nations of 'researchers'."*

## REFERENCES

Argyris, C., Putnam, R. & Smith, D., 1985, "Action Science: Concepts, Methods, and Skills for Research and Intervention," Jossey-Bass, San Francisco.

Checkland, P.B., 1981, "Systems Thinking, Systems Practice," Wiley, New York.

Conchelos, G. & Kassam, Y., 1981, A Brief Review of Critical Opinions and Responses on Issues Facing Participatory Research, Convergence 14:3.

Flood, R.L., 1990, "Liberating Systems Theory," Plenum, New York.

Flood, R.L. & Jackson, M.C., 1991, "Critical Systems Thinking: Directed Readings," Wiley, New York.

Flood, R.L. & Jackson, M.C., 1990, "Creative Problem Solving: Total Systems Intervention," Wiley, New York.

Hall, B., 1975, Participatory Research: an approach for change, Convergence 7:2.

Jackson, M.C., 1987, Community Operational Research: Purposes, Theory, and Practice, Dragon, 2:2.

Oliga, J.C., 1984, Methodological Foundations of Systems Methodologies, Systems Practice, 1:1.

Rosenhead, J., 1987, From Management Science to Workers' Science, in "New Directions in Management Science," M.C. Jackson and P. Keys, eds., Gower, Aldershot.

Tandon, R., 1981, Participatory Research in the Empowerment of People, Convergence, 14:3.

Tandon, R., 1988, Social Transformation and Participatory Research, Convergence 21:2/3.

Thunhurst, C., 1988, Community Operational Research, in "Systems Prospects: The Next Ten Years of Systems Research," Flood, Jackson & Keys, eds., Plenum, New York.

Ulrich, W., 1983, "Critical Heuristics of Social Planning," Haupt, Berne.

# THE SACRED AND PROFANE IN CRITICAL SYSTEMS THINKING

Gerald Midgley

Department of Management Systems and Sciences
University of Hull,
Cottingham Road, Hull, HU6 7RX, England

## ABSTRACT

This paper looks at what we mean by being critical about systems. In particular it seeks to expand our understanding of the process of making boundary judgments so as to explore the relationship these judgments have with values and ethics.

## MAKING BOUNDARY JUDGMENTS

In Critical Systems research, two needs in particular are stressed: first, the need to be critical about defining system boundaries, and second, the need to establish boundaries within which critique can be conducted. Ulrich (1983, 1988) and Flood and Ulrich (1990) are particularly fruitful sources of reference for thoughts on being critical about systems. Since placing boundaries around a problem, within which critique is to be conducted, automatically 'hides' aspects affecting it that have been defined as lying ;outside the scope of research, we inevitably have an incomplete view of the situation. As we can never get away from making boundary judgments, the best we can do is embrace critical flexibility with regard to such judgments.

This is very similar to the observation that making a classification sheds light upon a phenomenon while simultaneously casting the 'other' to that which is defined into darkness. The concept of 'otherness' can be traced back to the writings of Aristotle and Plato, and it has surfaced again and again in Western philosophy. As Fuenmayor (1990) points out using a similar metaphor of light and dark, by remaining aware that every concept has its other, we are able to retain the possibility of changing the boundaries of critique. In other words, awareness of 'otherness' is an effective remedy for 'hardening of the boundaries'!

## 'OTHERNESS' AND MARGINALIZATION

In the current paper I want to develop this generally accepted idea that there is always an 'other' to every system boundary that can be defined. Indeed, I want to suggest that when, of necessity, we cast light on a system by defining a boundary, the 'sharp line' between the region of light and its dark 'other' is, in a sense, an artifact: we also have the possibility of looking for grey areas in which *marginal* elements lie that are neither fully included in, nor excluded from, the system definition.

I wish to start the main body of this paper, then, by expanding upon the notion of marginalization. This will, in essence, be a re-presentation of some of the basics of the systems idea, albeit in different linguistic clothes. I would ask the reader to bear with me here, as the new language represents part of a conceptual reframing that will prove necessary for the coming analysis.

## TOWARDS AN INITIAL UNDERSTANDING OF MARGINALIZATION

To give an example of an obvious marginal element, we might look at the conventionally accepted organizational boundaries of a business. Customers, as traditionally defined, might not be seen as being 'within' the organization in the same sense as the workers, but the organization could not function without them so they cannot be placed wholly on the 'outside' either.

While customers are fairly obviously marginal to the way we traditionally define businesses, it is more difficult to identify 'hidden' marginalized elements. People who are unemployed are a typical example: local unemployed people would be excluded from most conventional organizational analyses, but they actually have a stake in the company's recruitment policy, might have a potential place within the traditionally defined organization, and are an integral part of the wider system of which the organization is also a part. Those elements of the wider system (including people who are unemployed) which are tacitly recognised as being pertinent to the organization, but are not explicitly taken in to the definition of the organization's boundaries, can be described as marginal to them.

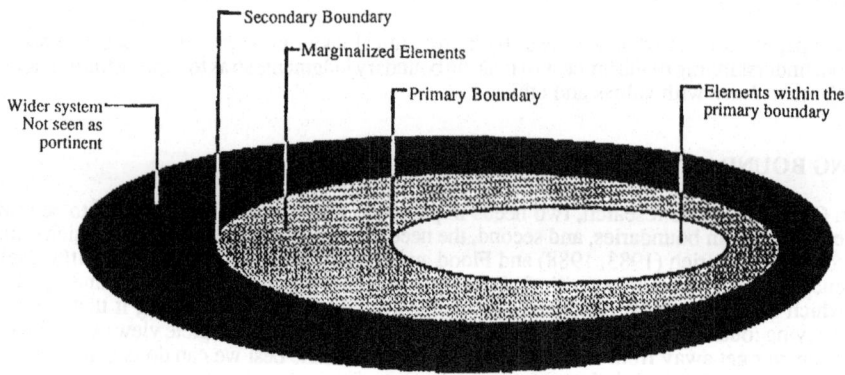

**Figure 1.** Marginalization

It is essential to note here that defining a hidden marginalized element involves recognising *an alternative system boundary*. To make this clear, let me use the analogy of a piece of paper. On the one hand the margin forms the edge beyond which there is no writing. A margin on a piece of paper is not one-sided however: *it is also defined by the boundary of the paper itself*. Now, it is a commonplace notion that a system boundary defines what is included in the system and, implicitly, what is excluded. The marginal area at the boundary, however, can only be defined with respect to *another boundary* because if there were no outside limits then there would be no way to differentiate what is marginal (but possibly hidden at first) from what is excluded. What is excluded is invisible - indeed, it is only seen to exist by implication given that we always acknowledge the theoretical presence of a wider system: for recognition of tangible and pertinent existence to take place, however, there must be boundaries defining a second system, and when there are, then we are no longer dealing with exclusion but with marginalization. See Figure 1 for a diagrammatic representation of this.

We can codify this analysis in the following way:

(1)    Marginalization implies the use of more than one system boundary, even if one or more of these boundaries is being employed tacitly or unconsciously in a given analysis.

(2)    We are therefore able to develop a systems language of *primary* and *secondary* boundaries. The primary boundary is that which is most obvious (it might be placed around a traditionally defined organization, a particular eco-system, a society, a planet, etc.), and the secondary boundary is that which allows recognition of the pertinent existence of elements outside the system being defined that nevertheless are seen to affect it. Elements seen to be lying between the two boundaries are *marginal* to the system.

(3)    It is therefore not merely 'otherness' which offers the potential for changing system boundaries, but recognition of marginalization. 'Otherness' refers to our theoretical understanding that any system that is defined is part of a wider whole, yet marginalization refers to those elements that are within the realm of our experience, are pertinent, but are seen to lie outside the primary boundary of analysis (nothing else is recognised as existing in the practical task of analysis).

## THE INSEPARABILITY OF THE CRITICAL AND SYSTEMS IDEAS

From here we can move on to integrate this understanding into Critical Systems theory so that we can expand our notion of what it means to be critical about systems:

(4)    Theoretical critique of system boundaries involves *making the secondary boundary explicit,* and *making choices between boundaries.*

(5)    These choices can be exercised in two ways. Firstly, having identified primary and secondary boundaries, we can choose to retain the primary boundary and deal with marginal elements in relation to it: e.g., we might wish to examine and/or change the relationships between a local community and a particular organization. Secondly, we have the choice of *disbanding* the initial primary boundary and making the secondary boundary primary: e.g., we might decide that wider issues raised by consideration of the needs of the local community are more important than those narrowly defined in relation to the organizational system boundary, and choose to place our primary boundary around the local community itself. There will then, of course, be the possibility of identifying a new secondary boundary. In practice, however, theoretical critique must have limits.[1]

(6)    From this we see that the systems idea and the critical idea are, as Ulrich (1983) and Flood and Ulrich (1990) have claimed, inseparable. In order to make practical choices between boundaries we must be guided by a sense of truth (i.e., what can be said to exist, either lying within boundaries or marginal to them), rightness (i.e., which boundaries it is right to employ) and subjective understanding (i.e., that it is possible to see things in very different ways).[2] It is the potential for rational consideration of any or all of these that defines Critical Systems thinking, and exclusion of one or more from rationality seriously impoverishes choice. Restricted choice will, of course, lead to an impoverished systems view because boundaries will become fixed through the use of tacit knowledges alone, without the benefit of being informed by rationally generated theory.

## VALUES AND BOUNDARY JUDGMENTS

The above is a reinterpretation of the systems idea that has, of necessity, been kept relatively simple in order to introduce a new language of *boundary* and *marginalization.* Now it is time to use this new language. What I want to do is look at the tension between what I have described as the primary and secondary boundaries. To do this we will first have to examine the relationships that exist between truth-orientated knowledge and value judgments.

There is, of course, a massive body of literature which challenges the assumption of the analytical philosophers that knowledge is value-free. In Critical Systems, the first writer to offer a detailed exploration of the idea that knowledge must be seen as value-laden was Ulrich (1983) who claimed that where the boundaries of analysis are drawn effects the ethical stance taken and the values pursued. To use our example of an organizational analysis, and the question of whether people who are unemployed should be included within the boundaries of it, it is obvious that the issues that can emerge within the primarpy boundary (i.e., when people who are unemployed are ignored) will be different to those that can emerge if they are included. If people who are unemployed are ignored, then (to generalise) it is most likely that issues of efficiency and effectiveness will emerge that take the status-quo value system for granted. If their concerns are admitted into the analysis, then the status-quo value system which allows the perpetuation of their unemployment is likely to come into question.

In this sense then, the boundaries of accepted knowledge define the values that can emerge. Similarly, the values adopted will direct the drawing of boundaries that define the knowledge accepted as pertinent. We are therefore equally justified in saying, on the one hand, that our moral choices have a basis in fact (what we see as objective reality) and, on the other, that the choices we make between objectively perceived boundaries are essentially ethical or moral choices.[3] According to Ulrich (1983) it is this dynamic relationship between truth and rightness, in which both have to be open to question (and in his view also have to be guided by an interest in emancipation), that defines Critical Systems thinking. Further, because Ulrich recognises this essential relationship between truth and rightness, he is able, in his 1988 work, to offer a systematic methodological guide to the practical business of drawing boundaries in which choice between alternative boundaries figures strongly.

## VALUES AND MARGINALIZATION

I began this paper with an attempt to build upon the basic idea of 'otherness' in order to bring out an understanding of marginalization as the process of consigning elements into the region between the primary and secondary boundary. Now I want to take one more step and use this analysis to build upon Ulrich's notion of the relationship between ethical reasoning and the making of boundary judgments.

My essential concern is to show that value judgments are not only related to what is or is not contained within given boundaries, but that they are also related to what lies in the margins. Indeed, we might postulate that the imposition of a *profane* status upon some marginal elements might reinforce or bolster the supposed objective necessity of the primary boundary, while imposition of a *sacred* status might protect the secondary boundary from dissolution.

The words "sacred" and "profane" might require a little explanation: essentially these mean 'valued' and 'devalued' respectively. This terminology has been borrowed from the tradition of structural anthropology, exemplified by the work of Douglas (1966), and it should be stressed that the words are not meant in a purely religious sense but refer to the development of a 'special status', whether positive (sacred) or negative (profane).[4]

Let us make it clear how a status of sacred or profane might be imposed on marginal elements by returning to Ulrich's understanding (1983, 1988) that choice between boundaries can also involve choice between different ethical concerns. I would like to suggest that, when the primary and secondary boundaries carry different ethical implications, a tension is set up. In our earlier example of an organizational analysis, for instance, we can see a tension between the concern for organizational effectiveness that is generated within the primary boundary of the business, and the concern surrounding employment rights that is generated by widening the analysis to the secondary boundary. At the risk of over-simplification, the two ethics in conflict might be characterised as "we should ensure our workers' survival in the market-place" versus "all people should have equal opportunities for employment".

Now, because most ethical issues and associated boundary judgments, both primary and secondary, can be said to have roots in culture (i.e., they are inter-subjectively accepted at either a conscious or an unconscious level), we are able to find evidence for cultural reactions to the ethical tensions that arise. These cultural reactions, I would argue, involve the imposition of value judgments on elements that are marginal to boundary definitions: i.e., marginal elements come to be characterised as either sacred or profane. Profanity supports the primary boundary by denegrating those elements which are marginal to it. In contrast, sacredness in the margins supports the secondary boundary. It works like this because sacredness is the 'other' to profanity, and profanity the 'other' to sacredness. Therefore, when marginal elements are seen as profane, elements within the primary boundary become sacred by implication and the primary boundary, along with its associated ethics, are reinforced. When marginal elements are seen as sacred, however, what is defined solely by the primary boundary becomes profane by implication, and the secondary boundary, with its associated ethics, comes to the fore.

This is not the end of the story however. Not only do ethical tensions give rise to sacredness and profanity, but this whole process actually comes to be overlaid with social ritual (Douglas, 1966; Leach, 1976). Ritual can be defined as behaviour, in whatever context, that contains certain stereotypical elements that involve the symbolic expression of wider social concerns (Douglas, 1966). An observation of the presence of ritual can give us a clue as to where sacredness and profanity might lie, and hence where ethical conflicts related to marginalization might be found.[5,6] In order to make all this clearer, the whole process has been represented diagramatically in Figure 2.

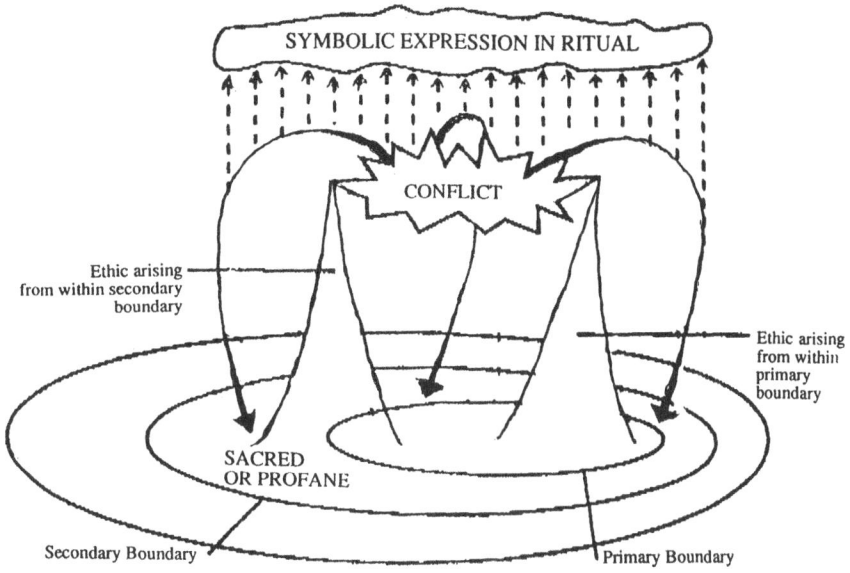

**Figure 2.** Margins, Ethics, Sacredness, Profanity and Ritual

To explain, in Figure 2 we see one ethic arising from within the primary boundary, and another from within the secondary. These come into conflict - a conflict which can only be dealt with by making one or other of the two boundaries dominant. This dominance is achieved by making elements in the margin (between the primary and secondary boundaries) either saered or profane. The whole process is symbolically expressed in ritual which, in turn, helps to support the total system. Here then we see some of the complexities of relationships between boundary, ethics and value judgements.[7]

## DYNAMISM AND COMPLEXITY

Although this is in itself not a simple set of relationships, there is a dynamism and complexity lying behind the relatively static model presented in Figure 2. To begin with there is no absolute *a priori* 'starting point' for analysis: if, for instance, we were to try to treat the system boundaries as entirely static starting points which give rise to ethical conflict, then we could have no notion as to whether marginal elements would actually become sacred or whether they would become profane. Sacredness and profanity, and the associated dominance and supression of system boundaries, only has meaning in relation to an understanding of *history* of movements within the system, and in relation to *interactions between the system and numerous others.* This last point is crucial. As Douglas (1966) has pointed out, sacredness and profanity (and associated ritual) only make sense if seen in the context of the wider system: the single system with its discrete primary and secondary boundaries is an idealised semi-dynamic model that helps us understand the principles involved in the relationship between boundary and value, but in everyday life we move from one system representation to another, and these will often overlap. We might gain insight into some phenomena by using this model, but like all models it is a means of reducing complexity. We should always remain critically aware that we live within a dynamic web of boundaries, marginalizations, ethical conflicts and value judgments, and never be tempted to regard any systems representation as an absolute.

Holding this warning in the back of our minds, it might nevertheless be useful to offer further clarification of the model by using it to interpret phenomena from the world of business.[8]

## THE PROFANE STATUS OF "THE UNEMPLOYED"

Let us start with the example of the organizational analysis I have been using periodically throughout this paper. We have already seen how both customers and people who are unemployed might be considered marginal to the conventionally accepted primary boundary of a business.

To give more detail, we'll start with people who are unemployed. As a group "the unemployed" have a profane status within the mainstream of society which can be said to relate to an ethical conflict between the need to preserve the living standards of the workers (within the primary boundary) and simultaneously offer equal opportunities to employment (within the secondary boundary that encompasses all members of the local community). Were unemployed people not marginalized and regarded as profane, the primary boundary would simply not be able to stand the ethical pressure, and our understanding of "business" would have to undergo a massive revision. Not surprisingly given its profane status, the experience of unemployment is beset with ritual, from form-filling to signing on. These rituals are at one and the same time functional in terms of the Social Security system, and symbolic in terms of reinforcing the marginal and profane status of being unemployed.[9]

The profane status ascribed to people who are unemployed is typical in showing how marginalization, profanity and ritual form a mutually-supportive systemic relationship. It also offers an example of how profanity reinforces the primary boundary: as we have noted, if unemployed people were not marginalized and regarded as profane, the primary boundary simply would not be able to stand the ethical pressure for equality of opportunity. Here, then, we see how the system works to perpetuate power relationships.

## THE SACRED STATUS OF THE CUSTOMER

Moving on to customers (as traditionally defined), we find that they sometimes have an ambiguous status in that they are most often seen as sacred, but can, in some circumstances, be looked upon as profane. The various elements of ritual associated with the staff-customer relationship, such as wearing 'smart' dress and offering stereotypical responses ("have a nice day"!), regulate and legitimise the customer's marginal, sacred status.

I say that the status of the customer is ambiguous, however, because workers on the shop floor sometimes demarcate public from private ritual: while they may offer the standard, ritual company image to the sacred customer, they may nevertheless operate a subculture in which the customer is viewed as profane, and this profane status is reinforced by rituals of gossip or even passive-aggressive behaviour that strays into the public domain.

To look at this in terms of marginalization, we notice that the primary boundary around a single business gives rise to an ethic of short-term profit-maximization, but lying behind this is the secondary boundary in which customers and other businesses are included. Within this secondary boundary there is a recognition that all profit-earners are also customers, so an ethic of equality of both consumption and profit-gain arises that is set to conflict with the ethic of short-term profit-maximization that results from the primary boundary. Once again we see that the sacred status of the customer is boundary-preserving. This time, however, it is the secondary boundary that is being bolstered: if customers were not respected as sacred, how could the ethic of short-term gain be offset? If customers could legitimately be fleeced, how would we be able to stop a frenzy of short-term profiteering that, in the long-term, would damage everybody?

Having noted the ultimate importance to businesses of maintaining the sacred status of the customer, we have to ask why counter-cultures arise amongst some staff in which customers come to be regarded as profane. One answer, it seems to me, is that in some circumstances these staff members might not be perceiving the same primary boundary around the business: they might see themselves, consciously or unconsciously, as lying in the margins outside the primary boundary and, as 'workers', being regarded as profane by powerful 'bosses' who sit within the primary boundary. Indeed, to generalise, they might see themselves as the relatively powerless victims of an ethical conflict between profit-maximization for others (a dominant ethic associated with the primary boundary) and the ethical principle of power- and money-sharing that is implied by the secondary boundary. Given that customers lie beyond what the staff identify as the secondary boundary, but their bosses see as the primary boundary, coercion and conflict are bound to arise. This is only one possible explanation for customers being seen as profane: others using the model spring to mind, but space restrictions prevent me from exploring them here.

Here we begin to get a small taste of the complexity of these systems in interaction. The above example also brings home the importance of being aware of a pluralism of viewpoints within cultures. Certainly in the West culture could never be described as monolithic: not only do some groups identify with one viewpoint while others identify with another, but individuals have the capacity to act through more than one system representation. Paradoxically, this not only allows individuals to hold conflicting viewpoints, but also introduces the possibility of critique as sets of system boundaries, ethics and value judgements meet and are played off one against the other.

## CONCLUSION

I began this paper by introducing a language of primary and secondary boundaries to help us develop an understanding of marginalization. I then tried to use these ideas to focus our attention upon what we mean by being critical about systems, and stressed the importance of being aware of the dynamic relationships that exist between subjective-, truth- and rightness-orientated inquiries. Having explored Ulrich's claim that truth and rightness judgments inevitably affect each other, I moved on to develop his understanding in order to show that, when primary and secondary boundaries give rise to ethical conflict, marginalized elements tend to be seen as either sacred or profane. Profanity supports the primary boundary, and sacredness the secondary. Furthermore, we find that marginalization and ethical conflict, leading to sacredness or profanity, is expressed symbolically in ritual. Here, then, we find a complex systemic web of primary and secondary boundary judgments, marginalization, ethical conflicts, power and knowledge relationships, value judgments and symbolic ritual.

We must remain aware, however, that what I have described is an idealized, semi-dynamic model, and that the 'real world' offers an incredible dynamism and complexity that is beyond the full comprehension of any individual. Nevertheless, this model will hopefully be of some use in helping us to understand some of the 'mechanisms' behind that elusive relationship between truth and value that is so important to grasp if we wish to aspire to being critical about sytems.[10]

## NOTES

1. This is very similar to the usual systems idea of defining the level of resolution of the system, only I am not assuming that system boundaries have an *a priori* concrete existence independent from human understanding.

2. This idea assumes that we act as though there are three essentially interrelated "worlds" of understanding: the objective natural world, the normative social world and my individual world. These correspond to the ideals of truth, rightness and subjective understanding respectively. Elsewhere (Midgley, 1990, 1991) I have developed this view of ontology, originally proposed by Habermas (1976), into a theory of rationality that underpins, not only Critical Systems, but also the need for methodological pluralism and a reunification of science.

3. The relationship between boundary (an aspect of truth) and value (rightness) is an essentially systemic one. It would be a nonsense to say that either one should be seen as an absolute *a priori*. Nevertheless, while we may paint a meta-theoretical picture of the interdependence of boundary and value, in the *actual process* of thinking we move between "worlds" of truth, rightness and subjective understanding (Midgley, 1990, 1991). In practice, then, we talk in terms of the origin of a particular value or ideology lying in 'the system', or a system boundary as having a particular ideological root: the notions of origins and roots in everyday thinking are essentially bound to the rationality of the moment and, following their emergence, should ideally become available for critique. When theories of origin become ossified, whether these be theories of the natural world (e.g., evolution, creation, the 'inalienable' laws of physics) or theories of epistemology (e.g., that any of the natural, social or subjective "worlds" has an absolute *a priori* status) then critical thinking inevitably becomes limited.

4. The choice of the words "sacred" and "profane" is deliberate. Although I could have stuck with a more 'neutral' or secular terminology such as 'valued' versus 'devalued', this would have left me open to the accusation of perpetuating an artificial Western distinction between the secular and religious (with language associated with the secular being better respected in academic circles). In terms of the relationships between boundary, value and ritual, I believe that the same processes operate whether they be classified as religious or secular. Also, the more emotive language of sacredness and profanity better conveys the urgency of addressing problems that might arise when human beings becoming marginalized.

5. Of course sacredness and profanity, with their associated rituals, can also be seen to flow from other sources not traditionally described in systems terms. For example, Douglas (1966) and Leach (1976) both offer interesting examples of danger that is seen to arise in the margins of categorisation systems: Douglas's analysis of the *Abominations of Leviticus* is particularly interesting for the way it explains the Biblical proscription of certain meats on the basis that the animals from which they are derived infringe the God-given categorisation system which animals, birds and fish should conform to in order to be considered holy. Although we often refer to a series of related categorisations as a "categorisation system", this is not systemic in the same sense as I have used the term, and hence the analysis presented in this paper should

be seen as complementary to structural anthropological analyses and not in competition with them. Indeed, exploration of the mutual reinforcement of categorisation systems and boundary judgments will hopefully prove a fruitful avenue for further research (see also note 8).

6.  It must be emphasised that it is ethical conflict in relation to marginalization that is the key to understanding sacredness and profanity here. Where consideration of primary and secondary boundaries do not give rise to obvious issues of rightness, then sacredness and profanity will not come to the surface of consciousness, although they might nevertheless be acted out unconsciously. This is perhaps why so many 'natural' scientists still claim that knowledge and boundaries can be regarded as value-free: they see no rightness implications in the choices they make between system boundaries in their areas of interest. There is both a validity and a legitimacy problem here. All choice involves judgments of right and wrong (Habermas, 1976), including choices between boundaries, so the argument for value-neutrality is invalid unless one dispenses with the notion of choice itself. Also, we have recently come to realise that the way we look at the ecology of the natural world has very definite implications for what we judge to be right and wrong socially (see for instance Midgley, 1990), so the continued legitimacy of the argument for value-neutrality in terms of being able to support a viable social system must also be brought into question.

7.  Flood (1990) has introduced Foucault's theory of power and knowledge into Critical Systems. Foucault (e.g., 1980) sees an intimate relationship between power and knowledge in that each is a meaningless concept without the other. Indeed, power is expressed in the rise of some knowledges into positions of dominance and the subjugation of others. In many ways, then, the processes identified in this paper complement Foucault's understanding by providing some further explanations of the 'mechanisms' by which some knowledges and ethics achieve dominance while others come to be supressed.

8.  Space restrictions confine me to examples from business (which I have already worked out), although I am also looking at various other environmental and social issues. One particular area of interest for me is the profane status ascribed to people with disabilities, and how definitions of what is normal and abnormal relate to the drawing of system boundaries.

9.  Incidentally, it is worth noting the fact that the marginal position of people who are unemployed is underscored by an analysis of the rhetoric of many left-wing and Green movements in the West which have elevated unemployed people to a sacred status. Many socialists talk about people who are unemployed almost as if they are martyrs to Capitalism, and many Green thinkers see those who successfully structure their lives outside the dominant concerns of mainstream employment as a vanguard of resistance to materialist conformity. Here we see a direct inversion of mainstream values which, we should note, does not challenge the marginal position of people who are unemployed, but reinterprets their status.

10. I would like to offer my thanks to Bob Flood, Gary Wooliston, Wendy Gregory and Pippa Carter, all of whom helped me sharpen up my ideas through conversation. I would also like to note that I intend to develop these ideas in a longer paper, probably for the journal *Systems Practice*.

REFERENCES

Douglas, M. (1966), *Purity and Danger: An Analysis of the Concepts of Pollution and Taboo*, London, Ark.

Flood, R.L. (1990), *Liberating Systems Theory*, New York, Plenum Press.

Flood, R.L. and Ulrich, W. (1990), Testament to Conversations on Critical Systems Theory between Two Systems Practitioners, *Systems Practice*, 3, 7-29.

Foucault, M. (1980), *Power/Knowledge: Selected Interviews and Other Writings 1972-1977*, Gordon, C. (ed.), Brighton, Harvester Press.

Fuenmayor, R.L. (1990), Critical and Interpretive Systems Thinking, 1. What is Critique? In, *Toward a Just Society for Future Generations. Volume 1: Systems Design*, Proceedings of the 34th Annual Meeting of the International Society for the Systems Sciences, held in Portland, Oregon (USA), July 8-13, 1990.

Habermas, J. (1976), *Communication and the Evolution of Society*, English edition published 1979, London, Heinemann.

Leach, E. (1976), *Culture and Communication: The Logic by which Symbols are Connected*, Cambridge, Cambridge University Press.

Midgley, G.R. (1990), Critical Systems and Methodological Pluralism, In, *Toward a Just Society for Future Generations. Volume I: Systems Design*, Proceedings of the 34th Annual Meeting of the International Society for the Systems Sciences, held in Portland, Oregon (USA), July 8-13, 1990.

Midgley, G.R. (1991), Unity and Pluralism, *Ph.D. Thesis*, City University, London.

Ulrich, W. (1983), *Critical Heuristics of Social Planning: A New Approach to Practical Philosophy*, Berne, Haupt.

Ulrich, W. (1988), Systems Thinking, Systems Practice and Practical Philosophy: A Program of Research, *Systems Practice*, 1, 137-163.

# CRITICAL SYSTEMS THINKING: REAL FICTION?

Gary Wooliston

Department of Management Systems and Sciences
University of Hull,
Cottingham Road, Hull, HU6 7RX, England

## ABSTRACT

Critical Systems Thinking has hit upon a dialectic, where post-modernity is beginning to challenge the traditional modernist, or late-modernist perspective. Through this impending dialectic we can propose further forms that allow greater freedom of analysis. Following Linda Hutcheon, two particular approaches to discourse are evident in Critical Thinking. First, an ultra-modernist angle that seeks to disentangle verbs and interest with a view to establishing opportunities and limitations. Secondly, a Post-modernist approach that angles the discourse at the reality of itself as a constitutive element within modernism. This dialectic may establish its reality as yet another fictional form (as Post-modernist novelists such as Eco and Rushdie create autonomous worlds) and, therefore, beg the question "Is Critical Systems Thinking real fiction?" The paper explores this question.

## INTRODUCTION

This essay seeks a conflation of 'critical' and 'creative' through Hutcheon's (1988) 'Historigraphic' novel. This conflation

"... finds its object neither wholly in the cultural sphere, nor wholly in the critical-institutional sphere, but in some tensely re-negotiated space between the two." (Conner, 1989, p. 7)

For the purpose of this essay the 'cultural sphere' will refer to post modernist fiction and the 'critical-institutional sphere' will refer to Critical Systems Thinking. Examples of post-modernist fiction include Eco (1983) and Rushdie (1981); while examples of Critical Systems Thinking include Flood (1989, 1980), Flood and Jackson (1991) and Ulrich (1983). In some senses post-modernism can be seen as a critique of modernism; in which case Flood (1989) develops a Foulcaulvian Archaeology criticising pragmatism isolationism and imperialism. In other senses post-modernism recognizes the importance of frivolity in Condillac's 'Essay on the Origin of Human Knowledge' (see Derrida, 1990); and here a gap exists between post-modernism and Critical Systems Thinking. This latter sense shall be considered in this essay.

A description of tradition-, late-, ultra-, and post-modernism opens up the debate for further possible re-interpretations of 'critical' and 'creative' positions. And the whole debate takes places within a Derridean re-interpretation of the Hegeleon dialectic.

## HITTING UPON A DIALECTIC

The "... coil within the great corpus of modernism" (Hassan, 1982, p.129) generates many paradoxes: the paradox of a self-reflective modernism allowing a post-modernist textual intensification seems to warrant critical study. The language of the dialectic itself needs questioning. For the purpose of this paper an Hegelian synthesis (as proposed in 'Phenomenology of Mind') is over-run by the complusive non-identity which momentarily identifies itself as a complicit synthesis. Adorno's (1973, p. 157) notion that

"Only in the accomplished synthesis, in the union of contradictory moments, will their [identity and non-identity] difference be manifested".

assumes the perceived necessary level of 'indifference' to allow the surfacing of contradictions. This holistic tendency will not be assumed, rather a Derridean 'difference'. Derrida's 'Margins of Philosophy' (1982) contains three dimensions of difference. The first and third dimensions seem relevant here: difference as non-identity of differents; and difference as 'temporisation'. The first dimension disallows Hegelian 'Aufhebung' to be seen as a determinable totality; the dialectic is thus an indeterminate action. The third dimension refuses to give privilege to the present (a "... challenge [to the] traditional philosophical belief in [the] immediate, indubitable features of mental life ..." (Hoy, 1990, p. 52)). The establishment of the dialectic, as being both indeterminate and temporary gives an energy to demonstrate developing positions relevant to Critical Systems Thinking.

## DEVELOPING POSITIONS

Two discourses (post-modernist and ultra-modernist) are developing from the traditional and late-modernist perspectives. The descriptions are deliberate. 'Perspective' relies upon the integrity of the individual at a particular time; while 'discourse' is a dispersed and discontinuous network of sites (post-modernist discourse, notably, decentres the transcendental knower) obeying neither the boundaries of humanistic knowledge nor the autonomy of time (see Foucault, 1988, 1980).

Definitions of these four positions (traditional-, late-, ultra- and post-modernist) will take a multi-disciplinary form. Since the basis of the debate upon 'realfiction' centres around Linda Hutcheon's (1988) notion of 'historiographic metafiction', a literary and philosophical form is adopted (though these boundaries seem blurred).

Connor (1989, p. 126) defines 'historiographic metaficition' as "... works of fiction which reflect knowingly upon their own status, forgrounding the figure of the author and the act of writing, and even violently interrupting the conventions of the novel, but without relapsing into mere technical self-absorption".

Historigraphic metafiction also challenges conventional discourse from within "... the economic (late capitialist) and ideological (liberal humanist) dominants of its time". (Hutcheon, 1988, p. xiii).

This area of discourse immediately calls for a crossing of disciplines, willfully tackling the inherent narrative structures across literature, history and theory allowing "... theoretical self-awareness of history and fiction as human constructs" (Hutcheon, 1988, p. 5).

Human constructs that emphasise differences not metanarratives (see Lyotard (1984) and Habermas (1983) for the debate upon the 'rationality' of post-modernism). Human constructs that are not beyond narrative legitimation because of some technolgoical bombardment of rationalisation. And human constructs that have moved truth to ficition and narrative (Lecercle, 1990, p.76): to the historiographic novel.

## TRADITIONAL-MODERNISM

The table gives a rudimentary indication of the area within which the debate upon 'realfiction' is taking place. Traditional-modernist and late-modernist perspectives create the possibilities of post-modernist and ultra-modernist discourses. Flavours of debate between traditional- and late-modernism are well documented in Wilde's 'Horizons of Assent' (1981). This text also develops a theme between modernism and post-modernism: disjunctive irony (where confusions of the world are controlled by an equal poise of opposites) and suspensive irony (an acceptance of the randomness of life through suspending divergent impluses). Wilde's thesis is a literary one, seeking to study the use of irony in its contriving

"... to declare its flexibility, openess to change, and to retain its essential identity as a characteristic response to the polysemic world we inhabit". (Wilde, 1981, p.1)

Another response suggests a comparison between the narratives employed in the works of Forster and Woolf, and the narratives employed in the works of the hermeneuticians Gadamer and Dilthey. The novelists and the philosophers both seem to adhere to unifying doctrines, and plausible hypotheses. The unification is considered an holistic ideal. For example, in Virginia Woolf's 'A Summing Up' (1953) Sasha Latham through a questioning of a vision (divine or mundane?) choses aesthetic closure, not resolution. Gadamer thinking along similar lines

"The reader is guided not only by the assumption of an immanent unity of meaning; his understanding is also constantly guided by transcendent expectations of meaning that arise from the relation of what is being said to the truth ... the prejudgement of completeness comprises not only the formal element that a text should fully express its meaning, but also that what it says is the whole truth." (Gadamer, 1975, p. 261 - 2)

Crudely, these two examples demonstrate a human will to 'transcend' the impossibility of unity in order to reveal 'aesthetic closure' and the 'prejudgement of completeness'.

Table 1

| MODERNISTIC PREFIX | TRADITIONAL- | ULTRA- | LATE- | POST- |
|---|---|---|---|---|
| Authors | Forster; Woolf; Gadamer; Dilthey | Federman, Sukenick, Gass, Robbe-Grillet. | Compton-Burnett; Wyndham-Lewis; Habermas; Adorno | Derrida, Barthelme, Lyotard, Barthes. (Death of the Author, Barthes; Kristeva, Foncault (what is an author?) Baudrillard, Eco, Rushdie. |
| Notion of Critique | Transcendental possibilities Hermeneutic consciousness. | Ironical critique (Wilde, 1981) Critique of distance, detachment, depth essentialism, anthropomorphism, humanism, privileging of sight. | Critique of re-ification; Critique of traditional Legitimation Crisis | Complications Critique of autonomous, transcendental subject (eg Hegel's Phenomenology of spirit) |
| Notion of Fiction | as "a reality of a kind we can never get in daily life" (Forster, 1983, p. 44). Nostalgia; to "Solace us" (Wilde, 1981, p. 107) | Novel inventing own reality, to absent itself from reality. Centrifugal, journey toward chaos (Federman, 1965) Cuts off referential points | "Appearances are not held to be a clue to the truth; but we seem to have no other" (C-B, 1959, p. 1) (3) Characters vary as a function of linguistic surface | Novels inventing historiographic reality, reference deferred or confused. Centre becomes a fiction (Foncault, 1977) Diachrony reinserted into Synchrony (Hutcheon, 1988, p. 99) |
| Notion of Reality | Unstable, unsatisfactory truth = depth (no truth in appearances) aesthetic autonomy. Holistic, complex, natural and social. | Becoming eventually unreal, surfiction "abolishes absolute knowledge and what passes for reality, it even states defiantly that reality as such does not exist" (Federman, 1978, 122) | Reading appearances correctly, stressing language. Dialectic between ultimate comprehensibility of the hidden and limits/complicationsof the visible. | 'Bricolage' randomness, decentred. Simulacra-process leading to break-down in media and panic-shaken (over) production of the real (Baudrillard, 1983) |
| Idea of Progress | Hegelian synthesis; (consensus) Penetration of phenomena to reach noumena. Pragmatic holism; Fusing of horizons. | Abstraction through structures of imaginative control. Refutation of modernist beliefs. To discard self-reflection(1) | Toward ideal speech situation. Relativism of surface contingencies. To re-unite theory and practice (Giddens, 1990). | Author " stripped of its creative role and analysed as a complex and variable function of discourse" (Foncault, 1977, p. 137) Problemaisation of activity of reference. |
| JUSTIFICATION SYSTEMS | Origins and finalities. Enlightenment, Emancipation from repression. Fusion (consensus) as healthy | The creation of fiction that abolishes reality (Federman, 1965) 'Simultaneous multiplicity'(2) | Communicative Competence. Retaining as early modernists have, privileges of a god like observer. "Class compromise" - Governments, business, labour movement. | History begging for deconstruction to question the function of unity of history (Parker, 1981, p. 58) Shift from Justification to Signification(4) |
| (Dominant Metaphors) HISTORICAL NARRATIVE | (opinion; transcendental; hermeneutic circle) Plausible stories, unifying linear development. | (Surfiction) Questions verbal textures | (transcendental pragmatics) Retention of Marxist notion of capitalistic transcendentalism. | (Mass Media) texts exposing the fictionality of history Narrational questioning, dissolving. |
| (Historical meanings/themes) | (Universal, consistent development) | (Self destroying stories; contradictions) | (Particular fighting Universal) | (Contextual; suspensive divergence) |
| PROTAGONISTS | Well made, fixed identity, stable set of social and psychological attributes (Federman, 1975). A being that can be understood is language (Heidegger) | 'word beings', changeable, unstable, unreasonable, illusory, fraulent, unpredictable, as the discourse allows. (Federman, 1975) | Speech acts Opaque, conceivably contradictory, verbal constructs; self-assured set against relativstic. (C-B, 1959) | "Challenge humanist assumption of a unified self and an interagrated consciousness by both installing coherent subjectivity and subverting it". (Hutcheon, 1988 p. XII) |
| (Epistemological) REPRESENTATION OF KNOWLEDGE | Depth analysis, hidden behind visual, an ability to penetrate phenomena "the skeleton [that] is wrapped in flesh" (Woolf, 1954, p. 162) | Superficiality and depth; philosophy of language dominating the concrete reality of things. Contradictory (through narration attempts to escape the narrator). | Obvious, appearance games truth in appearances. Battle for depth, surface winning. | Narrative; Semiotics; emphasis on ideology and fragmented democratic politics. Drive into questions of subjectivity (sexuality, desire, pleasure) (Newman, 1989). |

## LATE MODERNISM

Late-modernism compares Compton-Burnett with Habermas. Arguments to relate these two authors are maybe not as concise as the example in traditional-modernism; though the change in perspective would suggest a certain degree of opacity. Compton-Burnett's debate with traditional-modernism is perhaps best exemplified through the Weltenschauugen of her two protagonists, Horace and Mortimer (Compton-Burnett, 1959). Horace upholds a discourse that strains toward certainty, and a belief in verifiable knowledge. Mortimer strains toward relativisim (Wilde, 1981). The battle between Horace and Mortimer seems to be won by Mortimer's belief in linguistic surfaces, though conviction is lacking. For example, when Horace informs Mortimer that the fire is smoking, (Compton-Burnett, 1959, p. 10), Mortimer replies that while a fire is smoking it is hard to verify what it is doing (it is hard to verify anything).

During Habermas's (1984) consideration of Weber's (1953) views upon occidental rationalism, there is an attempt to decentre the framework of western culture. A 'culturalistic position' is temporily upheld

"Whether, and if so, how the relativism of value contents affects the universal character of the direction of the rationalisation process, depends then on the level at which the pluralism of 'basic points of view' is set. The culturalist position requires that for every form of rationality it is possible to specify on the same level at least one abstract point of view of which this form could at the same time be described as 'irrational'." (Habermas, 1984, p. 181)

Habermas's generation of this 'culturalistic position' complements Compton-Burnett's approach to relativism. Both authors agree that relativism offers the sorry state of 'imbalanced rationalisation' (Habermas, 1984, p. 183; Compton-Burnett, 1959, p. 153-4), however, it is left to Habermas to appeal to "civilised men" to relieve this imbalance and develop a system of validity that incorporates both value spheres and social systems. The ambiguities in both Compton-Burnett's and Habermas's works demonstrates the compexity of the issue of relativism.

## ULTRA-MODERNISM

Having described the traditional- and late-modernists perspectives, attention now focuses upon the ultra-modernist and post-modernist perspectives.

Ultra-modernism has as its protagonists 'word beings' that are changeable, unstable, unnameable, illusory, fradulent, unpredictable, as the discourse will allow (Federman, 1975). Literal chaos reigns through a thorough refution of modernistic beliefs. For Hutcheon (1988) ultra-modernism marginalises literature through its extreme treatment of the autonomy and supremacy of the artist. Graff considers Federman's surfiction as

"... little more than the obligatory hyperbole of avant-garde self-promotion." (1979, p. 172)

There is, however, some value with ultra-modernism. The extreme manner in which Federman dispenses with the 'verisimilitude' of the traditional novel illuminsates the relative complicity of the historiographic novel (Wilde, 1981). The historiographic novel demarginalises literature through confrontation with history (see for example Eco (1983) and Rushdie (1981)) in a formal and thematic manner (Hutcheon, 1988).

## POST-MODERNISM

The brief 'supporting role' of Ultra-modernism leads us onto post-modernism. Post-modernism is now: An attempt to conflate knowledge and experience. Any attempt to critique post-modernism must be achieved within (post-modernism), as the difference between the critical and cultural sheres seems irreducibly different (Montag, 1988, p. 91-2). Post-modernism, therefore, holds an ironical control over any comparative (or critical) study.

Ironical, since Foucaulvian notions of anti-totalising lead to an epistemic totalisation, and Lyotardian masterful denials of mastery (Hutcheon, 1988). This irony shows us that

"Is it not the characteristic of reality to be unmasterable? And is it not the characteristic of system to master it? What then, confronting reality, can one do who rejects mastery?" (Barthes, 1977, p. 172).

What does Derrida do? Taking Saussurian (see Saussure 1959) signifier (word) and signified (concept) the notion of the referent leaves the system. The referent still exists, but is not immediately accessible through knowledge. Derrida's (1981) metaphysics of presence (reality is directly given to the subject) denotes a 'transcendental signified' intruding into consciousness without 'discursive mediation' (Callinicos, 1989). This absence of 'discursive mediation' disrupts Lockean empiricism and gives the signifier intrinsic worth. Signifiers no longer signify objects, subjects no longer are linguistically constitutive, and meaning becomes intratextual. For Historiographic fiction there is a separation of

"... facts of history-writing from the brute events of the past" (Hutchian, 1988, p. 149))

Realistic fiction, however, (see Barthes, 1982) refuses to accept the split between signified and referent, and by cutting off the signified provides the illusion that the signifier relates to the referent (thereby suggesting that consciousness has no need for a discursive Intermediary). Historiographic metafiction, however, reinstates the signified alongside the signifier while allowing the referent. The positioning of the referent may show the temporary and discursive nature of knowledge. The positioning of the referent also allows fiction ability to question reality, since 'reality' is the "reality of the discursive act". (Hutcheon, 1988, p. 151).

## CONCLUSION

The question as to whether Critical Systems Thinking is real fiction has been given a four dimensional framework. This framework enables debate across many modernistic issues. Issues such as: notion of Critique, Justification Systems and idea of progress. Further essays will discursively develop this rudimentary framework with the aim of widening further the question: Is Critical Systems Thinking real fiction?

## REFERENCES

Adorno, T.W. (1973), "Negative Dialectics," Routledge/Kegan Paul, London

Barthes, R. (1977), "Roland Barthes by Roland Barthes," (trans Howard, R.) Hill/Wang, New York.

Baudrillard, J (1983), "The Precession of Simulcra," in Simulation, Semiotext, New York.

Callinicos, A.T. (1989), "Against Post-modernism: A Marxist Critique," Polity, Cambridge.

Compton-Burnett, I. (1959), "Manservant and Maidservant," Victor Gollancz, London.

Connor, S (1989), "Postmodernist Culture: An Introduction to Theories of the Contemporary," Blackwell, Oxford.

Derrida, J, (1980), "The Archeology of the Frivolous, Reading Condillac," trans Leavy J.P, (1989) University of Nebraska Press, London.

Derrida, J. (1981), "Dissemination," (trans. Johnson, B). University of Chicago press, Chicago.

Derrida, J (1982), "Margins of Philosophy," (trans Bass, A.) University of Chicago press, Chicago.

Eco, U (1983), "The Name of the Rose," Secker/Warburg, London.

Federman, R (1965), "Take it or leave it," Fiction Collective, New York.

Federman, R (1975) ed., "Surfiction: Fiction now and Tomorrow," Swallow Press, Chicago.

Federman, R (1978), "Fiction Today or the Pursuit of Non-knowledge," Humanities in Society, 1.2, 115-31

Flood, R.L. (1989), "Archeloogy of (Systems) inquiry," Systems Practice, 2(1), 117-124

Flood, R.L. (1990), "Liberating Systems Theory," Plenum, New York.

Flood, R.L. and Jackson M.C. (1991), "Critical Systems Thinking: Directed Readings," J Wiley and Sons, Chichester.

Forster, E.M. (1963), "Aspects of the Novel," Butter & Tanner, London.

Foucault, M. (1977), "Language, Counter - Memory, Practice," Cornell University Press, New York.

Foucault, M. (1980), "Power/Knowledge: Selected Interviews and other writings 1972-1977," ed. C Gordon, Harvester Press, Brighton.

Gadamer, H-G. (1975), "Truth and Method," Sheed & Wood, London

Giddens, A. (1990), Jurgen Habermas, in "The Return of Grand Theory in the Human Sciences," ed. Skinner Q, Canto, Cambridge

Graff, G. (1979), "Literature Against Itself: Literary ideas in Modern Society," University of Chicago Press, Chicago.

Habermas, J. (1983) Paris Lectures; published as the "Philosophical discourse of modernity," (trans. Laurence, F.) (1987), Polity, Cambridge.

Habermas, J (1984), "The theory of communicative action, Volume 1: Reason and the Rationalisation of Society," (trans, McCarthy, T). Heinenann, London.

Hassan, I (1982), "The Dismemberment of Orpheus: Toward a Postmodern Literature," Oxford University Press, New York.

Hoy, D. (1990), Jacques Derrida in "The Return of Grand Theory in the Human Societies," (ed Skinner, Q.) Canto, Cambridge.

Hutcheon, L. (1988), "A Poetics of Post-modernism: History, theory, fiction," Routledge, London.

Lecerle, J-J (1990), Post-modernism and Language, in "Post-modernism and Society," eds Boyne, R and Rattansi, A, Macmillan, London.

Lyotard, J-F (1984), "The postmodern Condition: A report on knowledge," (trans. Bennington. G), University of Minnesota Press, Minneapolis.

Montag. W (1988), What is at Stake in the Post-modernism debate? In "post-modernism and its discontents: Theories, Practices," ed. Kaplan, A. VERSO, London.

Newman, M. (1989), Revising Modernism, Representing Post-modernism, in "Post-modernism: ICA documents," ed. Appignanesi, L., Free Association Books, London.

Parker, A (1981), "Taking sides (On History): Derrida Re-Marx," Diacritics, 11, 2, p. 57-73.

Rushdie, S (1981), "Midnight's Children," Jonathon Cape, London.

Saussure, F. de (1959), "Course in General Linguistics," eds (Bally, C and Sechehaye, A) trans Baskin, W., The Philosophical library, New York.

Sukenick, R. (1969), "The Death of the Novel and other stories," Dial Press, New York.

Ulrich, W. (1983), "Critical Heuristics of Social Planning: A new approach to Practical Philosophy," Haupt, Berne.

Weber, M. (1958), "The Protestant Ethic and the Spririt of Capitalism," Routledge and Kegan Paul, New York.

Wilde, A (1981), "Horizons of Assent: Modernism, Post-modernism, and the Ironic Imagination," John Hopkins University Press, London.

Woolf, V (1953), A Summing up in "A Haunted House and other short stories," Hogarth Press, London.

Woolf, V (1954), "Jacob's room," Hogarth Press, London.

REFERENCES TO TABLE

(1)     Sukenick considers that if art does not reflect reality, then the final reflections should dispense with self-reflection. Sukenick is working on the dialectic between real-seeming artifice and reality. Proposing that the artifice is not an imitation (achieved through self-reflection) but an invention (achieved through the self-reflexive).

(2)     Sukenick's (1969) simultaneous multiplicity uses montage and parataxis (arrangement of clauses or propositions without connectives) to dissolve continuities and ideal structures; and reveal the sophistical view of existence.

(3)     Compton-Burnett.

(4)     Pluralist view of history.

THE CRITICAL SYSTEMS METHODOLOGIES

AND THE "TRUTH OF INVARIABILITY"

Ioanna Tsivacou

Research Laboratory of Samos
University of the Aegean
Athens, Greece

ABSTRACT

This paper deals with the prerequisite of active participation of
stakeholders in an organization, in order to apply critical systems
methodologies with the objective of arousing emancipatory interest.  It
discusses the meaning of the emancipation in the labour mouvement tradition,
and the emergence of apathy of the working people towards it.  The critical
systems thinking is obliged to face this apathy in its attempt to motivate
the emancipation interest.

INTRODUCTION

An attempt is made to confirm that critical systems methodologies,
due to their concern over the emancipatory interest, are obliged to pay
attention to the context of the system.
In order to accentuate the importance that critical systems methodologies
give to the preparation of the context, the paper annotates the meaning
that critical systems thinking gives to the word "emancipation" and the
connection of this meaning to that of social change.
This annotation leads to the emergence of the conceptual pair
"participation-emancipation" and to the interdependence between the two
terms.
The paper proceeds to the historical ancestor of the conceptual pair
and distinguishes the different kinds of participation according to the
means used for the pursuit of the emancipation.  But, it confirms also,
that due to the above mentioned interdependence, the observed today apathy
for the ideal of the participation is greatly explained by the weakness
of the will for emancipation.
The will for emancipation is the result of the production and circulation
of the statements of truth.  In our days the dominating discourses
disseminate the truth of "invariability"; the impact of all that being the
fading of the will for emancipation and social change.
Critical systems thinking, working in the context of the system and
seeking to apply a participatory democracy, is obliged to take into consideration the new truth.

## THE IMPORTANCE OF THE CONTEXT

In the most group problem-solving methodologies, the context of the methodology has not drawn the same attention as the content and the process of the methodology.  The reason probably must be attributed to the nature of the problems and to the kind of contexts where these methodologies have been usually implemented.

With the appearance of the critical systems thinkers and their concern to apply critical systems methodologies, the context goes to the front stage because their main interest is the coercive type context; this means that the priority in their analysis is given to the context itself.

This subject is very interesting, since the aim of the critical methodological approach is not limited to the structuring of a problem situation and to the designing of the way out from this situation; instead it attempts to develop a cognitive process in the system relevant to what Habermas calls "emancipatory interest" (Habermas, 1971).  By adopting the emancipatory interest, the systems methodologies transcend their limits: from the resolution of a problem situation, they are now examine the possibilities of a radical, qualitative change in the context of the situation.

## THE MEANING OF THE EMANCIPATION

The question that arises here is:  what does the Critical Systems Thinking (C.S.T.) mean as emancipation?  The answer, usually, is according to the meaning that Habermas attributes to the term.  Flood and Ulrich elaborate further with emancipation from hidden presuppositions (Flood and Ulrich, 1990).  In this case is the emancipatory interest motivated by self-reflection and not the emancipation itself.  Because the self-reflection liberates the subject from its false consience about the conditions that produce its material, social and psychological oppression, and not from the conditions themselves.

Habermas emphasizes the practical consequences produced by the process of "self-reflection".  But, meanwhile Habermas, as it is known, gives to the term "praxis" the meaning of a political dialogue on an ethical content, he does not deny nevertheless the factors labour-production which function automatically in the social tissue.  These labour-production forces seems to be more acceptable by M. Jackson.

Jackson (1988), clarifying the meaning of the term thus: "freeing themselves from constraints imposed by power relations and in learning, through a process of genuine participatory democracy, to control their own destiny".

With the answer of Jackson, the emphasis shifts from the "what" to the "how".  And one of the fundemental prerequisites of "how", is participatory democracy, or at the micro-level of the formal organizations, the discursive, problem-solving methodologies, which apply participation in action.

## THE INTERELATIONSHIP BETWEEN PARTICIPATION AND EMANCIPATION

The notion of "participation" has not yet been fully investigated by the C.S.T. in relation with the concept of emancipation.  Critical systems thinkers, have not yet clarified whether they accept the participative function for the systems methodologies because they accept the social character of the work, or whether they see it as a technique for problem-solving. This is a crucial point for a more assiduous study of the relations between productive and communicative action. But, either the critical systems thinkers stop at the point where the participants, through a collective self-reflection, will be released "from dependence on hypostatised powers" (Habermas, 1972); or they accept that beyond the emancipation from the ideology, the involvement of the actors in a game of social action, in order to change the social relationship and to make the man "a species-being" is also necessary; and only then, the man will recognise and organize "his own powers as social powers so that he no longer separates the social power from himself as political power"(Marx, 1963); The condition of the "collective" and of the "participative" are common. And that happens because both the marxist teaching and the analysis of Habermas, lie within the framework of critical dialectics, on the horizon of the modernity where the scientific paradigm, in order to be legitimate needs to be based on a meta-discourse; in this case, the meta-discourse of the emancipation (Rorty, 1985).

Following this tradition, the participation supported by C.S.T. cannot be replaced with the participation announced by the school of Human Relations. It is not a participation, of which the final end is to supply the participants with the necessary psychological motives, in order to increase their commitment towards the organization. Neither a participation which will function as a channel of information for the decision making of the organization. On the contrary, it is a participation derived from the participative value which inspired the requests of the labour movement. A value based on the desires of the working people to change the existing productive relations, to make the working man able to control the product of his labour, to bring an end to the exploitation of man over man.

POLITICAL AND SOCIETAL PARTICIPATION

This request for participation has been legally and practically implemented in many european countries by socio-democratic governments, under pressure exerted by labour and socio-democratic political parties. We call this kind of participation "political participation", because the means used for its fulfilment belong to the political sphere, in contradiction to the "non political participation" as first defined by the Hawthorne team.

The policy and executive issue are the main objective of the political participation (Guest, 1979). It is obvious that the accomplishment of this objective has been attempted through strategic action. But this action has taken place without a repetitive reflection on the causes of the action, so, very soon, it has been degenerated in instrumental action. This fact represents the consequences derived from the antinomy between "end" (emancipation) and "means" (limited to the manifestation of the technical interests of J. Habermas).

The critical systems methodologies, introducing the factor of the emancipation in their rationale, are represented as the evolutionary continuation of the "political participation", since they propose the same final objective, but completely different means. The "political participation" uses political means, as the representative participation and the created bureaucratic networks, for a purpose belonging not to the level of the politics, but to the level of the civil society. Therefore it continues the duality between political man and human being. The C.S.T. tries to transcend the contradiction by applying means, as the discursive methodologies, raised to the level of the civil society.

These constitute a process, developed not in the political and legal field, but in the communicative field. Especially when the field of the discursive systems methodologies tries to be a field of a non distorted communication. That is when the context of the methodologies manages to be culturally homogenous in order to be transformed in a communicative community. For this reason, we call this kind of participation "societal participation". It is a participation attempting to reconcile through discourse the "abstract man" with the "citizen" as member of the civil society.

## THE APATHY FOR PARTICIPATION

Individuals, who have been members of works councils, as distinguished researchers (Guest, 1979), confess to a lack of interest of the workers or employees to participate at any level in the organization. This indifference to participation has been extended today to wider layers of working people and has been transformed to an apathy for "political participation". Further small group, direct, non political participation, win the interest of the people. This is the participation which is only played in the domain of the civil society, and it is connected with the day to day work and the immediate benefit that the people derive from it.

Thus today the subject of the change of social relationship and the conflict among centers of power, the subject of emancipation, does not have the same priority in the scale of values of the working people as it did some years ago.

## THE WILL FOR EMANCIPATION AND THE TRUTH OF OVERTHROW

The organizations and especially the companies, are fields of conflict of interests and not places for psychoanalytical investigations. The acceptance by them of systems methodologies, animated by the emancipatory interest, means that they dispose primarily the will for emancipation. Or we accept that there is in the context of the system this kind of will, or we believe that this will, will be only a result of the emancipatory interest developed during the process of the methodology. In the last case, that means that the analyst of the methodology is the suitable and qualified person to estimate the needs of the system. This logic leads us back to the instrumentalistic model, therefore, we are obliged to accept that the will for the emancipation exists prior to the methodology. The methodology, motivating the emancipatory interest gives form to this will.
As the emancipation causes a breakdown in the existing situation, the will for it means the belief that this breakdown is possible. It means that the members of the company believe to the "truth of the overthrow" of the existing social relations.

We use the term "truth" according to the meaning that Foucault gives to the term. We believe that the perception that the overthrow is possible, which has dominated since French revolution, is a truth historically based and not an ideology. This truth has been produced under historical events, but has been spread and has overwhelmed with a great number of scientific statements. As Foucault (1980) says:"each society has its regime of truth; its general politics of truth: that is, the types of discourse which accepts and makes function as true;the mechanisms and instances which enable one to distinguish true and false statements...... the status of those who are charged with saying what counts as true".

The theory and the practice of the labour movement, which for brevity is known as "left", from French revolution till the decade of 60's, is the collective forces of power which imposed the truth of over-throw. In this regime of truth political participation has acquired its dynamics, because it has been derived as a result of power in the regime of truth of overthrow and of emancipation. This regime has been also the product of a network of power relationship between right-left, capitalism-socialism, conservative-progressive. This power relationship has regulated not only the relations in the realm of the production and circulation of the material goods, but also in the realm of the production and circula-tion of statements.

## THE WILL FOR EMANCIPATION AND THE TRUTH OF INVARIABILITY

From the beginning of 70's and after a different regime of truth has come about. The power relationship, which generate rules for the distinction between truth and falsehood, is modified; This alteration begins to be more apparent during the decade of 90's, especially after the depth of change occured in the social and ideological struggles and consequently in the political debate.

The "truth of the overthrow" is displaced by the "truth of the invariability" of the existing situation of things. New statements are found and transmit the new truth. From the economics of the new-liberalism till the philosophy of the post-structuralism, the mechanisms charged with the generation and transmition of the statements, the means and the texts which legitimate the truth, propagate the new scientific discourse: the vanity of the revolution; the self-enlightment of the individual conscience for the unveiling of its psychological constraints; the truth of the successive approaches for the regulation and equilibrium of the systemic oscillations.

The "truth of the invariability" cannot support a participative process animated by the emancipation. Because, this kind of truth operates for stability, not for change. In a world dominated by the "truth of the invariability", the C.S.T. must also investigate the con-text of the organization from this point of view. The confirmation that an organization is of a coercive-type context, must be followed by the analysis of the statements of truth that circulate in it. The systems methodologies which are going to be applied, will not aim only at the liberation of the participants from false conscience, but, also, enable them to detach "the power of truth from the forms of hegemony, social, economic, and cultural, within which it operates at the present time" (Foucault, 1960). Applying geneological analysis and seeking the rules of formation of discourses about truth, the C.S.T. will be obliged to "analyze the discourses articulation with and regulation by non discursive practices (social and institutional practices)" (Smart, 1983).
So, as under the communicative praxis of Habermas is concealed the labour-production relation, under Foucault's analysis of discourses structuration are concealed the mechanisms of power connected with historical and economic processes. If the critical systems methodologies, contribute to the emergence of the root-causes of the truth formation, they will reveal also the centers of resistance. And this revelation will reinforce the emancipatory interest of the participants.

## CRITICAL SYSTEMS THINKING AND GENEALOGICAL ANALYSIS

The C.S.T. desires the restoration of modernity. It tries to revive a normative theory for the legitimation of new forms of social organization. In this attempt, the acceptance of a thought as

417

that of Foucault, does not constitute a contradiction (Flood, 1990). Foucault's thought, and his genealogical interpretation, can be very useful in the micro-level of the formal organizations. Extending the meaning of emancipation according Foucault's suggestion, a societal-participative systems methodology offers the opportunity to the participants to be continuously aware of the systems methodology itself. To be continuously conscious that each statement, every enunciation, are pronounced in a scientific domain, in a discipline (Foucault, 1971). This fact by itself subverts the discourse, since makes obvious that also the methodology obeys to the rationality of the discipline. It extends the function of the emancipatory interest up to the level of the communicative praxis and gives the opportunity to the participants to be aware of their legitimation authority concerning the validity of the discursive terms. We show to them that the scientist who works on a critical methodology, as Jackson (1990) confesses, quoting the words of the Horkheimer, is a moral individual "never aims simply at an increase in knowledge as such. Its goals is man's emancipation from slavery". This paradigm of the ethical scientist, is, perhaps, more able to empower the actors to pursue the change of the existing social relations, than a new scientific paradigm.

CONCLUSION

It has been argued that the critical systems thinking is the successor of the scientific inquiry, based on the meta-discourse of the emancipation, which animated the labour movement. Therefore, the critical systems thinking must supersede the apathy of the working people towards the participative value and the ideal of emancipation. This apathy has been created from the antinomy of the labour movement inquiry between aims and means.

It has been also discussed that the will for emancipation is influenced by the "truth of the overthrow" of the existing social relationship, according to the meaning that Foucault gives to the term of the "truth". This "truth" today, has been replaced by the "truth of the invariability", through the new rules of the scientific discourse.

The critical systems thinking must enlarge its meaning of emancipation as suggested by Foucault distinguishing between the regime of truth and the regime of power. By doing this the participants of the systems methodology are helped to be critical of also the methodology itself. This position does not lead people to the relativism, but enables them to comprehend that the will for emancipation is an ethical decision connected with man and not with a scientific paradigm.

REFERENCES

Flood, R. L., Ulrich, W., 1990, Testament to Conversations on Critical Systems Thinking Between Two Systems Practitioners Syst. Prac., 3(1):21
Flood, R. L., 1990 "Liberating Systems Theory", Plenum, New York
Foucault, M., 1971, Orders of Discourse, Social Science Information, 10(2)
Foucault, M., 1980, "Power/Knowledge: Selected Interviews and Other Writings 1972-1977" C. Gordon, ed., Harvester Press, Brighton
Guest, D., 1979, A Framework for Participation, in:"Putting Participation into Practice", D. Guest and K. Knight, eds., Gower Press, Westmead.
Habermas, J., 1971, "Knowledge and Human Interests", Beacon Press, Boston

Jackson, M., 1988, Systems Methods for Organizational Analysis and Design,
    Syst. Res., 5 (3):202
Jackson, M., 1990, The Critical Kernel in Modern Systems Thinking,
    Syst. Prac., 3(4):363
Marx, K., 1963, On the Jewish Question, in:"Karl Marx Writings",
    T. B. Bottomore, ed., C. A. Watts and Co Ltd., London
Rorty, R., 1985, Habermas and Lyotard on Postmodernity, in:
    "Habermas and Modernity", R. J. Bernstein, ed., Oxford Polity Press
Smart, B., 1983, "Foucault, Marxisme and Critique"
    Routledge and Kegan Paul, London

INTERPRETIVE CRITICAL IDEOLOGIES FOR

RESEARCH METHODOLOGIES

Julie Binko
University of Minnesota
568 S. Smith
St. Paul, MN 55107

Introduction

To document successful transformation processes of the American
educational system, we need to explore and develop new paradigms of
research inquiry and methodology. Much of our current thinking of
research design is based on a static model of system dynamics. C.W.
Churchman explains that research for American socio-cultural systems
should become evolutionary rather than regulatory, with design for a
viable "mode for simultaneous and multitudinous interacting
relationships" (Churchman, in Jantsch, 1975). Furthermore, the
ideological underpinnings of scientific inquiry must be defined when
extrapolating from Churchman's paradigmatic proposition. The new
definitions will carve outlets for inquiry and move us towards a view of
evolutionary and transformational paradigms.

This paper will attempt to broaden the range of critical systems
inquiry by delving into three traditions of scientific interpretation:
hermeneutics; phenomenology; and epistemology. Critical systems theory
is the base of inquiry and encompasses three major concepts:
emancipation, pluralism and pure rationale. Emancipation and pluralism
are important for the transformation of any social system. They are
based on the critical social theory of communicative competence and
self-reflection. The interplay of these concepts, competence and self-
regulation, catalyzes a dynamic force, making humans aware of their
"false consciousness" and encouraging them to act in changing their
community or world (Argyris et al., 1985). According to critical social
theory, emancipation with a view towards the future, one can critique
freely and make choices (Carr and Kemmis, 1986). Pluralism prescribes
improvement of human existence through action. While emancipation and
pluralism are important catalysts of change, I have chosen only to focus
on the expansion of critique from pure rationale.

Pure rationale disables our consciousness to see beyond a current
situation. Humans become too boxed in definitions as well as self-
perceived and socially confirmed realities (Berger and Luckmann, 1961;
Carr and Kemmis, 1986). Pure rationale allows for a narrow and
sometimes false view of a situation (Carr and Kemmis, 1986). It will be
argued that pure rationale should be broadened to bring forward
explicitly other ideas from critical science. I am proposing to expand
the area of critique from pure rationale to other areas of critical
thought as an exploratory avenue into new paradigmatic thinking; for
example, critical pedagogical thought. In this paper I have chosen
emancipatory democracy (Giroux, 1988) as an expansion of critique.

*Systems Thinking in Europe*, Edited by M.C. Jackson *et al.*
Plenum Press, New York, 1991

Two phenomenological ideas will be used for methodological approaches: (a) moving towards a multi-dimensional human (Marcuse, 1966) and, (b) using the time dimension of the future. Finally, the epistemology of knowledge generated will be another critical foundation for methodological practices.

Extrapolating from Churchman's evolutionary paradigm, or at least stretching it further is necessary. There is ongoing educational research involvement at the state level that needs a transformational view of research. My current involvement in a state funded educational transformation project is one example. This project provides a rhetorical example of the desired future--transformation towards a "community of learners."

Juxtaposing explicit critical ideologies against the meaning of a "community of learners" enriches the meaning and possible impact of the transformation (McCutcheon, 1981). All three scientific interpretive approaches will be discussed as they relate to the transformation of an educational system, in this case, the state-funded research project's "community of learners."

Emancipatory Democracy

In emancipatory democracy all members take an active role in determining participation or involvement in their community. It is a movement to reinvent "a view of authority that expresses a democratic conception of collective life, one that is embodied in an ethic of solidarity, social transformation, and an imaginative vision of citizenship" (Giroux, 1988). All views of society are legitimized rather than assuming only one view is appropriate within the community. Legitimization encourages dialogue among community members as they work to understand one another's views and this "allows them to mediate, legitimate and function in their capacity" (Giroux, 1988).

Through reflection and self-actualization, any and all members of a future community would define roles in the community system. They would be able to serve as referents and critics of a vision for community change. Roles would be diverse, individuated, and empowering. Participants would be involved on their own terms and in their own capacities.

In the case of educational systems, these community members would include administrators, teachers, students, and parents. All together they would comprise the "community of learners." Administrators would support and facilitate empowerment, so that all other community members have ownership in their self-specified roles. Emancipatory authority would shift the traditional meaning of authority from a legitimating basis to " a view of community life in which the moral quality of everyday existence is linked to the essence of democracy" (Giroux, 1988).

Democracy would become the mediator of empowerment. Teachers would become two-way facilitators, empowered by administration to make their own instructional decisions using their intellectual capacities. The other direction would be towards the student. Learning would be facilitated and mediated in order to meet the needs of student situations and encourage dialogue among all participants in the community of learning.

Democracy in learning creates knowledge that is empowering (Giroux, 1988). The learner becomes referent to the society as a whole. This approach would lend itself towards individualized learning programs. Affirmation of life experiences enables learners to mediate and construct new social realities.

For parents, emancipatory learning frees them to be involved in the education community. Parental involvement includes participation on all levels from instructional to administrative. Administratively, parents could express their societal view that then could be translated into educational community policy. Conversely, by participating in their children's learning program, their different views of family and life experiences would make learning relevant for the child.

The state educational transformation rhetorically supports these emancipatory ideas within the "community of learners." Site-based management, locating power in the school, is central to administration in the learning communities. Decisions are made that correspond to needs of the "learning community," improving the quality of everyday existence.

Within the state-funded project, the role of the teacher is to facilitate and mediate learning through the carefully prescribed and iterative learning plans. Iteration allows for both student and parental involvement in the education process. This involvement reflects the student's context, interests, and abilities. From this plan, learning generates knowledge that validates experiences, making their learning relevant and empowering.

## Multi-Dimensional Man

Herbert Marcuse (1966) wrote extensively on the phenomenon of the one-dimensional man. He believes that technical aspects of our socio-cultural systems have bled into scientific inquiry. They have allowed humans to be sequestered into a minimal existence, socially, physically, and emotionally. Science promotes the dehumanization of humankind into a series of boxes and matrices. The result is that ideals are flattened so that social contradiction and opposition are absorbed (Binko, 1991). The outcome is that society no longer supports the diversity it needs to flourish (Marcuse, 1966).

If we refer back to the idea of emancipatory democracy, we need to extend our definition of humankind to become more inclusive of the traits that make us vital. One step to restore humans being to multi-dimensional beings would be to integrate the realities of human existence into scientific inquiry. A suggested methodological approach would be that of Habermas, who based his methodology on three systemic approaches that complement the theoretical concept of communicative competence, specifically the technical, the practical and the emancipatory (Habermas in Gregory, 1990). Communicative competence, reflecting back to emancipatory democracy, provides a dialogic stage between researcher and community. This dialogue would allow the socially constructed realities of all parties to be heard.

Habermas' technical systemic approach of control and prediction can be loosely associated with positivistic trends of scientific inquiry (Mingers, 1980). His practical approach synthesizes social construction and reality among participants (Gregory, 1990). His emancipatory systemic approach acknowledges power and organizational interests, but does not imply what "we ought to do" (Oliga in Gregory, 1990). The theoretical goal of inquiry for Habermas is communicative competence (Mingers, 1980; Bowers, 1984). It would seem that a combination of the technical, practical and emancipatory approaches would be more adequate than any taken alone. In total they address the multi-dimensional existence of humankind.

Trist's sociotechnical perspectives (1981) are more clearcut than Habermas' and more suggestive of a new paradigm of inquiry into systems. Some of Trist's elements are the following: community interaction, feedback mechanisms, lateral power structures, collaboration, innovation, ownership, and the optimization of people and machinery in a complementary fashion.

Trist's perspectives are more encompassing for effective research. This effectiveness is due to the provision of multiple perspectives and the "paradigm" or set of assumptions shared by the group (Argyris et al., 1985). Multiple perspectives are essential to the legitimization process of emancipatory democracy. This research paradigm is dialectic. From this dialogue emerges a unified relationship of science and community. This relationship would nullify the view of otherness exemplified often in positivistic research.

## Dimension of Time

Once a group comes to a shared set of assumptions, they must work towards those goals. This process occurred in the state-funded project, whereby transformational goals were established to guide participants in change.

In accordance with platonic thought a future condition must be articulated in order for transition to occur. As the teleos or goal is implemented, the future view evolves until each individual in the community "arrives at a mutual understanding in the conduct of public affairs" (Huebner in Pinar, 1974, p.37). This happened in the state project. The transformation goals changed often in number and focus. They came from input from the educational communities and included such modifications as rhetoric, focus, and simplification. Initially the future condition was contrived because of the initial top-down mandate of the project. Since then, there has been some attempts by the state to create ownership in the future direction of the project.

## Knowledge Generation

The final consideration towards establishing new methodological concerns is the type of knowledge that is generated in a research environment. The meaning of research knowledge generation will be shifted from "knowledge production" (Malchup, 1980) towards the meaning of knowledge "in situ" (Geertz, 1973) and praxis. These interpretive ideas will encourage professionals to generate useful knowledge for use in a community context. Knowledge generation from this research paradigm is in a "context in which beliefs are criticized, justified and evaluated" (Argyris, et al., 1985, p. 11).

Another epistemology of this research knowledge is based in practical meaning guided by the norms referenced by the group. Norms are negotiated and practically deliberated between the researcher and the community participants. This approach embraces the move towards developing the multi-dimensional parameters of scientific inquiry described above.

If knowledge is acted upon in a research capacity, it will set the stage for action research. In addition, knowledge into praxis moves towards the desired future state described by the community. As community members begin to describe their intentions, assumptions, beliefs, and values, the meanings of their actions become more apparent. Actions are really the connection between the means and ends of a transformational process. In order for a specific transformation to occur, intervention must take place within the context of the community. Reflection must be included to assure that knowledge be placed in action that incorporates the beliefs, intentions and frames the problems and solutions within the community. After an intervention occurs there should be also an evaluation and self-correction phase to redirect any misinterpretations.

This process that supports the proposed intervention strategies is named action research (Argyris, et al., 1985; Schon and Argyris in Whyte, 1989). Within the parameters set forth herein, this type of research environment is supported. The reiterative cycle of action research is compatible with the emancipatory and pluralistic qualities of critical systems theory. Within the critical idea of emancipatory

424

democracy the dialogic qualities of this research are supported. Also, the sociotechnical perspective of collaboration and community voice is also compatible and incorporated within this research and intervention strategy.

## Conclusion

This paper attempts to bring together three interpretive ideologies to attempt to create a new research paradigm. The transformational process being used is supported by the ideological parameters, but some modifications are taking place as the transformation evolves. More exploration in the area of foundational bases of inquiry will be needed as moves are made to implement critical systems theory into professional practice.

## References

Argyris, Chris; Putnam, Robert; and Smith, Diane, 1985, <u>Action Science: Concepts, Methods, and Skills for Research Intervention</u>, San Francisco: Jossey-Bass.

Argyris, Chris; Schon,Donald, 1989, Participatory Action Research and Action Science Compared, In <u>American Behavioral Scientist</u>,32,5, 612-623.

Berger, P.L.; Luckmann, T., 1966, <u>The Social Construction of Reality</u>, New York: Doubleday.

Binko, Julie, 1991, Using the Future to Create Community and Curricular Change, In <u>Changing Schools: Recapturing the Past or Inventing the Future</u>, Nancy P. Greenman and Kathryn M. Borman, eds., Ablex Press. (forthcoming)

Bowers, C.A., 1984, <u>The Promise of Theory</u>, New York: Longman Press.

Carr, Wilfred; Kemmis, Stephen, 1986, <u>Becoming Critical</u>, London: Falmer.

Geertz, C., 1973, <u>The Interpretation of Cultures</u>, New York: Basic.

Giroux, Henry, 1988, <u>Schooling and the Struggle for Public Life: Critical Pedagogy in the Modern Age</u>, Minneapolis: University of Minnesota.

Gregory, Wendy, (1990, December), <u>Approaches and Methods of Systems Design: Critical Pedagogy</u>, Paper presented at the NATO Advanced Research Workshop at Asilomar Conference Center, University of Hull, Hull UK.

Huebner, Dwayne, 1974, Toward a Remaking of Curricular Language, In William Pinar, ed., <u>Curriculum Theorizing</u>, (pp. 237-248), Berkley: McCutchan.

Jackson, M.C., 1982, The Nature of 'Soft' Systems Thinking: The Work of Churchman, Ackoff and Checkland, <u>Journal of Applied Systems Analysis</u>, 9, 17-29.

Jantsch, Erich, 1975, <u>Design for Evolution: Self-Organization and Planning in the Life of Human Systems</u>, New York: Braziller.

Marcuse, Herbert, 1966, <u>One-Dimensional Man</u>, Boston: Beacon.

Malchup,F., 1980, <u>Knowledge and Knowledge Production</u>, Princeton, N.J.: Princeton University Press.

McCutcheon, Gail, 1981, On the Interpretation of Classroom Observation, <u>Educational Researcher</u>, 10, 5-10.

Mingers, J.C., 1980, Towards an Appropriate Social Theory for Applied Systems Thinking: Critical Theory and Soft Systems Methodology, <u>Journal of Applied Systems Analysis</u>, 7, 41-54.

Trist, Eric, 1981, The Sociotechnical Perspective, In A. Van de Ven and W.F. Joyce eds., <u>Perspectives on Organizational Design and Behavior</u>, (pp. 19-73), New York: Wiley.

Ulrich, W., 1990, What is Called Critical Systems Thinking?, In <u>Proceedings of the Annual Meeting of the ISSS, Portland, Oregon, July 8-13, 1990</u>, B.H. Banathy, ed., New York: ISSS 1990.

COMMUNITY OPERATIONAL RESEARCH AND THE

NEW PUBLIC HEALTH MOVEMENT

Clare Townley, Colin Thunhurst and Charles Ritchie

Community Operational Research Unit
Northern College

and

Nuffield Institute for Health Services Studies
University of Leeds

INTRODUCTION

In this paper we will consider the characteristics of 'The New Public Health Movement'. We will identify 'community participation' as a fundamental underlying principle of that movement. We will consider the extent to which operational research, specifically the newly emerging community operational research, can be a vehicle for the strengthening of community participation within the health sector.

THE NEW PUBLIC HEALTH MOVEMENT

Health For All By The Year 2000

The last decade has seen the emergence of what has been called 'The New Public Health Movement' (Ashton and Seymour, 1988). The movement is international, but it has taken on specific characteristics and a specific formulation within the United Kingdom.

Globally, the movement was heralded by the Alma Ata Declaration, adopted by the World Health Organisation in 1978. This declaration reflected a growing awareness of the historical failure of a highly technologically based, very expensive, interventionist, medical model to meet the health needs of the majority of the populations in developing countries and a substantial minority of the population in developed countries. The declaration drew together the threads of a new strategy, designated Primary Health Care (Phillips,1990), which adopted a more 'holistic' conception of health ("a state of complete physical, mental and social well-being, not merely the absence of disease or infirmity") and which was to be built upon the growth and extension of basic health services as well as the development of the local community in terms of infrastructure, education, initiatives and resources. Primary Health Care (PHC) was defined as:

"essential health care made universally accessible to individuals and families in the community by means acceptable to
them, through their full participation and at a cost that the
community and country can afford.  It forms an integral part
of the country's health system, of which it is the nucleus,
and of the overall social and economic development in the
community" (WHO, 1978)

The concept was turned into more strategic format with the adoption, at the 34th World Health Assembly in 1981, of the Global Strategy
of Health For All by the Year 2000:

"the main social target of governments and WHO in the coming
decades should be the attainment by all citizens of the world
by the year 2000 of a level of health that will permit them to
lead a socially and economically productive life"

Such a goal was consciously utopian.  But it has served as a useful
vehicle for focussing attention on the very substantial inequalities
that exist in health experience within and between nations, and on the
extent to which this experience - and the inequalities - have been the
product of social, environmental and economic circumstance.

Throughout the world, the goal has been taken up at a regional
level within the WHO.  Regional Offices have been charged with formulating strategies of more local relevance.  Most have approached this
task by specifying a number of particular targets.  For example, the
European Region of the WHO adopted a series of 38 such targets in 1984
(WHO, 1985).

These European Targets are grouped under six headings: Health for
all in Europe by the year 2000 (Targets 1-12); Lifestyles conducive to
health (Targets 13 - 17); Healthy environment (Targets 18 - 25);
Appropriate care (Targets 26 - 31); Research for health for all (Target
32); and, Health development support (Targets 33 - 38).  The Targets are
based on the principles of equity, health promotion, community participation, multisectoral cooperation, primary health care and international
cooperation.  They are seen to be related pyramidically with Targets 1
to 12 building upon Targets 13 to 31, which themseleves build upon
Targets 32 to 38.  They incorporate timescales defined accordingly.

In 1986, the WHO European Region launched its Healthy Cities Programme whereby exemplar cities throughout Europe were selected to pursue
city wide health strategies in pursuit of Health For All By The Year
2000 moulded broadly around the Region's 38 Targets (see Ashton and
Seymour (1988)).  Within Britain an increasingly large number of cities
and towns, inspired by the WHO European Region, but working independently of it, have launched their own healthy city/healthy town
programmes.  Most have joined together under the umbrella of the UK
Healthy Cities Network, recently renamed the UK Health For All Network.

Health For All Within the United Kingdom

The British adaptation of the Healthy Cities model has been characterised by an accentuation of the dimension of inequality.  The spread
of Health For All at a global level coincided with a rediscovery of
inequalities in health within the United Kingdom.  This rediscovery had
been prompted by the publication of the Black Report in 1981.  It was
accelerated by the the clumsy and publicly damaging way that it was
received politically, and by the realisation both by health authorities
and by local government of the very far-reaching implications for their

own activities of its major findings (see Townsend, Davidson and Whitehead, 1988). The national Black Report was followed by local Black Reports (see, for example, Thunhurst, 1985) which explored the extent to which nationally reported inequalities were replicated locally and the extent to which these translated into geographical inequalities.

The formulation of Health For All within the U.K., seen by many as a natural mechanism for translating the rediscovery of inequality into more concrete policy, thus reflected the particular concern over the equity principle expressed in global Health For All. For example, the City of Sheffield, in their Healthy Sheffield 2000 Strategy, has taken the WHO's Target number 6, which concerns the extension of life expectancy and refomulated it:

'By the year 2000 the variation in life expectancy between the most and the least healthy parts of Sheffield should be reduced by 50 per cent' Healthy Sheffield 2000 (1987)

The overall strategy has been re-conceptualised and re-formulated to reflect this increased attention to inequalities, and also, within Sheffield, to provide more substance to the facilitating structures on which the strategy is constructed (see Healthy Sheffield 2000 (1989)).

The Health For All By The Year 2000 approach, incorporating, as it does, explicit written targets, draws attention to the need to be able to carry out an accompanying monitoring process. This has prompted exercises, conducted both at the European regional level and within the U.K., to establish a set of indicators which could be used to monitor progress towards achievement of the targets. And in the British context, the added emphasis on inequality has implied a need to monitor progress within cities and towns, as well as at a city wide level.

It has always been anticipated that a number of the monitoring indicators would be 'soft' indicators - requiring primary data collection and the use of 'soft' analytic techniques. But for regular monitoring, where the repeated conduct of primary surveys is prohibitive, the most effectively deployed data sources would need to be secondary ones, available only from routinely available data sources. In the light of this, the UK Healthy Cities Network commissioned a study on core indicators which highlighted, inter alia, steps that the members of the Network might take to improve the usefulness of secondary data sources for these purposes, Thunhurst (1989).

Equity is just one of the six principles underscoring 'Health For All'. Although expressed discretely, all six are recognised to stand in a symbiotic relationship. To address inequalities in health it is necessary to develop appropriate methods of health promotion employing approaches incorporating a maximal orientation towards community involvement. Thus, also, the decade has seen the development of a very extensive range of community health development projects (see Health Education Authority, 1990).

COMMUNITY OPERATIONAL RESEARCH

Community Operational Research Unit

The development of Community Operational Research was chronicled in a number of papers to the first international conference of the UK Systems Society (UKSS) held in 1988 (Flood, Jackson and Keys, 1989). Particularly, a contribution from Thunhurst (1989) outlined the processes of creation of the Community Operational Research Unit at the

Northern College in South Yorkshire and the methodological and philosophical principles upon which the work of the Unit was progressing. The Unit's commitment to methods of participatory research was spelt out and a number of promising areas of application were identified.

The last three years have seen much of this initial promise come to fruition (see Thunhurst, 1990, and Ritchie and Thunhurst, 1990). In the area of housing, the Unit is engaged in a number of projects, the longest standing of which is its work with the Thurnscoe Tenants Housing Co-operative.

The Thurnscoe Tenants Housing Co-operative was formed in 1988 in response to proposals from the National Coal Board to sell off NCB houses in Thurnscoe to private developers. The Community Operational Research Unit has been helping it strengthen its management capabilities in a diverse range of ways. In this it has employed a broad cross-section of OR techniques - conventional techniques, such as spreadsheet modelling, newer problem-structuring techniques, such as the Strategic Choice Approach, and OR-related techniques, such as survey analysis.

Housing ownership in the U.K. is currently undergoing very sub-stantial changes, with large Government-induced shifts from the public to the private sector. As an alternative to private tenancy, existing council tenants are exploring methods of co-ownership and self-management. Assisted by a grant from the Joseph Rowntree Memorial Trust, housing will continue to be a priority area for the Community OR Unit.

Community Operational Research with Community Health Groups

In parallel with the work with tenants and residents the Unit has been developing projects with community health groups. There are, it is estimated, some 12,000 community health projects operating throughout the U.K. They have already been most succesful in generating self-activity within communities aimed at their own health improvement or at social and economic changes with health consequences. But they have been less succesful in seeing their concerns and objectives reflected at a statutory level. It is here that it is hoped the Unit will be able to perform a role in assisting groups to respond to consultation documents and to formulate their own proposals which will ensure a wider rep-resentation of perspectives in statutory health planning processes.

Community Health Workshop. In June 1989, the Community OR Unit hosted a workshop at the Northern College which drew together a number of OR practitioners with people involved in community health activities in order to discuss and debate the potential for OR work with community health groups. The event highlighted the clear potential for OR involvements employing both conventional 'hard' quantitative OR tech-niques, as well as the newer problem structuring ones (Community OR Unit, 1989). It was agreed that it was necessary to undertake a number of pilot projects to investigate the potential and the problems encoun-tered.

Sharrow Community Health Project. As a direct result of the event, the Unit established contact with the Sharrow Community Health Project. Sharrow is an area of Sheffield characterised by very high levels of deprivation and correspondingly low levels of health. Studies of in-equality in health conducted at a ward level, (Thunhurst, 1985), show the ward to have the lowest life expectancy for men in the City and the second lowest for women.

The Health Project had been running for two years as part of the larger South Sheffield Community Project and employs two part-time development workers. Despite the efforts and clear intentions of the workers to do so, little progress has been made in involving members of the local community in the management of the project. It was agreed that this should form the focus of the Unit's initial involvement. Two sessions were held with the development workers during which they were encouraged to talk through their ideas about what community managers should and should not do. The ideas were clustered into a number of areas and a preliminary map of hopes and worries was produced.

Unfortunately, at this stage, one of the development workers, the primary contact, left the project. The other worker, and more particularly the chief development worker in the umbrella project, were less supportive of the Unit's involvement. Indeed, there was a feeling that this involvement was seen as a threat or insult to the capabilities of an experienced community development worker. Although not formally abandoned, work with the Sharrow Project has been discontinued.

Meanwhile, contacts were developed with other community health practitioners through the Sheffield Community Health Network. The Network was formed by local workers who were interested in discussing issues, providing mutual support and information, and campaigning on community health and alternative therapies. It is supported strongly by Healthy Sheffield 2000.

Heeley Community Health Project. Following a presentation of the Unit's work to an early meeting of the Network, the Unit was contacted by a general practitioner member of a steering committee, which was trying to establish a community health project in Heeley, another inner-city area of Sheffield.

Initially, the steering committee had seen the Unit's involvement centering on evaluation of the health project. However, after a first meeting, it was agreed that the Unit should also be involved in determining forms of community participation in the management of the Project and in fundraising. Shortly after our early meetings, the Project managed to raise sufficient funds to appoint a full-time development worker. During meetings, it was agreed that the Unit's involvement was welcomed by her, particularly in the areas of evaluation and community participation in management.

A wide range of methods and approaches have already been used in the evaluation of community health projects (see, Beattie 1990). It was agreed that the Heeley Project should develop a portfolio approach, using various methods for recording and tracking the progress and impact of the project. These include diaries, photographs and interviews. The Steering Committee has held regular evaluation sessions which are facilitated by staff from the Unit.

An important element of the evaluation was to try to establish aims and objectives for the project. Two sessions were held with members of the steering committee and local health practitioners and community workers. The sessions identified a number of possible activities and a range of aims for the project. More recently, Unit staff have worked with the project's development worker to look at ways of presenting and analysing the information which is being collected.

As in the Sharrow Project, some work has been done to help the steering committee to clarify their thoughts about community participation in the management of the project. Areas of decision-making and

Integral to the issues of evaluation and community invovlement is the need to determine health concerns in the local community. It was decided that it would be useful to develop a community health profile of Heeley.

Community Health Profile. In building a Community Health Profile, community operational research practitioners will work hand in hand with planners, community workers and the community at large, to produce a clear picture of the current state of health in an area.

A profile provides an ideal opportunity for the inter-agency, collaborative and participative approach advocated by the New Public Health Movement. Community Health Profiles can be an arena for the general population to express their views and concerns over health issues, thus empowering them both to decide what is important and to campaign for change in working towards WHO's 38 Targets.

Community Health Profiles are eclectic. They may include anything from the more traditional analyses of mortality, morbidity, immunisation rates, etc. through to audits of local amenities such as food available in shops, availability of public toilets, benches and local transport, as well as more qualitative analyses of data such as those factors which people feel contribute to their state of mental and physical health: pollution, noise, state of housing, danger from traffic, lack of knowledge on diet, unemployment, poverty, discrimination and childcare.

Not only is there a great potential for community operational research here to combine qualitative and quantitative data, but also to set standards for conducting community health profiles which, until now, have been individual and therefore not comparable.

## Strengthening Community Participation

The development of community health profiles provides clear example of the potential community participation strengthening role of community operational research. The repertoire and the balance of quantitative and qualitative analytic techniques provide a useful toolkit. However, as has been discussed more fully elsewhere, (Thunhurst, 1989) practitioners will need to consider issues of process with at least with the same priority as the enhancement and application of their technical base.

REFERENCES

Ashton J., and Seymour H., 1988, "The New Public Health", Open University Press, Milton Keynes

Beattie A., 1990, "The Evaluation of Community Development Initiatives in Health Promotion: a Review of Current Strategies", paper prepared for the HEA/OU Baseline Review of Community Development and Health Education (CDH), mimeo

Community Operational Research Unit, 1989, "Community OR and Community Health", CORU 89/1

Health Education Authority, 1990, "Community Participation in Health Promotion", Health Education Authority, London

Healthy Sheffield 2000, 1987, "Sheffield Health For All By The Year 2000 - Draft Targets", Sheffield City Council

Healthy Sheffield 2000, 1989, "HS2000 Strategy", Health and Consumer Services Department, Sheffield City Council

Phillips D.R., 1990, "Health and Health Care in the Third World", Longman, Harlow

Ritchie C., and Thunhurst C., 1990, How Useful Are Operational Research Methods and Techniques to Community Organisations, in: Centre for Voluntary Organisations (Eds) "Towards the 21st Century: Challenges for the Voluntary Sector", AVAS, London

Thunhurst C., 1985, "Poverty and Health in the City of Sheffield", Environmental Health Department, Sheffield City Council

Thunhurst C., 1989, "Core Health Measures for UK Healthy Cities", UK Healthy Cities Network, Liverpool

Thunhurst C., 1990, Community Operational Research - Thinking Globally Whilst Acting Locally, in: International Federation of Operational Research Societies (Eds) "Operational Research 90", Pergamon, London

Townsend P., Davidson N., and Whitehead M., 1988, "Inequalities in Health", Penguin, Harmondsworth

World Health Organisation, 1978, "Alma Ata 1978: Primary Health Care", 'Health for All' Series, No. 1., WHO, Geneva

World Health Organisation, 1981, "Global Strategy for Health for All by the Year 2000", WHO, Geneva

World Health Organisation, 1985, "Targets for Health for All: Targets in Support of the European Regional Strategy for Health for All by the Year 2000", WHO Regional Office for Europe, Copenhagen

# WHICH EVALUATION METHODOLOGY WHEN? A CONTINGENCY

# APPROACH TO EVALUATION

Amanda Gregory

Department of Management
Systems and Sciences
University of Hull
Hull, HU6 7RX

## INTRODUCTION

In the light of a project being undertaken by the National Association of Councils for Voluntary Service (NACVS) and Hull University, funded by the Leverhulme Trust, this paper seeks to show how an appropriate choice of evaluation methodology can be made by an organisation. To achieve this aim, the paper presents an analysis of the theoretical underpinnings of four types of evaluation and formulates a system of evaluation methodologies showing in what circumstances each may be most appropriately applied.

## THE FOUR FORMS OF EVALUATION

### Goal Based Evaluation

The goal approach to evaluation has its theoretical roots in the functionalist school of thought. Functionalism is characterised by belief that the organisation exists to serve some purpose and belief in the rationality of management. With regard to evaluation, incorporation of these principles results in a definition of effectiveness based upon the accomplishment of organisational goals. In practice several variants of the goal approach to evaluation can be identified, differing with respect to goal determination.

The most explicitly functionalist form of evaluation accepts the stated goals of management to be the legitimate goals of the organisation. However, the stated goals of management tend to be at a high level of abstraction, hence it is more usual to base a goal evaluation on operational goals as specified in management work-plans.

*Systems Thinking in Europe*, Edited by M.C. Jackson *et al.*
Plenum Press, New York, 1991

The stages in a goal based evaluation might include:
1. Determine the overall desirable goal state of the organisation.
2. Translate the goal state into a coordinated set of objectives and specify the period in they are to be achieved.
3. Select indicators by which achievement may be measured.
4. Express the goal state in terms of the chosen indicators.
5. Measure the end of period situation with regard to the goal indicators.
6. Compare the state of the indicators at the period end with the ideal expressed at the beginning of the period and thus assess the extent of goal achievement.
7. Investigate any mitigating circumstances which might have influenced organisational performance.
8. Formulate the next vision of goal state to be pursued.

## Systems-Resource Based Evaluation

The systems-resource approach to evaluation is based around the principles of structural functionalism in general and cybernetics in particular. Management cybernetics is the study and construction of organisational models which specify those relationships and functions necessary for an organisation to be viable.

Following the cybernetic approach of idealised model building, the systems-resource approach to evaluation involves the identification of multi-variate criteria by which the organisation's effectiveness can be judged. Thus this approach involves an expert rating the organisation, based upon his personal perception of the organisation's functioning, with regard to the characteristics specified in the ideal model.

The stages in a systems-resource based evaluation might include:
1. Appoint an appropriate 'expert' to undertake the evaluation.
2. Provide the evaluator with such information as he requests to enable him to judge the organisation upon the specified criteria.
3. Arrange a meeting at which the expert can reveal his findings and at which discussion might take place. The evaluator should be prepared to justify his ratings and advise the organisation on how it might improve its performance.
4. The organisation should seek to incorporate the evaluator's advice into its work plans and invite the evaluator to undertake a follow-up exercise six months hence to determine whether the organisation has improved.

## Multi-Actor Based Evaluation

The interactionist paradigm, in which the multi-actor approach has its theoretical foundations, implicitly assumes a subjectivist stance to social theory. Hence for interactionists, the social world is predominantly processual in nature. Interactionists concentrate on identifying differences in peoples' beliefs about the social world and facilitating a higher level of understanding between all those parties having an interest in the situation.

The multi-actor approach is based upon a definition of effectiveness which seeks to examine the organisation's ability to satisfy interested parties' needs over time. Thus in practice this approach is quite simple, for all one has to do is ask interested parties about what they want the organisation to do in the next period and about their degree of confidence in the organisation. However it is not quite that simple, for the satisfaction of constituent preferences may be unrealistic in the light of constraints, hence a certain amount of compromise must be agreed and take place.

The stages in a multi-actor based evaluation might include:
1.  Design a time-recording system aiming to capture what the organisation does currently.
2.  Maintain time-sheet records for an appropriate period based upon the decision-making cycle of the organisation.
3.  Identify all those groups having an interest in the organisation. Invite a significant number, by means of questionnaire, to express their opinions on their level of satisfaction with the organisation and their expectations of the organisation.
4.  Analyse the opinions expressed in the survey regarding expectations of the organisation and compare with actual time-sheet data for the period. Based upon the expectations of interested parties expressed in the survey formulate an ideal work-plan.
5.  Amend the ideal work-plan in the light of resource constraints.
6.  Plan a similar exercise to be undertaken in the next period to determine whether the organisation has improved and moved closer to the interested parties' ideal.

## Culture Based Evaluation

In addressing the question of what distinguishes living systems from non-living, Maturana and Varela (1980) propose that the fundamental characteristic of living systems is autonomy which, they state, is realised through the process of autopoiesis.

Based upon Robb's (1989) exploration of autopoiesis in the organisational sense, it may be claimed that organisations maintain a distinctive culture by autopoietic processes. Indeed, it is culture which distinguishes one organisation from another and may be said to bound the organisation with respect to its environment. At the component level, organisational culture is operationalised by role-taking. Upon entry to the organisation, individuals are fitted into impersonal roles which serve to ensure the organisation's survival beyond the involvement of the individuals with the organisation. Further, the dominant norms and values of the organisational culture are sustained by their being internalised by role occupants through continuous reinforcement via interaction with other role-takers.

Based upon this view of the organisation, an approach to evaluation can be formulated which concentrates on the bargaining relationship determining the level of individual moral commitment to the organisation.

The stages in a culture based evaluation might include:

1. Instruct each member of the organisation to answer questions relating specifically to such matters as whether they feel adequately rewarded for their work, how well they get on with colleagues, whether they feel at odds with their role within the organisation, and so on.
2. Arrange for groups of colleagues to meet to discuss the opinions surfaced in the first stage of the exercise and the perceived negative/positive aspects of organisational membership. Instruct each group to record any statements they find significant or about which there is disagreement.
3. Present each group with all other groups' statements of opinion.
4. Arrange for and lead discussion between groups on their views of the organisation, how that view differs from the views and expectations of others, and on how services and relationships might be improved in the light of the opinions expressed to generate more satisfaction for the members of the organisation.
5. Plan a similar exercise to be undertaken six months hence to determine whether the organisation has improved and moved closer to members' ideals.

CHOOSING AN APPROPRIATE EVALUATION METHODOLOGY

Given that there are several types of evaluation methodology available, some sort of contingency model is needed which matches the types of evaluation to the situations in which they might be employed.

The classification adopted rests on the observation that the need for different types of evaluation is a consequence of the differing characteristics of the Evaluation Group (EG). The EG is the party of people directly concerned with carrying out the evaluation and it may consist of representatives of various bodies having an interest in the organisation, for example - staff, management, funders, clients, etc.

Two key variables of character have been identified as being crucial to the form of evaluation adopted, the EG's orientation to evaluation and the organisation and the EG's variety, we shall consider these in turn.

The Evaluation Group and it's Approach to Evaluation

When formulating a classification of evaluation contexts, a distinction between subjective and objective outlooks, based upon Burrell and Morgan's (1979) work, was originally employed. However, in talking to people it has been found that this is quite a difficult concept to grasp, so in practice this variable has been replaced by a quantitative-qualitative dimension, detailed in Table 1, which people find easier to understand.

Table 1. Approach to Evaluation Data

|  | Quantitative | Qualitative |
|---|---|---|
| 1. Oriented towards investigating: | tangible reality | individual perceptions |
| 2. Organisation is to be seen as: | one of a type | a unique entity |
| 3. The evaluation is to serve the purposes of: | external parties | internal parties |
| 4. Evaluation results are to appear: | value-free | value-full |
| 5. The evaluation is to be: | an isolated exercise | a recurring exercise |

Based on the above variables an EG can be classified as best adopting either a quantitative or a qualitative orientation.

## The Evaluation Group and its Variety

The second variable we are taking as a determinant of the appropriate mode of evaluation in a given situation is the variety of the EG. Variety is a concept first introduced by the cybernetician Ashby (1956) to refer to the number of possible states a system can exhibit. The variety of the EG is made up of several factors, such as those detailed in Table 2.

Thus, EGs can be classified as being either of high or low variety and as per Ashby's 'Law of Requisite Variety' the evaluation undertaken must reflect the variety of the EG. In practical terms this has been taken to mean that a low variety EG should seek to attenuate the variety of the evaluation project by restricting its focus to internal matters, whilst a high variety EG's frame of reference should embrace the organisation's environment.

Table 2. Variety of the Evaluation Group

|  | High Variety | Low Variety |
|---|---|---|
| 1. Size of group: | medium | small/large |
| 2. Level of knowledge about evaluation: | high | low |
| 3. Importance of contingent decisions: | considerable | minor |
| 4. Availability of resources: | abundant | scarce |
| 5. Decision making ability of the group: | good | poor |

|  | | Orientation to Evaluation: | |
|  | | Quantitative | Qualitative |
| Variety<br>Of The<br>Evaluation<br>Group | High | Quantitative-High<br>Variety | Qualitative-High<br>Variety |
|  | Low | Quantitative-Low<br>Variety | Qualitative-Low<br>Variety |

Fig. 1. Classification of Evaluation Contexts

## A CLASSIFICATION OF EVALUATION CONTEXTS

Having defined what is meant by the term EG and having examined the two key variables of its character, a framework for classifying evaluation contexts can now be drawn-up. Combining the two classificatory dimensions of 'orientation to evaluation' and 'variety' results in the production of a four-celled matrix as in Figure 1.

## CLASSIFICATION OF APPROACHES TO EVALUATION

Based upon the classification presented in Figure 1 and the analysis and definition of the different forms of evaluation, the evaluation types can now be fitted into the classification to give a contingency approach to evaluation as shown in Figure 2.

The goal approach was felt to be most appropriate in low variety EG situations because the organisation in isolation from its environment is taken as the system in focus for the evaluation. This approach is quantitative because numerical indicators are usually employed in assessing the extent of goal achievement.

The systems-resource approach is appropriate in high variety EG situations because it exploits the ability of the EG to explore the environmental context of the organisation and the restraints it has imposed upon it by other organisations with which it interacts. This approach is quantitative in orientation because the evaluation results are based upon an expert's numerical rating of the organisation against certain critical characteristics.

|  | | Orientation to Evaluation: | |
|  | | Quantitative | Qualitative |
| Variety<br>Of The<br>Evaluation<br>Group | High | Systems-Resource<br>Based Approach | Multi-Actor<br>Based Approach |
|  | Low | Goal Based<br>Approach | Culture<br>Based Approach |

Fig. 2. Classification of Approaches to Evaluation

Multi-actor evaluation fits into the high variety class because the interested parties involved will be predominantly from bodies external to the organisation and together they will embrace the whole of the environment. This form of evaluation's emphasis on the surfacing of different interested parties' level of satisfaction and expectations gives the approach a distinctive qualitative orientation.

It was felt that culture based forms of evaluation might fit into the low variety EG class due to their internal focus. Furthermore, culture based evaluation has been placed into the qualitative orientation category due to its focus on matters pertinent to the individual and its emphasis on communication and relationships.

CONCLUSION

In this paper, the principles guiding the project being undertaken by Hull University and NACVS have been set-down. Firstly, we looked at the theoretical foundations of four forms of evaluation. We then reviewed the different quantitative/qualitative and high variety/low variety orientations capable of being exhibited by an Evaluation Group. From this point, we were then able to form a classification of evaluation contexts. Finally, we related the different evaluation methodologies to the different evaluation contexts.

REFERENCES

Ashby, W. Ross, 1956, An Introduction to Cybernetics, Methuen, London.

Burrell, G., & Morgan, G., 1979, Sociological Paradigms and Organisational Analysis, Heinemann, London.

Maturana, H. R., & Varela, F. J., 1980, Autopoiesis and Cognition: The Realization of the Living, Reidel, Dordrecht.

Robb, F. F., 1989, The Application of Autopoiesis to Social Organizations - A Comment on John Mingers' "An Introduction to Autopoiesis....", Sys. Pract., 2:343.

# THE SYSTEMATIZATION OF PRACTICE

René Victor Valqui Vidal

The Institute of Mathematical Statistics and Operations Research
Build. 321, The Technical University of Denmark
DK–2800 Lyngby, Denmark

## INTRODUCTION

All over the world exists a great variety of oppressed groups' practice in connection with their struggles to survive and participate in the historical process of creating an alternative society. These activities are usually developed in closed interaction with professionals of different qualifications. The point of departure is usually a concrete problem to be solved. This practice usually contains many activities related to organization, planning, analysis, engineering, decision–making, evaluation, etc. These practices are usually in the academic world denominated as alternative consulting work, anti–capitalistic practice, or emanicipatory research, to emphasize their critical, radical and socialist perspective.

We will point out that such kind of work has been going on in Scandinavia, in what concerns the work of computer scientists with trade unions, and in Latinamerica, in what concerns the work of social workers with popular sectors in shanty towns, for many years. Later, such kind of alternative practice has also arised in England under the name of Community Operational Research.

Our discussions in this paper will be kept at two different levels one that is related to consulting work in general as an alternative practice and a second that is more related to systems work.

The purpose of this paper is two–fold:

a)  to motivate a discussion within the system's community around problems that will impulse the search for a conception (theoretical, methodological and practical) of an adequate <u>systematization</u> of alternative practice of our profession, and

b)  to communicate to the system's community the experiences and achieve– ments from Latinamerica, in what concerns the alternative practice of social workers, that are transferable to our own discipline.

In this paper general ideas and concepts will be discussed and many questions will be formulated with the objective to show that to perform a systematization is a very complex task but a necessary one to avoid "practicism" and "empiricism".

*Systems Thinking in Europe*, Edited by M.C. Jackson *et al.*
Plenum Press, New York, 1991

# THE ELEMENTS OF ALTERNATIVE CONSULTING WORK

The starting point of any consulting work is a client having a problem, and a group of professionals (consultants) willing to suggest a solution. More specifically in alternative work the client is usually denominated as "an oppressed group" as for instance a local trade union, a local community, a popular sector, etc. The problems to be solved are of different nature and content, usually there are problems of diagnosis, planning, design, or organization. The formulation and solution of such problems usually demand some expertise that will be provided by the consultants (OR (Operational Research)—workers, System Analysts, Computer Scientists, Social Scientists, Social Workers, Economists, Lawyers, Doctors, Engineers, etc.). For the sake of concreteness, let us first see three real—life activities that are regarded as alternative work.

## The Scandinavian Approach to Systems Design

This kind of alternative consulting work has been going on for the last 20 years. The clients in these projects have been trade unions both at the local and the central level. The main problems have been: how to increase the workers' in—fluence on new technology, how to develop independent union activities in design and use of computers and other kinds of new technology, and how to perform systems design within the context of democratization. The consultants have been primary computer scientists having research positions at universities or alike research institutions. This kind of activity can be considered as an alternative in two senses. First, it is an alternative to the well—known "socio—technical approach" that it was regarded as anti—trade union and even anti—democratic in its practice. Secondly, it is an alternative to the traditional way of designing computer systems. In this respect a "negotiation model" has been developed for independent trade union investigatory work and participation in management pro—ject groups.

As far as I know a "systematization" has never been performed within this tradition. Many papers, reports and books have reported practical experiences or methodological achievements that are elements of a systematization but they have not reflected their social praxis as consultants and socialists, specially in what concerns their work with social—democratic trade unions.

## Community Operational Research

Critical analysis of the limited role of OR in Society emerged in England and Denmark in the early 70's. Critical OR was limited in its practical activities. In 1988 the Community OR Unit was launched in England by the Council of the Operational Research Society. It has been established as the result of an initiative taken by Jonathan Rosenhead during his Presidency of the OR Society in England, over the period 1986–87, and as a part of a re—evaluation of the direction and role of OR. The clients are typically organizations with several of the following characteristics: relatively small financial resources, not possessing a managerial hierachy, not producing goods or services for sale, operating through consultans or internal democracy, etc. The consultants in these projects are operational researchers and systems analysts belonging to the "soft" side of these disciplines.

Probably it is too early to begin with a systematization of this praxis, but a very important point to notice is the broad motives of the consultants to partic—ipate in this work. Some have a humanistic approach while others have a clear formulated critical and radical approach.

One of the interesting aspects of Community OR is its relation to the traditions of Participatory Research as a way of working with deprived communities in developing countries. Participatory Research has as principles the collective investigation, analysis and action with the active participation of the constituency in the entire process. One of the most important area in which work has already begun has been in the selfmanagement of housing and community health groups.

## Alternative Social Work in Latinamerica

In many countries in Latinamerica, professionals have been working together with oppressed groups in connection with their struggles to survive and participate in the historical process of creating an alternative Society. CELAT (Centro Latinoamericano de Trabajo Social) is a center for Latinamerican social workers performing this kind of alternative work. There is a lot of practical work done during the 80's, some examples are:

- community health in Honduras
- small firms and community development in Costa Rica
- popular education in El Agustino, Peru
- women organisation in popular sectors in Quito, Ecuador
- community health and popular education in a marginal sector, Lima, Peru.
- etc.

The clients are here oppressed groups, living in shanty towns under bad conditions and having not many basic facilities. Thus there exist in Latinamerica a great deal of popular praxis to which professionals are connected essaying to contribute in the popular projects. One of the most important activies of CELAT has been the organization of such praxis in a systematic way with the purpose to give some elements of analysis to enrich the development of the popular organizations and the qualifications of the professionals that are working with the popular sectors. It is within the work of social workers in Latinamerica that discussions and experiences with systematization of practice has reached their highest level. One of the purposes of this paper is to transfer these knowledge from the under-developed world to the industrialized countries.

## Some Questions

After these short discussions of alternative projects, we will formulate some fundamental questions, that are seldom discussed, dealing with the political, sociological and epistemological aspects of alternative consulting work. The answer to these questions will force the praticioners to become more aware of the nature and purpose of alternative work, this process of "consciousness" is a necessity because alternative work is much more complex and demanding than traditional consulting work.

Some of these questions are the following:

- When a client can be considered as "an oppressed group"? What level of political awareness is demanded from the oppressed group?
- Which kind of qualitifications are demanded from the consultants? What level of political awareness is demanded from the consultants?
- What are the real motives of both the clients and the consultants to go through such an alternative problem solving process? What are the potential and the real conflicts in such a social process both of political, sociological and methodological nature?

It is usually assumed within the Social Sciences, OR and related Systems Sciences, and in common talk that consulting work is primarily related to practice, a kind of social engineering. In a process of reconceptualization it is very important to emphasize that in any consulting work we should have reflexion and theory about the actions taken within the actual problem solving process. Consulting work is for us not only pure action, it is also, consciously or unconsciously, the result of a conception and a diagnosis of the reality where one wants to intervene (theory). There is no practice in a profession, even the most routinized or institutionalized, that it does not have a therotical conception about the situation one wants to affect. Within the traditionally consulting field this theoretical conception is neither explicitly formulated nor consciously reflected by the professionals performing the consulting work.

Consultants together with a proposal of how a given problem ought to be solved, they also have a view–of–the world and a rationality of how to tackle the situation that they communicate in direct or undirect form to the clients. These two dimensions are always present in consulting work. In addition, the consultants are usually professionals that are members of institutions or organizations and their work are restricted within the limits and possibilities of their institutions or organizations. Thus usually different objectives and motives are present and many contradictory processes might be generated. The systematization has to reconstruct and make explicit these to dimensions of the consultants practices, as well as the process and contradictions that are developed in the interrelations and interpretations of the different objectives that are present in the reality to be transformed.


SYSTEMATIZATION: POTENTIALITIES AND DIFFICULTIES

A fundamental criticism that can be formulated to the intents to systematize practice can be connected to the logical difficulties when essaying to go from a particular case to a generalization without falling in the trap of empiricism.

A given practice of a community or a professional has two fundamental charateristics:

a)   Each pratice has its own particular combination of features in its development, thus two similar situations within a similar context might be approached in different ways.

b)   On the other side, the consulting practice with a given group is not social praxis. They can only give a report of the activities and advances of research but not a theoretical document. Theoretical work can only be alaborated based on the action of the classes, not as isolated persons or groups of individuals. The practice of no–classes (f.ex. consultants) are related to theory and praxis as far as their actions are understood, in their articulation and significance with the praxis that the class is developing.

Therefore, it is impossible to pretent a theoretical reflexion only based in a particular practice. This means that the first task in a systematization is the articulation of this practice with the social praxis of a class. This will permit to relate this particular practice with the theory that is directly related to the class–praxis to which it corresponds. For instance, a consulting work with popular organizations to obtain food is a practice related to the strategies for survival of the popular class in a Society in serious economic crisis. Only the practice of the fundamental groups of Society (the classes) are the ones that might have social praxis. It is to these activities in Society that the fundamental texts on dialectics are referring to and it is from the handling of these classes that theory can be obtained and knowledge generated, if we find the appropriate methodology.


WHY SYSTEMATIZATION?

We believe that it is very important to support the systematization of alternative practice within our profession due to two main reasons:

a)   The consultants working with a popular sector are usually rather absorbed by the daily problems of the practical work. In most cases, there are very few possibilities to perform a reflexion about the problems mentioned in the last sections. The systematization gives the possibility to reflect more theoretically about a practice, locating it in its context, analyzing it, and re–thinking the developed work, the methods applied, the problems and contradictions, and how they were solved, etc., and from these to plan and reorientate the future actions. Consequently, the systematization has a value for each particular practice, because it permits to give a retrospective look to the past, and taking departure on what has been learned to reorientate the future. It is in this way that it permits the improvement of practice of the consultants, doing it more efficiently and effectively in relation to the formulated objectives.

b)   There is very little literature about the above mentioned problems and the obtained experiences are not broadly diffunded. This means a limited reflection about consulting practice. The systematization pretents to contribute to the superation of this weakness. It is desirable to have a communicative and sharable practice that will increase the possibilities of collective learning and reflection.


WHAT IS SYSTEMATIZATION?

Now, we have sufficient background to give a more precise conception of what we mean by systematization. A systematization describes, arranges and reflects analytically the development of a practical experience of an alternative consulting work with respect to the following aspects:

a)   Theoretical and methodological
b)   Contextual, both institutional and historical–social
c)   The interactions among the various actors that are participating in the practice: professionals and popular sectors
d)   The social processes that are carried through, including an analysis of the elements that facilitate and/or support their development as well as those that make difficult the planned actions
e)   The results of the experience
f)   The development of some generalizations that can be extracted from the practice.

It is important not to confuse a systematization with an assessment. An assessment is the evaluation of the actions taken in a practice, as for instance the evaluation of achieved goals, the suitability of the methods, etc.. A systematization will usually also include an assessment but what is more important is the other mentioned elements. Some of the essential questions are:

- What was done? How? Why?
- What influence had the political–economic–social context in the actual practice?
- What was the relation between the professionals and the popular sectors?
- Which social processes were generated by the practice?
- Which contradictions influenced the practice or were generated by the practice? How were they solved?

## CONCLUSIONS

Focusing in alternative consulting work in both Scandinavia and Latinamerica we have argued for the necessity of a systematization of alternative practice within the systems movement.

In this connection we have a lot to learn from the experiences of social workers in Latinamerica while dealing with the problems of popular sectors. A task of this paper has been to transfer some of this knowledge and experiences from Latinamerica to Europe.

Paradoxically, while traditional system's workers in Europe and USA are talking and discussing about how to transfer their discipline to the Third world, here we are doing the opposite that is transferring experiences with alternative consulting work from Latinamerica to Europe.

It is very important to diffuse the performed systematizations at three different levels:

- to the popular organizations that have participated, so that the adquired experiences can be used in their future struggles,
- to our colleagues of the system's community to enrich our profession and collect experiences about alternative work, and
- to other processionals of different disciplines to promote similar activities and interchange experiences.

An extended version of this paper with the title: "OR and Social Praxis", has appeared in "Liber Amicorum for Arne Jensen, 16th February 1990", IMSOR, DTH, 1990.

## REFERENCES

[1]   Bjerknes, G. et al, Computers and Democracy – A Scandinavian Challenge, London 1986.

[2]   Thunhurst, C., Community Operational Research, In Systems Prospects: The next ten years of Systems Research, Flood R.L. (Ed.) Plenum, 1989.

[3]   Jackson, M.C., Community Operational Research: Purposes, Theory and Practice, DRAGON, Vol.2, No. 2, 1987, pp 47–73.

[4]   CELAT, La Sistematización de la Práctica, cinco experiencias con sectores populares, Lima, 1986. (In Spanish).

[5]   Vidal, R.V.V. (Ed.), OR–epistemology: some essays, IMSOR, DTH, p. 120.

TIMEO DANAOS ET DONA FERENTES: A PHILOSOPHICAL-CUM-

EPISTEMOLOGICAL CRITIQUE OF THE CRITICAL SYSTEMS PERSPECTIVE

Haridimos Tsoukas

Warwick Business School
University of Warwick
Coventry, CV4 7AL

The purpose of this paper is first to outline a synopsis of the main tenets of the critical systems perspective (CSP), and second to provide a philosophical-cum-epistemological critique of its claims. The thesis that will be advanced here is largely unsympathetic towards CSP. The latter is beset by a number of philosophical-cum-epistemological, logical, and sociological errors which render its central claims problematic. However, in the extremely limited space of this article only the philosophical-cum-epistemological aspects of CSP will be considered.

## THE CRITICAL SYSTEMS PERSPECTIVE: A SYNOPSIS OF ITS CLAIMS

What principally differentiates CSP from the positivist systems perspective (PSP) and the interpretive systems perspective (ISP) is its self-proclaimed "critical" nature. The term "critical" has been construed in two ways. The first interpretation refers "to an explicit state of awareness which reminds us, in the face of incomplete knowledge, that we must never accept knowledge or methodological output as total or absolute" (Flood, 1990a:331). In this sense "critical" is identical to critical attitude, self-reflection, awareness of hidden presupositions, disclosure of assumptions of various perspectives.

The second interpretation is more radical. Critique here means "oppositional thinking, an instrument for fighters and resisters" (Flood, 1990b:66) that denies the purely instrumental rationality of positivist science, in favour of the emancipatory rationality of critical science. An emancipatory rationality strives for, and operates within, an "ideal speech situation" (Habermas, 1987) within which free, undistorted communication between equally empowered agents takes place, and consequently a consensus is rationally arrived at. In this sense "critical" means liberation from repression, emancipation, concern with equality and justice, fullfilment, empowerment, absence of false consciousness and alienation (Fay, 1975; Flood, 1990b; Jackson, 1987,1990a,1990b).

The preceding terminological dualism is reflected in the parallel development of two views within CSP: the contingency view and the fundamentalist view. A contingency view of CSP sees the latter as being chiefly concerned with either the disclosure of the assumptions of various problem-solving methods, or with the classification of problem situations in such a manner so that an informed methodological choice can be made (Jackson and Keys, 1984; Keys, 1988; Jackson, 1990b). For this purpose various typologies have been suggested.

A fundamentalist view of CSP consists mainly in a radical critique of both PSP and ISP. It has been argued that positiv- ist science, best exemplified by hard systems thinking and traditional O.R., conceives of systems and problems as things in themselves with an objective existence and unproblematica- lly transparent functions, causes, or purposes. Consequently, problem situations are depoliticised and dehistoricised, and means-ends relationships are neutralised. The status quo is implicitly accepted and alternative social arrangements are not considered (Mingers, 1980; Oliga, 1988; Jackson, 1990a, 1990b; Flood, 1990a, 1990b).

The interpretivist perspective, exemplified by soft systems thinking, interactive management planning and problem modelling, is similarly castigated for favouring regulation and the status quo instead of advocating radical social change. In addition, the alleged inability of ISP to deal with actors' potential false consciousness and distorted communica- tion is highlighted. Systematically distorted communication, it is argued, stems from socially embedded coercive contexts and structurally determined power differences among actors, which the interpretivists find unable to explain, let alone to change (Jackson, 1990b; Flood, 1990b, Flood and Ulrich, 1990; Oliga, 1988; Mingers, 1980).


A PHILOSOPHICAL-CUM-EPISTEMOLOGICAL CRITIQUE OF CSP

The critical systems perspective is beset by a number of problems. Below we shall provide a critique of the central assertions of the CSP from a philosophical-cum-epistemological angle.

1. Insofar as CSP is couched in terms of scientific dscourse, its proponents attempt to advance an arbitrary ideological position, namely that of "emancipation", which can be articu- lated (and evaluated) only in terms of a meta-language which is unavailable within the domain of science. Modifying Godel's Incompleteness Theorem, it can be said that any language always contains a number of undecidable propositions -- hence the need for a meta-language (Hofstadter, 1979; Beer, 1985). As Toulmin et al (1984) have demonstrated, all arguments are ultimately based on some sort of "backing" which serves the purpose of grounding and closing the argument; but a "claim" can be evaluated relatively independently from the "backing". The latter serves the purpose of a meta-language; it is a

fundamental worldview which is taken for granted but which in itself cannot be rationally justified.

The point made here is that while science is certainly not grounded transcendentally, it can function as a quasi-autonomous activity without concerning itself with the foundation of its norms. Science developed by enlarging the area of knowledge which was not subject to arbitrary judgement, thanks to the Platonic distinction between knowledge and opinion, episteme and doxa. As Kolakowski (1978:394) has remarked in his criticism of Habermas, "decisions as to good and evil and the meaning of the universe cannot have any scientific foundation; we are bound to make such decisions, but we cannot turn them into acts of intellectual understanding. The idea of a higher reason synthesising these two aspects of life can only be realised in the realm of myth, or remain a pious aspiration of German metaphysics".

2. CSP supporters want to see the means-ends distinction abolished and are highly critical of the lack of a rationally achieved consensus with regard to norms and values (Habermas, 1987; Fay, 1974; Mingers, 1980; Flood, 1990b). They fail to understand two things. First, the institutionalisation of central societal values occurs in a manner that is not totally transparent to the members of society. Rationality itself cannot be rationally justified any more than other valuable ideas such as autonomy, democracy, freedom, property, justice or the rise of monotheistic religions can. All these are significations, cultural creations defying rational explanation, and not the discovery of reality behind the appearences (Castoriadis, 1975; Rorty, 1989). Central cultural significations demise and emerge in an arational manner, although the process of their formation can be reconstructed.

Second, the idea that human beings are able to rationally shape the world around them according to their wishes is wrong and dangerous. It is wrong because our reason is as much the result of an evolutionary selection process as is our morality. Like other traditions, the tradition of reason is learnt and culturally transmitted, is not innate. Reason cannot be the panoptical Zeus, and therefore need not be privileged, since reason itself is an endowment conferred on us by cultural evolution (Hayek, 1988).

The idea that one can redesign morality is fatally dangerous. Morals stand between instinct and reason; they are not the rational creation of human beings and they are transmitted by cultural evolution. As Hayek (1988) has remarked, learnt moral rules have been institutionalised not because human beings recognised by reason that they were better, but because they made possible the growth of an extended order exceeding anyone's vision, in which more effective collaboration enabled its members, however blindly, to maintain more people and to displace other groups. Critical systems thinkers, along with positivists, rationalists, and -alas!- socialists, fail to appreciate the enormous importance of moral practices because their epistemology is such that whatever is not scientifically proven, or is not fully understood, or lacks a fully specified purpose, or has some unknown effects, is unreasonable (Hayek, 1988). Although morality shapes scientific practices at any point in time, morality itself cannot be rationally redesigned --not, at least, without disastrous consequences. It seems that the price for Promethean scientific arrogance is Socratic moral humility.

3. Even if the ideological premise of "emancipation" is accepted, proponents of CSP have yet to answer the question "by what criteria are we to judge whether "emancipation" consists in one state of affairs or in another"? One possible answer is that the acquired knowledge is in itself emancipatory since it allows someone to uncover what presence conceals, and consequently free someone from previous biases, prejudices, or psychic perturbations (Fuenmayor, 1990).

Undoubtedly the above can be said of psychoanalysis, but is it valid to transpose this argument to units of analysis above the level of the individual? If the answer is 'yes', it presupposes a coherent, unitary subject willing to be "liberated". Can social institutions be considered as subjects in this sense? Who is to judge whether they have been "liberated"? According to what criteria? Given that a result of CSP's viewing of science as an ideologically motivated discourse is the absence of formal criteria for the evaluation of knowledge claims, one is left with the use of extra-scientific quarantors for the settlement of scintific disputes. If one places the latter against a larger social background, it would not be a far-fetched conclusion to say that "there must be some compulsion other than the rules of thought, and that must take the form of social repression" (Kolakowski, 1978:418). Any similarities of such a scenario with totalitarian regimes around the world is certainly not a coincidence. There is an intrinsic link between totalising discourses and totalitarian solutions (Popper, 1961, 1966).

4. Critical social science, and CSP in particular, suffer from an essentialist conception of human beings. Interestingly, they conceive of the latter in a peculiarly individualist and asocial manner: humans would exist in a primordial state of autonomy, fullfilment and communication competence if it was not for the corrupting effects of society. The structure of the narrative is essentially religious: the fall of Adam and Eve from paradise to the unequal, unjust and distorting society. But why is it the case? Through what mechanisms do the distortion and repression of an otherwise pure "communication domain" take place? How do we know what is the essence of human beings and of their "communication domain" anyway? What constitutes "true consciousness" which is somehow distorted into "false consciousness"? By virtue of what criteria does "false consciousness" constitute one state of affairs rather than another? How do we have access to such knowledge?

What CSP proponents assume is their privileged linguistic access to an essentially speechless world. The idea of human beings having a certain language-independent nature is difficult to sustain. Even if human beings had such nature there would be no way we could know of it. As Rorty (1989:7) has observed "if we could ever become reconciled to the idea that most of reality is indifferent to our descriptions of it, and that the human self is created by the use of a vocabulary rather than being adequately or inadequately expressed in a vocabulary, then we should at last have assimilated what was true in the Romantic idea that truth is made rather than found. What is true about this claim is just that <u>languages</u> are made rather than found, and that truth is a property of linguistic entities or sentences".

5. The totalising discourse of the CSP renders it unable to distinguish between different types of social research having their own legitimate boundaries of inquiry. Following Whitley (1989), social research can be classified along two dimensions. The first dimension is the proximity of social research to curent organisational practices and rationalities, and the second is the degree to which social research is concerned with the generation of causal explanations.

In the preceding matrix, PSP falls in the low proximity to current practices-low concern for causal explanation cell since it attempts "to identify regularities between relatively abstarct properties of idealised objects which are remote from commonsense descriptions" (Whitley, 1989:21). ISP falls into the high proximity to current practices-low concern for explanation quadrant since its concern is with the initiation and establishment of organisational practices which seem to be successful in current circumstances without systematically attempting to explain why they are sucessful. Finally, CSP falls into the low proximity to current practices-high concern for explanation cell given its interest in the identification of causal mechanisms responsible for the generation of particular outcomes in particular circumstances. However, CSP deals with highly abstract entities which makes its immediate application in current organisational contexts more difficult.

Considering the preceding matrix and the classification of the various systems perspectives in it, it is clear that much of the criticism directed by proponents of CSP at the interpretivists is clearly misdirected. The boundaries of the interpretivits' inquiry simply do not include a high concern for explanation, just as the CSP's boundaries do not include a concern with immediate application in current organisational contexts. There can be no apriori argument in favour of the one or the other; it's just the way it is!

CONCLUSIONS

All the abovementioned weaknesses of CSP stem from two fundamental premises. First, CSP accords reason an omnipotent and omniscient character which it does not possess, since reason itself is a cultural endowment and not an innate human characteristic. The idea of reason as the ultimate arbiter gives rise to the projection of supposedly rational ultimate values (i.e. emancipation) which cannot be evaluated within the domain of science. Secondly, CSP is a totalising discourse unjustifiably asserting its privileged knowledge of human nature, axiomatically accepting the homogeneity of social institutions in need of "emancipation", and unable to see the analytical differences between different types of social research. For all the preceding reasons, the contribution of CSP into systems thinking and problem management is highly questionable.

REFERENCES

Beer, S., 1985, "Diagnosing The System For Organizations", J.
Wiley, Chichester
Castoriadis, C., 1975, "L' Institution imaginaire de la socie-
te", Seuil, Paris
Fay, B., 1975, "Social Theory and Political Practice", G.
Allen & Unwin, London
Flood, R., 1990a, A new decade a new spirit, Syst. Pract.,
3:331:335
Flood, R., 1990b, Liberating systems theory: Toward critical
systems thinking, Hum. Rel., 43:49-75
Flood, R. and Ulrich, W., 1990, Testaments to conversations on
critical systems thinking between two systems practitioners,
Syst. Pract., 3:7:30
Fuenmayor, R., 1990, Systems thinking and critique: I. What is
critique?, Syst. Pract., 3:525-544
Habermas, J., 1987, "The Theory of Communicative Action",
Polity Press, Cambridge
Hayek, F.A., 1988, "The Fatal Conceit", Routledge, London
Hofstadter, D.R., 1979, "Godel, Escher, Bach: An Eternal
Golden Braid", Penguin, London
Jackson, M.C., 1987, New directions in management science, in:
"New Directions in Management Science", M.C. Jackson and P.
Keys, eds., Gower, Aldershot
Jackson, M.C., 1990a, The critical kernel in modern systems
thinking, Syst. Pract., 3:357-364
Jackson, M.C., 1990b, Beyond a system of systems methodol-
ogies, J. Opl. Res. Soc., 41:657-668
Jackson, M.C. and Keys, P., 1984, Towards a system of systems
methodologies, J. Opl. Res. Soc., 33:473-486
Keys, P., 1988, A methodology for methodology choice, Syst.
Res., 5:65-76
Kolakowski, L., 1978, "Main Currents of Marxism", Oxfod Uni-
versity Press, Oxford, Volume 3
Mingers, J.C., 1980, Towards an appropriate social theory for
applied systems thinking: Critical theory and socft systems
methodology, J. Appl. Syst. Anal., 7:41-49
Oliga, J., 1988, Methodological foundations of systems method-
ologies, Syst. Pract., 1:87-112
Popper, K., 1961, "The Poverty of Historicism", Routledge &
Kegan Paul, London
Popper, K., 1966, "The Open Society and its Enemies", Routledge & Kegan Paul, London, Vol.1
Rorty, R., 1989, "Contingency, Irony, and Solidarity", Cambridge University Press, Cambridge
Toulmin, S., Rieke, R., and Janik, A., 1984, "An Introduction
to Reasoning", Macmillan, New York, 2nd Edition
Whitley, R., 1989, Knowledge and practice in the management
and policy sciences, Working Paper, No.174, Manchester Busi-
ness School

INFORMATION SYSTEMS

INFORMATION SYSTEMS: INTRODUCTION

What is an information system?    Is it possible to define precisely
an information system?    Is it necessary or even possible to have one
definition of an information system which everyone can agree with?

It would seem reasonable that any editorial introducing papers
concerned with the nature and purpose of information systems, would
begin by answering the questions posed above.    It is immediately
obvious when commencing, however, that these questions are impossible
to answer.    There is no one definition of an information system which
is agreed by academics and practitioners working in the area and the
papers presented here reflect some of the different perspectives taken
when studying their nature and purpose.    Most commercial organisations
using what they would refer to as management of information systems tend
to view them as being essentially technical systems comprising of
computer hardware and software and one of the major criticisms of the
design of such systems has been that the design has been treated as a
technical exercise normally giving low priority to user requirements
and interface issues.    The papers here recognize that information
systems should be viewed from a wider perspective leading to the
acknowledgement that an information system consists of not just
information technology but also people.    Given this recognition,
issues concerning effectiveness and communication will involve
consideration of the use of natural language, semantics and logic.
One area of considerable interest, and reflected in the papers
presented here, is the consideration of how knowledge and expertise
can be modelled for use in information systems.

Knowledge elicitation and representation are generally recognized
as the stages in the development of knowledge based systems which
cause most problems and have stimulated great interest in recent times.
They reflect the stages crucial to the development of the boundary and
interface between the people and the technology and the approaches and
methods advocated in the papers presented here illustrate the benefits
of using systems ideas and paradigms.

The papers naturally take a systemic approach in attempting to
solve current issues concerning the nature and purpose of information
systems and this leads to taking a much wider definition of information
systems which includes not only constituent technology but also the
people involved with them.    One of the most disquieting advances
recently, however, is the advocation and adoption of formal methods
for specification of systems.    This could make the designer-user gap
even wider than previously by producing and implementing specifications

which the user will not understand and in the quest for formality
could lead the designer to overlook, even more, the user requirements
of the system.

Users need information to support them in the performance of tasks,
and the content and structure of data held in technical systems must
correspond to the users' requirements and expectations.

THE NATURE OF INFORMATION SYSTEMS AND

THE ROLE OF PRACTICE

D. E. Avison* and G. Fitzgerald+

*Department of Accounting & Management Sciences
Southampton University
Southampton SO9 5NH, UK

+Templeton College
Kennington
Oxford, OX1 5NY, UK

INTRODUCTION

This paper explores the nature of information systems,
in particular the academic discipline of information
systems. Its purpose is to further the debate on the
academic basis of information systems. We highlight the
role of practice on information systems education and
address some of the issues that this raises.

The discipline of information systems has been defined
as the effective analysis, design, delivery and use of
information and information technology in organisations and
society. The implication of such a definition is that
information systems concern not just the implementation of
computer systems but also, for example, manual information
systems and informal information systems, and issues
relating to people, organisations, and society are
relevant.

However, the view that information systems is largely
a technology-oriented domain predominates in information
systems education and research, and this is partly due to
the influence of practice.

THE FOUNDATION DISCIPLINES

From the starting point of the above wide definition
of information systems, there would seem to be a number of
foundation disciplines relevant to information systems. In
their paper, Liebenau & Backhouse (1989) stress sociology
and semiology. The former, they argue, constitutes the main
thrust of research into social organisation, institutional
dynamics, group interaction, working conditions and social
policy, whereas semiology brings together the range of
studies associated with contexts of language and

communication, meanings, grammars signs and codes. To quote
from Stamper (1987):

> 'An arrangement of computers linked by a
> telecommunications system does not constitute
> an information system. Such a technical system
> only processes information in the sense that
> it operates upon structures of symbols that
> are intrinsically meaningless..... Meanings
> are only conferred upon the symbols
> manipulated by computers when they are
> interpreted by the social system within which
> the technical system is embedded. When an
> organization uses a computer it must in some
> way or other establish the meanings of the
> signals being processed electronically, which,
> in a nutshell, is why semantics is one of the
> central research areas in our nascent
> discipline'.

Thus it would seem that the study of semantics and
semiology would be a crucial reference discipline for
information systems. However we disagree with those who
argue that these are the only fundamental bases of
information systems.

Placing too rigid boundaries on information systems
reduces its richness and effectiveness. A stress on
sociology and semiology may be just as narrowing as the
stress on technology. There are other foundation
disciplines. As well as sociology, semiology and computer
science and information technology, they include systems
theory, politics, ethics, psychology, anthropology, applied
psychology, ergonomics, linguistics, economics and
mathematics.

In a later paper, Backhouse, Liebenau & Land (1991)
follow-up the original article and include comments from those
who had responded to their original call for views.

Some respondents felt that the technology is the
fundamental basis of information systems. Although
Backhouse, Liebenau & Land do not go to this other extreme,
they do seem to have taken a position somewhat nearer the
technological end of the continuum than in their earlier
article.

Again, we argue that information technology is an
important part of information systems but by no means the
only aspect of importance or indeed the most important, as
is often implied. We prefer to avoid describing information
systems in terms of a position on a continuum, but as a
position on some kind of multi-dimensional matrix.

The influential factors in information systems do not
relate directly to the technology itself but the ways in
which it is used and to organisational and human factors.


THE ROLE OF SYSTEMS THINKING

Although Liebenau and Backhouse mention some of the
other foundation disciplines listed above as contributing
to information systems (amongst these they emphasise
economics and anthropology), and are well aware that
information systems should not be primarily concerned with
the technical components such as computer based networks,
it is strange, perhaps, (and particularly disappointing to

460

the present audience) that no mention at all is made of
systems thinking. Despite the influence of systems thinking
in many disciplines, it has obviously not made an impact on
all members of the information systems community.

Avison & Wood-Harper (1990), Wilson (1990) and others
are attempts to show the relevance of systems thinking and
soft systems to information systems. However, many
information systems practitioners in particular, but also
some academics, still see systems thinking as irrelevant to
information systems. This view is, perhaps, slowly
changing, as evidenced by recent issues of practitioners'
magazines such as Computing (Miles, 1991) and the British
Journal of Healthcare Computing (Vidgen & Hepworth, 1990).


THE INFLUENCE OF PRACTICE ON TEACHING AND RESEARCH

There is a view that 'what practitioners do is the
discipline' and it is certainly true that practice has greatly
influenced education and research. This influence is by and
large positive, but one of the results of this emphasis is that
the theoretical basis for the discipline has not been well
debated or defined.

If we look at the history of information systems, we see
the influence of practice and the emphasis on technology.
Originally systems analysts developed computer applications for
isolated operational activities. Practice has advanced so that
applications addressed the totality of operational needs of the
organisation, usually supported by databases, and thence to
management information systems (MIS), decision support systems
(DSS) or executive support systems (EIS) which are, perhaps,
not too dissimilar concepts, though the technological support
may be different.

Computer networks, satellites, graphics and other aspects
of the new information technology have been part of the
widening technological basis, though problems of the
application backlog have remained. There have been attempts to
try alternative methods, such as attacks on the life cycle as a
method of developing information systems, prototyping supported
by various tools, and movements towards decentralised computing
and end-user development.

It is noticeable from this that practice has concentrated
on the technology, and this bias has been reflected in much
information systems education. Courses in data processing and
systems analysis have been followed by courses in databases,
management information systems and then to those in information
systems methodologies, techniques and tools. These have often
been supplemented with courses in particular technologies. Thus
information systems education has been more concerned with
computer science courses reflecting this technologically driven
view of the world and such courses have been running since the
mid 'sixties. It would seem that practice has been setting the
course content and in general practice has been leading
education.

Education is required not just for computer
scientists, but also for 'those concerned with the analysis
and design of computer-using systems' (Buckingham et al,
1987b). There is also a need to support those whose primary
interests are in business and administration and to 'whom
computer technology would be one management tool among
others' (Davis, 1987).

461

Some information systems courses, particularly in polytechnics, have placed emphasis on systems analysis and design and they aim at training systems development practitioners for the large number of organisations struggling with systems development. In the UK at least, the polytechnics have tended to embrace the new and emerging discipline of information systems. The universities were much more reluctant and less influenced by the growing demands of industry and commerce.

This polarisation of responsibilities was unfortunate because the universities tended to control the research, and information systems as a discipline has not been well supported in terms of funded research. Indeed it was sometimes viewed by the universities as not a proper subject for academic institutions to embrace. Perhaps the main reasons has been those mentioned above: the difficulty to define, or bound, information systems, a lack of a theoretical basis, and it could be argued that the subject has been over-influenced by practitioners.

In 1987, IFIP and the British Computer Society (BCS) published a suggested curriculum which reflected the much broader scope of information systems as influenced by the fundamental changes in technology and the recognition that many of the issues concerning practitioners were no longer technical. As Buckingham et al (1987b) point out:

> 'the information system designer, using this
> title now in its broad sense and at the
> highest level, is concerned with the
> application of information technology in
> organisations and society and with the design,
> development and use of information systems
> which add effectively to their welfare and
> successful activity.'

However, despite this attempt at widening, the role is seen as still focused on the information system designer and not on the manager who seeks to utilise information and technology as part of the armoury of skills to enhance the business of organisational effectiveness.

This person is beginning to be catered for, less in information systems or computing science departments, but more in business schools. Some cater for a business information systems specialist but most advocate a more seamless integration of information systems/information management into the mainstream business courses, for example, into business strategy. This seamless integration is often not evident in practice, but nevertheless it characterises the notion that management is now about information systems as much as it is about finance or marketing or other functional specialisms.

Turning from teaching to research, a similar pattern is seen. Keen (1987), in an overview of research in management information systems, critically examines particular areas of research and researchers. He argues that the mission of information systems research is to study 'the effective design, delivery, use and impact of information technologies in organizations and society' and he goes on to spell out some of the implications of this. For example, he uses the term effective to imply that practice should be improved through research. He regards systems design, described as a craft, as the core area of information systems. It concerns the possibilities and constraints of applying technology. The interest is not with the information technology itself, he argues, but with its application.

He gives a potted history of information systems research in the following form:

- The early 1970s - managing systems development, design methodologies, economics and computers.
- The mid 1970s - decision support, managing organisational change, implementation.
- The early 1980s - productivity tools, database management, personal computing, organisational impacts of technology, office technology.
- The mid 1980s - telecommunications, competitive implications of IT, expert systems, impact of IT on the nature of work.

Again, there is a high proportion here of technology-oriented information systems research (for example, design methodologies, computers, implementation, productivity tools, office technology, telecommunications). Some of these aim to provide 'solutions' to perennial problems (for example, productivity tools, database management, personal computing and expert systems, have all had claims, some argue wild claims, made in support of them). There is also significant research looking for particular gains for businesses in an economically competitive environment (for example, economics and competitive implications).

The early 1990s may include technological research, but we hope that there will also be emphasis placed towards social, organisational and human aspects. The paper by Lyytinen et al (1991) is one paper that reflects this movement.

The set of papers contained in Mumford et al (1985) sprang from a Colloquium in 1984 called 'Information Systems Research - A Doubtful Science'. This was a provocative title because attempts were made then (and more recently in a follow-up conference in Copenhagen (Nissen et al, 1990) to suggest that research in information systems should move away from scientific methods to concern itself with human activities, and may therefore be more in the realm of a social science.


CONCLUSION

Practice has been very influential on information systems education and research. Indeed it is often said to be leading the academic discipline, and this view is tenable. particularly when viewed in terms of the technology. However, this has to some extent worked to the detriment of information systems education and research. Some disciplines, in particular, systems approach, semiology, sociology and anthropology have not made the impact that would seem appropriate to information systems. Information technology, on the other hand, has been over-emphasised. Further, the impact of practice has also distorted information systems research, so that technological advances, in particular, are emphasised. This has also meant that practice is dominating information systems research in that commercial organisations have the necessary resources to advance more quickly.

By redressing the balance, research and teaching in academic institutions may develop, and lead to the greater acceptance of information systems as an academic discipline and influence practice as much as practice influences research and teaching.

BIBLIOGRAPHY

Avison, D. E. & Wood-Harper, A. T. (1990) *Multiview - An Exploration in Information Systems Development*, Blackwell Scientific Publications, Oxford,

Backhouse, J., Liebenau, J., & Land, F. F. (1991), On the Discipline of Information Systems, *Journal of Information Systems*, **1**, 1.

Boland, R.J & Hirschheim, R.A (eds) (1987) *Critical Issues in Information Systems Research*, Wiley, Maidenhead.

Buckingham, R.A., Hirschheim, R.A., Land, F.F. & Tully, C.J. (1987a) *Information Systems Education: Recommendations and Implementation*, Cambridge University Press.

Buckingham, R.A., Hirschheim, R.A., Land, F.F. & Tully, C.J. (1987b) *Information Systems Curriculum: A Basis for Course Design*, in Buckingham *et al* (1987a).

Davis, G. B. (1987) *A Critical Comparison of IFIP/BCS Information Systems Curriculum and Information Systems Curricula in the USA*, in Buckingham *et al* (1987a).

Keen, P. G. W (1987) *MIS Research: current status, trends and needs*, in: Buckingham *et al* (1987a).

Liebenau, J & Backhouse, J (1989) A Need for Discipline, *Times Higher Education Supplement* (31.3.89).

Lyytinen, K., Klein, H, & Hirschheim, R (1991) The Effectiveness of Office Information Systems: A Social Action Perspective, *Journal of Information Systems*, **1**, 1.

Miles, R (1991) Fundamental Guidelines for using Soft Systems Methodology, *Computing* (24.1.91).

Mumford, E., Hirschheim, R., Fitzgerald, G & Wood-Harper, A. T. (eds) (1985) *Research Methods in Information Systems*, North Holland, Amsterdam.

Nissen, H-E, Klein, H. K & Hirschheim, R (1990) *The Information Systems Research Arena of the '90s*, Volumes I and II, Lund University, Sweden.

Stamper, R (1987) *Semantics*, in: Boland & Hirschheim (1987).

Vidgen, G & Hepworth, J. (1990) Yesterday's Philosophy, *British Journal of Healthcare Computing*, 7,7.

Wilson, B. (1984) *Systems: Concepts, Methodologies and Applications*, Wiley, Chichester.

INTELLECTUAL INFORMATION SYSTEMS

IN ORGANIZATIONAL MANAGEMENT

Dmitry Chereshkin

Institute for System Studies (VNIISI)
Moscow, USSR

Growing complexity of management processes connected, in particular,
with larger production scales and greater impact of technical and economic
decisions on the ecological system and society at large, calls for
enhancing capabilities of information systems (IS) used in organizational
management.  Their intellectualization is one of the trends of such
enhancement of IS capabilities within the user - IS system as to provide
for:
 - transition from solving rigidly determined problems to poorly or
less rigidly structured ones;
 - continued development of the user - system friendly interface;
 - extension of the structure of information used, knowledge
integration as a basis of logical conclusions made;
 - use of a wide variety of models for solving users problems.
 Intellectual information systems for organizational end economic
management are nowadays defined as systems integrating in their structure
traditional information retrieval and computer information (host and/or
distributed) systems, decision support systems (DSS) and expert systems
with knowledge bases concerning a particular subject domain.
 Integrated systems uniting the aforementioned components taken in
different combinations are beginning to came into the markets of many
Western countries.
 In a real IIS intellectualization is secured by:
 - utilization of the user-to-system language in the natural or close
to natural language;
 - generation of the knowledge base of the appropriate subject domain,
where it is used in situation analysis and alternative generation and
evaluation;
 - enhancement of capabilities of logical information processing;
 - systems inner sophistication and enhancement of its "convenience"
for the user;
 - development of special software capable of identifying situations,
evaluating them and finding key problems;
 - development of (mathematical, structural, logic, etc.) simulators of
possible situations and their components and model adaptation to situation
changes;
 - this system's integration into general computer network (external
integration) and integration of IS of different types within its structure
(internal integration);

- development of means of information representation to user in the way most convenient for him.

It is obvious that all the intellectualization components mentioned are implemented by integration of various hard- and/or software into IIS structure, utilization of same systems engineering solutions, etc.

Hence, there arises rather complex system task of choosing different intellectualization components and ways of their implementation. This task exist both at the designing phase and in the process of development of a real IIS.

An approach to IIS designing, based on detecting the relationship between the universal functions within the system and intellectualizing components constituting it is suggested.

The universal functions are composed of:

- user-system communication in the course of solving the problem (information input-output, interactive regime, etc.);
- necessary links created between the system elements in the course of solving the problem (system structure adaptation to a specific problem);
- situation analysis and alternative generation;
- alternative evaluation and choice of decision.

All the aforementioned functions are implemented by means of combining following components:

Systems engineering - new IS architectures; distributed systems; teleaccess; implementation of functions by means of hardware and software.

Technological - future computer types; special processors; expanded peripherals; communication channels and means of commutation.

Program - interaction procedures; system/system interface; recognition of speech, patterns and schemes; modeling, obtaining and generation of knowledge; database and knowledge base management; expert systems.

Information - data; knowledge; preferences of the decision-maker.

Model - models of computation, situations, logical conclusion, preferences, etc.

Personal - adaptation to change of situation, instruction, definition of new problems and interactive conditions.

Hence, we may say, that the following general factors, characteristic of IIS, are known:

IS intellectualization components;

universal functions within the system;

components implementing the system functions.

Two more groups of factors defined in conference with the user are indispensable for real designing. They are, firstly, the initial requirements of the user (i.e., what system and for what subject domain he would like to have) and, secondly, a set of evaluation criteria (i.e. the acceptable cost, possibilities of utilization of the already existing system, etc.).

All these factors being available make it possible to begin with the designing process. In the most typical case the necessary decision may be arrived at by consistent form and studying the number of three dimensional matrices. E.g., matrix combining user requirements, informatization components and universal functions is formed. This matrix study helps to formulate particular and concrete requirements for all the system functions. Then the matrix composing the relationship between the specified function requirements, for the components implementing the functions and the evaluation criteria is built.

But manipulating three-dimensional matrices, even with the help of the computers constitutes rather complex computational, informational and logical problem so it is more advisable to use a sequence of interactive two-dimensional matrices. Here, all the (interim and final) decision procedures may be computerized with the help of a special programming complex developed and a database concerning specific options of different components established.

The AOAFS (Aim-Oriented Alternative Forming System) programming complex ensures consistent forming and interactive regime study of two-dimensional matrices. Here the AOAFS provides for information support (i.e., specific data filling of information) of all the matrix components.

IIS designing process begins with building, specifying and studying "initial user requirements for IIS - universal system functions" matrix. At this phase user requirements are formulated and specified in conference with user himself, i.e. "general form" user-attractive IIS is elaborated. Specifications are maintained consistently beginning with the most general ones ( What is your perception of the system as a whole? ) and ending with the ones refining the requirements for each universal function. It is necessary to note that the AOAFS comprises "Questionnaire" block to obtain knowledge including user interrogation.

"Requirements - functions" matrix study helps to specify with accuracy initial user requirements for IIS within the universal functions.

At the second phase "intellectualization requirements within the functions - intellectualization components" matrix is formed and studied. As a result possible solutions to meet user requirements are elaborated. All the solutions are formulated in the terms used for IS intellectualization components.

Within the third phase "possible solutions - components implementing the system functions" matrix is build. A set of system, technical, program, information and other components is specified as a result of matrix study. All this phase existing and constantly actualized database comprising specific system, technical and all other components is widely used. E.g., this database comprises the whole range of the known designs of the new IS architectures, future computer characteristics, etc.

The final phase of the design process is connected with "system components set - evaluation criteria" matrix formation and study. The choice of evaluation criteria is the most complex task of this matrix formation. The most obvious criterion is the cost of IIS development. But as is shown in practice the criterion of maximizing the user hard- and software exploitation is quite important. Besides the criterion of simple (convenient) user-system interaction is of greatest importance for the user. So it is necessary to choose an appropriate criteria set in each particular case. The second complex task includes criteria ranking according to the user preferences, as well as taking them into account. The AOAFS complex comprises expert evaluation processing system ensuring unbiassed ranking of the criteria set chosen.

The whole designing process is of iterative character, where it is possible at each subsequent phase to return to the previous one, introduce any changes and continue the process further.

Unfortunately, there are not as many intellectual information systems in this country (as we should like to have) used in organizational management. The AOAFS complex developed in practice is used for creating certain Decision Support Systems (DSS), though we believe that changes taking place in our economy and social life will call for cardinal changes in IIS development and application.

INFORMATION SYSTEMS AS LINGUISTIC SYSTEMS:

A CONSTRUCTIVIST PERSPECTIVE

R. A. Stephens and J. R. G. Wood

Bristol Transputer Centre,
Department of Computer Studies and Mathematics
Bristol Polytechnic, Bristol, UK.

## 1. Introduction

Information systems are usefully defined by Lyytinen (1987) as organisational communication systems that are just technically implemented and which rely to an increasing extent on information technology. This perspective directs attention to the language context; technology is always subordinated to, and supports language change. By treating information systems as fundamentally linguistic systems involving signals that are used by people in relationships, we will share Checkland and Scholes (1990, p55) concern for the attribution of meaning by the individual and the purposeful action which the information system serves. This requires attending to those meaning generating systems which individuals are involved in and attending to the meanings which make those particular actions meaningful to particular groups.

In this paper we examine the perspective on language adopted by Maturana and Varela in their theory of autopoiesis, and consider its implications for information systems. Language is not a vehicle for the transfer of knowledge or information in the conventional sense, but a manner of co-existence, the means of participation within a consensual domain. It is a social phenomena and its properties are connotative rather than denotative. Similarly, the process of 'knowing' is more than simply observing the world as if it were external to us, it involves our material activity as one of nature's forces, operating within nature ("all knowing is doing and all doing is knowing", Maturana and Varela, 1987, p26). What is at issue is the success or failure of this transformation - this active 'objectification' of knowledge - rather than a passive 'mirroring' or 'representation' of the world. Accordingly, knowledge is not a reflection of the world, but a transformation of experience into linguistic ontology. In this way, knowledge is not something people possess in their heads, but rather something people do together; meaning is contingent on the context of human linguistic activity.

## 2. Linguistic Systems

Characterising information systems as linguistic systems or social action systems is not new (Lyytinen, 1985, Lyytinen and Hirschheim, 1988), yet the study of language as a means for effective co-ordination of human action is a neglected topic. This may seem surprising considering the importance of language within information systems, that language is the principal medium of information systems and that many of the people information systems are intended to serve, spend most of their working time as language practitioners (listening, talking, convincing etc.). Lehtinen and Lyytinen (1986) argue that information systems design should begin by addressing the communicative character of information systems in the sense that information systems are formal linguistic systems that serve limited communication between social agents to support their action.

The organised provision of information in organisations is always linkable in principle to action (Checkland p54-55), and we may consider this action to be, for the most part, linguistically motivated. Supporting the notion of information systems as "social systems only technologically implemented" (Lehtinen and Lyytinen, 1986), Lyytinen and Hirschheim (1988, p20) give ontological status to information systems "as involving a set of human practices which exhibit regularity and impose constraints on people's behaviour, but which can also be transformed by knowledgeable social actors".

Lyytinen (1985) identifies five views of language informing information system design, but only one of these, the interactionist, would acknowledge an active social component in language use. The four others, the denotational, generative, cognitive and behaviouristic, in one way or another treat language as passive and in many ways transparent: experiences emerge independently of language and we simply apply words to reality.

In contrast, the interactionist version, which is based on ordinary language philosophy, regards the primary function of language to be performative or rhetorical, and only secondarily and in a derived way referential and representational. It works by people materially moving one another by its use to behave in certain ways. Support for the interactionist version can be found in the theory of autopoiesis developed by Maturana and Varela who have examined the biological origins of language.

### 3. Autopoiesis and Language

At the risk of over-simplification, autopoiesis may be defined as the homoeostatic operation of an organisation; all living systems are autopoietic and all autopoietic systems are living systems; as autonomous systems, the main guideline for their characterisation is not a set of inputs, but the nature of their inner coherences, which arise out of their interconnectedness. The theory is defined by Maturana and Varela (1980 and 1987), and requires a lengthy appreciation. Informative discussions, however, are provided by Winograd and Flores (1987), and Mingers (1989).

For Maturana and Varela, language is prior to conscious thought. Consciousness is generated through the operation of the consensual domain within which language is generated, but the fundamental act is that of distinction. In this we specify a unity as an entity distinct from a background, characterise both unity and background with the properties with which this operation endows them, and specify their separability. However, the individual/environment dualism, characteristic of most behavioural, cognitive and systems approaches, is denied by Maturana and Varela. It is only at the meta-level of the observer that the organism and its environment are separated into clear- cut components. In this way, the subject-object relation, so characteristic of Western thought, is only valid within the domain of language.

The domain of language arises in the co-ontogenic co-ordination of action (the coupled histories of organised beings), which Maturana and Varela call the linguistic domain.

These co-ordinations of action bring forth different entities. In the flow of recurrent social interactions, language appears when the operations in a linguistic domain result in co-ordinations of actions about actions that refer to the linguistic domain itself (Maturana and Varela, 1987, p 209). However, because of the nature of the autopoietic organisation itself, every change that an organism undergoes is necessarily determined by its own organisation, communicative interactions are intrinsically non-informative. The observer, by neglecting the internal determination of the autopoietic systems which generate it, may describe communicative interactions as if they were informative, but phenomenologically the linguistic and autopoietic domains are exclusive. Although one generates the elements of the other, they do not intersect (Maturana and Varela, 1980, p120).

## 4. The Communication Process

What then can we say about language? If there is no transmission of information through language, then its function is connotative, not denotative (Maturana and Varela, 1980, p32). Consensus arises only through co-operative interactions in which the resulting behaviour of each organism becomes subservient to the maintenance of both. Any denotative function lies only in the cognitive domain of the observer and not in the operative effectiveness of the communicative interaction. Furthermore, while the primary function of language lies in the arena of communication, language also plays an important role in shaping the individuals' mental processes. Thus we can expect such mental processes to be shaped by forces that originate in the dynamics of communication.

Recent theoretical developments associated with the social constructionist movement in social psychology have attempted to establish an ontological status to the communication process (see Gergen, 1982). In the social constructionist view, all behaviour is socially and historically situated, making social psychology the essential foundation for all psychology and indeed epistemology. This contrasts with the dominant paradigm of the philosophy of science in which transituational, transhistorical verities are sought (termed as exogenic by Gergen). This does not point to solipsism, however; ontology is grounded in the consensual domain, language, communication process or social community. Reality is neither objective, nor subjective, but intersubjective, generating the subject/object complementarity. As with hermeneuticists, the criteria for knowledge is the validity of a given interpretation within the prevailing rules of communication within the culture. In effect, interpretations may be rendered acceptable or unacceptable to the extent that they meet currently adopted standards of intelligibility.

The suggestion here is that language is chiefly a set of practices employed by people for purposes of successful interchange (Gergen, 1985). Forms of discourse emerge as a response to certain practical problems encountered in human relationships. One of the major uses of language in social interchange is as a signalling device. Linguistic terms can be rapidly and effectively deployed as a means of designating the presence or absence of various objects, entities, or states of affairs. Within the arena of human affairs, however, reference is frequently made, not to physical attributes or properties, but human action itself.

Gergen (1985) suggests that in the attempt to solve the pragmatic problem of referring to or signalling about human activity in the sphere of daily relations, a language of psychological dispositions (intentionality) is born. Similarly, Maturana and Varela (1987, p129-32) note that as observers we have access to both the system and the structure of its environment. By describing the behaviour of the system as a goal-orientated process, we are not referring to operation of the nervous system itself, but are describing the movements of a unity in an environment that we (as observers) indicate. Such statements only serve the purpose of communication among ourselves as observers.

However, as Gergen (1985) suggests, in order to account for human action we must speak as if people possess motives, needs, drives, intentions, wants, preferences, etc. - all terms that reinstate at the internal level what persons appear to accomplish in their behaviour. In a similar way Maturana and Varela argue that consciousness and mind belong to the realm of social coupling (1987, p234). One crucial aspect here is the recognition that distinctions lie in a consensual domain - that they presuppose some kind of social interaction in which the observer is engaged. Networks of conversations (Maturana, 1988, p71), whether in the realisation of a social system or of a non-social community, are also operationally realised in language as coherent systems of descriptions and explanations that constitute a domain of reality. While humans will operate in may different domains of reality which, as different conversations and explanations intersect, change in any particular social or non-social community will take place as a conversational change; i.e., as a change in the configuration of the network of co-ordinations of actions and emotions that constitutes it and defines its class identity (Maturana, 1988, p72).

Since we exist in language, domains of discourse that we generate become part of our existence and constitute part of the environment in which we conserve our identity and adaptation. That is, they fall into the same position vis-a-vis the language as do 'real world' events. For Gergen this produces a reified language of psychological events; Maturana and Varela's account would take us a step further. Theories in which concepts like the self have a place so structure our experience as to create these very concepts; different theories, construct different 'mental' organisations. Everything that appears to each of us as the intimate structure of our personal being has its source in the regularities imposed by human social dynamics (Maturana and Varela, 1987, p246).

In this respect, knowledge will appear vitally dependent on the vicissitudes of social negotiation; its constraints not directly experiential but social or linguistic. It is not the internal processes of the individual that generate what is taken for knowledge, but a social process of communication (Gergen, 1982, p207). It is within the process of social inter-change that rationality is generated. Thus Maturana and Varela can maintain that reason is not an unanalysable property of the mind, but an expression of our human operational coherence in language, and that as such it will have a central and constitutive position in everything that we do as human beings. We are thus invited to adopt the principle that social structures and mental structures are in reciprocal relation to each other because one is involved the the creation of the other.

## 5. Implications for Information Systems

Within Maturana and Varela's work the term information is used only in the restricted sense of forming from within. Communication is not seen simply as a matter of informa-tion transfer from one location to another, but as ontologically formative, as a process by which people can, in communication with one another, literally in-form one another's being. By defining the primary human reality as conversation, albeit in some extended sense, Maturana and Varela encourage us to consider all human action in terms of its contribution to a developing discourse; communication can then be considered as a shared domain of (inter)action, as the conservation and construction of meaning systems. As such these meaning systems will be characterised by their flexibility and evolutionary nature. This will be in sharp contrast with computer-based information systems so it is not surprising that semantic breakdowns occur. The connotative nature of such linguistic systems is then reflected in the amount of programming time spent on maintenance and largely attributable to 'shortcomings' in the analysis.

Subjectivist approaches in systems thinking share with contemporary psychology, AI and expert systems a view of man in which mental events, mental activities, mental operations, mental organisation, and mental transformations are of greater importance than events, activities, operations, organisation, or transformations of the external world. Not only are these mental operations cut off from their roots in social and historical practice (the consensual domain), but also, in being located within the mind of the individual, they cut off people from effective action to change their circumstances rather than their subjective understanding of these circumstances. Maturana and Varela's analysis, however, recognises the intersubjectivity of language; that communication presupposes structural coupling, that it is the co-ordinated behaviours mutually triggered among members of a social unity. Since *we* exist in language, the domains of discourse that *we* generate become part of *our* domain of existence and constitute part of the environment in which *we* conserve identity and adaptation.

Accordingly, we must acknowledge of the vital role of the investigator/analyst/researcher in the construction of these knowledge or meaning systems. The reality of meanings cannot be grasped at either of the extremes of subjective idealism or functionalist behaviourism, but must be achieved in the dialectical interpenetration of subject and object in which neither has primacy. As an alternative we should interest ourselves in semantic rather than causal relations, that is, actions should be studied by reference to the meanings which actors take them to express, that is how the actor intends it, and how the others will take it, rather than by reference to its supposed causes. Knowledge is not individually subjective, but is inter-subjective and interpretation reproduces an existing structure, without which there can be no communication (conservation of adaptation).

The work of analysis will then be properly seen as the exploration of these meaning systems via a linguistic domain, involving participating observers attaining their respective poieses. Thus research/analysis cannot have a subject/topic in the conventional sense, but must involve the emergence of meaning from within. Moreover, the inquiry (analysis) will become an end in itself and information systems become evolving social forms of sense-making; a change process founded on an intersubjective recognition of phenomena (linguistic co-ordination of action).

## 6. References

Checkland, P. and Scholes, J., 1990," Soft Systems Methodology in Action", Wiley.

Gergen, K.J., 1982, "Towards a transformation in social knowledge", New York: Springer-Verlag.

Gergen, K.J., 1985, Social pragmatics and the origins of psychological discourse, in "The Social Construction of the Person" K.J. Gergen and K.E. Davis, (eds) Springer-Verlag

Lehtinen, E. and Lyytinen, K., 1986, Action Based Model of Information System, Inform. Systems, Vol. 11, No. 4, pp 299-317.

Lyytinen, K., 1985, Implications of Theories of Language for Information Systems, MIS Quarterly, March/1985, pp 61-74.

Lyytinen, K., 1987, A Taxonomic Perspective of Information Systems Development: Theoretical Constructs and Recommendations, In "Critical Issues in Information Systems Research", R.J. Boland and R.A. Hirschheim (eds.), Wiley.

Lyytinen, K. and Hirschheim, R. 1988, Information Systems as Rational Discourse: An Application of Habermas's Theory of Communicative Action, Scand. J. Mgmt., Vol. 4, No. 1/2, pp 19-30.

Maturana, H.R., 1988, Reality: The search for objectivity or the quest for a compelling argument. Irish Psychol. 9(1), 25-82.

Maturana, H.R. and Varela, H.R., 1980, "Autopoiesis and Cognition: The Realization of the Living", Reidel, Dortrecht.

Maturana, H.R. and Varela, F.J., 1988, "The Tree of Knowledge", Boston: Shambala.

Mingers, J., 1989, An Introduction to Autopoiesis - Implications and Applications, Systems Practice, Vol. 2, No. 2.

Winograd, T. and Flores, F., 1986, "Understanding computers and cognition, A new foundation for design", Addison-Wesley.

# THE ENGLISH LANGUAGE AS A SYSTEM AND

# ELEMENTS OF ITS LOGIC

Michael Yu. Chernyshov

Irkutsk Computing Centre
USSR Academy of Sciences
Irkutsk, 664033, U S S R

## INTRODUCTION

Natural Language (NL) is the most amazing of the phenomena worthy of attention of scholars. It is the might  and
the indicator of weakness of one's intelligence, the truth
messenger and liar. We hope we know it, and still, when
speaking, we fear to look clumsy. It seems, in investigations
we sometimes sooner guess its properties and laws than recognize them by analysis on the basis of some accepted method.

The problem of recognition of an adequate NL formal grammar, logic and semantics is one of the most interesting and
complicated. And its solution will require new and new approaches. The logic system of NL is substantially richer than
the apparatus of the present-day formal logic. This gives a
stimulus to continue investigations in elucidation and representation of NL logic and, consequently, NL semantics(its basis being constructed by logic). Adequate KR languages must
include some sufficiently complete sets of reasoning functions
based on some refined logic. But presently, reasoning by
analogy, conditional and temporal reasoning cannot be completely modelled on the old basis. We made a shy attempt to
represent some interesting aspects of NL logic without going
down into spheres of nonmonotonic logic.

In our investigations, NL (English) is considered as a
homeostatic (i.e. closed, multi-level, hierarchical, selfcor-
related) system of structures and functions, i.e. of NL components, relations (properties, characteristics, interconnections, etc.), meanings, senses, accents. Structural elements
that control the sequential order of text synthesis, and also
those controlling the directed attention of communicants in
dialogue, are identified in NL (Chernyshov, 1990). We postulated the existence of a special class of NL signs called
constructives and elucidated their role, nature and place in
in the system of NL (English).

Semantics of relations in a NL text representing relations in a described  world cannot be reduced to only referen-

tial world (RW) relations $R^R$ and, hence, is supplemented with relations added by the speaker and represented in text structures in the form of structural semantic relations $R^S$. NL semantics may be subdivided into : (i) semantics of relations in RW; (ii) semantics of relations and relations of relations in the denotative world (DW) of concepts; (iii) semantics of NL lexical signs in DW of linguistic concepts (DWLC); (iv) semantics of relations in DWLC between NL text lexical signs; (v) semantics of NL text structures (formulas: phrases, clauses, sentences); (vi) semantics of relations in and between NL text structures.

Logic may serve to represent only semantics of relations. In his Extended Standard Theory Noam Chomsky (1972) postulated that semantic interpretation of a sentence is held to be determined by the pair (deep structure + surface structure); grammatical relations play role in determining meaning; such matters as scope of logic elements and quantifiers, coreference, focus and presupposition are determined by rules that take <u>surface structure</u> into account. N.Chomsky subdivided 2 levels of semantic representation: (a) the level of logic form (LF) (semantic interpretation determined by rules of grammar - SI-1); (b) the level of "fuller" semantic interpretation (LF + other cognitive representations - SI-2)(Chomsky, 1977). LF includes: the scope and interpretation of logic constants and quantifiers; indications of coreference and disjoint reference, of what elements are in focus, etc. Chomsky contends: "...surface structure determines LF." Consequently, by analogy with semantics, it is reasonable to subdivide logic of intensional relations in NL texts into the extralinguistic logic (bound up with RW and DW) and the linguistic logic (for surface text structures that are some comfortable universal apparatus for representing knowledge on relations and their logic, furthermore, this apparatus is not bound to specificities of RW/DW semantics). We distinguish between (a) logic of relations between objects in RW $(l_1)$; (b) logic of relations between objects in DW $(l_2)$; (c) logic of relations between NL text signs (represented in the form of a certain structure) $(l_3)$. Together $l_1$ , $l_2$ , $l_3$ may form a joint system for representing knowledge on logic of relations for NL.

## AN APPROACH TO REPRESENTATION OF LOGIC AND SEMANTICS OF RELATIONS IN NATURAL LANGUAGE TEXTS

Let us consider one of such approaches for some NL text T, wherein each following sentence $S_i$ step-by-step forms a progressing situation in a world $W_1$ (one of a set of possible worlds corresponding to interpretation of T), can introduce in consideration some additional or parallel world $W_2$, into which some objects $a_j$, j=1,n already described in T for $W_1$ and relations $R_i$, i= 1,m in $W_1$ (relations being generally interpreted as predicates or predicate formulas) may be mapped. The fact of existence of relations in $W_k$ can be represented by propositional sentences of the form $P_i^k(X_1,X_2)$ (for property predicates) or $R_i^k(P_1,P_2)$ (relation predicates).

Let us define a simplified version of language that will describe NL sentence logic and will allow to express the basic concepts and propositions.

<u>Notation. Nomenclature.</u> p - phrase; <u>from the structural</u>

476

viewpoint: p is a finite sequence of lexical-level NL signs, $p ::= \langle w \rangle / \langle kw \rangle / \langle p \rangle \langle w \rangle / \langle p \rangle \langle kw \rangle / \langle p \rangle \langle K \rangle \langle p \rangle$ , where: w - word; k - <u>elementary phrase constructive</u>, i.e. a NL sign employed to construct a phrase as a syntactic sign by making it semantically complete (i.e. by constructing a valuable, realizing the speaker's elocutive intention, part of sentence semantics); it does not possess any proper meaning, but can either acquire some meaning in a context or take part in semantization (initial/supplementary) of contextually dependent NL signs; K - <u>sentence constructive</u>, i.e. a NL sign employed to construct a sentence as a text sign by making it semantically complete (i.e. by adding a valuable part of sentence semantics); it does not necessarily possess any proper meaning, but can either acquire some meaning in a context or take part in semantization (initial/supplementary) of contextually dependent NL signs; <u>from the logic viewpoint</u>: in the general case, p may form a propositional variable or an n-placed k-th order (embedded) predicate $P^k(x_1,...,x_n)$; <u>from the semantical viewpoint</u>: p has at least one synthesized meaning corresponding to a synthetical concept. S - set of sentences; $S^\alpha \in S$ - subsets of sentences that belong to an $\alpha$-th grammatical type. $S_i$ - sentence, which: <u>from the structural viewpoint</u>: $S_i$ is a finite sequence of phrase-level NL signs, which is characterized by structural (syntactic) sufficiency and obligatory logic and semantic completeness; $S_i \rightarrow \Phi ::= \langle p \rangle \langle K \rangle / [\langle p \rangle \langle K \rangle]^3 / [\langle p \rangle \langle K \rangle]^* / \langle K \rangle \langle p \rangle \langle K \rangle /$ etc. (constructed for nominative and complete NL simple sentences including: affirmative, negative, exclamatory, imperative, interrogative types), where: $\Phi_i$ , i=1,N form a set $\Phi^k$ ($\Phi^k$ - set of admissible sentence structures obtained by some k-th method of analytical decomposition of sentences); <u>from the logic viewpoint</u>: we shall speak that S can be mapped both into a set of structures ($\rho_k : S \rightarrow \Phi^k$) and into a set of logic relations $l_1 + l_2 + l_3$ ($h_1 : S \rightarrow l_m$ , m=1,3); <u>a sentence</u> with some set of relations in it can form different types of propositional formulas to represent (i) a set of represented objects, (ii) a set of properties of these objects, (iii) diverse relations between objects and relations of objects and properties of objects.

Let us describe the concept of sentence logic. From now on: $L^R / L^D / L^s$ - logic relations characteristic of, respectively, RW, DW, SW (denotative world of structural relations for NL structures); $R^R$ - relation (REL) in RW (on objects, sujects situations, etc); C - propositional connective (basic: $\neg$ , $\vee$ ; derived: $\wedge$ , $\rightarrow$ , $\cup$ , $\leftrightarrow$ ($\equiv$) ); c - sentence surface connective (and/or/,/etc.); Q - NL logic quantifier; q - NL surface quantifier; æ - NL logic sentence constructive of a NL sentence logic formula (an operator employed to construct sentences semantically); $\mathcal{L}$ - set of logic relations possible for S; where $l_1 \in \mathcal{L}$ is a set of logic relations between events and objects of RW (desribed by a text, i.e. mapped into DW relations of concepts, that can always be represented structurally), that can be represented in some $S_i$. Then we may speak that the function $L_j^R(S_i)$ puts $l_1$ (logic formulas of the following form) in correspondence to any $S_i$. For example : "He has some money." $L^R(S_i) =$ HAV(SUB1, OBJ2). But many relations cannot be expressed in RW terms, e.g., "He needs some money." "The money slaved, ate him." There are no necessary referents in RW. Let $l_2 \in \mathcal{L}$ be a set of DW logic relations intended to represent objective relations of RW from some subjective viewpoint. $l_2$ enriches the world of possible relations at the expence of concepts of mind and personal (speaker's) intentions. The function $L^D$ ($S_i$) puts $l_2$ in correspondence to any $S_i$. So, some RW situations and relations are represented in terms of $l_2$ as a result of (i) mapping $l_1$ into the set of functional DW logic concepts and (ii) reinterpretation of these situations (/relations/events) from the viewpoint of a speaker who would like to say, e.g., not "He loves her.", for which $L^D(S_1)= R^R{}_1 = $ LOV(SUB1,SUB2); $L^D(S_1)=R^D{}_1=\overline{R}_1(\widehat{P_1},P_2)$; but rather "She is beloved by him.", for which the RW situation is the same: $L^D(S_2)= L^R(S_1)$; but $L^D(S_2)= R^D{}_2= \overline{R}_2(\widehat{P_2},P_1)$. Here $R^D{}_1$ and $R^D{}_2$ have definite logic (and semantic) "vectors" (tensors, in the general case); SUBi - subjects of RW; $P_i$ - concepts of SUBi in DW.

Practically, such relations in NL logic are ever more complex, especially as far as "relations between relations" (events/conditions/etc.) are concerned. Not only relations of the type "condition-event", "cause-consequence", which are characterized as consequence relations (directed relations) in time, can be expressed in $L^D$, but also relations of the type "object of directed relation (action/state/perception/etc.)-subject of directed relation", for which the direction lies in the abstract world of relations. Obviously, in NL speech structures one represents DW logic (rather than RW), the linguistic ("surface") structure being dependent on one's elocutive intention. Moreover, NL occurs to be not only the carrier of logic, but in its turn adds to the total world of possible logic relations (a) the semantics stored in NL signs (lexical units, phrases, sentences of various special types), which is inevitably imposed on NL logic; (b) the logic of NL formulas, which is more complex than any formal logic known; (c) the logic stored in NL constructives, which practically represent a sort of symbiosis of connectives, quantifiers and lexical units in one form (Chernyshov, 1990). So, we postulate also $l_3 \in \mathcal{L}$ as a set of lingua-logic relations between sentence structure (set of possible formulas) components ($L^S$); these components and the structures on the whole being the result of mapping the logic $l_2$ for $S_i$ into a set of functional-logic (operator) NL structures representing the principal universal part of the NL logic apparatus. The functional (predicate) filling of relations is the other, nonuniversal part of this apparatus.

Since NL (especially an analytical NL, such as English)is a universal tool to represent any logic of relations, no special representation formulas are needed, "...surface structure determines LF." (Chomsky, 1977). Hence, for example,

(1) "We put on the lamp, because it grew dark in the room."

$L^R = R^+(R_2(P_3,P_4),R_1(P_1,P_2))$; $L^D = $ because$(R_2,R_1)$;

$L^S = R_1 \ae^{cx} R_2$

(2) "Since it grew dark in the room, we put on the lamp."

$L^R = R^+(R_1(P_1,P_2),R_2(P_3,P_4))$; $L^D = $ since $(R_1,R_2)$;

$L^S = \ae_1{}^{cd} R_1 \ae_2{}^{cd} R_2$

(3) "It grew dark in the room, and the lamp was put on."

$L^R = R^+(R_1(P_1,P_2),R_2(P_3,P_4)); \quad L^D = since(R_1,R_2)(conditional relations?) / when(R_1,R_2)(temporal relations);$

$L^S = R_1 æ_1 R_2 \Rightarrow R_1 æ_1 æ_2^{cd} R_2 \leadsto æ_3^{cd} R_1 æ_4^{cd} R_2 ,$

where: $æ_2^{cd}$="consequently"/"then"; $æ_3^{cd}$="since"; $æ_4^{cd}$="then/,"

Obviously, in each case, $L^R$ describes relations on a set of described objects of RW , while allowing to tell between P and R, for example, "John is a boy, who has a car." - $L_1{}^R = R^+(R_1(P_1,P_2),R_2(P_2,P_3));$ "John is the boy, who knows that she has a car." - $L_2{}^R = R^+(R_1(P_1,P_2),R_2(P_2,R_3(P_3,P_4)));$ but does not allow to express specific properties of relations. Furthermore, in the most actual problem (of text analysis), $L^R$ is the logic subject to elucidation. $L^D$ is the result of subjective reinterpretation of $L^R$ , which includes additional abstract relations. $L^S$ is the map of relations $L^D$ into the logic of surface operator structures, which adds its specific information (contained in the linear sentence structures). For example, it is hardly possible to express such NL logic relation as "the more.., the better..." in terms of the notation adopted for $L^D$. But $L^S = æ R_1 C æ R_2$ , i.e., the NL linear structure is a comfortable tool to represent NL logic relations. The logic correctness of the accepted notation in terms of $L^S$, and also the fact that this notation is a correct NL logic formula for any $S_i$ containing logic relations, must be demonstrated.

## REPRESENTATION OF RELATIONS BETWEEN EVENTS

### ( RELATIONS BETWEEN RELATIONS )

Whereas the operators representing relations between objects of RW or between ordinary concepts of DW (such as: "is", "is greater (>)", "is smaller (<)", "is equal (=)",etc.) or else between objects and their properties possess the obvious objective sense, the representation of relations between relations ($R^+$) is possible only in terms of DW relations and in terms of relations expressed through surface structures of NL. These are intensional relations of the type: "consequence (result), since - cause"; "those..., who..."; "(result), when (cause)"; "so (characteristics), as..."; "if..., then..."; "the (condition), the (result)"; etc., which are not characteristic of RW. These relations represent the result elucidation of laws of reality by an individual, express the knowledge of an individual about these laws and (or) about the state of affairs in the world. The logic of these relations (which is assumed to be subdivided into (a) logic of universal laws and (b) logic of realized "events(due to condition/cause)/(in time)...", "logic of admission of an event/condition/...", "Logic of characterization of an object/event/...", "logic of comments on facts/...", etc.) is an important part of logic of relations in DW ($L^D$), which cannot be represented in terms of $L^R$ , but may be represented in terms of logic of surface structures, i.e., $L^S$.

Now let us demonstrate that $L^S$-formalisms form well-formed formulas, which contain obviously logic components including $æ$ . The following proposition is proved for a particular class of sentences with NL logic to show the nature of $æ$ .

PROPOSITION 2. Let $\mathcal{S}^k \subset \mathbf{S}$ be a subset of well-formed logic based conditional NL sentences, for which

$$\Phi^k (S_i) = K X_1 \; c \; K X_2 \tag{1}$$

where $X ::= d \; (p/S)$ is a relation predicate; $d$ is an adverbial phrase, and such that

$$L^S (S_i) = \mathcal{F}_1 \; c \; \mathcal{F}_2 \; ; \; \mathcal{F}_m = \text{æ} R_m \tag{2}$$

and $L^D(S_i) = R^D_i(R_1, R_2)$, where $R_1, R_2$ express logic quantificative relations between situations (/conditions/events), then any such $S_i$ there exists $S_j$ such that

$$\Phi^k (S_j) = \Phi^k (S_i) \; , \quad \text{where } X ::= Q \; d \; (p/S) \; ,$$

$$L^S (S_j) = Q \; (\mathcal{F}_1 \rightarrow \mathcal{F}_2) \; ; \; L^D (S_j) = L^D (S_i) \; ; \tag{3}$$

furthermore, $\quad L^S (S_j) \curlyvee L^S (S_i)$ . $\tag{4}$

PROOF : In (2), C is a binary connective; the nature of æ is logically inobvious (predicate?/operator?), although we assume that elements æR construct well-formed propositional formulas. Suppose the contrary: æ is not a NL logic operator and any formulas with æ are not well-formed. Hence, (4) is not satisfied, i.e., there are no formulas that are structurally equivalent to (2). Now, let us try to construct an obviously well-formed formula with obvious conditional logic, which contains arbitrary predicates $X_i$ and possible binary connectives $(\rightarrow /\curlyvee)$. (a) Case 1: C:=" $\rightarrow$ " , then commutativity for $X_1 \rightarrow X_2$ implies

$$a \; X_1 \rightarrow a \; X_2 \tag{5}$$

(5) is structurally similar to (2), when "a" and "æ" are either both quantifiers or both variables. If "a" (resp., æ) is a variable then in (5) we may rewrite aX as a new predicate (/formula) Y, and (5) is equivalent to (2). Now, suppose, "a" is such that similarity of (2) and (5) seems impossible, i.e. both "a" are operators applied to $X_1, X_2$. Hence, a new complex operator may be constructed

$$X_1 \xrightarrow{\;*\;} X_2 \quad \text{(or in other terms: IFA X1 \; THENA X2)} \tag{6}$$

(6) is similar to (2) iff "a" is a quantifier Q, i.e., $L^S = Q \; (X_1 \rightarrow X_2)$, which is an obviously well-formed formula that is structurally equivalent to (5) independently of the type of "a", and hence, due to transitivity is equivalent to (2) containing æ . (b) Case 2: C:="$\curlyvee$". The formula $X_1 \curlyvee X_2$ represents the inverse logic order of the antecedent and the consequent with respect to $X_1 \rightarrow X_2$, hence cannot represent the order in (2) and may be exempted from consideration.

REFERENCES

Chernyshov, M., 1990, Elements of NL logic in AI systems, in: Proc. 8th Int.Congress of WOSC (to appear).
Chomsky, N., 1972, "Studies on Semantics and Generative Grammar," Mouton, The Hague, p. 134.
Chomsky, N., 1977, "Essays on Form and Interpretation," Elsevier, North-Holland, Amsterdam. - 216 p.

DATA BASES, KNOWLEDGE BASES AND

INTELLIGENT INFORMATION SYSTEMS

Boris E. Polyachenko, Filipp I. Andon, Oleg L. Gunko

V.M.Glushkov Institute of Cybernetics
USSR Academy of Sciences
252207 Kiev, USSR

## INTRODUCTION

Issues affecting modern information system design are considered based on Data Base(DB) and Knowledge Base(KB) technologies. Mathematical models of such systems are suggrested, DB and KB synthesis and optimization are explored. Techniques and tools for DB and KB software and intelligent systems design are examined. These tools are oriented to parallel computing system environment and used for various computer architectures. Proposed means provide controlled level of parallelism based on knowledge and use object-oriented approach to support DB and KB design and programming. The structure of sofware packages are described. Performance characteristics are improved several times as compared to using standard system facilities. The architecture of highly parallel DB and KB machines are suggested, various applications of suggested tools are disscussed.

## PARALLEL KNOWLEDGE-BASED SYSTEMS

Parallel knowledge-based systems(KBS) for specific problem domain (PD) will be considered later. This intelligent information systems support new information technology for problem solving. There are following stages of knowledge representation and processing in KBS: conceptual modelling, formal logical and procedure-oriented knowledge presentation(Mylopoulos and Brodie, 1988).

There proposed an implementation schedule of the KBS query, which describes its main processing stages in the KBS. On any stage there is a corresponding system supporting applicable technology implementation.

1. The query specification as a system job for required knowledge acquisition is based on the formal PD desciption. The usage of graphical possibilities and means for query specification is widespread, a friendlyman-machine interface being provided.

2. The solution schedule is generated according to the specified query. It is based on logical inference techniques for the knowledge representation models used in the DB.

3. The task solution is performed in the computing system ,on the base of DB information. On any technological stage and corresponding support system there should be determined a user, language, object set and operations, allowable query set, inference tools, equivalent

transformation and query optimization means. The KBS query implementationschedule mentioned above includes three main technological stages and three corresponding control levels (CL) of the query implementation.

CL1: task specification by means of the formal lanquage, problem statement qualification in the interactive mode.

CL2: declarative task description conversion into procedural (operational) representation.

CL3: task solution by the abstract computer is based on the algorithm constructed. The task is presented as the query set described as expressions allowable for this level, i.e.expressions of the corresponding algebra which will be considered later. The instruction set of abstract computer is in agreement with the operation set allowable on this level. Any of the queries is the DB processing program. The specialized( with hardware, software-hardware or software implementation) processor with DB processing orientation and a particular data model usage is used as an abstract computer, e.g. a parallel DB machine with hardware support of relational statements. The query processing schedule on the CL3 level with fixed (constructed) algorithm of the solutiones is in agreement with common technology of the task solution in DB processing system on the DBMS basis.

Mathematical models of the multiprocessor computing system(MCS) are discussed assuming the following MCS architecture: distributed memory, distributed control (distributed operating system) and universalswitching system. Let M processor (M≥1) be available, each of these possesses its own main memory; the shared external memory is available and the shared main memory can be available. Any hardware components of MCS can communicate with each other by means of switching system (processor). Such MCS obviously spreads over the wide range of MCS architectures as static as well as dynamically reconfigurable ones (Evans, 1982; Glushkov, 1974). Here we'll describe new techniques for parallel computations during DB and KB processing program parallelizing and parallel processing at MCS with various architectures. These techniques are rather general approaches, can be used for wide range of problems and provide high degree(controlled) parallelism, referred here as vertical, horizontal and relation binding paralleling techniques(Polyachenko et al., 1989). The vertical paralleling technique controls the computations in asynchronous way, horizontal paralleling technique provides data structure decomposition and its parallel processing in multiprocessor environment. These methods are oriented to DB and KB parallel processing (Polyachenko et al., 1990) within intelligent information and other computer-aided systems. Program and computation process model transformations are reduced to the following ones:

1.Partitioning: graph cutting into subgraph satisfied to precedence and resource constraints.

2.Buffering: to each array of considered subgraph will set into correspondence the buffer in the main memory of computing system. The buffer size can be static or dynamic value.

3.Single assigning: each subgraph is assigned to execution by exactly one processor. Subgraph nodes(programs) can be processed sequentially or in multiprogramming way for independent programs in relation to their informational dependencies or asynchronously by means of vertical programming technique.

4.Multiple assigning: each subgraph is assigned to execution by several processors. Subgraph programs are chosen to be processed with correspondence to traditional or vertical paralleling techniques.

5.Synchronous processing: supergraph is constructed in which the node set corresponds to the subgraph set. This supergraph will be processed as usual graph by traditional tools with precedence constraints defined by their informational relations. Computations optimization is performed by traditional scheduling theory models.

6.Asynchronous processing: supergraph is processed by means of vertical paralleling technique at graph level. If vertical paralleling technique is used at graph node(program) level, such approach is referred as two-level vertical paralleling technique.

Above mentioned transformations give the possibilities to choose efficient(optimal) method of computations organization with respect to computing environment and current situation.

Two dual optimization computation problems are formulated and solved for distributed query processing in computing network.

PROBLEM 1. To find such DB, KB distributions and computation organization technique, which provides the optimum of the query processing overhead(effeciency) function for given computing network, DB set and query set.

PROBLEM 2. To choose computing network(system) configuration, DB relation distribution and computation organization technique, which optimized computing system efficiency for the DB and query set given.

For these problems researching and solving mathematical models and methods are suggested. To note that these problems are NP-hard.


PROGRAM MODELS

For formal DB processing program representation we'll use the algorithm's algebra(Andon, Polyachenko, 1987) under the memory with the set of elementary conditions, the set of elementary functional statements, the set of elementary statements for communications with external informational environment. This model is based on object-oriented paradigm and described objects, their classes, properties and object inheritance, interobject communications and interface with para-llel MCS.

Let us consider following statements for communication with external informational environment which corresponds to MAJAK language for our original supercomputer ES-1766 (Wolcott, and Goodman, 1988): WRITE array element, READ array element, LOCK array element, UNLOCK array element, SEND data, RECEIVE data, CALL program. The multimodular program(MMP) $\Gamma$ with arrays, which consists of the set $P$ of sequential components(programs), is described as the set of pairs $\Gamma=\{(K_1,P_1),\ldots,$ $(K_N,P_N)\}$, where $K_i$ - component names, $P_i$ - corresponding components' programs. Let us $V(K_i)$ and $W(K_i)$ are the set of source arrays(DB, relations) and the set of resulting arrays of the $K_i$ component correspondingly. The set of source arrays(relations) resides on external informational environment (external devices). The execution of the MMP will be followed to reside resulting arrays at external informational environment. The equivalent transformation means that from equality of source array follows the equality of resulting array. Later we'll consider MMP equivalent transformations only.

Dyscrete dynamic system(DDS) is defined as the state set S and the system of transition rules $\langle==\rangle$ which are given on the set S. The sequence $pr=s_0,\ldots,s_n,\ldots$, where $s_j==\rangle s_{j+1}$ for each $j=0,1,\ldots,n-1,\ldots,$ will be referred as a process in DDS given, and $s_0$ - its initial state.

There are the subset $\{s*\}$ of final states in S, to which, by definiton, the transition rules not applicable. The deadlock states will be referred as states which are not final states and to which the transition states are not applicable. The process which is terminated by final state will be terminatable and the process which to go to deadlock state will be deadloack one. We have described in (Andon, Polyachenko, and Gunko, 1987) traditional computation process organization with synchroniza-tion by means of LOCK, UNLOCK statements and/or 'rendez-vous'-like principle (Evans, 1982), as well as asynchronous (vertical and horizontal paralleling) computations, proved their features for different important data processing program classes and examined deadlock conditions.

Following stages of DB program life cycle are described: specification, synthesis, optimization, parallelizing and realization. Corresponding support tools will be presented.

The synthesis of parallel intelligent systems are based on representation and usage of suggested models of computing system, programs, databases, computation organization techniques methods and methodology as well as the technology of complex DB program production and maintenance which consists of all technological steps of software life.

## INTELLIGENT SYSTEM DESIGN IN PARALLEL ENVIRONMENT

Intelligent system design will be considered on the example of parallel intelligent decision system(DSS) synthesis by means of proposed models usage for system description and development. Such class of systems are based on expert systems and parallel knowledge based computing systems and introduced to provide support to the human decision makers at the management level in distributed manner. This work provides theoretical and practical guidelines in the development and management of such knowledge based DSS on the basis of above suggested models, techniques and tools. Multiprocessor computing systems with local/distributed DB and KB are capable of monitoring and controlling subsystems while exchanging information via data networks with other systems and providing the interface with human desicion makers that acts as supervisors and managers. These tools provides intelligent shell and interactive design process for knowledge based DSS in parallel environment of multiprocessor sytems with various architectures.

## SOME APPLICATIONS AND PERFORMANCE EVALUATION

Described models and paralleling techniques have been used in design of some projects conducted in the Institute of Cybernetics of the Academy of Sciences(USSR), and at the International Basic Laboratory for Artificial Intelligence(Czechoslovakia), namely, in software package for computing process organization at ES(IBM/360,370) computers and their multimachine systems (Andon, Polyachenko, and Gunko, 1987), in the operating system for multiprocessor supercomputer ES-1766 with macropipeline computation organization (Andon, Polyachenko, 1987; Wolcott, Goodman, 1988), in the relational database manipulation system, in the software package for optimization program generating, for pattern recognition problem solving and in the project of higly parallel multiprocessor data and knowledge base machine in transputer environment(Polyachenko, 1989; Polyachenko et al., 1989). Now last project is designing within International cooperation with Universities of USA, Italy and firms of France, Hungary, Czechoslovakia.

The real and test tasks of DB and KB handling on MCS are examined. The processing time of such tasks is reduced 1.5-7 times as compared to the usage of the traditional computations organization facilities. Hardware implementation of suggested methods can improve the efficiency of parallel computations organization. For these reasons we designed the architecture of higly parallel multiprocessor DB machine with associative parallel processing, active hardware buffering and parallel input/output processor. Some of described vertical and horizontal paralleling techniques, program transformations and optimization at this MCS are provided by means of hardware tools. Performance evaluation analysis and experimental studies conducted on both real and test tasks show that execution time has been reduced 2-9 and processors workload has been increased 2-3 times due to application of proposed facilities as compared with standard operational tools.


REFERENCES

Andon, F.I., Polyachenko, B.E., and Gunko, O.L., 1987, Models for Organization of Asynchronous Computations in Multimodular Data Processing Programs, Cybernetics, 23:5.

Evans, J., 1982, "Parallel Processing Systems", Cambridge Univ. Press, London.

Glushkov,V.M., 1974, Recursive Machines and Computing Technology, IFIP Congress-74, 2:65, North-Holland, Stockholm.

Mylopoulos, J., Brodie, M., 1988, "Readings in Artificial Intelligence and Databases", Morgan Kaufman Publishers, London.

Polyachenko, B.E., Andon, F.I., and Gunko, O.L., 1989, in: "Analysis, Synthesis and Parallel Processing of Large Data and KnowledgeBases. International Symposium on Data Analysis and Learning Symbolic and Numeric Knowledge. Invited Papers", Nova, New York.

Polyachenko, B.E., 1989, Parallel Computation Optimization in Multiprocessor Environment, in: "14 IFIP Conference on System Modelling and Optimization", VEB, Leipzig.

Wolcott, P., and Goodman, S.E., 1988, High-Speed Computers of the Soviet Union, Computer, 21:9.

# THE HUMAN-CENTRED PARADIGM IN KBS DESIGN THEORY

John G Gammack and Robert A Stephens

Bristol Transputer Centre
Department of Computer Studies and Mathematics
Bristol Polytechnic, UK

## 1. Introduction

In this paper we outline two radically different approaches to Knowledge Based Systems development and suggest that progress has been hampered through uncritical adoption of the methods and metaphors of hard science, leading to a 'techno-centric' conception of knowledge. The technical advantages of this approach however, tend to be gained at the expense of the overall human- machine functionality by denying the value of the user's construction of knowledge and subserving the user's problem solving ability to the dictates of pre-conceived software. The contrasting, human-oriented, approach to developing systems views knowledge as socially distributed and owned, implying methods for agreeing common understanding coupled with a devolution of responsibility for knowledge construction onto those whose world it affects.

Case studies of KBS development are used to illustrate both the techno-centric and human-centred approaches; in the course of which, we identify acquisition methods and their appropriate usage within the broader theoretical paradigms.

## 2. Meta-theoretical standpoints

Classically there are two ontological traditions which provide the philosophical assumptions underlying systems development. These have been variously identified by Winograd and Flores (1986) as rationalistic and hermeneutic, Burrell and Morgan (1979) as functionalist and interpretivist, and S.P. Gill (1988) as rationalistic and humanist. Similarly within human inquiry, social scientists such as Reason and Rowan (1981) distinguish positivist and post- positivist approaches; Lincoln and Guba (1985) distinguish positivist and naturalistic approaches. Although these dichotomies are not exactly equivalent they may be broadly encompassed within the traditions Gregory (1986, 1987) identified as realist and idealist. As such they tend to share a family of assumptions which lead to different approaches to the practice of system development, as the following synopsis based on Gregory (1986) reveals.

Realism assumes that objects have a true structure and relate lawfully to one another, and discovering this is the task of science. These laws are useful in prediction and control, and transcend the particular moment and circumstances of their elucidation. Objects are

*Systems Thinking in Europe*, Edited by M.C. Jackson *et al.*
Plenum Press, New York, 1991

not transmuted in the investigation of them, i.e. the truth of an experimental result must not be affected by the observer or participants in the experiment, but exist independently. Idealism (the tradition of cybernetics) assumes that observers are fundamentally connected to the world by their acts of observation. This shapes what is observed so that meaning is elucidated by reference to oneself or to others, rather than to a canonical external referent. Idealism emphasises meaning- assignment in the world, and the conception of objects in the world is a subjective interpretation. Different observers may conceive 'the same material object' in different ways, but what is objectively so is a matter of consensual agreement.

Knowledge is viewed differently under these two perspectives. Realism considers knowledge to be reducible to primitive (atomic) components. It is logically separable from the knowing of it by a knower: i.e. to be representable in a knowledge base it is considered as a commodity that can be traded, retrieved, and transferred. It is also reasonable to think of knowledge as a set true facts bound together with rules. Idealism views knowledge as a process, which involves interpretation in a context, not as an enduring state of a static object. Knowledge consists of coherent relationships connecting knowers with the world, not facts with each other. Primitivity is also relative, a part of a process of describing relationships between knowers and their world. "Facts are anchored not in (absolute) truth, but in particular contexts" (Gregory, 1986, p 839).

Underlying the present practice of knowledge engineering is the assumption that there exists a corpus of domain facts independent of the way they are known to the knower and that they can be manipulated according rules of a prescriptive nature. We therefore question whether 'engineering' is always an appropriate metaphor for KBS development, and suggest that many of the problems apparent in KBS stem from an unexamined realist conception. Numerous academic concerns such as validation, rule consistency and ambiguity, lose dominion upon adoption of an idealist ontology.

Recognising the centrality of being in the world to the interpretation and application of knowledge leads to KBS designs based on human- centred or anthropocentric principles such as human-machine symbiosis (see K.S. Gill (1990) for a list of foundation ideas and concepts in human-centredness). Human, context sensitive, interpretation replaces interminable lists of rules, their qualifications and exceptions, thus eliminating spuriously precise formal specification. Similarly, general limits on machines imply that at a functional level consistency must be built into the system (White, 1988). This however inevitably limits the scope of the system and tends to restrict KBS to rather trivial well-specified domains. The human-centred alternative extends the overall scope by removing interpretive responsibility from the machine, since the context sensitivity unavailable to machines allows humans to deal effectively with formally inconsistent propositions.

The concern with formal correctness (e.g. rule consistency) and correspondence manifests itself in the current (functionalist) preoccupation with KBS validation. Validation becomes a problem when the machine is viewed as an autonomous reasoning artefact. Within a human-centred design framework this problem can be appropriately located at the level of the machine. The idealist ontology implies a concern, less with correspondence than coherence: with qualitative adequacy appraised by performance measures, user satisfaction and consensus. Since a machine is expected to perform its intended function consistently, functional validation is an important aspect of KBS development. This, however, is not in itself enough in a KBS, where the symbols inevitably relate to the world through meaning imputed by active human interpretation. As

Checkland and Scholes (1990 p 55) remind us "the boundary of an IS, if we are using that phrase seriously, will always have to include the attribution of meaning, which is a uniquely human act." Accordingly, not just KBS validation, but design also, must begin by recognising the inevitability of human interpretation.

## 3. The role of the user in KBS Development

Viewing knowledge as a disembodied transferable commodity has led practitioners to separate design from use and focus attention on the technical aspects of system reification. The deception implicit in this step has been exposed by Klein and Lyytinen (1985) as leading to the convenient view that the problems of information systems lie outside the scientific domain of inquiry. They go on to explain how in the orthodox IS literature data are defined as meaningful measurements and information as data which help to achieve organisational goals. The contingent view is "an example of reification by treating humanly defined goals as if they were functional requirements which are beyond human control. This conceptual trick works by first construing separate technical system goals and by then only admitting objective measurements which relate to these (and no other) goals as meaningful data. This produces system requirements which take on an objective appearance, but lack human support." (Klein and Lyytinen, 1985, p143).

The techno-centric approach often accompanies a condescending attitude towards the user, denying them a real design influence, constraining behaviours to an imposed conceptual model and knowledge representation. Systems developed this way often prove irrelevant or unworkable, incurring expensive post-hoc HCI.

By contrast the idealist perspective recognises the user as essentially a constructor and interpreter of symbolic knowledge and it is because the user's involvement is ultimately with the knowledge-base that design and acquisition issues cannot be considered independently of the system's use. A human-centred perspective focuses designers' attentions on the particular settings, activities, and interactions between people and relations between people and machines.

The reality of knowledge use implies participatory strategies in a human-centred approach to knowledge acquisition. Participative methodologies are not new in systems thinking, but to date have had little practical impact in KBS development. Recognising that "users cannot be disembedded from their social and cultural contexts" K.S. Gill (1989) pioneered a model for a participatory approach to knowledge transfer. In his Parosi project the users' construction of knowledge plays an active part in the design process, leading to mutual learning and shaping of knowledge. Discussion of participative approaches has identified such problems as group think, user uncertainty and resistance, differential power and experience, and pseudo-compliance. We recognise these difficulties but maintain that adopting a participatory or collaborative acquisition paradigm is required given the conception of knowledge outlined above. New directions and methodologies in human inquiry are described in Reason (1988) and Lincoln and Guba (1985) which may provide practitioners with models for the collaborative design of information systems.

## 4. KBS Development Paradigms - Two Case Studies

Practical examples now illustrate some of the theoretical points made above, and show how the assumptions underlying the development paradigm affect the overall system functionality.

## 4.1 Case study 1 - Life Underwriter training system

For largely commercial reasons, the project managers made several design decisions prior to the knowledge engineering itself. These were: a single expert (a senior underwriter) from the company would be allocated to the project; an expert system (to train juniors to detect risk in application forms for life insurance) would be implemented using an already purchased shell, and that the knowledge acquisition would be done by the researchers working to deadlines. In preliminary interviews, the expert suggested that underwriting comprised risk identification in 6 distinct areas. A decision was therefore taken to build 6 modular expert systems, linked together, imposing a division of labour which subsequently appeared unconsidered.

Each area was taken in turn, and, because of the requirement to represent knowledge in rules, used elicitation techniques based on identifying concepts, their relationships to categories of risk, and the discriminating information required of the user that narrowed down the search space. In the module for assessing geographical risk for example, the pattern of elicitation began with *Concept Generation* taking terms from a reassurance manual (usually names of countries chosen by the expert to cover a wide range); followed by *Repertory Grid* to get constructs that discriminated those terms (war zones or otherwise; prevalence of disease, or differential risk according to nationality of applicant); then using the terms to construct a tree of rules using the *Context Focussing* technique. The rule tree was then implemented more or less directly in the shell's production system architecture. Substantially similar procedures were followed in building the other modules such as medical and financial risk, (though see Gammack, Battle and Stephens (1989) for details of an alternative approach to acquisition and system development in this latter domain). The expert then decided a prioritised ordering of the modules, so the six expert systems were sequentially linked and embedded within a larger program, which was then presented to juniors for testing.

It emerged that the categories identified by the expert did not always map onto the categories understood by the juniors, and insufficient account had been taken of their point of view in the design of the modules. In particular, the initial menu presented to the juniors invited them to categorise medical conditions into those provided by the reassurance manuals, with which the expert was familiar. The first screen required an untrained junior to choose among such terms as 'pulmonary', 'cardio-vascular' and 'endocrine' before answering more discriminating questions on diseases. However, because neither the designers nor expert had appreciated that trainees would not generally know these medical terms, the system as originally designed was unusable, and required modification.

In relation to this point about identifying the user's conception, a small study (Gammack, 1990) was done using the repertory grid, in the constructivist mode in which it was originally intended (Kelly, 1955). Presenting the junior underwriters with the same list of diseases as had been culled by the expert from manuals, their constructs made no use of the traditional medical categories of the reassurance manual, but rather reflected a practical concern with information on the application form directly relevant to making an assessment. Is the condition treatable or terminal? Has the person been recently hospitalised? Is the applicant currently on drugs, and if so then which? These issues cut across the medics' theoretical conception of bodily systems, and imply an expert system design grounded in user-relevant issues.

Following the system development, S.P. Gill (1990) conducted a participatory inquiry into the phenomenology of underwriting working with one of the juniors involved earlier. This revealed aspects which would have affected the design had they been appreciated at the outset. For instance underwriters do not sequentially look for types of risk, ticking them off some checklist, but instead appear to build up a picture of the client's life-style, which may emerge as life threatening risks over the policy's duration. A 'farmer' may be a low risk occupation, whilst 'Australia' may be a low risk country, leading to a conclusion of low risk when processed in sequential modules, yet an Australian farmer might be a high risk category (due to flying small planes in remote areas, or skin cancer). This project is more fully detailed in Bolger et al (1989) and Gammack (in press).

## 4.2 Case Study 2 - IDIOMS - an anthropocentric KBS development

The IDIOMS (Intelligent Decision-making In On-line Management Systems) project (Fogarty et al, 1990) is committed to a human-centred interface design strategy. The aim is to build a tool for gathering management information from very large databases, made possible by a powerful parallel hardware platform. The design aims to integrate 'knowledge' with 'data', with the human at the centre of the overall system. Briefly, a qualitative, humanly understandable representation of patterns in the database (e.g. historical) based on human supplied (rather than merely statistical) categories is abstracted, which the decision maker can then augment with information representing subjective judgements, and awareness of contextual information, business rules, new legal constraints and other non-historical considerations. Since such information is cognate with the representation of the data, the inferencing takes place on an enhanced basis, with the computational assumptions and the reports relatively transparent.

The IDIOMS philosophy can most easily be seen by contrast with some more conventional designs. In an expert system the rules are *prescriptive*, brittle, and rarely effectively combined probabilistically. By contrast, IDIOMS sees rules as a high level *description* of a tendency emergent from the database by virtue of selective viewing. The history of their use affects their strength, but the surrounding circumstances also affect their interpretation. This interpretation into the world must come from outwith the rule base itself, though due respect is paid to historical usage. An expert system's configuration of rules in any dynamic domain is likely to become increasingly obsolete and inappropriately dogmatic: IDIOMS constructs a representation afresh, based on the very latest data in the database. In banking applications, where patterns of transactions (or indeed workplace practice) may change in time, this design ensures up to date rules tailored to the context of the application.

Expert system rules are usually specified in terms of one particular goal variable, and are incapable of advising on other potentially interesting variables. IDIOMS aims to offer a much more flexible querying facility, supporting the range of queries a user may want to ask. This however is not achieved by anticipating a wide set of possibilities, and coding those in, but rather by designing the system to support queries designed by the user. Relatedly, new rules or constraints which are subjective in origin play no role in traditional expert systems, where the user is often a glorified data entry device. IDIOMS recognises not just the inevitability, but also the importance of including subjective and contextual information, and the user interface is designed to support this functionality. As an example, a change in the mortgage law regarding owners in common, means that history may not be a good guide to decising the riskworthiness of singles, and the automated system requires qualification. This is often done by crude post-hoc amendments to statistical output, but in IDIOMS, relevant information and judgements may be directly incorporated in the representation of the database PRIOR to any inferencing.

Although this work is at an early stage, we hope it will provide a vehicle for exploring a number of issues in human centred systems design, as well as KBS development more generally.

## 5. References

Bolger, F., Wright, G., Rowe, E., Gammack, J. and Wood, R., 1989, LUST for life: developing expert systems for life assurance underwriting, in: "Research and Development in Expert Systems VI", N. Shadbolt, (Ed.), CUP.

Burrell, G. and Morgan, G., 1979, "Sociological Paradigms and Organizational Analysis", Heinemann.

Checkland, P. and Scholes, J., 1990," Soft Systems Methodology in Action", Wiley.

Fogarty, T.C., Gammack, J.G., Battle, S.A., Miles, R.G., 1990, Intelligent Decision Support Using On-line Database, in: Proc. BCS Parallel Processing Specialist Group Conference on Commercial Parallel Processing, Unicom Technology Transfer Series, Uxbridge, p193-99

Gammack, J.G. Battle, S.A. and Stephens, R.A., 1989, A knowledge acquisition and representation scheme for constraint-based and parallel systems. in: Proc. IEEE conference on Systems, Man and Cybernetics, Cambridge MA, p1030-1035

Gammack, J.G., 1990, User Participation in Systems Design, in: 1990 Workshop on Human-Centred Systems, Brighton Polytechnic, UK

Gammack, J.G., in press, Knowledge Engineering Issues in Decision Support, in: "Expertise and Decision Support", G. Wright and F.M.I Bolger, (eds.), Plenum, New York.

Gill, K.S., 1989, Reflections on Participatory Design, AI and Society, 3: p297-314.

Gill, K.S., 1990, "Summary of Human-Centred Systems Research in Europe", SEAKE Centre, Brighton Polytechnic, UK.

Gill, S.P., 1988, On Two AI Traditions, AI and Society, Vol.2, p 321-340.

Gill, S.P., 1990, A Dialogical Framework for Participatory KBS Design, in: Tenth European Meeting on Cybernetics and Systems, University of Vienna, April 17-20, 1990.

Gregory, D., 1986, Delimiting Expert Systems, IEE Transactions on Systems, Man and Cybernetics, Vol SMC-16, no. 6. p 834-842.

Gregory, D., 1987, Philosophy and Practice in Knowledge Representation, in "Human Productivity Enhancement-Organisations, Personnel and Decision-Making", J. Zeidner (ed.), NY, Praeger.

Kelly, G., 1955, "The Psychology of Personal Constructs". New York: Norton.

Klein, H.K. and Lyytinen, K., 1985, The poverty of scientism, in: Research Methods in Information Systems, E. Mumford, R. Hirschheim, G. Fitzgerald and T. Wood-Harper, (eds.), p131-162, North-Holland.

Lincoln S.Y. and Guba E.G., 1985, "Naturalistic Inquiry", Sage.

Reason, P., 1988, "Human Inquiry in Action: developments in new paradigm research", Sage, London.

Reason, P. and Rowan, R., 1981, "Human Inquiry, A Sourcebook for New Paradigm Research", Wiley.

White, I., 1988, The Limits and Capabilities of Machines - A Review, IEE Transactions on Systems, Man and Cybernetics, Vol SMC- 18, no. 6. p 917-938.

Winograd, T. and Flores, F., 1986, "Understanding computers and cognition, A new foundation for design", Addison-Wesley.

# THE APPRECIATIVE INQUIRY METHOD

## A SYSTEMS-BASED METHOD OF KNOWLEDGE ELICITATION

F.A. Stowell

Faculty of Science
Nuffield Centre
Portsmouth Polytechnic
Portsmouth
Hampshire
United Kingdom

D. West

Business Information Systems
Bournemouth Polytechnic
Talbot Campus
Fern Barrow
Bournemouth
Dorset

## INTRODUCTION

In this paper the authors offer a brief outline of a practical approach to knowledge elicitation (KE) for expert systems based upon the notion that human expertise consists of both "objective" knowledge (i.e. factual, rule-based, text-book, logical, tangible, deterministic knowledge) and "subjective" knowledge (i.e. knowledge resulting from experience, judgement, intuition, prejudice, rules-of-thumb).

An argument is put forward concerning the potential benefit of viewing KE as an inquiring system as a means of trying to address the problem inherent to many current KE techniques, namely, the difficulty in eliciting the more "subjective" areas of human expertise. The authors' argument for a change in emphasis from knowledge elicitation to inquiry reflects their attempt to develop an approach which is capable of providing a framework within which, prior to elicitation, an expert's knowledge can be explored and made explicit. A practical KE approach which has been developed out of an "interpretive" theoretical foundation is presented. The approach, which has been developed upon Vickers' notion of "appreciation", draws upon systems thinking, and in particular "soft" systems thinking, as a means of facilitating what will be referred to as "appreciative inquiry" (i.e. emphasis is placed on learning and understanding a domain through the process of inquiry).

## KNOWLEDGE ELICITATION

Knowledge elicitation (KE), can be described as the process of extracting "knowledge" or "expertise" from a human expert with a view to organising it in such a way that it can be manipulated within a computer-based "expert system". This process is of fundamental importance during the development of an expert system since, arguably, the power of the expert system relies upon the richness of the knowledge-base. Knowledge elicitation is more than asking an expert to relate all he knows - it involves the processes of learning, questioning and

Tel. (0705) 843017;
Fax. (0705) 843335;
E.M. StowellF@ uk.ac.port.cv

*Systems Thinking in Europe*, Edited by M.C. Jackson *et al.*
Plenum Press, New York, 1991

493

above all understanding, or *appreciating* (see Vickers, 1965) both the experts knowledge *and* the way in which it is used. Within the KE process there are a number of problem areas, relating to those actors involved in the process, which are briefly outlined below in order to give some illustration of the type of problems that need to be considered when undertaking KE.

## The Expert(s)

One of the first questions that needs to be asked when starting an expert system project is whether there is a suitable expert available who is willing to participate. The amount of the expert's time needed during the project should not be underestimated. Consequently, it is important to ensure that there is management support for the project as a whole.

There are a number of important questions that need to be addressed at the start, including "Who is the expert?", "Is there more than one expert?", "Is the expert suited for undergoing the KE process?", "Can he/she be spared from his work long enough to be useful?" "Is he/she willing and able to divulge his/her expertise?".

## The User(s)

The role of the prospective "user" of the expert system is arguably the most important actor in the development process. "Users" are often treated as a consequence of the design of an expert system and, therefore, play a minor role in the KE process. However, it is argued here that the "user's" requirements should dictate the complexity of the expert system which, in effect, should be "user-driven".

It is suggested here that the user of an expert system can, justifiably, be regarded as an expert in his/her own right since as they are deemed to *need* an expert system to help them in their problem solving presumably they have some ability in recognising the problems that they face. For this reason, it is the authors' view that during KE the user's view of the domain of expertise should also be explored as an integral part of the KE process.

## The Knowledge Elicitor

One of the main problems for the knowledge elicitor is that of avoiding the introduction of his/her own bias and prejudice into the development process as he/she interprets what has been learnt from the expert. In domains where the elicitor has some expertise there is a danger of distorting the expert's knowledge by the elicitor imposing his own ideas upon the information elicited. Furthermore, the knowledge elicitor may bias a KE session by asking leading questions that anticipate or direct the expert's response. This difficulty may become particularly noticeable when conducting unstructured interviews or undertaking group activities, such as brainstorming, where a dominant character may impose his view upon the group as a whole and thereby directing the knowledge elicited (See Cleaves, 1987 for an interesting discussion of the problem of bias).

It is emphasised here that the knowledge elicitor's job cuts across many different types of skills and tasks, technical knowledge and inter-personal skills (eg. Hart, 1986). One problem that we wish to highlight is that many of those who are assigned the task of developing an expert system are either, by discipline, computer scientists or have little IT knowledge at all.. It is often the case that the educational process of computer scientists has little room for the "human" aspects of problem-solving (Stowell, 1990), whilst those without IT training find little real practical guidance in the literature about the KE process and how it needs to be undertaken within their own organisation. Few people have been trained as knowledge elicitors and, instead, seem to learn their skills through a process of trial and error - a situation which has tended to result in the lack of a unified and comprehensive approach to knowledge elicitation.

Recent thinking about the elicitation process has placed increasing emphasis upon the need for the knowledge elicitor to play a less dominant role during the KE process and to give the

494

expert and the user the opportunity to express their requirements and their knowledge (eg. Bell and Hardiman, 1989; Stowell and West, 1990; Gill, 1990). This change in emphasis of the knowledge elicitor's role requires, we argue, an equally radical change in the approach to knowledge elicitation. To this end a method which we call "Appreciative Inquiry" has been developed.

## THE APPRECIATIVE INQUIRY METHOD (A.I.M.)

The "Appreciative Inquiry Method" offers a framework within which the expert's expertise can be explored and modelled so as to provide the knowledge elicitor with a *subjective appreciation* of the domain. Furthermore, the method encourages the expert to learn about his expertise in that appreciative inquiry facilitates an exploration of "tacit" areas of knowledge which are otherwise difficult to quantify.

It is not intended to provide a detailed account of the method of the Appreciative Inquiry Method since this is available elsewhere (Stowell and West, 1990; West, 1991), but to provide an overview of the stages involved and a description of the process together with the activities of the expert.

AIM comprises four stages which can be as: Identify, Specify, Model and Extract. The process is iterative and intended as an investigative framework in the spirit of action research. The first stage of AIM is intended to help the expert to determine the main elements of his domain of expertise and to define his boundary of expertise. This is achieved through the employment of a "map" based upon the Venn diagram convention. The expert is shown an example and then asked to produce a map of his/her domain of expertise which is then used as a form of agenda to discuss the domain. This activity has been shown to provide a great deal of information but more importantly it provides a potential learning exercise for both expert and knowledge elicitor: the elicitor and the expert to learn about the domain: the elicitor learns from the explanations of the map that are given by the expert about the map and there is an opportunity for the expert to learn about his/her own view of the domain through the process of drawing and explaining the map (West, 1989). Experience has shown that the map appears to offer a quick and simple means for the expert to express a view about a given domain and in a form which can be readily communicated to others (West, 1989).

The second stage of the method is to specify each element defined in the map. It is here that we borrow, unashamedly, ideas from Soft Systems Methodology (SSM) (Checkland, 1981; 1990). The expert is asked to explain each element of the map through answering questions relating to a derivation of the so-called "CATWOE" test of SSM. The aim is to ask questions that are content-independent which do not presuppose a given answer nor which are formed by virtue of the elicitor's understanding of the domain. Once completed the information gained from the specification stage enables a concise definition of each of the elements of the map to be produced by the elicitor. The resulting description of the map element, therefore, is the product of the elicitor's *interpretation* of the information given by the expert during the previous stages of the approach.

From the definition of each map element a type of "conceptual model" is developed by the elicitor (Checkland, 1981), showing the activities which need to be present in order to operationalise the activity that the expert has defined. This model (or models) provides a pictorial representation of the elicitor's interpretation of the expert's domain of knowledge. The model is the result of information verified and tested through the iterative process provided by the first two stages of the approach and, therefore, represents a logical description of the expert's view of the domain. The model can be used to help structure relevant questions into the described situation so as to be able to extract data to build the expert system. An "agenda" developed from the model's activities provides the elicitor with a useful framework for the next level of discussion with the expert. Development of the element descriptions and conceptual models can take place away from the interview room which reduces the time that the expert needs to devote to being interviewed and also lessens the amount of pressure placed upon the knowledge elicitor. It is important to stress that this process of KE is iterative and, thus, the method itself provides a framework for inquiry that can be continued until sufficient understanding, or detail, has been achieved.

# REFLECTIONS

The process described through the three stages of the method enables the elicitor to identify "facts" about the domain that can be developed in a form suitable for the expert system, and, information which defies being structured as rules but which may be considered as providing a *context* in which the "facts" reside. This latter *contextual* type of knowledge, which is often difficult to define in terms of facts and rules, may be identified and reinvestigated by a further application of the proposed KE method in order to bring about an increased appreciation of the domain.

The significance of viewing KE as an "Inquiring system" lies in the way in which emphasis is removed from the notion elicitation, with its concentration upon *extraction* of knowledge. Instead, AIM focuses upon the need to understand, or *appreciate* the domain prior to extracting knowledge suitable for the expert system. The increased understanding which is sought through the inquiry process supported by AIM may also be seen as a way of providing a framework with which to utilise the more "traditional" approaches to KE, eg. structured interviews, repertory grids, prototyping.

The objective of developing AIM was an attempt to develop a means of eliciting the more heuristic aspects of human expertise. Vickers' ideas about "appreciation" (Vickers, 1965), were seen as being of potential use since they seemed to offer a useful description of the way in which experts can be said to gain and practice their expertise (West, 1990a, 1990b). The strength of adopting such a model of expertise is that it takes into account not only the gaining of factual, formal knowledge but also offers an explanation of the way in which subjective knowledge is developed (Stowell and West, 1988, 1989, 1990; West, 1990a). The method of KE advocated here, which is conceptually an hermeneutic approach, is underpinned by the method of action research (Susman and Evered, 1978). Attempts are made not to devalue the interpretive ideas upon which AIM is based by producing a method that is deliberately no more than a framework for the guidance of the knowledge elicitor and expert. The simple tasks that are required of the expert at each stage of the method are intended to avoid imposing themselves upon the KE process but to encourage the expert to divulge his expertise without inhibition and without the need for long, complex and tiring procedures. The results of using this approach so far suggest that the general framework of AIM appears to provide "guidance" for the elicitor and expert during KE rather than imposing a step-by-step technique which determines the KE process.

One additional advantage of the method described above is that it provides documentation of the elicitation process (ie. by means of the expert's map, the production of the description of the different map elements, the conceptual model and the agenda produced from this model). Documentation has been referred to as a neglected area of knowledge elicitation (Burton et al., 1987), but AIM provides, as a natural product of the method, a set of comprehensive documentation which, for the most part, has been produced by the expert or else has been verified, corrected and restructured by the expert. The practical activities involved in the different stages of the method provide the basis for the expert, user and elicitor to gain an "appreciation" of the problem domain. Since AIM can be used to help explore a user's understanding of the problem domain, or to investigate the views of several domain experts, it can be seen to provide a means of making these different views explicit. This is achieved by promoting discussion about the domain, modelling the different views expressed, which in turn, helps to structure the elicitation process as well as providing appropriate documentation.

## REFERENCES

Bell, J. and Hardiman, R.J. (1989) The Third Role - The Naturalistic Knowledge Engineer in "Knowledge Elicitation - Principles, Techniques and Applications", (Ed. D. Diaper). Ellis Horwood Ltd.: Chichester. pp49-85.
Burton, M., Shadbolt, N.R., Hedgcock, A.P. and Rugg, G. (1987) A Formal Evaluation of Knowledge Elicitation Techniques for Expert Systems: Domain 1 in "Proceedings of the First European Workshop on Knowledge Acquisition for Knowledge-Based Systems", Reading, UK. Section D3.
Checkland, P.B. (1981) "Systems Thinking, Systems Practice". Wiley: Chichester.

Checkland, P.B. and Scholes, J. (1990) " Soft Systems Methodology in Action". Wiley: Chichester.

Cleaves, D.A. (1987) Cognitive Biases and Corrective Techniques: Proposals for Improving Elicitation Procedures, International Journal of Man-Machine Studies, Vol. 27, pp1555-166.

Gill, S.P. (1990) A Dialogical Framework for Participatory KBS Design in "Cybernetics and Systems", (Ed.: R. Trappl), Part 2. Kluwer Academic Publishers: Dordrech. pp1039-1046.

Hart, A. (1986) "Knowledge Acquisition for Expert Systems". Kogan Page Ltd.: London.

Lewin, K. (1948) Action Research and Minority Problems in "Resolving Social Conflicts". (Ed.: G.W. Lewin). Harper: New York. pp201-220.

Stowell, F.A. (1990) Designing Information Systems: Towards Client-based Design. 8th International Congress of Cybernetics and Systems Conference, June 1990, City University of New York.

Stowell, F.A. and West, D. (1988) Expert Systems: Knowledge Acquisition and the Systems Epistemology in "Cybernetics and Systems", Edited R. Trappl, Kluwer Academic Publishers: Dordrech. pp941-948.

Stowell, F.A. and West, D. (1989): Expert Systems: Ramifications for the Knowledge Engineer, Systems Analysis Modelling Simulation, Vol. 6. No. 9, pp673-678

Stowell, F.A. and West, D. (1990) The Contribution of Systems Ideas During the Process of Knowledge Elicitation in "Systems Prospects: The Next Ten Years of Systems Research". Plenum Press: New York. pp329-334.

Susman, G.I. and Evered, R.D. (1978) An Assessment of the Scientific Merits of Action Research, Administrative Science Quarterly, Vol. 23, pp582-603.

Vickers, G. (1965) "The Art of Judgement: A Study of Policy Making". Chapman and Hall: London.

West, D. (1989) Field study documentation. Unpublished.

West, D. (1990a) 'Appreciation', 'Expertise' and Knowledge Elicitation - The Relevance of Vickers' Ideas to the Design of Expert Systems, Journal of Applied Systems Analysis, Vol. 17, pp71-78.

West, D. (1990b) Knowledge Elicitation as an Inquiring System: Towards a Knowledge Elicitation Methodology in "Proceedings of the 34th Meeting of the International Society for Systems Science", 8-13th July, 1990. Oregon, USA.

West, D. (1991) "Towards a Subjective Knowledge Elicitation Methodology for Expert System Design". PhD Dissertation. Unpublished.

# SYSTEMS PERSPECTIVE FOR KNOWLEDGE REPRESENTATION

Maryvonne Longeart

Département d'informatique
Université du Québec à Hull
Hull, QUE, CANADA

Gilbert Boss

Faculté de philosophie
Université Laval
Québec, QUE, CANADA

## 1. INTRODUCTION

Knowledge representation has been associated with the development of computer systems exhibiting some degree of intelligence. However, it has been recently acknowledged that capturing knowledge about the world is also essential to systems engineering and the elicitation of the users requirements regardless of the so called "intelligence" of the system being implemented (Bordiga, Greenspan, Mylopoulos, 1985). Any system specification effort is partly a matter of deriving a world model and deciding which type of objects should be represented and which of their properties and relations are relevant. this world model represents the context within which the computer based system will operate.

The need for a general purpose representation tool compatible with any "world view" or ontology has stimulated researches on specification languages and knowledge representation environments for systems analysis. The assumption that has dominated most research in this area is that the basic structure of conceptual systems could be represented as a taxonomically organized network. A taxonomy is a hierarchy of concepts such that more general concepts are accessible from less general ones to which they are related. In such a hierarchy, a type is associated with a set of necessary and sufficient conditions that must be satisfied for something to be of that type. A subtype is defined by its genus which identifies a more general type under which it is classified and its differentia which distinguishes it from its genus. This is typically an Aristotelian way of defining a concept and Aristotle's theory of category is the paradigm of this kind of hierarchy. The importance given to taxonomy is a general feature of virtually all representation tools based on the structured object form of knowledge representation, even though it has been generally acknowledged that provision should be made for different types of relation in the system.

Experiencing with one of those knowledge representation environment, we tried to represent several philosophical systems as networks of concepts (Longeart, Boss, 1990). The working hypothesis on which our research is based is that philosophical systems are models of the world *par excellence* and that, given the conscious and systematic way in which a philosophy is developed and constructed, the world view inherent in a philosophical enterprise would be more easily grasped than the implicit one present in any ordinary commonsense knowledge.

We shall first present the structured object paradigm of knowledge representation. Then, an example of partial representation of a philosophical system will be presented. This example has been worked out using the CODE representation environment (Skuce, 1989). Finally, provisional observations will be made on the use of structured objects to represent abstract conceptual constructions.

*Systems Thinking in Europe*, Edited by M.C. Jackson *et al.*
Plenum Press, New York, 1991

## 2. STRUCTURED OBJECT REPRESENTATION FORM

Following Nilsson's terminology (1980), we shall use the term "structured object" for any representational scheme whose basic elements are analogical to nodes and arcs of graph theory. This includes semantic networks, frames and objects in the object oriented approach. Basically, objects are equivalent to frames and the frame description form of representation is mainly an elaboration of the semantic network one. Semantic networks can be viewed as a collection of conceptual graphs representing mutual connections between concepts and their insertion within a context which constitute the domain of discourse. (Sowa, 1984, 1989)

A conceptual graph is a connected graph formed from concept and relation nodes such that each relation node is linked only to its requisite number of concepts and each concept node is linked to one or more relation nodes (apart from the special case of a graph consisting of a single concept).

Suppose Book_3 were a specific sample of Hobbes' *Elements of Philosophy*. The individual object referred to by Book_3 is an instance of the abstract concept, or type, book, referred to by the generic word "book". Let us use the notation InstOf (for "instance of") to represent a relation node between a concept node representing an instance and a concept node representing a type.

Philosophy books are books. Let us use the notation IsA (for "is a") to represent a relation between two types, one being eventually more general than the other.

Types and instances can be organized in hierarchies as in graph 1 below representing Book_3 as an InstOf T_Philosophy_book which IsA T_book.

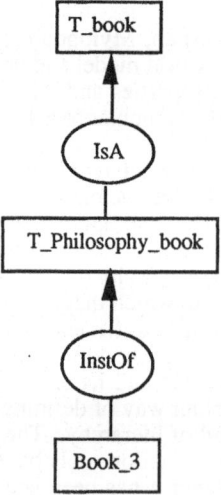

**Graph 1** *A type hierarchy*

A binary predicate can be represented by a triplet (Object, Attribute_j, Value_j). If all triplets associated with an object are united in a single structure, what is obtained is a unit of information called a "frame". The object or frame representation form is an elaboration of the semantic network form where the emphasis is on the structure of concept nodes themselves. Those structured nodes are called objects or frames. Each couple (Attribute_j, Value_j) in a frame is called a "slot". The frame notation for Book_3 above would be:

Book_3:
    Author        :       Hobbes of Malesbury
    Title          :       Elements of Philosophy
    InstOf        :       T_philosophy_book

500

# 3   EXAMPLE OF REPRESENTATION USING THE CODE SYSTEM

Using the CODE environment, we partially represented several philosophical systems as networks of concepts.

## 3.1   CODE As a Structured Object Representation Form

CODE, for "Conceptually Oriented Design/Description Environment", is a general purpose knowledge description and acquisition environment intended to assist in the process of developing descriptions, definitions or specifications of systems. It belongs to the family of frame based languages of representation.

Three basic tools are the keys to understanding the potential of the CODE system for any conceptually oriented analysis. They are the CODE implementation of Concepts, Properties and Inheritance mechanisms.

CODE organizes knowledge into small units termed conceptual descriptors (cds). A concept is described by its associated cd. Cds are arranged in inheritance hierarchies so that more specific concepts may inherit properties from more general ones.

A property is a unit of information associated with a concept in a cd. Properties are divided into system properties and user properties.

The most important  systems properties are:

| | | |
|---|---|---|
| cdName | : | The name of concept |
| supers | : | One or more superconcepts |
| kinds | : | Subconcepts which form disjoint partitions |
| subconc | : | A list of all other subconcepts that are not part of some partition |
| instanceOf | : | Cds that this one is an instance of |
| instances | : | Instances of this cd |

User properties are defined by the user. They are grouped in categories.

A suggested application of CODE is in the area of software engineering. Our own goal in using the CODE system was to evaluate the expressive power of concept hierarchies for representing conceptual systems in general and to explore possible applications in as well as outside the field of computer science.

## 3.2   Example of Transposition of an Abstract Conceptual System into a Structured Object Form Using Code

The philosophy of Thomas Hobbes (1588-1679) and more specifically, his logic and methodology as exposed in the first section of his *Elements of philosophy*, written in 1655, served as the raw material for this study (Hobbes, 1966).

We concentrated our efforts on Hobbes' theory of names and proposition. Names play a major part in Hobbes' philosophy. Hobbes believed that knowledge, the end of which was power, could be obtained only by adopting a method analogous to computation defined as "either to collect the sum of many things that are added together, or to know what remains when one thing is taken out of another". We start with concrete names which denote bodies and we connect them into propositions to form abstract names which denote the causes of concrete names.

The graph below shows part of a conceptual network constructed with CODE to represent chapters 2 (Of names) and 3 (Of proposition) of Hobbes' *Elements of philosophy*.

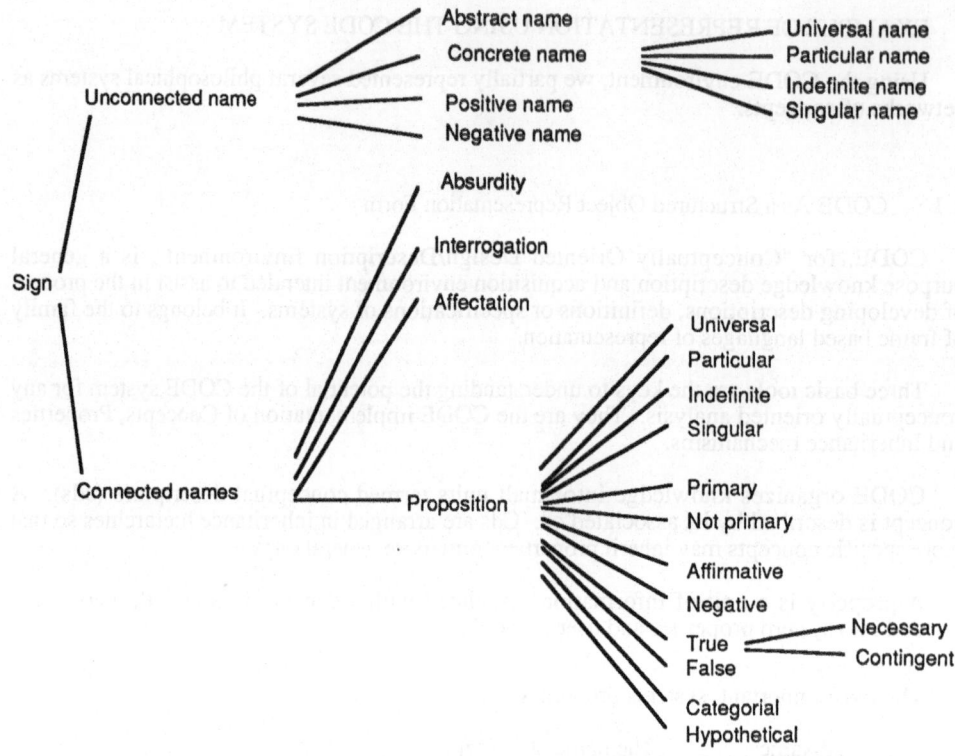

**Graph 2** *Hobbes' hierarchy of names and proposition*

A node in the network is a complex structure describing the corresponding concept. The network representation of Hobbes' hierarchy of names and proposition together with the information encapsulated in each node via user properties helped to capture some important insights of Hobbes' theory of knowledge.

For example, Hobbes distinguishes between the denotation of a word and its signification. Denotation and signification could be introduced as user properties in the node **Sign**.

| **Sign** | Denotation | : Things in the world |
| | Signification | : Conceptions |

These properties are inherited by all nodes dependent on the node **Sign** in the hierarchy unless otherwise specified. Not all names have a denotation (some names denote nothing in the world). This can be dealt with using the flag mechanism in CODE to block inheritance of that property.

Concrete names and abstract names do not denote the same type of things. Again, this can be represented by adding a constraint on the denotation.

| **Abstract name** | Denotation | : Things in the world (inherited) |
| | Constraint | : DenotationType: Accident of bodies |
| **Concrete name** | Denotation | : Things in the world (inherited) |
| | Constraint | : DenotationType: Bodies |

502

It is also interesting to note that **Universal name** is under **Concrete name** in the hierarchy and as such, it inherits its denotation from concrete name. This makes it clear that for Hobbes, universal names denote bodies and not essences as the scholastics believed at that time.

The representation also helped in disclosing ambiguity or even apparent contradiction in the original text or in a possible interpretation of it. For example, the node **Connected names** inherits Denotation from the node **Sign**. The constraint on the denotation for the node **Proposition** is "Truth value". However, Hobbes insists that truth and falsity are not things in the world and should be considered respectively as synonymous to "true proposition" and "false proposition". We shall not try to solve that problem here. The point is that the representation was a very useful tool for knowledge elicitation.

However, some important relations could not be represented in any satisfactory way. Here are some examples.

1.  A proposition is **composed of** names (the name of a subject and the name of a predicate) and a copula. Each component plays a specific part in the resulting composition and knowledge is nothing but this **combination of names**. The process is analogous to computation. Just as we add numbers or quantities, we add names drawing conclusions that are contained in the premisses but of which we were not aware.

2.  Concrete names were invented **before** proposition but abstract names were invented **after** because you need a concrete name (e. g. "body") and a copula ("to be") to form an abstract name (e. g. "to be a body" or corporiety).

3.  Absurdity arises if names of accidents (abstract names) are confused with names of bodies (concrete names).

These informations could only be introduced in the representation as comments on the model. In other words, temporal relationships, part/whole relationships, processes, events, to mention only a few, are not the kind of things a taxonomically organized network is meant to represent, but they are nonetheless essential to any world view whether the model of the world to be represented is explicitly stated in a philosophy or implicitly contained in some ordinary commonsense knowledge.


4. CONCLUSION

Requirements limited to the hardware/software specification leave out the crucial fact that a computer system, as an information system, includes not only programs and machines but also human beings with their conceptions of the context of the system as a whole. The requirements specification of a system should include a model of the world representing the user knowledge of the environment of the system under design.

Using a knowledge acquisition/representation tool based on the taxonomical approach to knowledge representation, we tried to represent several conceptual frameworks as they are explicitly stated by philosophers.

Our research shows that taxonomically organized networks are inadequate for representing most of the conceptual relations found in philosophical systems, and one should say in commonsense representation of the world. There might be taxonomical perspectives in any such system of concept, but only incidentally does the taxonomy sometimes constitute the core of the system as a whole. In any case, even though inheritance mechanisms have been at the centre of discussion about semantic nets and the structured object form of representation in general, it does not appear to be an essential feature as far as the expressive power of the representation is concerned. Unfortunately, this issue has hardly been addressed as such in the conception of knowledge representation tools, due to the taxonomical bias of most of the research in the field.

In the future, we plan to

1.      pursue the process of transposing Hobbes' epistemology using as much as possible the resources of CODE as a knowledge representation tool, pinning down limitations as we go along;

2.      specify theoretically what would be an adequate representation tool that would overcome the observed limitations. (Boss, Longeart, 1991)

Acknowledgments: We thank Jean-Marie Comeau for useful comments on an early version of this paper  We thank Douglas Skuce and the AI Laboratory of the University of Ottawa for the opportunity of using the CODE system. This research project has been supported by a grant from the Department of Philosophy of the University of Ottawa.

REFERENCE

Bordiga, A., Greenspan, S.and Mylopoulos, J., 1985, "Knowledge Representation as the basis for Requirements Specifications". IEEE Computer (April 1985)

Boss, G., 1988, "Rupture et Système chez Hobbes", Dialogue, Vol. 27, No 2, pp. 215-231.

Boss, G., Longeart, M., "Représentation philosophique par réseau sémantique variable". To be published in Laval théologique et philosophique, spring 1991.

Brachman, R.J. and Levesque, H.J., eds, 1985, Readings in Knowledge Representation. Los Altos: Kaufmann,.

Hobbes, Thomas, 1655, English Works of Thomas Hobbes of Malmesbury, Molesworth, William, ed., London: J. Bohn, 1839. New edition: Aalen: Scienta (1966)

Longeart, M. and Boss, G., 1990, "Structured Object Representation of Abstract Conceptual Systems". Rapport de recherche RR 90/11-17. Département d'informatique, Université du Québec à Hull.

Nilsson, N.J., 1980, Principles of Artificial Intelligence. Palo Alto, CA: Tioga.

Skuce, D., ShenKang, W., Beauvillé, Y., 1989, "A Portable Generic Knowledge Acquisition Environment that Understand Basic Logic and Language". Technical Report TR-89-11, Computer Science Department, University of Ottawa.

Sowa, J.F., Conceptual Structures: Information Processing in Minds and Machines. Reading, MA: Addison-Wesley, 1984.

Sowa, J., ed., Proc. of the Workshop on Formal Aspects of Semantic Nets, Catalina, Feb. 1989.

SYSTEMIC METAPHOR ANALYSIS FOR

KNOWLEDGE BASED SYSTEM DEVELOPMENT

R. C. Paton, H. S. Nwana,
M. J. Shave and T. J. Bench-Capon

The MEKAS Project
Department of Computer Science
The University of Liverpool
P. O. Box 147, Liverpool, L69 3BX

INTRODUCTION

One of the key phases in the development of a Knowledge Based System (KBS) is that of knowledge acquisition. All KBSs must go through this stage and it presents a major hurdle to be overcome when building such systems. The focus of attention at this stage of a KBS development is the domain which, in simple terms, is that knowledge needed to solve a set of real world problems. Research on the MEKAS project has indicated that knowledge about a domain not only includes the tasks and objects which it addresses, but also the wider cognitive and real world environment in which it is set (Paton & Nwana, 1990). As such, human knowledge about domains is so complex that without an *analysis* stage that probes the underlying nature of a real world domain and how experts may conceptualise it, the knowledge that is eventually incorporated into a KBS remains shallow and incomplete. Just as conventional systems development requires a *systems analysis* phase to analyse the problem situation, so KBS development requires a *knowledge analysis* phase which characterises the domain by organising the information acquired from a variety of knowledge sources into a coherent and unambiguous whole.

We propose seven fundamental characteristics that should be analysed for the purpose of describing a domain (Paton & Nwana, 1990):
• Theory - conceptual framework used in construction and maintenance.
• Metaphor - language of domain description; especially global.
• Metatheoretical constraints - concepts such as time and causality.
• Relations with other domains - similar areas of knowledge.
• History - the evolution of the domain in time.
• Structure - parts, relations and organisation of the domain itself.
• Purpose - problems the domain addresses, in terms of their solution.
These seven characteristics do not represent mutually exclusive areas of knowledge and are highly interrelated. Together they constitute what we call a domain characterisation.

Discussion in this paper will be limited to certain details concerned with the metaphorical nature of domains and how an understanding of this can assist the knowledge acquisition process. The requirement for this seven-featured approach to analysis has been incorporated into an iterative knowledge analysis methodology called SAAGS (Paton et al., 1990). Although there is insufficient space to give details of the operation of SAAGS it should be noted that it requires an analyst to formulate anticipations and assumptions about a domain prior to acquisition from knowledge sources such as books and experts. These anticipations may then be confirmed or refuted at a post-acquisition stage with the generation of models.

## SYSTEMIC METAPHORS AS ANALYTICAL TOOLS

The basic assumption of our approach is that humans construct models which allow them to refer to and come to an understanding of complexities of the real world. These models are dependent on theories (ie., general models) for their construction, maintenance and evolution. Models and theories are not hodge-podges of cognitive fragments; they are integrative in nature. The approach to model and theory development we elaborate is summarised in Figure 1. It is based on a realist interpretation of the world as consisting of observable and non-observable entities (Harré, 1970). Its application to domain characterisation is discussed more fully in Paton et al., (1990).

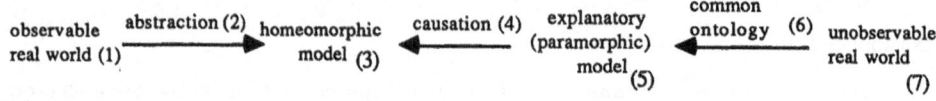

Note:
(1) and (7) are the real world domain.
(3) and (5) are cognitive constructions, the products of cognitive processes.
(2), (4) and (6) are cognitive processes involved in model construction.

Figure 1 - Model and Theory Construction

In this scheme, theory or model formation may be entirely dependent on an abstraction from the observable real world or, more commonly, will also involve other models which provide explanatory details about the non-observable real world. These explanatory models make use of metaphors as they seek to describe the unknown in terms of what is known. In this way metaphor provides the context by which analogies and similes can be made. This specifies a common ontology. (For completeness, note that some theories make use of mathematico-aesthetic ideas such as order, symmetry, transformation and harmony).

There are basic kinds of metaphor that pervade everyday language about complex domains and so are important for knowledge analysis. Some metaphors are related to a domain as a whole and provide global details, others are related to specifics. Two examples of global metaphors that are important to knowledge analysis are systemic metaphors and spatial metaphors (see Paton et al., 1990).

"System" is a very common word in everyday language. Systemic metaphors are characterised by a set of basic properties (called systemic M-properties (see Paton et al., 1990)):

Interacting parts,
Organisation,
Collective behaviour and whole system functionality.

In order to provide the widest discrimination between systems, the systemic M-properties are associated with a set of systemic metaphors which inherit all the properties given above and include machine, organism, society, circuit, game, text and culture. There are further properties which typify particular systemic metaphors in the language used when talking about them, for example:

```
Circuit:    flow, transfer, conduit, cycle, transferred thing(s), drain.
Machine:    efficiency, input, process, output, goal, purpose, power.
Organism:   growth, organised complexity, level, adaptability, openness.
Text:       interpretation, context, translation, meaning, style,
Culture:    tradition, expression, ritual, belief, symbol, idea.
```
The M-properties of the systemic metaphors listed above, though not exclusive to a particular type, are associated most clearly with that type. For example, in saying that a culture is adaptable we are likely (though not necessarily) referencing the organismic metaphor (ie., culture is like an organism). The caveat for a knowledge analyst is that metaphorical analysis provides possible confirmation of a hypothesis about particular language. It is an approximate technique. Its advantages are that it relates the semantic richness of language to domain models of the real world in a meaningful way.

Metaphorical analysis can proceed to identify language used in analogies and similes within the context of a particular systemic metaphor. These M-properties have further properties associated with them. For example, the properties which an expert associates with the circuit metaphor (ie., circuit M-properties) can be anticipated by a knowledge analyst, such as:

```
Flows:      fluid, electrons, energy, information, materials, ideas.
Conduit:    pipe, wire, channel, link, topology.
Topology:   tree, network, chain.
Transfer:   from...to..., transferred thing, temporal relations.
```
It is possible to go further and note that properties of the circuit M-properties (ie., the right hand side of the list above) are related to each other in several ways. Some examples would include:
(i)  Conceptual interrelations between things that flow can be located in a generalisation hierarchy. This can be important if an expert describes things with different kinds of language but within the context of the same systemic metaphor. A simple example is the description of electrical phenomena in terms of the movement of water.
(ii) The idea of transfer carries with it certain verbs which convey information about time, causality and mechanism. The verbal associations can be revealed in the case relations between noun phrases. For example transfer verbs will take source, destination, object and agent cases.
An appreciation of these and many other properties allows a knowledge analyst to make disciplined anticipations about a domain in terms of underlying theory, ontology and the language used to talk about it. Knowledge acquired from various sources, and especially from people, can be investigated on the basis of explicit analytical techniques.

The systemic M-properties of the real world domain are unlikely to reference a single systemic metaphor. For example, in an investigation of an artificial neural network applications domain, several systemic metaphors were used to describe a network including, machine, circuit, organism and society. This mixture of metaphors is partly related to the complexity of the domain and the need to use different metaphors to emphasise particular features. Metaphorical analysis provides a way of managing the vast amount of verbal data that emerges from interviewing an expert in a coherent model.


INTERPRETING EPISTEMIC STRUCTURES

It is possible to anticipate likely conceptualisations of domains through an understanding of how people construct and maintain them. As a result, the initial stages of knowledge acquisition should focus on the real world domain and the cognitive domain (models) which experts use. In our approach, a knowledge analyst attempts to come to share a similar epistemic horizon on a problem domain as an expert. We describe this as a hermeneutic approach (see Figure 2a).

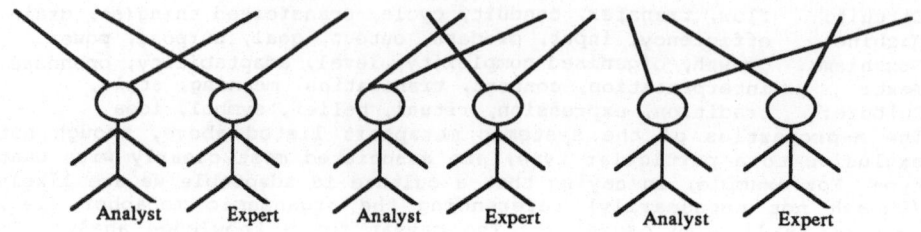

2a Sharing the same epistemic horizon.  2b Too narrow a focus can miss important details.  2c Domain characterisation begins with a broad perspective.

Figure 2 The Hermenuetic Approach to Domain Characterisation
(adapted from Salmond, 1982)

An understanding of the metaphors used in a domain facilitates the sharing of similar perspectives. The language used suggests the kinds of models of the domain that have been constructed. An appreciation that language, including scientific language, is metaphorical in nature, provides a knowledge analyst with valuable insights concerning the expert's perceptions of the domain. This will be especially valuable when there is no established or comprehensive theory of the domain. For example, metaphors will have been used to fill gaps in the vocabulary. The language used in these metaphors will provide details about the conceptual context in which models are formed.

At a general level, the hermeneutic approach provides a means for harmonising perspectives on a domain. It emphasises the cognitive nature and identifies key human constructs (cognitive objects). It also emphasises the referential nature of a domain (ie., cognitive constructs are used to refer to entities in the real world). The approach has breadth: without sufficient breadth to begin with the knowledge analyst may miss part of the domain altogether (see Figure 2b & 2c). Knowledge analysts are not passive, dumb, non-interactive agents. They bring assumptions and preconceptions with them when embarking on knowledge acquisition. It is crucially important to make this background information explicit. This is achieved by requiring that initial approaches to the characterisation are broadly based.


RESULTS FROM INVESTIGATIONS

The occurrence of systemic metaphors provides analytical details about a variety of aspects of a domain. In this section some outcomes from two investigations are discussed: the installation of HP-UX software and applications of neural networks in industrial domains. The material that is reported is a small selection from metaphorical analyses and further details can be found in (Nwana et al., 1990 and Paton et al., 1990).


The Presence of Key Lexical Types

Theoretically relevant and controlled anticipations prevent time-wasting, and establish important expectations about the vocabulary of discourse at an early stage in a domain characterisation project. In order to manage anticipations about the nature of the domain and make the assumptions of the knowledge analyst more explicit, it is necessary prior to any elicitation of knowledge from an expert that an analysis of key texts/references is made. This provides information about the possible vocabulary of discourse. Two examples are now briefly discussed.

In the software installation project, we anticipated that "data" would be an important word and used the techniques of metaphorical analysis to anticipate details about the possible metaphorical nature of the domain. Three systemic metaphors were located: machine, circuit and text. Language associated with each was identified and can be prefixed before the following M-properties (eg., data-driven):
   Machine (eg., input, output, processing, driven).
   Circuit (eg., stream, transfer, piping_of, pipeline, filter).
   Text (eg., database, storage, dictionary, interpreter).
These anticipations were then used as passive probes (ie., based solely on what the expert said rather than through interaction with the knowledge analyst) in the elicitation sessions with the expert.

Analysis of selected literature concerned with neural networks revealed that this domain is a hybrid from several substantial bodies of knowledge with the transfer of such concepts as:
   Neurobiology (eg., brain structure, network architectures).
   Psychology (eg., learning, vision, associative memory).
   Mathematics (eg., transfer functions, non-linearity).
   Computer science (eg., non-algorithmic solutions, program).
   O.R. and statistics (eg., optimisation, cluster analysis).
   Physics and electronics (eg., spin-glass problems, feedback, noise).
This preliminary broad perspective on the potential importation of concepts and problems from related domains helps anticipate the kinds of systemic metaphors that are likely to be used in the domain being investigated. In this case it was anticipated that machine, circuit and organismic metaphorical contexts were likely to be found.

## Relational and Iconic Features of Models

The nature of systemic metaphors in a domain shows the kinds of theory used in its construction and maintenance. Relational properties will tend to be related to domain structure and organisation.

One example from the HP-UX domain, concerned the expert's use of the notion of the "system". Analysis of the interview text revealed that he used the term in three different ways: the hardware system, the software system and the software as executed in the hardware system. It was not always easy to differentiate the three or identify the boundaries between them. Analysis of the discourse indicated an iconic aspect to the expert's knowledge based on the relational language used (eg., transfer verbs and verbs associated with kernel functionality such as "drags", "handles", "controls"). This allowed us to anticipate the organisation of possible systems and so, at the next interview session, we showed him a set of simple graphs which are related to the systemic metaphors. His choice helped us clarify the nature of the different systems he had described. Clearly care must be taken in using a technique to probe a expert for an iconic representation as it would be easy for the analyst to introduce bias.

A separate investigation within the neural networks project was to test a set of simple techniques that probe an expert for the systemic nature of a domain. This kind of approach is an active probe in that new knowledge may emerge from the interaction of the expert with the material and the active participation of the knowledge analyst. It was used after explicit accounts were made about its purpose and relation to prior analyses. In this case, they were made after two previous iterative cycles had suggested a need to confirm or refute hypotheses about the domain. The procedures described are still under development and form part of a larger analytical package.
(i)    The expert was given a list of systemic M-properties and asked which applied to the domain under discussion and if there were any others. The list that emerged was:
Interrelated parts      Levels of Organisation          Complexity
Collective behaviour    Structure of units (parts)      Purpose
These helped to clarify the systemic level of description.

(ii)    The expert was given a list of systemic metaphors and asked if an artificial neural network could be described by each in terms of component parts. This provided structural details about the network:
Machine      " can be divided into units and connections ",
Circuit      " can be considered as flow of activity ",
Organism     " not an organism but an integral part thereof ",
Society      " society of interacting individuals...".
This provided the possibilty to probe the expert to investigate the dominance of machine language in this domain.

## Managing Metaphors used for Didactic Purposes

Experts often communicate what they know in terms of what they think the knowledge analyst will understand. The expert may intentionally manufacture a metaphor in order to promote understanding by the analyst. The analyst must be able to understand these metaphors and the referents they address. Consider the following example from a transcript about local feedback for a kind of unsupervised learning.

> " ...the analogy one can draw is imagine you've got a number of channels and imagine water running down them. If it only gets into one channel its going to erode and therefore is going to be deeper and more likely to catch the water the next time...we found that when water went down the hillside just little abnormalities might start off a little rivulet. The next time the water went down it gets deeper and it gets deeper."

Here, the expert is seeking to explain the problem by referring to a situation he anticipates the analyst will understand. For the purposes of analysis, the analogy has two contextual dimensions, the circuit metaphor and landscape metaphor. These metaphors provide details about feedback and analysis of the response revealed a theoretical relation between global and local feedback mechanisms and unsupervised learning.


## CONCLUSION

Domain characterisation is a very important activity that can help knowledge analysts overcome some of the pre-design problems associated with knowledge acquisition. The pervasive nature of systemic metaphors can be used to provide important insights into the analytical processes that are carried out.

ACKNOWLEDGEMENTS
Thanks to our experts Ken Chan and Brian Ward and to Professor Rom Harré for his advice and comments. The work of the MEKAS project is sponsored by Shell Research Ltd., (Thornton Research Centre) and Unilever Research (Port Sunlight Laboratory).

REFERENCES

Harré, R. (1970), *The Principles of Scientific Thinking*, London: Macmillan.
Nwana, H.S., Paton, R.C., Shave, M.J.R & Bench-Capon, T.J.M. (1990), "The Relevance of Domain Characterisation to Knowledge Acquisition for Second Generation Expert Systems", MEKAS Report 17, Department of Computer Science, University of Liverpool, December 19. Submitted for publication.
Paton, R.C. & Nwana, H. S.(1990), "Domain Characterisation through Knowledge Analysis", *Proceedings of AAAI-90 Workshop on Knowledge Acquisition: Practical Tools and Techniques*, Boston, Mass., July.
Paton, R.C., Nwana, H.S., Shave, M.J.R. & Bench-Capon, T.J.M.(1990a), "Foundations of a Structured Approach to Domain Characterisation", MEKAS Report Number 13, Department of Computer Science, The University of Liverpool, November 24. Submitted for publication. This is an extended version of a paper presented at the Eighth Workshop of the European Society for the Study of Cognitive Systems (ESSCS), Oxford. September, 1990.
Salmond, A.(1982), "Theoretical Landscapes", in Parkin, D.(ed), *Semantic Anthropology*, London : Academic Press.

Jun-Kang Feng

Department of Computer Science
University of Manchester, UK

Hugh Rooms

School of Information Science
Portsmouth Polytechnic, UK

*ABSTRACT*

*Knowledge based processing becomes increasingly important for a new class of information systems. The process of knowledge based system construction is termed knowledge engineering, in which the transformation of the knowledge about a domain from experts to a computer system is known to be a difficult task. Appropriate techniques are therefore required. During our research on the use of knowledge based techniques for database design, a systems approach was adopted, on which a conceptual modelling technique was developed to tackle the problem. This paper will present this technique.*

# 1 KBS Construction and Available Methods

In the field of knowledge based systems (hereafter called KBS), we are in the stage that is very similar to that which took place in the field of software engineering in the sixties [GM89]. The need for appropriate approaches and well-defined methodologies has arisen. The construction process of a knowledge based system is termed knowledge engineering. The issue of developing techniques for knowledge engineering is now particularly important, because a new generation of computer systems is expected to have powerful knowledge based processing capabilities among others [BY86] [ACS85], moreover [Myl86] even pointed out that it is time to consider the software development paradigm in favour of one that views software development primarily as knowledge base construction rather than traditionally as the construction of programs and/or databases. Some research has been done in order to formalise a theory for knowledge engineering, and some principles and systematisation have emerged [Buc82] [Hay84].

Building a knowledge based system can be viewed as a mapping of domain knowledge into a computer system. Domain knowledge refers to the sum or range of what has been perceived, discovered, or learned about the domain [Ram89], together with an understanding sufficient to apply the knowledge to suitable problems within that domain [Fil88]. Domain knowledge is usually composed of judgemental rules, procedures, algorithms, and the structure of the domain, whereas the computer system used to implement a KBS can be viewed as having three components, namely, its architecture, its working principle — pattern-directed inference, and its formalisms of knowledge representation. It is based on a computational model with its particular forms of program, data memory, and executor. Knowledge engineering consists of three strands [Gre88]: knowledge elicitation, knowledge analysis, and implementation of a KBS.

A number of methods have been suggested for knowledge engineering.

P.Grimaldi and A.Marcelli [Gri89] presented an approach aiming to establish the feasibility of a KBS, and to collect the necessary data for building a prototype. But this approach does not offer any intermediate representation of knowledge and it only cover part of the whole process of knowledge engineering.

Keravnov and Johnson's Techniques [KJ86] is a set of software tools including *epistemic nets,status diagrams,relational networks, task-trees* and *production rules*. These techniques do not seem to form an integrated methodology, and they are basically centre on objects.

Brouwer-Janse and Pitt's Technique [BP86] constructs a model of domain knowledge by extracting the problem-solving behaviour of experts, rather than by looking at objects and their relations. This technique seems more useful in "knowledge elicitation" than "knowledge analysis", because it looks at the domain from the viewpoint of an expert's behaviour. It does not produce a domain structure directly, therefore it does not facilitate the implementation of a KBS either.

The **Interpretation Model Technique** is a tool kit put forward by Breuker and Wielinga [BW84]. It structures a domain with a hierarchy of categorised objects. But the starting point of this method is to work out a typology of basic elements, including processes and almost everything of interest. The number of these elements may be inconveniently large. In addition, it does not distinguish processes from objects, which would be necessary in handling many domains.

All these techniques centre on the identification of domain objects and modelling the reasoning processes. They all share certain problems: that they only cover part of the whole process of building a KBS; that they are not particularly integrated; and that they are not explicitly based on the systems approach.

## 2   Basic Considerations for a Modelling Technique

It has been proved that in knowledge engineering, the transformation of the knowledge about a domain from experts to a computer system is a difficult task. During our research, in order to tackle this problem, a conceptual modelling technique was developed according to systems theory and from the practical viewpoint of KBS construction.

This is a modelling technique. Due to the diversity and interdisciplinary nature of modern complex problems, the modelling approach is widely adopted [Khe88]. It was pointed out [BMS84], in particular, that the growing demand for systems of ever-increasing complexity and precision has stimulated the need for higher level concepts, tools, and techniques in every area of computer science. Some attempts have been made to meet the demand by defining a new, more abstract level of system description. These attempts and related activities are termed conceptual modelling. Our technique is to model the domain knowledge so as both to achieve a better insight of the domain and to ease the transformation of knowledge to a knowledge representation language or a KBS development environment. The relevant process can be called knowledge modelling, and it results in a model of the domain knowledge, which fills the gap between a domain knowledge and its implementation medium.

A modelling technique is creditable only when it has proved to have theoretical bases, well-defined objectives and construction procedures, and identified ways to validate the results of its applications.

Our modelling technique is based on the systems approach. Modelling refers to the study of the mechanisms inside a system, and through using basic physical (biological, economic, etc) laws and relationships, a model, ie a representation of the system, is inferred [Khe88]. Therefore systems theory is naturally adopted as an appropriate approach. A system is a set of elements together with relationships between elements and their attributes related to each other and to their environment so as to form a whole [SSK90] [Khe88]. A system is a way of viewing the reality in the world. Both the domain knowledge and a KBS are thus looked at with systems view and a structured methodology for KBS construction was developed.

With this technique, a domain knowledge is modelled with a data flow paradigm and an object-oriented viewpoint, and a KBS is looked at on three levels [BL84]. The first is the system engineering level, which focuses on the organisational aspects of a KBS using the techniques such as successive refinement and version management. This is a human's view, namely, how a human deals with the information content in a KBS. The second is the symbol level, which focuses on data structures, algorithms with the mechanisms such as locking, backward-chaining etc. This is a machine's view, namely, how a machine deals with the information content in a KBS. The third is the knowledge level. It considers exactly what knowledge is represented in a scheme and what a KBS tells us about the world.

The objective of the present technique is to encapsulate the three different perspectives identified above. It models the domain knowledge with a human's perspective, ie the domain structure - what an expert deals with, and the reasoning process - how an expert deals with problems in the domain. This directly results in a way in which a KBS is organised. It then transforms the model of domain knowledge into a form which can be implemented by a machine, ie a machine's perspective. The resultant KBS should be capable of offering a user the domain knowledge and explanations about its reasoning when required. In other words, this technique will help the production of a user interface that works on the knowledge level.

When the methodology was developed, the following additional factors were also taken into consideration: (1)The degree of analysis performed in the preceding stage, namely, knowledge elicitation; (2)The nature of the domain and the typical way of representing its structure; (3)The complexity and representation power of the implementation medium; (4)The formalism adopted should facilitate the understanding of the domain by using graphic and symbolic analogy, and should facilitate the transformation of knowledge to an implementation medium.

# 3    A Structured Methodology

The technique developed can be expressed as a well-structured top-down methodology, which consists of five stages.

**Stage 1**: Function analysis with *data flow paradigm*

In terms of static structures, a system can be viewed as composed of parts, but from the functional (ie dynamic) point of view, the functions performed by the system's parts are the elements that constitute the system [SSK90]. For example, in the domain of database design, there are functions such as "tasks" and "stages" in a design, which serve as elements of the domain. The functional viewpoint is more essential, because the static structure of a system can change from time to time while its basic functions remain as long as its nature does not change.

Elements (ie functions) interact with each other, therefore they have relationships. The relationships are embodied in the processes in a domain, and the processes can be structured as a hierarchy. Thus through a top-down function analysis, one can model the structure of a domain with a hierarchy of abstraction levels in terms of functions and their relationships.

The data flow paradigm is employed to carry out the analysis. Originally the data flow concept was developed to model data processing systems [GS79]. Due to its power, simplicity and familiarity, the data flow paradigm has been employed in many other domains, such as CAD framework development [HT90]. The point is that many domains can be viewed as systems and a system can be modelled as a hierarchy of functions that are composed of inputs, processes, and outputs. The data flow paradigm fits such a hierarchy of functions. Moreover, a system has boundaries, by illustrating sources of inputs and destinations of outputs as well as processes, a set of data flow diagrams for a system specifies its boundaries.

Through a top-down function analysis, both the sequence of the processes in one dimension, and the depth of the processes in another dimension are revealed. In other words, the relationships between functions are modelled. Moreover, the domain knowledge is modelled with different levels of abstraction. The representation of domain knowledge will therefore also be of different level of abstraction. This will help identify meta-level knowledge, such as the knowledge about the contexts in which lower-level knowledge is used. In fact a function hierarchy is also a context hierarchy, which is a way of viewing the structure of a problem space [Fil88]. Context is a widely used mechanism for controlling the reasoning process in a KBS, for instance, it is used to realise so called "rule-filtering" in a KBS. Another example of meta-level knowledge which can be identified is the knowledge about the procedure in which the tasks are performed. This helps structure a goal-tree.

As a result of the function analysis, the knowledge in a KBS is organised according to the hierarchy of functions. Obviously, this stage is conducted in terms of the human level of a KBS as mentioned earlier.

**Stage 2**: Identification of domain objects

Within each function or process, there are domain objects. They are the elements on which the functions are performed, or that they produce or change. Objects and the relationships between them constitute the domain and are manipulated by experts to approach their goals. In this stage, objects, their attributes, and their relationships are identified and stored in a KBS as data structures. For example, within the function of "Judge the modelling stages of database design", we have object class "modelling-stage" with attributes "stage", "order", and "status"; and object class "model" with "name", "order" and "status" as its attributes. Among these attributes, "order" is used to manifest the relationship between different modelling stages or models.

Stage 1 and 2 establish a knowledge model of the static aspect of a domain. This model is implemented in a KBS in terms of the knowledge base organisation and data structures for object classes and their attributes and relationships, which form the basis of the structure and functioning of a KBS. In a KBS, knowledge is represented in the forms of *facts*, *production rules*, *objects* and *procedures*. Facts are the instances of object classes, which are usually the results of the firing of rules; rules are the relationships between one status of a problem solving process in terms of facts as well as rules and another status. Their firing advances the status of a problem-solving process by changing facts in the working memory or even rules in the rule base(s); and procedures are ordered operations to be performed on objects. Therefore up to this stage when domain objects have been identified, a vital basis of the machine level of a KBS has been established.

**Stage 3**: Identification of goals

A system exists as a whole for its particular goals. The third stage is to identify the goals and sub-goals in each function. A hierarchy of goals and sub-goals can be identified in a top-down and a backward inference way. For instance, to sort a heap of bricks according to their sizes, the nearest sub-goal whose realisation would make the goal achieved is "put down" the largest brick among those in ones hands or in the heap; whereas to achieve "put down", one needs to "pick up" the largest brick from the heap, so "pick up" is a sub-goal of the next level down, and in this case is a leaf-node of the goal tree. The structure of the goal tree serves as a model of an expert's reasoning process and problem-solving strategy. A goal hierarchy reveals the dynamic aspects of a system. In terms of implementation, goals can be treated as a special type of object. Therefore with a data structure, a goal and its attributes can be represented, and with the organisation and functioning of rules, a goal tree and its function can be implemented.

**Stage 4**: Identification of actions

To achieve goals and sub-goals, actions are performed on objects by an expert. In this stage, actions within each function will be identified.

Actions are either merely the knowledge about what to do to achieve goals, in which case the job is actually done by a human, or activities which are usually done by a human and which will now be carried out by a KBS. The first category may be called *virtual actions* from the point of view of a KBS implementation, such as the actions of "put down" and "pick up" in the example of sorting a heap of bricks of different sizes. The second category consists of *real actions*. For example, in the domain of database design, a number of jobs such as "entity-relationship modelling" and "schema analysis" that are usually performed by a human will now be carried out by a KBS.

To distinguish the two categories of actions is important from the viewpoint of the implementation of a KBS. When a virtual action is identified, we will not consider its implementation, the only thing needed is to produce advice for the user of the KBS; however once a real action has been identified, the way to implement it must be worked out. In other words, the natures of actions indicate the way of implementing the system. For example, the action "create criteria" which are used for identifying the components of a database schema is better implemented by a program written in a conventional language such as COBOL, whereas the action of "judge which modelling stage can be carried out at a time point" is relatively easily implemented with a production system language such as OPS5. In addition to its classification, the pre-conditions and post-conditions of each action should also be listed.

An action may be taken by an expert with incomplete knowledge, therefore uncertainty is involved in the reasoning process. A KBS should capture this kind of knowledge as well. Along with the actions being identified, how certain it is taken under a particular circumstance needs to be identified too.

**Stage 5**: Identification of Productions

The final stage of the methodology is the identification of productions (ie rules in a production system), which are the main form of domain knowledge representation in a production system. The inference engine in a KBS works on the productions and working memory elements to achieve the goal of the system.

With a systems approach, the identification of productions can be carried out in a top-down and backward inference way. The strategy is to examine the goals and design productions to carry out necessary actions to achieve these goals.

Condition elements in the left hand side (hereafter called LHS) of a production (rule) are derived from the pre-conditions of an action. These condition elements either ensure that the action can be carried out, or ensure that the results of the action do not already exist in the data memory, or ensure that this production is allowed to participate in a matching process under the right context. Actions in the right hand side (hereafter called RHS) of a production are derived from the post-conditions of an action. In order to satisfy the post-conditions, actions such as "modify", "remove", and "add" are used in the RHS of a production to make changes to the data configuration of the data memory or to the productions themselves in the rule memory.

A *procedure* for the identification of productions is as follows:-

*Step 1* Choose an action related directly to a goal.

*Step 2* Write the LHS of a production, using one particular condition element for specifying the goal which is the context in which the production is used, and one condition element for each of the pre-conditions.

*Step 3* Write the RHS of the production, using a statement to give a message to the user that the action has been accomplished and using commands such as "remove", "add", or "modify" to change appropriate elements of the working memory in order to achieve post-conditions. In addition, the goal element should be removed or marked as satisfied.

*Step 4* Check to make sure that the production does leave the working memory in a legal configuration.

*Step 5* Check the interactions of the production with already existing productions for this goal, modifying as necessary.

*Step 6* Each of the conditions has one of the three following purposes: [a] specifying the goal as the context in which the production can be searched and fired; [b] ensuring that the action can now be carried out; [c] ensuring that the result of the action does not already exist. Choose a condition with the purpose of type [b], create a new condition element specifying that the chosen condition is **not** satisfied, and use this in a new rule in place of the old condition.

Check to see whether or not any other conditions need to be modified as a result (They may have become superfluous or incorrect). Create a new RHS element that establishes a sub-goal of satisfying the chosen condition.

*Step 7* If Step 6 results in a new rule, go back to Step 4; otherwise create a production by modifying the condition of type [c]. This production's LHS should test all the conditions resulting from carrying out the action. The RHS of the production should output a warning message that the goal is already satisfied and remove the goal or mark it as satisfied. Check this production with Step 4 and 5. Then go back to Step 1 for another action that is probably needed to achieve this goal.

This way, for each goal, there will be two levels of productions that should usually be identified. The first level has only one production, which is derived directly from the action whose post-condition(s) satisfy the goal. The second level productions are derived from the conditions of the first level production. For each of the conditions whose purpose is of the [b] type, a new rule is created in order to set up a sub-goal or to make the KBS do something in order to satisfy the condition. And for each of the conditions of the [c] type, a new production is created to make sure that the goal has not been satisfied already.

The underlying logic is: in each function, a goal — actions required to reach the goal — first level production — second level productions — new goals as sub-goals ... this process is repeated till no more sub-goals need to be generated. This process results in a well structured hierarchy of production clusters.

# 4 Summary

In this paper, after the discussion of knowledge engineering and problems with available methods, a systems based modelling technique for KBS construction was presented as a structured methodology. This technique covers most of the process of knowledge engineering and its applications result in a knowledge model and a well organised knowledge based system.

# References

[ACS85] Albert,T., Charn,B. and Sears,J. 1985 in "On Knowledge Base Management Systems", Springer-Verlag, 1986.

[BL86] Brachman,R. and Levesque,H. "The Knowledge Level of a KBMS" in "On Knowledge Base Management Systems" pp.9-11, Springer-Verlag, 1986.

[BM86] Brodie,M. and Mylopoulos,J in "On Knowledge Base Management Systems", Springer-Verlag, 1986.

[BMS84] Brodie,M., Malopoulos,J. and Achmidt,J Preface in "On Conceptual Modelling", Springer-Verlag, 1984.

[BP86] Brouwer-Janse and Pitt "Knowledge Acquisition: Methodological Issues and Problem Solving Profiles", Proceedings of ECAI, Vol.2, pp.120-127, 1986.

[Buc82] Buchanan,B. "New Research on Expert Systems", in: Hayes,J., Michie,D. and Pao, Y. (eds.), "Machine Intelligence", Vol.10, Edinburgh University Press, Edinburgh, pp.269-299, 1982.

[BW84] Breuker,J. and Wielinga,B. "Interpretation of Verbal Data for Knowledge Acquisition", Report 1.4 Espirit Project 12 Memorandum 27 of the Research Project "The Acquisition of Expertise", University of Amsterdam, 1984.

[Fil88] Filer,N. "The Use of Knowledge Based Techniques for Electronic Computer Aided Design." PhD thesis, University of Manchester, January 1988.

[GM89] Grimaldi,P. and Marcelli,A "Toward a Structured Approach to Expert Systems", Nort-Holland, in "Microprocessing and Microprogramming", 25, pp.27-32, 1989.

[Gre88] Greenwell,M. "Knowledge Engineering for Expert Systems". Chichester, Ellis Horwood, 1988.

[GS79] Gane,C. and Sarson,T. "Structured Systems Analysis: Tools and Techniques", Improved System Technologies INC., New York, N.Y. 1979.

[Hay84] Hayes-Roth,F. "The Knowledge-Based Expert System: A Tutorial", COMPUTER Vol.17 n.9, pp.11-28, 1984.

[HT90] Hamer,P. and Treffers,M. "A Data Flow Based Architecture for CAD Frameworks", Proc. ICCAD-90 — IEEE Conference on CAD, pp.482-485, 1990.

[Khe88] Kheir,N.(ed) "Systems Modeling and Computer simulation", Marcel Dekker, Inc. New York, 1988.

[KJ86] Keravnou,E. and Johnson,L. "Competent Expert Systems", Kogan Page, London, 1986.

[Myl86] Mylopoulos,J. "On Knowledge Base Management Systems", Springer-Verlag, pp.3-8, 1986.

[Ram89] Ramamoorthy,C. "Knowledge and Data Engineering" IEEE Trans. on Knowledge and Data Engineering, Vol.1 No.1 1989.

[SSK90] Schoderbek, Schoderbek, and Kefalas "Management Systems", Business Publications Inc. Texas, 1990.

INFORMATION SYSTEMS DEVELOPMENT

INFORMATION SYSTEMS DEVELOPMENT : INTRODUCTION


PLANNING

A number of issues are considered important in the development of organisational information systems, and the papers in this section reflect these. Firstly, there is a need to develop business plans within organisations and then plans for the enhancement and development of information systems that support the business plans. Business planning logically precedes information systems planning, and enables applications to be selected for investment and developed with appropriate priority. Decisions must be taken during information systems planning that have a global effect on the organisation, such as centralisation or decentralisation of data processing and systems development, and standardisation on hardware and software. Increasingly, information is regarded as a resource within organisations, that needs professional management, in the same way that financial and material resources are managed. Investment in information technology must be carefully justified in terms of individual and organisational efficiency and effectiveness.

METHODOLOGY

A second issue that is of concern to the information systems developer is the choice of methodology used to define requirements for information systems, and specify solutions based on the use of information technology. A number of analysis and design techniques exist, such as entity modelling and data-flow diagramming, and these techniques have been packaged in various ways and marketed as development methodologies such as Information Engineering and SSADM. The systems movement has contributed to debate over the effectiveness of information systems development methodologies and the debate is reminiscent of the critique of systems engineering and operational research conducted by P B Checkland in developing Soft Systems Methodology. Information Engineering and SSADM can be seen as 'hard-systems' approaches, equivalent to systems engineering, and open to the criticism that they are primarily concerned with means rather than ends and are unduly influenced by a functionalist view of social reality inappropriate for understanding and intervening in human activity. A number of papers in this section are concerned, in one way or another, with the need to open up information systems development to soft-systems thinking.

PROJECT MANAGEMENT

Finally, there is the issue of monitoring and controlling information systems development projects. Such projects have a poor record for timely completion within budget to satisfactory quality standards. A number of reasons have been offered to explain this failure. For example information technology has been used to automate what were considered to be routine human tasks, and insufficient account has been taken of the human discretion that was previously applied to keep the system running smoothly. In general the organisational impact of introducing advanced technology has been under-estimated and the potential benefits over-stated. The quality of project planning, costing and scheduling of systems development projects seems to have been lower in information technology than in other branches of engineering. Papers in this section indicate the concern of systems thinkers to address the issue of effective project management.

SOFT SYSTEMS AND STRATEGIC PLANNING

JOHN L THOMPSON

HEAD OF MANAGEMENT STRATEGY
HUDDERSFIELD POLYTECHNIC, ENGLAND

"Planning is one of the most complex and difficult intellectual
activities in which man can engage. Not to do it well is not a sin; but
to settle for doing it less than well is."
                                                    (Russell Ackoff; 1970)

This paper examines the nature of strategic planning and considers the
valuable contribution that can be made by soft systems thinking and methodologies. My aim is to illustrate the potential for matching the holistic
thinking contribution of soft systems with selected views on the management
of strategic change in organizations. I conclude with the argument that
soft systems concepts can provide a useful contribution to courses on
business policy and strategic planning.

Planning the future, that is thinking about the most appropriate
strategies, and changes of strategic direction, is essential for organizations, particularly those experiencing turbulent environments. Thompson
(1990) argues that an organization is being managed strategically effectively when its resources and values are closely congruent with its external
environment. Furthermore resources should be managed sufficiently flexibly
to ensure that necessary changes of policy and direction are both determined
and implemented at the appropriate time. For many organizations this
implies and results in a decentralized structure with considerable authority
and responsibility delegated to managers in charge of business units
(general managers) and individual functions.

The effective management of strategic change must be planned, but this
need not necessarily imply centralized and formal planning systems. Porter
(1987) contends that such systems, popular in the 1960's and early 1970's,
have fallen from favour because they failed to generate effective strategic
thinking. The plans were often drawn up by planners with insufficient
involvement by the managers who would be charged with implementing them, and
moreover the plans were treated as rigid and inflexible statements.
Robinson (1986) argues that the role of planners should not be to plan, but
to enable good line managers to plan.

Planning, then, should concentrate on understanding the future, which
is, of course, uncertain and unpredictable, and helping managers to make
decisions about strategic changes. These changes can relate to either or
all corporate, competitive and functional strategies. Corporate strategic

change concerns the overall scope and perspective of the organization -
what business or businesses it competes in. It may be diversified and
multi-national; equally it may be concentrated on one type of product or
service and limited geographically. Competitive and functional strategies
support the corporate strategy and they relate to the essential search for
competitive advantage in each chosen line of business. Each and every
functional area - marketing, operations, finance, people and information -
can be the source of a distinctive competitive edge which generates a
positive response from consumers. Corporate strategy is the responsibility
of the overall strategic leader, typically the chief executive, but
general managers and functional heads may be empowered to change, modify or
improve competitive and functional strategies.

Changes can be the outcome of a formal planning process which embodies
tried and trusted strategic planning techniques, and which rely on systematic information gathering and analysis. On the other hand a visionary
strategic leader, aware of strategic opportunities, and convinced that they
can be capitalized upon, may decide by himself (or herself) where the
organization should go, and how the strategies are to be implemented. Very
little needs to be recorded formally. Equally a broad strategic direction
might be clarified and defined and managers then encouraged to experiment
with gradual changes, learning by trial, error and reflection. Detailed
strategies evolve through this learning process, emerging from a pattern of
decisions taken over time. Mintzberg (1973) has referred to these three
modes of strategy creation as planning, entrepreneurial and adaptive
respectively. The modes are not mutually exclusive and most organizations
are likely to utilise at least two of them simultaneously. Planning is
involved in all of the modes, albeit that the cerebral and intuitive
aspects take precedence over formal and systematic procedures in entrepreneurial and adaptive changes.

STRATEGIC CHANGE AND SYSTEMS THINKING

This section of the paper re-examines the notions of planning, entrepreneurial and adaptive strategy in the light of ideas from soft systems
thinking and in particular Checkland's methodology (1981) for tackling
complex problem situations. It is assumed that readers of this paper are
already familiar with Checkland's work.

Planning Strategic Change

Checkland's methodology, beginning with the drafting of a rich picture,
can provide an extremely useful framework for structuring the thinking
process involved in analysing strategy. The rich picture helps evaluate
the current state of an organization: where it is strong; where it is weak;
external opportunities it might seek to seize; possible environmental
threats which might affect it - constituting a SWOT (strengths, weaknesses,
opportunities, threats) analysis; together with how well it is performing
financially and competitively, and any important problem areas which need
addressing. From this knowledge base the methodology can assist in the
search for ways of changing things and planning how these strategic changes
might then be implemented. All this against the background of clearly
understanding the long-term intentions (the mission) for the business.

Following the seven stages of the methodology can be a type of formal,
or at least disciplined, planning, for it provides a way of looking ahead
from where the organization is to where it might seek to go, linked to an
exploration of constraints and options.

The root definition for the relevant system, and the subsequent conceptual model, would reflect the essential purpose of the business. If the

organization has a documented mission statement it should ideally correspond closely to the root definition. If it becomes apparent that current strategies could be changed or improved upon the methodology allows for a series of iterations which basically work upwards from current strategies and/or downwards from the first conceptual model. The strategic options for evaluation are likely to be input by a group of managers working together – logically those who will implement the changes. The outcome should be proposed changes which are feasible in terms of the environment and the organization's resources, and acceptable in relation to the values and expectations of the strategic leader and other stakeholders.

In many respects this is not unlike the planning model proposed by Ackoff (1986). Here organizations are encouraged to project the future assuming that strategies remain unchanged and the environment stable, and then postulate an ideal future position. Means for closing the gap are then formulated together with an implementation and control programme.

## Entrepreneurialism

Checkland's methodology can also be applied to the entrepreneurial (or visionary) mode of strategy creation, as an awareness of problem themes, say limitations of present strategies, and strategic opportunities, can lie behind the strategic leader's vision for the future. In this instance the root definition and conceptual model reflect this vision. Ways of fulfilling the mission must be sought, and again this could be an iterative process which links capabilities and skills with strategic opportunities. This could be almost wholly cerebral, the quality and robustness of the proposed changes depending substantially on the strategic leader's awareness of the environment and his or her appreciation of the organization's true capabilities. Whereas the planning mode application is likely to be part of an annual cycle, the entrepreneurial analysis might take place any time.

These ideas are illustrated in Figure 1 on the following page.

The top part of the diagram represents the present strategy, structure and relative performance of the organization. Historically this will have been affected by issues of leadership and culture, and by the influence and power of stakeholders. The key concern is how successfully the organization is pursuing its basic purpose or mission, which might be implicit, or more ideally, stated explicitly. Under the planning mode this analysis would be carried out systematically and the results quantified; with the entrepreneurial mode on-going awareness and insight by the strategic leader is quite likely to provide the assessment and evaluation.

The middle section of Figure 1 is built around a vision of the future. This conceptualization of the future strategy and structure will relate directly to the organization's purpose or mission, be it implicit or explicit. Deliberations then address whether changes of strategy – which of course must meet the tests of feasibility and desirability – would improve the effectiveness of the organization in respect of its mission. Whilst an entrepreneurail strategic leader might be expected to be more visionary, dynamic and risk oriented in the selection and evaluation of options for changing the strategic perspective, the decision could be to make no fundamental changes, and instead continue with present strategies, seeking modification and improvement whenever appropriate.

## Adaptive Strategy

Soft systems thinking recognises that organizations have a purpose, but not set, defined, agreed, inflexible objectives. Rather, objectives and strategies emerge from organizational processes and management decisions.

# PLANNING STRATEGIC CHANGE - A SYSTEMS APPROACH

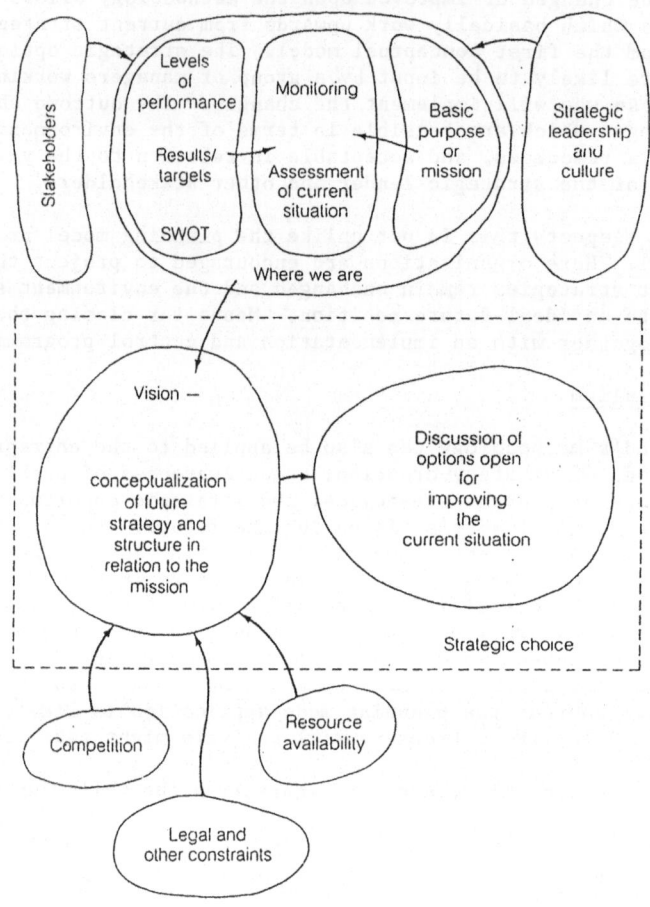

FIGURE 1

Adaptive strategic change occurs when managers in decentralized organizations make decisions which alter functional and competitive strategies. Policy guidelines can limit their flexibility, containing the extent of the decentralized power, or allow them considerable freedom. Organizations are seen as learning systems, and the value of "bottom-up" change, as well as centrally-led change, is appreciated. These managers, however, are likely to be specialists in particular functions, and loyal to specific business units or divisions, and affected by their personal meaning system or "weltanschauung" concerning the organization and their role and contribution. Sometimes other managers will be consulted about proposals; often they will not. Conflict can result from decisions which ignore other interests. How, then, can organizations ensure managers take a more holistic perspective and do not ignore those implications of their decisions which impact upon other parts of the organization?

Firstly I would argue the organization structure itself (together with management systems and processes) must not act as a barrier to holism, whereby only the most senior personnel are able to take a corporate perspective. This can be accomplished by focussing on the inter-dependencies between functions and divisions, and by ensuring that both rewards and sanctions reflect organization-wide contributions. Secondly managers must

524

be aware of the mission and corporate objectives of the organization, be supportive towards this, and appreciate how their contribution can be effective as well as efficient. Training in systems ideas can provide an ideal vehicle for this, although it is not the only way. The objective: shared values supported by effective communications and information systems.·

Figure 2 below postulates that decision making should encapsulate measures of both efficiency and effectiveness, and thereby increase the likelihood that outcomes and emergent properties will be synergistic.

Functional strategies should firstly be evaluated regularly, and proposed changes analysed, in relation to agreed measures of efficiency – are things being done well? Could improvements be made? In addition the effectiveness of these strategies should also be considered. If the right things are being done functions are not simply being run efficiently, they are contributing towards competitive advantage and supporting the activities of other parts of the organization.

MONITORING STRATEGIC PERFORMANCE

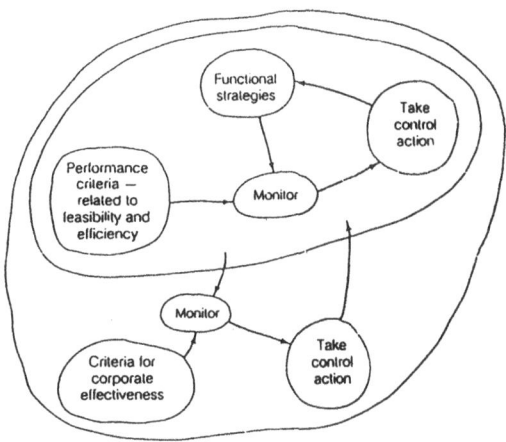

FIGURE 2

CONCLUSIONS

Organizational effectiveness requires that managers at all levels become and stay aware of the nature and changes in the external environment, and act appropriately. Peters (1987) says environments are becoming increasingly chaotic and that effective managers thrive on, rather than amidst, the chaos. They are appropriately proactive as well as necessarily reactive. Changes in corporate, competitive and functional strategies can occur in a variety of ways; central to the effectiveness of all of them is wide strategic awareness.

Soft systems ideas can make a major contribution here:

* Soft systems thinking ensures that attention is focused on the long-term purpose of the organization, and that decisions which involve deviations from the mission are not taken without due consideration.

* The thinking process can and should take place at all levels of the organization, with managers encouraged to think seriously about interdependencies and synergy.

* Proactivity and change orientation can be encouraged, rather than there be an emphasis on "more of the same" and reaction to external pressure.

* The focus can be on improvement rather than a search for ideal strategies, which is likely to prove frustrating. Moreover revolution can be constrained.

* Finally the thinking, and Checkland's methodology, can be applied retrospectively to test the quality of decisions that have already been taken.

For these reasons I believe that the linking of soft systems ideas and strategic change is potentially rewarding and that there is a strong case to be made for incorporating soft systems thinking and methodologies into courses on business policy and strategic management.

REFERENCES

Ackoff R L (1970), A Concept of Corporate Planning, Wiley.
Ackoff R L (1986), Management in Small Doses, Wiley.
Checkland P B (1981), Systems Thinking, Systems Practice, Wiley.
Mintzberg H (1973), Strategy Making in Three Modes, California Management Review, XVI, No. 2, Winter.
Peters T (1987), Thriving on Chaos, Pan.
Porter M E (1987), The State of Strategic Thinking, The Economist, 23 May.
Robinson J (1986), Paradoxes in Planning, Long Range Planning, Vol. 19, No.6.
Thompson J L (1990), Strategic Management: Awareness and Change, Chapman and Hall. The ideas contained in this paper have been developed from Chapter 16 of this text; and acknowledgement is made to Chapman and Hall for permission to reproduce the two figures.

BREAKING THE MOULD - THE EFFECTIVE USE OF INFORMATION

TECHNOLOGY IN ORGANISATIONS

Stuart Maguire Vas Prabhu

Newcastle Business School
Newcastle Polytechnic
Newcastle upon Tyne  England

## INTRODUCTION

It has become clear that the present status quo in relation to the provision of Information Technology within organisations does not allow the flexibility needed to cope with a rapidly changing environment (1,2).  In many instances organisations are trying to respond to changes through the provision of static information systems. Historically most medium and large organisations have relied on a centralised form of computer services department.  For the most part the systems that were implemented were of the operational, transaction-processing type(3).  They tended to be internal to the organisation and to reinforce the issue they were mostly of a historical nature.

It is very unlikely that the type of structure and personnel within traditional computing departments will have the range of skills necessary to react to environmental changes.  This paper focuses on a number of researched organisations where they have taken a different approach to the provision of information within their companies.  They were chosen because their organisations were going through a major change process.  This change had different forms and included both internal and external disruptions.

## TECHNOLOGICAL CHANGE

Traditional computer departments reflected the technology of the time. Few data processing managers had much feel for the business for which their computers were supposed to serve. Too many took refuge in impenetrable jargon as a means of protecting their departments from interference. Even fewer made any efforts to sell the positive commercial benefits of information technology to their less technically minded peers.

Large mainframe computers pointed the way towards a centralised computing structure(4).  The argument would always be that the centralised approach was the most

efficient in terms of economies of scale(5). An updated
mainframe computer might be 80% more powerful but only 30%
more expensive. In the last 10 years the onset of personal
computers has allowed User departments to develop their own
stand alone systems. The improvement in the effectiveness
of communications hardware and software has made the dream
of distributed computing a reality. Local area networks are
prevalent within most corporate environments. The work that
has been done recently in the area of Open Systems has meant
that compatibility is not the issue it was even 5 years ago
(6).

ORGANISATIONAL CHANGE

More organisations are now deriving benefits from
changing their existing organisation structure. Research
has shown that organisations which move away from the
traditional hierarchical structure to a 'flatter', organic
way of working reap dividends in terms of greater efficiency
and effectiveness.(7). This obviously puts into question
the traditional development of computer systems from the
centre. Organisations are finding that more effective
information systems are developed when the systems staff
understand the User area and are sympathetic to their
specific needs (8). These information systems cut across
traditional functions and organisational structures and
result in new communication paths and methods (9). The
flatter organisation and the use of inter-disciplinary teams
have already proved their versatility in the new management
philosophies and practices encompassed by the acronyms such
as TQM (Total Quality Management), TCS (Total Customer
Service) (10.11). The development of information systems in
a distributed way can only prove effective if Users and
Systems staff work as a team.

TECHNOLOGY PUSH VS.MARKET PULL

In the same way that the customer in the market-place
is becoming more demanding the User has higher and higher
aspirations in terms of what the information systems in
their area should consist of. Customers are expecting
shorter lead times, a wider variety of products and services
and better quality at a lower cost. The age of the
discerning User is at hand. They are demanding real-time
systems, shorter system development times, flexibility in
terms of their information needs, effective software
support, and no hardware downtime. In terms of costs more
and more organisations are moving away from the traditional
'top-slicing' that has taken place in the past in relation
to computer services. This refers to the way that User
departments have automatically paid for the central
computing facility through being charged directly as a
function of size or usage of the system. Many User
departments are now in a position to use this particular
'cost' in a more effective way themselves.

Each of these particular issues is important in its own
right in terms of improving the effectiveness of information
systems within organisations(12). With the onset of the
Single European Market and the increasing use of
telecommunications, organisations that don't adapt to the

changing environment may find themselves out of business. The following section shows how different organisations have reacted to change in different ways.

Case 1: This is an international company based in the north of England which produces a wide range of high quality pressure sensitive materials made specifically to customer requirements. The main reason for the introduction of new technology was to support its business strategy of reducing customer lead times . It found that this could only be attained by changing its fundamental manufacturing strategy to one based upon the concept of maintaining part processed stocks.

As the existing computer systems were found to be obsolete, an IT or automation strategy for the 1990's was formulated. Technical objectives were coordinated by a 'automation' steering committee to match the global objectives of the business strategy.

The process adopted was one of prototyping with dynamic specification and iterative development of user led projects. Several changes have been undertaken over the past few years. These cover the areas of statistical process control, materials handling, stock control and production planning systems, bar code reel tracking and time and materials recording with shop floor data capture. This is being facilitated by the introduction of a distributed data base technology with various communications packages and eventually networking.

The early visibility provided by this process greatly enhanced ownership and the iterative development gave management tighter control and understanding of individual projects. To date the company has delivered considerable success using this approach and developed and implemented several modules of a manufacturing information system.

Case 2: This is a heavy engineering company producing mining equipment mainly for the domestic market with a small but growing interest in export sales as well. Its over-dependence on one large customer had resulted in the past in severe consequences. To make matters worse, it was now being forced to compete more fiercely with others, for that business. The only way it felt it could survive was by cutting costs, improving quality and cutting lead times.

The plans which were developed involved the re-location and re-structuring of the plant. This was a major organisational adjustment based on the concept of the 'focussed factory' with clear responsibilities assigned to the management of the different areas and with greater co-ordination by top level management. It also installed a CAD system to speed up the design process and to improve customer response time. A number of FMS cells have also been introduced to speed up its throughput by cutting changeover times and the number of stages involved in production.

As the central D.P. department was quite small to begin with and several departments were already 'dabbling' in IT, the company decided to form a steering committee to look into all aspects of information technology. Each individual department was allowed the freedom to develop its own applications within the overall framework set up by the steering committee. A high degree of computer literacy amongst its engineering and manufacturing personnel allowed this devolved development process to operate reasonably smoothly.

Case 3: This company is one of Europe's largest manufacturers of PVC compounds and resins. A few years ago it decided to develop an integrated manufacturing and planning and control system for its 'compound' business.

Several major projects have been completed within this overall scheme. These include the development of manufacturing strategies (via the use of computerised simulation methods) to be adopted under varying market and environmental conditions; the development of forecasting models for product demand, raw material requirements, plant capacity etc; the development and implementation of plant loading and sequencing systems which interface with the company's underestimates MRP II system.

At the start of the exercise the central D.P. department was extremely small, as all support services were provided by the software house. All the above changes were introduced by a 'team' of user managers and academic staff via a 'teaching company' scheme. The work not only involved the design and testing of prototype systems but also its implementation. A great deal of expertise on the use of such systems was developed by the company managers involved who were eager to exploit their developments as quickly as possible.

This initial work has matured to such an extent that the company now has an IT manager and department with a clear strategy for implementing company-wide systems. A relational database system is currently being installed with several local area networks covering all aspects of its activities on site.

Case 4: This is a small sheet metal fabricating company in the North of England which is reasonably well established and run by a very dynamic and knowledgeable managing director. Even though the company is quite small in terms of size it is the firm intention of the M.D. to make the company into a world class organisation. The company is operating in a very competitive market and in order to survive during the early eighties took the decision to invest heavily in computer controlled manufacturing equipment which they felt would give them the necessary competitive edge and at the same time which would solve their difficulty of obtaining skilled labour. The actual IT investment has been in the purchase of CNC turret punches and guillotines with direct numerical control facilities for the production and transmission of process plans.

The company did not have any centralised DATA PROCESSING department in existence and took the view that it did not want to have one. Instead, it trained its engineering staff and the skilled shop floor workers to use

these new technologies. This has been done quite
successfully and is reflected now in the performance of the
company. It is intending to continue along this path by
developing IT skills amongst clerical and administrative
staff with the (part-time) help of outside consultants.

Case 5: The company in this example is a large
multinational metal manufacturer.  The particular division
studied has 1,200 employees. There were many opportunities
for the organisation to develop information systems as they
had not yet taken full advantage of the possibilities of
computer integrated manufacturing.  The company did not
impose corporate-wide systems on the various plants across
the world.

This Company does not have a large computing department
but historically the structure has been very centralised.
For the most part the central computing facility was the
prime mover in any major system development.  They tried to
keep tight control over any hardware purchases.

This led to frustration within the various User
departments.  Many of these areas didn't get the systems
they wanted to be developed.  This was exacerbated by the
long system development timescales.  There was also a direct
chargeout policy to the User departments.  This meant that
even a relatively small section would be asked to pay over a
hundred thousand pounds per annum for use of the computing
facility.  Most User departments would be prepared to pay
this amount if the information they were receiving was
accurate and up to date.  In many instances, however,
reports were produced from a master file which could be up
to eight weeks out of date.

The discontent showed itself in two particular ways.
Firstly, certain departments would go outside for specialist
consultants to develop particular projects.  Secondly,
various departments purchased their own stand-alone
computers which they used for their information needs.  Very
often this was made more difficult by the fact that they did
not receive any hardware or software support from Central
Computing. (Central Computing had a software development
programme organised for those staff who remained within
their area of influence).  This did not lead to the
effective introduction of Information Systems into Company
5.

Case 6: This company is a large British-based
manufacturer the bulk of whose business is in the area of
defence.  With the changing environment in Europe and the
apparent thawing of the Cold War this organisation had to
rethink its strategy in terms of winning business.  The
foreign competition had increased dramatically over the last
10 years and they were no longer guaranteed lucrative
British Government contracts.

This Company had a history of centralised data
processing with computing costs charged out as overheads.
As in many other organisations the accounting function was
computerised first followed by personnel.  There was no
information technology strategy and this led to systems

being developed in a piecemeal way. The rationale for getting systems implemented had more to do with expediency than with a clearly defined business plan. This "patchwork quilt" approach led to a lot of ill-feeling in the various User Departments.

The only change that had taken place in twelve years was that the data processing department had grown dramatically in size and with it all ensuing costs. The company found itself in a rapidly changing business environment. Their customers were looking for greatly reduced lead times. They were also looking for greatly diversified specifications on a range of products. They wanted to be able to change specifications in a short space of time. The existing computing department did not have the infrastructure in place to react to such demands. The firm had arrived at an impasse. A dramatic decision was made. The computer department in its present form was to be disbanded. The various sections within Company 6 were given autonomy in the purchase of hardware and the development of systems. The computing department staff were given the role of User support. This was alien to them in terms of their department's previous aims and objectives.

IMPLICATIONS:
Even though there are several differences in the approaches adopted by each of the case companies described briefly above, they all have one thing in common. They have broken the mould and have developed their own style, but nevertheless an effective style, for introducing information technology into their organisations. It should be borne in mind that meeting customer needs or often just the desire to survive is the most powerful motivator for introducing change. All IT and information management strategies must be seen in this light and in order to be effective they must be geared to achieve specific corporate objectives of the organisation.

Secondly, the phenomenal growth in the computing power to cost ratio over the past few years and the new developments in application software has provided enormous freedom, choice and power for the End user. Thirdly, the changes in organisation structures being imposed by the new management philosophies is clearly pointing towards the need for changes in the way in which we run our D.P. functions. Several important lessons can be drawn for the company wishing to adopt an effective implementation strategy for information technology.

o    A clear business strategy should form the basis for an appropriate information systems or information technology strategy.

o    The creation of a corporate climate which is conducive to seeking new information systems and information technology opportunities to support and enhance the business.

o    The creation of an organisational structure and culture which matches current business needs and which is

flexible enough to accomodate changes from effective
information systems.

o     Appoint people in both the business and information
      technology areas who are sympathetic to each others
      needs and capabilities.


Companies that have taken these steps are already moving
ahead of the competition. In the future the consequences
could be even more dramatic.

1.    Caulkin, S.            Crippled by Computers, Management
                            Today, July 1989, pp. 84 -88.

2.    Wheatley, M.           Taking Stock of the System,
                            Management Today, December 1989,
                            pp. 93 -96.

3.    Alloway & Quillard     User managers' system needs"
                            MIS Quarterly June 1983

4.    Checkland P.B.         Information Systems and Systems
                            Thinking : Time to Unite?
                            International Journal of
                            Information Management Vol 8 1988

5.    Beer S.                Brain of the firm
                            Wiley 1978

6.    Industrial            Survey of Computer Use in U.K.
      Computing              Industry, Industrial Computing,
                            October 1989, pp. 35-48.

7.    Burns & Stalker        The Management of Innovation
                            Tavistock Publications 1961

8.    Maguire & Hammond -    The systems life cycle:
                            inflexibility in an ever changing
                            environment - Systems Prospects -
                            1989 Plenum Press

9.    Heap, J.               Information Systems &
                            Organisational Engineering,
                            Management Services, November 1989,
                            pp. 6-10.

10.   Oakland, J.            Total Quality Management, Heinemann
                            1989.

11.   Schonberger, R.        Building a Chain of Customers,
                            Hutchinson Business Books, 1990.

12.   Milton Jenkins         Research Methodologies & MIS
                            Research - in Research Methods in
                            Information Systems -Elsevier
                            Science 1985

# TOWARDS AN EVOLUTIONARY THEORY OF INFORMATION SYSTEMS PLANNING

Margi Levy

School of Information Systems
Curtin University of Technology
Perth, W. Australia

## INTRODUCTION

ISP has evolved over the last twenty or so years in a
somewhat ad hoc fashion.  To begin it was seen as a way to
manage competing demands for computer systems by data
processing managers.  However, now it is increasingly seen as
an integral part of strategic planning i.e. there is as much
need to plan for the information systems resources in an
organisation as there is for the financial and human
resources.  While this view prevails in theory, in practice
ISP is not always regarded as an essential part of strategic
planning.  In some organisations it is still perceived as
merely part of the process to acquire computer resources, or
to spend allocated funds.  It is clearly not fully understood
by managers as has been seen from the rise of end-user
computing, where the strategic direction of the organisation
is often not considered in the decision to develop local
systems.

In practice, the case for ISP has largely relied on the
intuitive argument that it is probably a good thing to embrace
rather than reject information management.  However, the need
for organisations to manage their resources in the worsening
economic climate has highlighted the need for a more precise
statement of the contribution of ISP to the organisation.  It
is not sufficient to know that ISP works, but why it works and
the benefits which can be gained from doing ISP.  In other
words, what are the theoretical foundations of ISP.

## THE THEORETICAL LIMITATIONS OF ISP

Unfortunately, the literature has failed to adequately
show the link between ISP and organisational effectiveness.
As Galliers (1988) shows, most of the published research is:

1.  an explanation of a particular approach based on personal
    experience and case study evidence
2   intuitive argument leading to conceptual frameworks or
    proposed approaches

3.    a comparison of approaches (p182)

However, there have been few attempts to illuminate the underlying theory of ISP.  Indeed the preponderance of different approaches to ISP which draw on different themes particularly from the management literature would seem to support this hypothesis.  We see approaches based on Rockart's (1987) critical success factors, Porter's value chain analysis (Scott Morton 1988), and Porter's competitive strategy approach (Rackoff et al 1985).  This list is by no means exhaustive but is intended to give an indication that ISP is unsure of its roots and is attempting to draw on other fields to give it relevance.

Galliers (1988) has suggested that the rash of different approaches stems from the different perceived needs of practitioners. In his study, Galliers (1988, P188), concluded that there is a range of opinion why information technology strategy  studies are undertaken .  He suggests that this is why different approaches to ISP are required.  While it is undoubtedly true that different circumstances require different approaches to ISP (Earl 1988, P171) it is not clear why these particular management theories have been selected in specific cases.

If ISP is to be regarded as a professional and academic field in its own right ad hoc reasons for doing it are not sufficient.  Instead ISP must be able to demonstrate the theoretical basis for the solutions that it proposes for real world problems.

## ESTABLISHING A THEORETICAL BASIS FOR ISP

In modern society concepts of knowledge tend to be based on the scientific paradigm, leading to theories of universal laws governing the physical universe (Checkland 1981, Pp72-73).  Dominant theories of management and organisational analysis implicitly acknowledge that universal laws also govern social behaviour and that discovery of these laws enables the engineering or optimisation of human situations.

In the natural sciences it is usual for an idea to emerge which leads to a hypothesis which can be tested repeatedly by investigating a small number of variables and ultimately putting forward a theory if the tests are consistently successful (Checkland 1981, P56).  Pursuit of universal laws has also dominated the study of society (Habermas 1974, P254). Checkland (1981, P73) gives the example of the rise of operational research which attempts to scientifically model management systems.  The clear implication here is that the expert makes the local situation fit the universal law.  The theory is assumed correct, while it is the local situation which is out of step.

If it could be shown that ISP adhered to universal "laws", its emergence as a respected discipline, both professionally and academically would be assured.  However, ISP appears to be  derived from a different epistemological route than the traditional one of positivism.

ISP is about understanding the information needs of an organisation and the way that those needs are best addressed through the use of appropriate information systems. Hence it draws heavily on the experience of people who influence and are influenced by the organisation. ISP knowledge is therefore derived from human experience and validated in practice, a view which runs counter to the view that knowledge is the interpretation of universal laws. Knowledge in the social domain is invariably a matter of meaning and not just physical action. Meaning cannot be determined unless the context is also known. For example, our reaction to people marching in the street will be coloured by the purpose for which the march is taking place.

This is one reason why Checkland's (1981) soft systems approach has been applied so successfully to ISP by Galliers (1987) and others e.g. Galliers et al (1991). Soft systems does not deny the validity of individual perceptions, but seeks to understand the connections between different actors in what each believes to be a shared enterprise. Therefore, the soft systems approach seeks to make the links between the various parts of the organisation more explicit and in the processs enhance the overall enterprise.

The strength of the soft systems approach is that it has no preconceptions about the nature of the organisation or the people within it. Checkland & Scholes (1990, P6) describe the process as looking at an issue of concern, then selecting and modelling some relevant human activity systems, using the models to question the real world situation and propose some purposeful action to improve the problem situation. Soft systems clearly looks at the relevance of the local situation to produce an outcome which is relevant to that particular organisation. A holistic view of the organisation is developed which incorporates individual systems which are appropriate to the needs of that organisation.

## ISP AS LOCAL KNOWLEDGE

This leads us to question whether it is possible to view ISP in any other way than through the consideration of local knowledge. All organisations have different sets of data and different sets of values, hence the information derived from these will be different. Consequently the knowledge of the organisation and the use to which it puts the information will vary between organisations. In other words, different local realities are defined.

Metcalfe (1990) draws on evolutionary economics to show that the use of technology within organisations is dependent on their ability to use it at that time. Evolutionary economics draws an analogy, albeit in the broadest sense between biological evolution and survival of firms (Metcalfe 1990). What Metcalfe shows is that it is the capacity to learn and develop i.e. evolve, which distinguishes the successful from the unsuccessful business unit. He shows that variety of behaviour in organisations leads to major differences between decisions taken by organisations even within the same industry (Metcalfe 1990). In other words he shows that it is at the local level, not at the sectoral level that knowledge is applied by successful organisations. Metcalfe & Boden (1990) go on to show that even though the signals and information received by different organisations will be similar, individuals' experiences lead them to a variety of interpretations and hence different decisions.

Having recognised that ISP knowledge may be considered local knowledge, the issue becomes one of defining it at the local level. Habermas' concept of communicative action is used to suggest a means to gain knowledge in the local environment.

In the theory of communicative action rules are determined within the social context of the individuals. Rules are determined based upon dialogue between individuals until they are aware of recognising the meaning behind the words used. The context of the language is all important (Habermas, 1987, Pp17-20). As Lyytinen & Klein (1985) say "in communicative action people reach understanding through having a common background of assumptions about the world. Communicative action is based on a knowledge of norms, conventions, habits and accepted world views, all of which can be expressed in ordinary language" (P221).

Lyytinen & Klein (1985) proposed the use of the theory of communicative action for information systems research. They argued that it is not sufficient in information systems development to be concerned with "technical knowledge" issues. Information systems development is also concerned with enabling people to "create new meanings and concepts to cope with new situations". They perceived information systems as "formalized systems of communication built as a language game" (P226). The theory of communicative action offers the opportunity to address these issues in addition to those which can be addressed by the traditional scientific approaches to information systems development.

Earlier it was noted that the soft systems approach has been effectively used for information systems planning. There are considerable similarities between the approaches (Mingers (1980). Both the theory of communicative action and the soft systems approach endeavour to classify human activity as purposeful activity. The major objective of both approaches is to enable people to gain knowledge and understanding of their particular situation through a process of elucidation rather than imposition.

## CONCLUSION

We began by noting that ISP developed in a somewhat eclectic manner. What has been shown is that ISP knowledge is derived from the local situation. This means that the skills of the ISP practitioner are to do with understanding the importance of this local knowledge rather than the application of universal laws. Appropriate information systems can only be determined dependent on local circumstance and knowledge. ISP can only be understood in this context. ISP is therefore learnt in an evolutionary manner (Earl 1988 P171) dependent upon previous knowledge of the local situation. Hence it has been argued that the epistemology of ISP is that ISP knowledge is locally derived.

A consequence of this is that the practitioner of ISP needs to have techniques which enable them to distil the local knowledge. The practitioner must bear in mind that they may have to constantly update and validate what is meaningful information rather than rely on universal theories.

It has been shown that ISP has legitimacy other than through the adoption of management approaches. It can therefore claim theoretical validity determined through recognition of its place within the context of communicative action and evolutionary development in a local environment.

## REFERENCES

Checkland P B    (1981)    Systems Thinking, Systems Practice
                           John Wiley & Sons Ltd

Checkland P B    (1990)    Soft Systems Methodology in Action
Scholes J                  John Wiley & Sons

Earl M           (1988)    Formulation of IS Strategies
                           in Earl M (Ed) Information Management
                           The Strategic Dimension, Clarendon
                           Press Oxford

Galliers R D     (1987)    An Approach to Information Needs
                           Analysis in Galliers R D (Ed)
                           Information Analysis, Addison Wesley

Galliers R D     (1988)    Information Technology Strategies
                           Today: The UK Experience
                           in Earl M (Ed) Information Management
                           The Strategic Dimension, Clarendon
                           Press Oxford

Galliers R D     (1991)    Effective Strategy Formulation Using
Klass D                    Decision Conferencing and Soft
Levy M                     Systems Methodology, forthcoming
Pattison E                 paper for IFIP TC8 Working Conference
                           COSCIS - 91, 24-28 August 1991,
                           Helsinki

Habermas J       (1974)    Theory and Practice
                           Heinemann, London

Habermas J       (1987)    The Theory of Communicative Action
                           (Vol 2), Polity Press

Lyytinen K J     (1985)    The Critical Theory of Jurgen
Klein H K                  Habermas as a Basis for a theory of
                           information systems in Mumford E et
                           al (Eds) Research Methods In
                           Information Systems, Elsevier Science
                           Publishers B V (North-Holland)

Metcalfe S       (1990)    Introduction to Evolutionary
                           Economics, lecture to the Economics
                           Society, Curtin University

| | | |
|---|---|---|
| Metcalfe S<br>Boden M | (1990) | Strategy, Paradigm and Evolutionary Change, paper to "Processes of Knowledge Accumulation and the Formulation of Technology Strategy" Workshop, "Rosnaes", Denmark, 20-23 May 1990 |
| Mingers J C | (1980) | Towards an appropriate social theory for applied systems thinking: critical theory and soft systems methodology _J. Applied Systems Analysis_, Vol 7, 1980, 41:49 |
| Rackoff N<br>Wiseman C<br>Ullrich W A | (1985) | Information Systems for competitive Advantage: implementation of a planning process, _MIS Quarterly_, December 1985 |
| Rockart J F | (1987) | Chief Executives Define Their Own Data Needs _in_ Galliers R D (Ed) Information Analysis, Addison Wesley |
| Scott Morton M | (1988) | Strategy Formulation Methodologies and IT _in_ Earl M (Ed) Information Management - the strategic dimension Clarendon Press, Oxford |

# PLANNING IMPLEMENTATION STEPS FOR CO-OPERATIVE PROCESSING SYSTEMS

W. Gregory Wojtkowski, Margaret Barrett, Judith M. Barton, and
Wita Wojtkowski

Lancashire Polytechnic
Information Management
Preston, U.K.

Boise State University
Computer Information Systems and
Production Management
Boise, Idaho, U.S.A.

## INTRODUCTION

Co-operative Processing (COP) is a method of processing in which communication is an integral part of the mechanism of executing an application. COP processing model permits the application to be broken into pieces, to be executed on diverse hardware platforms, in order to derive the benefits of each. In this paper we discuss primary planning factors required to successfully implement COP applications.

To stay competitive in the 90's the business organization, between many other things, needs to:

- Improve enterprise efficiency
- Enhance customer service
- Improve flow of information
- Enhance decision making capability
- Understand information technology
- Employ information systems that reflect business plans

These competitive pressures and the evolution of technology allow the users to change the way they define their applications requirements. Currently the following precepts are emerging:

1. With a decentralized operating structure and a hierarchical reporting structure, application processing can be decentralized, but a considerable portion of the organization's data must be centralized.

2. Since the efficiency of many parts of the business depends on the use of computerized systems, terminal response time must be as low and as consistent as possible.

3. With an increasing demand for computerized solutions to business problems, application development should proceed as quickly as possible while development costs should be as low as possible.

4. As a greater percentage of the organization's staff becomes computer user, the interface for applications must be as easy to use and as easy to learn as current technology allows.

5. Since the business often operates 24 hours per day, provisions must be made for standalone, local processing for those times when the local processor cannot communicate with the host.

6. As pieces of the application are distributed to remote locations in response to the pressures of business and technology, a mechanism must exist to automatically ensure that all remote processors always have the most current version of the data files and programs necessary to run the system.

7. Despite the increasing interdependence of distributed processors, communications costs must be controlled to the extent current technology allows.

Changing old applications to a COP mode or building new COP applications offers solutions to many of the requirements listed above. For example, COP applications can add considerable functionality to the computer human interface, and this is very affordable in the COP environment. Functions such as context sensitive help, adviser, or list selections, when executed in the mainframe only environment, are typically quite limited due to the processing overhead. In systems designed for COP these types of functions, when implemented on the PCs, can be quite robust as well as commonplace.

In the broadest sense, one can define co-operative processing as a system design or architecture that employs two or more computers to complete a process (Tibbets and Randesi, 1990). Specific implementations of co-operative processing vary in the degree of integration of these computers and in the specific functions that each processor performs.

## CO-OPERATIVE PROCESSING IMPLEMENTATION PLATFORMS

Co-operative processing can be implemented in a number of ways (Mahnke, 1989, Mullen, 1988, Scherr, 1988). Most typically, it is based on a linkage between PCs and hosts. But, that is not the only approach that is viable. There is a good case that can be made for using a minicomputer of some kind as the dispersed computing resource, delivering the user interface via a dumb terminal. While that seems to compromise the chief benefit of co-operative processing - its enhanced user interface - there are considerable advantages to this approach, that may offset its disadvantages for certain situations.

## IMPLEMENTING CO-OPERATIVE PROCESSING

### Software Distribution

In order to implement a co-operative processing application, some mechanism must be built to ensure the use of the right software at the right time on each user's PC. Without such a mechanism, there can be no real level of control and the integrity of the data maintained by the system will, potentially, suffer.

The software distribution process involves three components. In the first component, the application must perform an audit of the user's application at the beginning of each session in order to determine whether the user has access to the right version of all program components and editing tables. This audit should also determine if any of the critical files has become corrupted since the previous audit.

Secondly, based on the results of this audit, the software distribution program must be able to take appropriate action. If the user has access to an out-dated or corrupted version of a program or table, then the software distribution application must be able to prevent the user from continuing until the correct files are loaded into the user's work area. Typically, the software distribution application will also write a transaction log entry at the host to indicate the results of the audit and, when necessary, what corrective measures were taken.

The third function of the software distribution application must be to support the scheduled roll-out of a new version of an application. Because of the amount of data that will typically be distributed in these circumstances, the distribution will normally be scheduled for late at night or over the weekend. In either case, the software distribution application must either dial in from the PC or link up with the PC from the host and establish the communications session. Then, the application must transfer the new files, verify that they have been downloaded successfully, and write a transaction log entry at the host to indicate the results of the transfer.

### The Selection of Application Development Tools

The planning process for the implementation of co-operative processing applications requires two steps. First, you must determine which cooperative processing models or topologies will be used in your organization. Then, you have to evaluate the tools that are available to assist in the development of your co-operative processing applications.

This second process itself will involve two steps. First, one must determine whether one wants to use third generation or fourth generation languages and low-level or high level communications interfaces in programming applications. Then, based on that decision, one must proceed with selection of the specific tools that best fit application's requirements.

There is also concern that must be considered when discussing implementation strategies and program development approaches. This concern is related to the gap between the users' functional requirements and the ability of MIS to deliver applications that fully meet those requirements.

Many applications in use today were written ten or more years ago. Over the years, the way the business functions has often changed. Certainly, as a result of the widespread use of PCs, people now have higher expectations concerning the user interface of their applications.

Despite changes in users' expectations and user requirements, the MIS departments and system developers in many organizations have not been able to enhance the existing suite of applications. The priorities of developers are often set according to the age-old philosophy: "if it isn't broke, do not fix it". So, most existing applications will probably not be enhanced and the users will have to cope with user interfaces that are functionally inadequate and out of date from an ergonomics standpoint.

With that in mind, it is likely that building new interfaces for existing applications and building new co-operative processing applications can be very high on any manager's list of MIS priorities.

## Learning Curve

Regardless of the productivity gains that may be realized by implementing a new development approach or a new set of tools, the improvements of those innovations may be forgone if managers feel that the new tools and techniques will take too long to master (Wojtkowski and Wojtkowski, 1990). So, it becomes important to consider the question: How long takes to learn how to use this new technology?

Indeed, this issue goes beyond the simple question of how long does it take to learn how to use a new screen painter or database. It also gets at the issues of individual and institutional inertia and 'office politics'.

Many times a new development system requires a different approach to the System Development Life Cycle (SDLC) in order to generate maximum benefits. Other times, highly skilled personnel in a development team feel threatened by a new piece of development software. If a developer is highly regarded and highly paid because of his proficiency in Command Level CICS, for example, why should that developer welcome the use of a tool that will allow any junior programmer to readily develop on-line applications without learning Command Level coding?

Once office politics and institutional inertia are taken into account, though, the critical issues concern the amount of time required to convert all or most of the programmers in a shop to the new system. Unfortunately for advocates of productivity software and methodologies, this isn't necessarily a short and sweet process.

Since a new technology will often require changes to the shop's SDLC in order to produce maximum benefits, a thorough re-evaluation of that SDLC will be required when installing any productivity tools of consequence. Tools that have little impact on the SDLC of the typical shop (e.g. a debugger, editor, or testing system) will, by definition, have a small impact on the overall productivity of that shop. And, as one might expect, the tools that will require the largest learning curves (e.g. CASE tools) may often have the greatest impact on productivity.

With these comments in mind, the selection of an appropriate set of development tools and techniques must offer:

1.    The Application of considerable increase in programmer productivity for as many different co-operative processing topologies as possible.

2.    A considerable increase in productivity in as many segments of the System Development Life Cycle as possible.

3.    A learning curve that is short relative to the productivity gains that are offered.

4.    The ability to support the migration of applications from a conventional architecture to a program-to-program communications architecture.

The closer a toolset or technique comes to meeting all four of these criteria, the better it may be for supporting the development of co-operative processing applications.

REFERENCES

Benbasat, I., and Dexter, A.S., 1982, Behavioral Aspects of Information Processing for the Design of Management Information Systems, IEEE Journal of Systems, Man and Cybernetics, Vol. SMC - 12, No. 4, pp. 439-450.

Lisker, P.,1987, "It's Time to Get Ready for New Co-operative Processing," PC Week, Vol.4, No. 10, p. C6.

Lyne, R., Iida, J. B., von Simson, C., 1988, "Co-operative Processing: Solving the Network Problem," Information Week, No. 195, pp. 66-68.

Mahnke, J., 1989, "IBM Adds a Suite of Packs for Co-operative Processing", MIS Week, Vol. 10, No. 21, pp. 25-28.

Mullen, J., 1988, "Advanced Program to Program Communications Facility Will Help Usher in the Co-operative Processing Era," MIS Week, Vol. 9, No. 49, p. 41.

Nilekani, N., 1989, "It Will Fly With a Little Help From Its Friends: Despite Apparent Obstacles Co-operative Processing Can  Take Off," Computerworld, Vol. 2, No. 7, pp. 75-79.

Scherr, A.L., 1988, "SAA Distributed Processing," IBM Systems Journal, Vol. 27, No. 3, pp. 370- 383.

Tibbets, J., and Randesi, S., 1990, "Co-operative Processing," Systems/3X and AS World, Vol 18, No. 1, pp. 62-64.

Wheeler, E.,F., and Ganek A.G., 1988, "Introduction to Systems Application Architecture," IBM Systems Journal, Vol. 27, No.3, pp. 250-263.

Wojtkowski W., and Wojtkowski W. G., 1990, " Co-operative Processing: An Agenda for Research. Proceedings of the 23rd Hawaii International Conference on Systems Sciences, Emerging Technologies and Applications Track, Vol.4, pp. 307-315.

# CLIENT PARTICIPATION IN INFORMATION SYSTEM DESIGN

F.A. Stowell

Science Faculty
Nuffield Building
Portsmouth Polytechnic
Portsmouth PO1 2EG
Hampshire, United Kingdom  Tel:(0705) 843017

## INFORMATION SYSTEMS

An information system can be viewed as an assortment of intelligence gathering devices used in a way specific to an individual, or individuals, to aid in the management of his/her environment. The key element of the information system is the human who, ultimately, will interpret the output of these devices. A computer is clearly only one method of gathering and processing data and is, as Winograd and Flores remind us, fundamentally a "..tool for human action.." (Winograd and Flores. 1987 p78). It is important, therefore, when we are considering the design and development of information systems that we put into context what is meant by *information technology* and its relationship to *information systems*.

The application of technological devices make a fundamental contribution to the dissemination of information but these devices alone are not information systems (Scarrott. 1989). The technical process is concerned only with the translation, storage and transmission of signals. It is the interpretation (literal and contextual) of the resulting code by a human that makes it information and the way in which that information is "shared" into an information system.

Information systems existed long before the invention of the microchip (Buckingham et al 1987) but paradoxically those "information systems" which utilise a computer in some way are often considered by some to be the exclusive province of the technologist. It is only relatively recently that serious attempts have been made to include the customer in the actual development procedures of computer-based data-processing systems (Mumford, 1979; see Hekmatpour and Ince, 1986). Attempts to produce "more user friendly" methodologies have, it is argued, not achieved their stated objectives (Stowell. 1991). There seems to be few computer system design approaches capable of taking into account the wider considerations involved in the generation and dissemination of information. Arguments that have been raised in the literature for a more client centred approach to information system design do not appear to have yielded any practical results. There seems to be two main difficulties to achieving this intent. First we need to attempt a *subjective* understanding of the problem-situation in order to appreciate *what* the information systems are in the social and cultural terms of the organisation and, second, provide a functional translation of the problem which satisfies the practical *"how-to-do-it"* requirements of the technology.

The underlying concept of most computer system design methodologies is essentially reductionist and many of the methods that have been developed were developed to satisfy a perceived operational need with no need for consideration for matters conceptual. The result of this tradition has been a stream of tools/techniques/methods which are geared to translating existing organisational operating procedures into a data specification. The analysis of organisational action is interpreted in a strictly instrumental fashion (Lyytinen and Klien. 1985). This seems to be based upon the assumption that the company information needs can be improved by improving the method of data processing. In 1979 Methlie expressed a similar view and defined the thinking into two categories, datalogical and infological (Methlie. 1979). Ideas about information design do not appear to have progressed over the past decade as the 1970's Mainframe pay-roll system thinking still seems to persist. Most design approaches can be described in terms of Methlie's datalogical definition but, it is argued, in the 1990's we require thinking which is more concerned about <u>information</u> systems.

## IDENTIFYING THE NEED FOR A SUITABLE IS DESIGN APPROACH

If we are to rethink the role of Systems Analysis there would seem to be an advantage to be gained in disregarding the methodologies and techniques currently used for the design of "information" systems and concentrate instead upon finding a way to help the analyst to express the "form of the problem form" (Vickers. 1981).

Hirschheim and Klien propose that information systems developers approach the task of designing information systems from a set of assumptions about the nature of the task. These assumptions relate to a particular epistemology about information system and will determine the way in which the system is developed. They suggest that most I.S.D. research focuses upon the functionalist paradigm and that alternative conceptions are warranted (Hirschheim and Klein, 1989)

## SUBJECTIVISM AND I.S.D.

Organisations are in a constant state of change and it is reasonable to suggest that the information systems too will reflect this situation (Stowell, 1990). The adaptive nature of organisations, which is part the result of a reaction to operational and social exigencies, suggest that an understanding of organisational change would be best achieved by the involvement of the analyst, the problem-solver, in the problem situation. This proposition could be addressed through the adoption of the interpretive paradigm as the intellectual theory underpining information design methodology. The interpretive paradigm is concerned with the recognition of the "real-world" as it is at a level of subjective experience. Personal experience of some attempts to incorporate this thinking into information systems design has occasionally resulted in the misapplication of the concept in practice.

One difficulty facing the analyst is the practical one that technology requires a functional explanation but information needs understanding. Despite advances in technology the human-being is still the essential element in an *information* system. An Information System is a mixture of technical and social events causally explained by an observer through a perception of actions. The information systems designer attempts to unpack and then re-pack "reality" as the process of design takes its course. The shallower the analyst's appreciation of the situation then the more likely the client is to be dissatisfied. To gain a better understanding the analyst needs to become involved in the problem.

The adoption of the subjective paradigm is not without difficulty for the analyst. Subjective involvement in the problem situation solicits criticism of analyst bias and lack of professionalism and care should be taken to avoid this criticism through the rigourous manner in which the concept is applied. Such criticism is not the exclusive province of subjectivism as Hoyer records when writing of traditional computer system design methodologies ".. the system designer may adjust their development practice to prevent the consequence they consider undesirable... the end users have no formal influence or power." (Hoyer, 1980 p130). Soft Systems Methodology may provide the means of achieving subjectivity but often the essential phenomenological element is lost when the design of the information system stage is reached.

The application of "Soft" systems thinking coupled with an appropriate means of producing a technical specification might provide one means of satisfying the requirements of ISD discussed above.

The whole method of investigation, analysis, development through to the design of an information system could be viewed as an Inquiring System (West, 1990). If the forgoing arguments are accepted the analysts first task is to gain what Vickers calls an "appreciation" of the problem situation (Vickers, 1983 p69). The adoption of this notion should be extended throughout the study as a whole which means that the development of the IS should be treated in the same way. It is here that the epistemological difference between Information Systems as understood by the "Systems" practitioner and technologist is most pronounced.

In many instances in the recent past where a subjective inquiring method has been used as a front end to information system design when the practitioner reaches the technical development stage the subjectivism which underpinned the inquiry up to this stage is engulfed by functionalism. For example, where Soft Systems methodology has been used as the method of inquiry (see Benyon and Skidmore, 1987; Wood-Harper, 1985) the translation of the information system as described by the Conceptual Model is taken over by the technical specialist and used independently to develop the computer supported "information" system. This approach often becomes an exercise in generality and the benefits gained from the development of the information system through the use of SSM are lost. This apparent functionalist attitude to subjective information system designed is suggested to be an error and defeats the whole point of adopting a subjective approach to design. The design of the information system must continue to be developed by the client and the analyst in the same fashion as the problem was addressed. The project should be treated in a similar fashion to a piece of action research. The appreciation gained about the situation by the "team" should help to guide the information system development process. The development of the information system can then be considered by both analysts and the group in its widest context rather than considered as a set of discrete steps in which the information system is predominated by the technology.

The I.T. practitioner should seek to become "facilitator" and consider himself as one member of the team who has some knowledge about I.T. The role of the analyst should be no different to another member of the team who may have knowledge about Accountancy or Production Control and his input should reflect that position and reduce the tendency for the I.T. specialist to predominate the information system design process. If the development process is conducted in this way then the possibility for real change increases since the recognition and subsequent acceptance (or rejection) for change will come from the group rather than imposed from outside (Stowell, 1990).

Using a soft approach for I.T. design, such as SSM, is not new but using it to enable clients to design their own information system is rare and there are critics who point to the incompatibility between the Conceptual Model and a Structured data Flow Diagram (SDFD) (Mingers, 1990; Doyle and Wood, 1991). However, there is only an incompatibility if the development of the SDFD is viewed as a separate exercise to be undertaken by a technical expert. For example, the use of SSM for information system design can facilitate the transfer of responsibility of design from the I.T. practitioner to the development group as a whole. If the development of SDFD is continued in the same spirit as the development of Conceptual Models then the difficulty of translating C.M. to SDFD diminishes and client led I.T. design becomes a real possibility (Stowell, 1990; Prior, 1990). It is important, however, that the hermeneutic concept which underpins SSM is maintained throughout the I.T. specification stage.

The essence of this paper is to accentuate the importance of the ethos of participation. Client involvement is not about finding ways to *bolt the client(s) to the technology* (as is with CASE tools) but to provide them with a means by which they can determine their information needs and to assist them in defining a technical specification. The argument expressed here is for the further development of methods truly suitable for client-led design.

Client-led design is not to be considered as de-skilling the role of the Systems Analyst but to help this role to evolve into that of an Information System Designer. Indeed the skills required for client-led design are argued to be greater than those for the more traditional methods of design. It is suggested that the Information Systems Designers technical expertise is used only at the end of the study to indicate to the client where I.T. may be usefully employed in an information system which s/he has aided the client to design themselves. The prime task of the designer should be to ensure a result which produces a description of an information system which includes the use of I.T. rather one dedicated to an I.T. solution.

Client-Led design emphasises that the degree of change can be regulated by the group itself and that the change may be quite radical. Consequently, the problems of individual and of management "*power*" are capable of being addressed within the development process itself (Stowell and Allen, 1988; Stowell, 1990; Checkland and Scholes, 1990). The process suggested here proposes that radical change, of which some commentators suggest SSM to be deficient (e.g. Jackson, 1988), rests with those who are involved in the problem situation. It may be that by facilitating the control of the development process to those who are most affected by the inherent changing situation a greater magnitude of change can be achieved than would be possible by an "*outsider*", no matter how expert.

It is proposed that a methodology for client-led information systems design should enable: (i) Subjectivity (ii) Client Participation (iii) Discussion by the client and practitioner about "their" information system (in its widest sense) (iv) Evaluation of Information Technology in terms of its potential aid within the defined information system and, finally, (v) Change.

## SUMMARY

An attempt has been made to consider some aspects of Information System Design. The contention has been made that the development of an Information System is better undertaken by those who are to use it but that because most design methodologies are developed from positivist ideas such a possibility is precluded. Moreover, those design methods which do attempt to include client participation seem to be more concerned with the application of computing to improve the way in which existing practices operate and rarely attempt to examine if the current working practices are appropriate to the environment in which the company presently exists. Despite the weight of literature that advocates the need to take into account the "people" part of an Information System there is little practical advice about how this might be achieved. Some key elements, considered relevant to Information System Design, are proposed as one means of evaluating the a design methodology before practitioner use.

## REFERENCES

Benyon, D. and Skidmore, S.(1987) Towards a Toolkit for the Systems Analyst Computer Journal Vol. 30 No.1 pp2-7

Checkland, P. and Scholes, J. (1990) "Soft Systems Methodology in Action" Wiley: Chichester.

Churchman C.West. (1971) "The Design of Inquiring Systems: Basic Concepts of Systems and Organisation". Basic Books: New York.

Doyle, K. and Wood, R.(1991) Systems Thinking, Systems Practice: Dangerous Liasons, Systemist, Vol 13, No.1,pp28-29

Hirschheim, R.A. and Klien, H.K. (1989) Four Paradigms of Information System Development, Social Aspects of Computing, October Volume 32 No. 10 pp1199 - 1214

Hoyer, R. (1980) User Participation - Why Is Development So Slow? in "The Information Systems Environment", edited H.C. Lucas, F.F. Land, T.J. Lincoln and K. Supper. Proceedings of the IFIPTG 8.2 Working Conference in Information System Environment. Bonn West Germany 1979. North-Holland: Amsterdam. pp129-138.

Jackson, M.C. (1982) The Nature of "Soft" Systems Thinking: The Work of Churchman, Ackoff and Checkland, Journal of Applied Systems Analysis Vol.9 pp17-29

Lyytinen, K.J. and Klien, H.K. (1985) A Critical Theory of Jurgen Habermas as a basis for a Theory of Information Systems in "Research Methods in Information Systems" (edited Mumford, E., Hirschhiem, R.A., Fitzgerald, G. and Wood-Harper, A.T.) Elsevier Science Publisher B.V.: Amsterdam. pp219-231.

Methlie, L.B. (1980) Systems Requirements Analysis - Methods and Models in "The Information Systems Environment", edited H.C. Lucas, F.F. Land, T.J. Lincoln and K. Supper Proceedings of the IFIPTG 8.2 Working Conference on Information Systems Environment. Bonn, West Germany 1979. North-Holland: Amsterdam. pp173 - 185

Mingers, J. (1990) The What How Distinction and Conceptual Models: A Re-Appraisal, Journal of Applied Systems Analysis, Vol.pp21-28.

Mumford, E. and Henshall, D. (1979) "A Participative Approach to Computer Systems Design"

Prior, R. (1990) Deriving Data Flow Diagrams From 'Soft Systems' Conceptual Model, SYSTEMIST Vol.12, No.2, May 1990 pp65-75.

Scarrot, G.G. (1989) The Nature of Information, Computer Journal Vol.32, No3, pp262-266.

Stowell, F.A. and Allen G.A. (1988) Co-operation Power and the Design of Information Systems, Systems Practice Vol. 1 No.2. pp181-192.

Stowell, F.A. (1991) Towards Client-Led Development of Information Systems, Journal of Information Systems, (in press)

Vickers, G. (1981) The Poverty of Problem Solving, Journal of Applied Systems Analysis Vol. 8 pp15-22.

Vickers, G. (1985) "The Art of Judgement. A Study in Policy Making" Harper and Row: London.

West, D. (1990) 'Appreciation', 'Expertise' and Knowledge Elicitation: The Relevance of Vickers's Ideas to the Design of Expert Systems, Journal of Applied Systems Analysis Vol.17, pp71-78.

Wood-Harper, A.T., and Fitzgerald, G. (1982) A Taxonomy of Current Approaches to Systems Analysis, The Computer Journal, 25 (1) pp12-16

THE USE OF A SYSTEMS APPROACH TO DEVELOP AN IMPROVED METHOD FOR

THE DESIGN OF COMPUTER-AUGMENTED WORK SYSTEMS

D C Sutton

SYSTEM SIX LTD
PO Box 107, Macclesfield,
Cheshire SK11 9SX, England, UK

## INTRODUCTION

If technology is ever to augment Man's intellect and information processing capability to levels that could seriously be considered symbiotic, advances are required on many fronts. In broad terms:

- Technology itself must become far more adaptable and exhibit some of the characteristics of an 'intelligent assistant',

- The current approaches to the design and provision of technology must be radically improved,

- There must be an improved understanding and utilisation of technology in the user organisations themselves.

These distinctions derive from a framework developed in the course of work upon issues relating to the design and supply of information systems. Attempts to address known deficiencies in clients' development approaches led to the recognition of a need to reconsider systems design from first principles. This article focuses upon the second of these areas.

## THE CORE FEATURES OF SYSTEMS DEVELOPMENT

Classical systems development tends to assume a linear model which progresses from the definition of user requirements to the delivery of a working system. Authorities vary as to the precise number of stages into which the process should be divided. Such approaches have been found deficient over the years, for example Belotti (1988), and modernised methods recognise that the process has feedback and feed-forward between the phases.

Nevertheless, formal methods still tend to force a linear progression upon events. The news reports a constant succession of systems design failures. These tend to produce a reaction calling for more strictly enforced standards. However, other authorities consider these responses to be 'more of the same' and that a more fundamental rethink is required. One client, highly placed in the systems provision industry, stated the position more forcefully:

"In the modern context of RAPID CHANGE:

- Traditional design methodologies approaches cannot cope.
- The latest design methodologies approaches cannot cope.
- Current systems procurement approaches cannot cope with ANY change, let alone rapid change."

*Systems Thinking in Europe*, Edited by M.C. Jackson *et al.*
Plenum Press, New York, 1991

553

Given a brief to seek ways to address this problem, we went back to first principles. Previous development work had highlighted importance of communication in all areas of man-machine interaction and so we started from basic models of organisation and communication.

## SYSTEMS DEVELOPMENT AS COMMUNICATION

Technology is an integral part of enterprise. Nevertheless technology people have languages and values that can be radically different from those in other organisational functional areas. One of our fundamental standpoints is to recognise the primacy of communication in all transactions with and in systems and there is a model which can be applied to organisational conversations in general. The model is called the ISC model (Inter-Systems Communication), it is described in greater detail in Sutton (1979), Sutton (1988) and is derived from Pask (1975).

The essence of the ISC model is that any communication, to be effective, needs to cater for three classes of concern. The obvious concern is the Topic of the conversation, what it is 'about'. However, the choice of what is said and how, depends upon additional factors.

The capacities and interests of the sender or initiator of the signal are a second class of concerns which naturally influence the choice of topic and the conduct of the conversation. The third set of concerns are those which address the capacities and interests of the receiver.

These additional concerns are crucial to successful interaction. They are addressed by psychologists as 'non-verbals', by sociologists as 'meta-messages' and by ethno-methodologists as 'context marking'. The capacity of any channel or process to carry information relating to these issues has a strong bearing upon the nature and effectiveness of any conversation that utilises it.

The ISC model has been found useful in evaluating the adequacy of technical systems including computer-based training systems. The main assertions of the model relevant to the area of technical systems' design and use are that:

- Systems design can be represented as a conversation between two parties, the 'user' and the 'provider'.

- Effective communication requires the three classes of information exchange to be supported, they address primarily the:
    - Intentions of each party
    - Expectations of each party
    - Topic under discussion.

- The Systems design process must balance the requirements and capabilities of each party in each of the three classes of concern.

When the systems design process does not give the appropriate weight to each of these issues, or allows the interests and expectations of one party to dominate those of the other, it fails to be effective or, all too often, simply fails.

## A FRAMEWORK TO DEVELOP A SYSTEMS DEVELOPMENT METHODOLOGY

We sought to specify a systems development framework that would support the full requirements for co-evolution dialogue. This necessitated interpreting the ISC model into the systems development space at a more detailed level.

The language of Chemistry is not the language of Dancing, or Accountancy and, similarly, the languages of Enterprise are not all immediately translatable into the languages of Technology. To capture this distinction, and follow up the implications, the ISC model distinguishes between languages for Self and languages for Other. For example, accountants talking amongst themselves will use a very different language from that they use to talk to clients, or to the tax inspector!

554

Figure 1 illustrates the ISC framework with the stages of classical systems development positioned upon it. We refer to a User system and a Provider system,. Although systems development casts Technology in the role of Provider, the framework is depicted in its more general form.

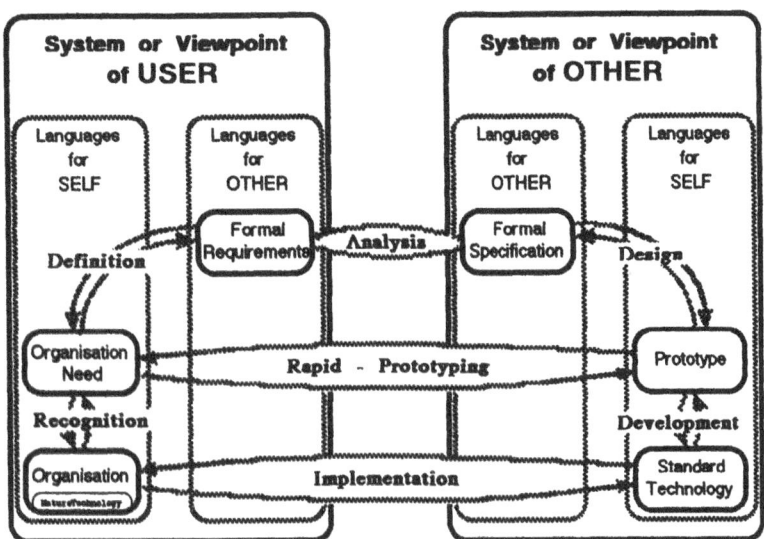

Figure 1    The Major elements of Systems Development mapped onto the Inter-Systems Communication model - showing the incomplete coverage.

Key points to note are:

- The vacant cells in the framework,

- The communication channel given **least** emphasis by formal methods is precisely the one **central** to the real meeting of the need.

- The only dialogue that is explicit and arranged to take place between Languages for each Other is the Analysis conversation.

- Rapid-prototyping is subordinate to the Analysis channel of communication and so any modifications must be expressly incorporated into the formal models via the 'correct procedures'.

- Rapid-prototyping is grounded in what Standard Technology can do.

Note that all three channels of communication are featured. It is possible to identify three 'Paradigms of Systems Development' that can be characterised in ISC terms:

- Formal Analysis and Definition of User Requirements
    (User expectations and intentions)

- Demonstration and Alteration to Satisfy User Needs.
    (Topic oriented to the needs of the total system)

- Stimulation of User Requirements by Available Technology
    (Provider expectations and intentions)

The third mode is characterised by the attitude "you can have any color (sic) as long as it's black". It predominates in the world of information technology where the IBM of old is the supreme stereotype. In the first paradigm the "customer is King" and procurement is characterised by invitations to tender and great buying power, stereotypes would be the MOD, and the major retailers.

The second paradigm is characterised by "user centred design" and strong market orientation. In the world of IT it is honoured more in the breach although "rapid prototyping" and "participative design" address, in part, the features of this mode. The move towards "Co-evolution" seeks to attain the balance this model identifies as essential.

From the perspective of the communication model, even the best of the modernised systems development tools and methodologies do not support adequate communication. The balance is, in fact, virtually the opposite of that which ISC depicts. The formal communication is at to abstract a level and Implementation is too irreversible to be a conversation channel.

The inherent shortcomings of current formal methodologies explains the rise of advocates of 'Rapid-prototyping' (RP) and 'Rapid System Development' (RSD). In ISC terms, RP/RAD attempt to open up a direct channel of communication at the Topic level by early demonstration and initiate discussion of the effects provided before it is too late to change the system.

RP/RSD evolved as a 'fix' for a deficient methodological framework. In its current form, it is still subservient, some argue subversive, to the 'formal' development procedures. Also, RP/RSD does not have the capacity to support the level of Co-evolution dialogue we know is necessary or achievable.

## SOME OF THE FEATURES OF THE NEW SYSTEMS DEVELOPMENT METHODOLOGY

All the cells depicted in the ISC framework have practical relevance and to support Co-evolution. Our aim was to specify a system development methodology that would support all the cells and all their interpretations and interactions. These are represented in Figure 2.

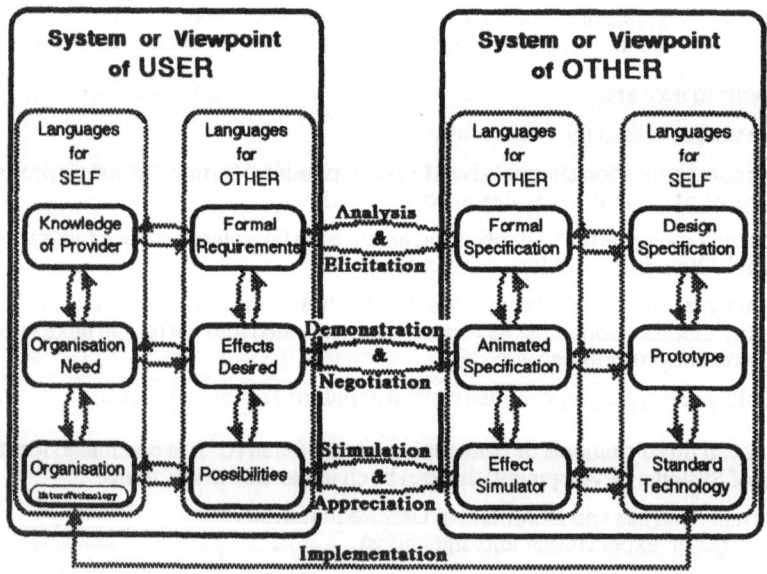

Figure 2    The elements a new Systems Development will have to support, coordinate and integrate, a specified by the ISC model.

We address first the way that an organisation Need develops into a Formal statement of Requirement. This strongly depends upon users' Knowledge of the Provider. Firstly, users will define a sub-set of their needs that they consider the provider capable of helping with. Secondly, users will then translate their needs and the chosen sub-set into a statement of requirements in the language they think, or are told, the provider will find acceptable.

It is the users who, initially, form an impression of what the provider is competent to do. It is especially important therefore that providers take care to see what the customer thinks they can do, before offering help. This is even more important when the customers' beliefs are based upon many past transactions (bitter experience?). Thus the conversations labelled Analysis and Elicitation refer not only to Systems Analysis and Knowledge Elicitation about the declared Need but about the User's beliefs about the Provider and how those beliefs might have shaped, narrowed or distorted the User's view of the need in the first place.

The above point is addressing a contextual element. Our stand-point is considerably more radical than a gap-filling exercise. We contend that the dominance of the Formal Requirements - Formal Specification dialogue is harmful and contrary to good communication principles.


## THE CORE OF THE NEW METHODOLOGY

The major communication and contractual emphasis must be shifted to the discussions about concrete Organisational Need and the capability and opportunity offered by Technology to satisfy those needs. The attempt to use rapid-prototypes to see if they meet the need, whilst laudable are not sufficiently well grounded. The first step is to bring the discussions on this level closer together.

To this end we see the users and providers requiring assistance to, jointly, translate from the users' language of organisational Needs to identifying and describing the Effects needed in terms with which the providers can work.. This change must be complemented by shifting the providers from trying to Prototype Technology to trying to Generate Effects. Both of these changes are making the translation from language for Self into language for Other.

On the users side this firstly brings them closer to actually thinking and talking about their work rather than translate their requirements into the abstract and artificial world of contract terms and conditions. On the providers side this removes the constraint of having to ensure compatibility with the Mature System and restriction to the use of Standard Technology.

However, when we attempted this, we found that the technology available was incapable of even simulating the effects required to meet the users' needs. The Effect Generator was, in the end, implemented as a tool kit able to use anything from any source in order to generate or simulate the effect the user asks for.

An Effect Generator is required to achieve three major things. Firstly it assists the provider by enabling the production of an 'Animated Specification' which simulates the effects the user has defined. In a carefully facilitated process, the provider has a conversation with the user(s) using the Effect Generator as a tool to 'mock-up' and demonstrate any user-described effect. This allows the users instantly to see the technology effect and accept or change their mind as the conversation proceeds.

The second contribution of the Effect Generator is more subtle but just as important. The user and provider have a conversation in which Effects are described, demonstrated, revised etc. This enables the user(s) to extend their appreciation of what the provider/technology) can do for them.

Thirdly, as they try out system ideas with the Effect Generator, the users are able to experience directly the potential of the technology to provide effects. The conversation may extend to role playing and users can then see how the availability of these effects will impact upon their other work. They can redesign and change that other work and modify their required effects in an integrated way.

The Effect Generator therefore stimulates the user so that they can appreciate technology possibilities that go beyond the limitations of the mature, and even the standard technology. (This is in direct contrast to the practice of conventional systems developers, who see 'users changing their minds' as a deplorable phenomenon.)

The configuration of the Effect Generator at the end of such a conversation is then capturable and formalisable and constitutes an 'Animated Specification'. This dynamic representation of requirements then constitutes the 'target' for the systems developers. This form of specification is far more accessible and concrete to both sides than the conventional formal statement of requirements which forms the basis of classical procurement contracts.

The above facility and the approach to exploit it requires a radical rethink by both sides of the systems development process as to their roles and contributions. Which brings us to the consideration of current institutionalised approaches to the procurement of technology. They too are out of touch with technology capability and organisations need to drastically rethink their approach to procurement.

## CONCLUSIONS

RP/RAD are to be seen, not as short-cuts, vaguely reprehensible because they do not fit into the standard development and procurement customs and practices, but as fore-runners, born of necessity, of a radical rethink of the approach to DAP of IT.

The key feature of the transitions required will be that the 'driving force' of development and procurement customs and practices will shift from the formal procedures for requirements, analysis and delivery to the real-time servicing of task needs.

I can best conclude by quoting the words of another forward thinking client who argues that the changes needed to address this problem area will require the following features:

"..Successful development and exploitation of future Information Technology will be dependent upon:

- Shifts from a mind-set based upon the 'Specification of Technological Requirements', to one aiming to provide: **'(Re)Configurable Technological Capability'**.

- Shifts from a mind-set based upon the 'Delivery of Modularised units of Costed Benefit', to one aiming to provide: **'On-going 'Attested Value Added' to User units'**..."

## REFERENCES

Belotti, V (1988) , *Implications for Current Design Practice for the use of HCI Techniques* In: People and Computers IV, Jones, D M & Winder, R, (Eds), Cambridge University Press

Central Computer and Telecommunications Agency, *Structured Systems Analysis and Design Methodology: Training Manual*

Kolb, D & Fry, R, (1975), *Towards an applied theory of experiential learning*, In: Theories of Group Processes, Cooper, C, (Ed), Wiley, Chichester

Mumford, E (1983), *Designing Human Systems,* Manchester Business School, Manchester.

Pask, G (1975) , *The Cybernetics of Human Learning and Performance*, Hutchinson, London

Sutton, D C (1979), *Some promising areas for research in the field of computer aided management education*, Centre for Business Research, Manchester Business School

Sutton, D C (1988), *Cognitive Transactional Analysis, a model of purposeful communication* In: Cybernetics and Systems '88, R Trappl (Ed), Kluwer,

Sutton, D C & Sutton, M M (1990), *Wheels within Wheels - A development of traditional Socio-technical thinking*, Journal of Management Education and Development, 21, 2

Wood-Harper, Antill & Avison, (1985), *Information Systems Definition: The Multiview Approach,* Blackwell, Oxford

Zuboff, S (1987), *Automate/Informate: The Two faces of Intelligent Technology.*, Organisational Dynamics

INVESTIGATIONS TO PREPARE FOR THE PROVISION OF INFORMATION
SERVICES FOR DANCE, WITH SPECIAL REFERENCE TO COMPUTERISED
INFORMATION HANDLING

Heather Williams

University of Surrey
Guildford, Surrey, U.K.

INTRODUCTION

To prepare for the provision of computerised information
services for dance, investigations were made with special
reference to the information systems of potential users. This
was to promote development of both accessibility to and
usability of such information services.

In this paper, various issues about changing situations
in the availability and accessibility of computerised
information handling are explored. Then dance information
users in situations other than secondary or higher education
are considered. Following this, a model of the situation of a
dance information user is related to a model for enquiries
about dance. A particular example of an attempt to monitor a
situation of dance information flow on an ongoing basis is
described.

CHANGING SITUATIONS OF EXPERIENCE OF AND KNOWLEDGE ABOUT
INFORMATION TECHNOLOGY FOR INFORMATION RETRIEVAL

The situation of a dance information user was modelled
using 'soft systems methodology' in combination with a
conceptual framework from cognitive information retrieval
research. The aims of this modelling were to represent the
factors that may limit effective contact of potential users
with information resources. For instance, knowledge and
notions about information resources, what use they may be and
how to approach and use them could be considered. The aspects
explored included those such as the 'ease of use' of the
'information channels', as suggested by Allen (1977) and the
extent to which information requirements can be specified by
the enquirer, as expounded by Belkin in (1977 and 1980).

Soft systems methodology was used because its theory and
practice involves models and their outcomes and formalises and
describes what occurs in each situation. Systems models enable
projections to be made about the future of a studied
situation. The methodology can involve the users of an

information system while overcoming the problem that users may not be able to recognise what they might require from an information service or what an information system could provide.

Fig. 1 represents information retrieval by a hypothetical user, who could be either a teacher or a student. Three examples of factors that could initiate information needs are exam syllabi, other curricula and dissertation topics. This diagram was arrived at after the construction of rich pictures, a root definition and a conceptual model. It is however, no longer any of these. The information needs are described in the diagram as a problem situation. Formulated queries arise out of information needs. Now, as hypothesised by Belkin (1977 and 1980) and Belkin et al (1982), at least a part of an information need cannot be expressed and is represented here as 'unspecified information needs'. There may be various sources of information available to a potential user of a dance database; e.g. the National Resource Centre for Dance at the University of Surrey (acronym NRCD), dance company archives, colleagues, public libraries. Also, although Local Education Authorities resources might not be of significance to the private dance school sector or to lecturers or students in higher education; they may be regarded as very useful by teachers within primary and secondary education. The enquiry, if made, is represented in the diagram as being directed to any, several or all of these sources of information. Any response that is obtained is shown here as influencing the situation of the information need experienced by the potential user of a dance database. As stated by Belkin (1990), if enquiries were expressed to an ideal interface to information resources, this expression could be in terms of the user's goals and existing knowledge rather than as a description of the information that would be useful. This is not the norm for computerised interfaces at present.

This person also has a level of knowledge about and access to information channels, including computerised databases and the circumstances requiring information handling and management. Both this level of knowledge and the nature of the information channels may affect the state of their information needs with respect to the balance between specified and unspecified information requirements due to envisaged or actual interaction with information channels. If these influences were represented on this diagram, they could be shown as enhancing the 'effective contact' with the information channel. By means of this cycle, the formulation of enquiries compared to the 'unspecified information requirement' might be affected.

These influences tend to be shifting. For instance, investigations into information provision and use for humanities scholars by surveys and other projects indicated a state of rapid change with respect to information handling by people within Higher Education. Developments in information technology and their uptake have been particularly noted by Katzen (1986) and Hirschheim (1988) amongst others. Similarly, the situation of acquaintance with and access to computerised information handling may be rapidly changing for dance information users.

In a questionnaire survey of potential users of dance database. Respondents were invited to write comments about what they believed to be the most significant problems for them regarding the online searching of a computerised database. In analysis, the comments were categorised as those pertaining to technical aspects of using online computerised databases, matters of work situation or finances and 'user experience and knowledge'. The number of different comments received for each category were 9 'technical', 13 'situation' and 10 'experience and knowledge'. So along with other features of their situations, user experience and knowledge were one of the pertinent area for dance information users, by their own indication.

The situation of a dance information user may not only be affected by changes in information technology for personal information handling, but also by changes in the means for recording and analysing dance. Tools for notation of dances and for choreography are becoming available for use with personal computers such as the 'Macintosh'. Also, an interactive video training package for a GCSE dance study has aroused admiration and interest from some involved in dance education.

Additional to the use of features of Belkin's proposals about the Anomalous State of Knowledge in the representation of the situation of a dance information user; several complementary aspects of cognitive information retrieval and soft systems methodology were identified. For instance, the Anomalous State of Knowledge (ASK) concept of Belkin and co-workers represents ill-defined information problems and 'soft' problems can also be initially difficult to define. Both approaches are intended to contribute to systems design.

SITUATIONS FOR DANCE INFORMATION USE OTHER THAN SECONDARY OR HIGHER EDUCATION

In a questionnaire sent to people working within secondary and higher education, a few respondents gave information about their work with community or theatre dance as well as dance studies in education. Despite this, Griffith (1988) commented on insularity of special-interest groups involved with dance, and on low levels of communication

With regard to both of these states of affairs, then factors identified as involved in computerised information handling could be investigated in order to complete the following matrix. In this, dance information users are considered to included those working in education, community situations, the professional theatre and media or publishing.

MONITORING OF SOME ASPECTS OF DANCE INFORMATION FLOW USING A SITUATION OF THE NRCD AND ITS USERS AS AN EXAMPLE

The 'formulated enquiries' part of the model of the situation of a dance information user was used as the link between the modelling and practical monitoring of dance information flow. By the combination of the model of the situation of a dance information user with one of normalised data from enquiries, another model was derived that was based on information obtained by analysing enquiries to the NRCD.

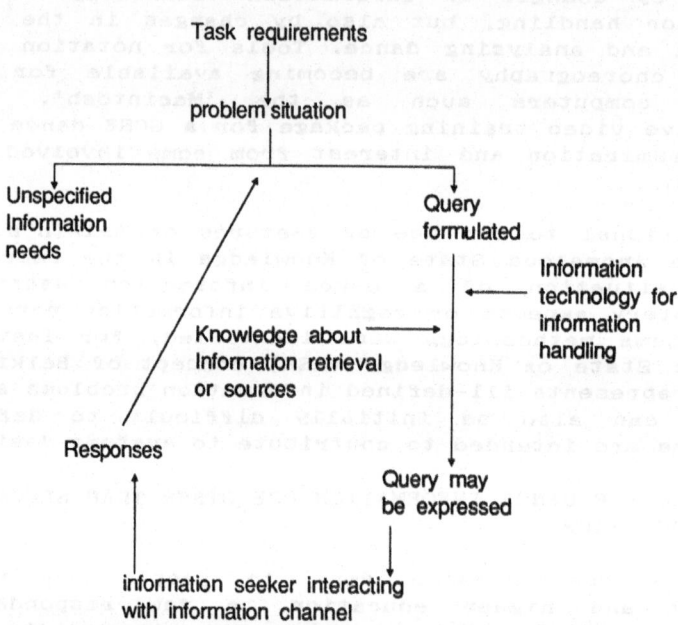

**Fig 1. Representation of the situation of a potential user of a dance database based on pictures and models obtained by ssm**

Table 1.  Matrix to indicate investigations that could be
carried out on information handling by dance information
users in various contexts

|  | IT affects question (ASK) | IT facilitates monitoring | change in information handling | recording & analysis of dance |
|---|---|---|---|---|
| Education | | | | |
| Community | | | | |
| Theatre | | | | |
| Media & Publishing | | | | |

but widely applicable.  A database was compiled, using the
records of over 1000 enquiries received by the NRCD between
1981 and 1988.  The outcomes of these studies should enable
mechanisms to monitor dance information services to be set up
and opportunities for computerisation to be recognised and
met.  Schemes for and developments in computerised access to
dance information will need to be adaptable and flexible to
meet the changing situations of the information users.  Hence
mechanisms to monitor enquiries about dance could result in
feedback loops to assist this. The next part of this section
is about just one attempt to set-up a monitoring scheme for an
information service.

     As  an  ongoing,  simple  and  effort-effective  means  of
monitoring enquiries; there could be a link between word-
processed responses and data collection about enquiries. For
instance, ongoing data input could be carried out at time of
writing responses, by use of moving relevant sections of reply
from letter 'window' to a data entry form 'window'. Then the
database  could  be  searched  on  reception  of  enquiry  for
previous similar enquiries or for other aspects of records of
enquiries.

     This procedure was not taken-up by the NRCD. However, the
identification of enquirer types and subjects has been welcome
by those offering this enquiry service. These findings could
be referred to during future developments of mechanisms to
monitor information retrieval systems of interest to those
working with dance. In addition, they might inform interface
design for such systems.

BIBLIOGRAPHY

Allen, T. J., 1977, " Managing the flow of technology:
     Technology transfer and the dissemination of
     technological information within the R&D
     organisation". The MIT Press, London.

Belkin, N. J., 1977, A concept of information for information
     science. PhD thesis, University of London .

Belkin, N. J., Oddy, R. N. and Brookes, H. M., 1982, ASK for
     information retrieval: Part I: Background and theory,
     Journal of Documentation, 38, (2), pp 61-71.

Belkin, N. J., 1990, The cognitive viewpoint in information
     science. Journal of Information Science, 16, (1),
     pp 11-15

Checkland, P. B., 1981, " Systems thinking, systems practice",
     John Wiley and Sons, Chichester.

Griffith, S., 1988, An analysis of information services
     available and critique of their response to user'
     needs. Unpublished essay from final year BA project on
     library and information services for dance,
     Ealing College of Higher Education.

Hirschheim, R. A., Smithson, S. C. and Whitehouse, D. E.,
     1988, "Microcomputer use in the humanities and social
     sciences: a UK university study". Oxford Institute of
     Information Management, Oxford

Katzen, M., 1986, The application of computers in the
     humanities: a view from Britain. Information
     Processing and Management 22 (2), pp 15-30

Naughton, J., 1980, "Soft systems analysis; an introductory
     guide", Open University, Milton Keynes.

GRAFTING A "SOFT FRONT END" ONTO A HARD METHODOLOGY -

SOME QUESTIONABLE ASSUMPTIONS

Brian Petheram

Department of Computer Studies &
Mathematics
Bristol Polytechnic
Bristol, U.K.

INTRODUCTION

This paper is primarily concerned with the application
of systems based methodologies to the development of
organisational information systems that involve the use of
computers. Checkland(1981) identified two different
paradigms of systems thinking, hard systems and soft
systems. This framework has been found useful in clarifying
the aspects of problem solving addressed by different
methodologies. Those which are based on hard systems
thinking are mainly concerned with designing solutions to
problems, the nature of which is taken for granted. Soft
systems thinking regards the nature of a problem situation
itself to be problematic and is focussed on learning rather
than design.

In the field of information systems development there
has been an increasing recognition that the overall aim of
producing a system that is well matched to the
organisation's needs is more likely to be achieved if both
hard and soft approaches are applied. Efforts to gain the
benefits of both have taken several forms, possibly the
most common being an attempt to graft a "soft front end"
onto a hard information systems methodology such as
Structured Systems Analysis & Design Methodology(SSADM)
described in Downs et al.(1988). The form of the soft
element is normally a close derivation of Checkland's Soft
Systems Methodology (SSM), the most recent account of which
is in Checkland & Scholes(1990). The idea of grafting soft
and hard systems methodologies is described more fully in
Miles(1988).

This paper will critically examine some of the
assumptions underlying this grafting process. The aim of

the discussion is to determine the extent to which the two paradigms are compatible when combined in this way. Both philosophical and methodological issues will be considered, and conclusions will be drawn on the likely impact of these issues on information system development practice.

PHILOSOPHY

There are fundamental differences between the ways in which hard and soft systems thinking view organisational reality. These have been more fully explored in Checkland(1981), but a key issue is the contrast between the positivist stance of hard systems and the interpretive stance of soft systems.This dichotomy manifests itself on a more pragmatic level both in the views of organisations as a context for information systems development, and in the modes of enquiry thought appropriate for bringing about change.

The positivist nature of hard systems thinking leads to a functionalist account of organisations. The underlying assumption is that the structures and processes that occur can be identified and modelled in such a way that the model is seen to be "correct". Such models can be constructed by gathering facts concerning the situation. If there is disagreement about the model, either the model or the questioner must be wrong. This reliance on the existence of an objective truth extends to the ways in which hard approaches deal with problem solving. There is assumed to be a problem existing, the nature of which is agreed on by the problem owners. The task of the problem solver is to determine how to solve it and to implement that solution. There may well be investigation to clarify the problem and to identify constraints, but the existence of the problem is taken as given. The mode of enquiry is normally in the form of a one off project dedicated to the solution of a particular problem.

The interpretivist nature of the soft approach leads it to regard the essence of organisational reality as lying in the meanings assigned to it by the participants. Those meanings are accepted as being likely to vary amongst the stakeholders in terms of their views of structures and processes. Before action is taken, it is thought necessary to investigate these perceptions of the nature of the situation and move towards building a consensus on the changes, if any, which should be carried out. Variances in models relevant to a situation are held to be a reflection of these differing perceptions rather than mistakes. Clear cut problems for which a solution can be engineered are regarded as the exceptional case, not the general. The model of organisational problem solving adopted by soft approaches is based on Vickers' concept of appreciative · systems(Vickers,1965). This sees management as an ongoing process over time, involving the formulation of perceptions and ideas about the organisation which may occasionally lead to a decision to act. Problems are not necessarily "solved", but effective action may be taken from time to time to alleviate them. Problem categories such as cost

control, profitability, and adapting to a changing
environment are compatible with this view.

These differences between hard and soft have
implications for information system development. Instead of
being a special and separate activity, it is seen as one of
a number of change processes that recur in the life of an
organisation. The hard systems project, with a beginning
middle and end, contrasts sharply with the soft systems
ongoing activity. Rather than search for facts, the analyst
seeks to establish the meanings ascribed to situations by
the participants. The role of the analyst in a soft systems
investigation is more of a facilitator than a leader and
the aim is to learn about what should be done rather than
how to do it.

The differences in the development process, rooted as
they are in the underlying philosophies, are likely to lead
to practical problems when attempts are made to combine
them. The notion of a soft front end defines the soft
systems activity as a stage of a linear project oriented
approach. Apart from the difficulties that an analyst might
have in performing a complete and sudden paradigm shift,
the nature of the relationship between the analyst and
those in the problem situation may well be disturbed by a
sudden change in roles and in level of participation. A
linear approach to problem solving also demands that the
outputs from one stage should form the inputs to the next.
The next section of the paper explores the problem of
achieving effective linkage between such fundamentally
different approaches.

METHODOLOGY

The starting point for a hard systems project is a
well defined problem. If a preliminary stage is added to
the project, such a problem definition becomes a necessary
input to the hard systems activity and should therefore
constitute the output of a soft front end. SSM is the most
widely known and used soft methodology and is the one
envisaged for the role of such a front end. Because SSM
seeks to address ill defined or "fuzzy" problems, a
perception seems to have developed that the result of an
SSM investigation is a well defined or clear cut problem.
Those who have studied and used SSM will know that this is
not the case.

The output of SSM, assuming a hard methodology is
substituted for Stage 7(taking action), is a set of changes
that are held by the participants to be systemically
desirable and culturally feasible. This set of changes is
not a well defined problem in the sense required as a
starting point for hard systems development, and if taken
as such may lead to unforeseen consequences. The desirable
and feasible changes may include some that would be
suitable for implentation via a hard systems approach, but
there are also likely to be "soft changes" - attitudinal or
organisational. It is the author's experience that these
are frequently mutually dependent. For example, a new

computer based information system may be feasible given that attitudinal or organisational changes are made. A switch to a hard approach may well lead to those changes not being addressed and consequent failure of the system.

There are two other assumptions which make this linkage problematic. The identification of a set of changes suitable for hard systems development does not guarantee that those changes in themselves constitute a viable system. Wilson's Information Systems Methodology (Wilson 1990) addresses this issue via the extensive and iterative activity involved in producing a Consensus Primary Task Model, but it is not clear how a grafting approach would reconcile this more creative approach with the hard systems' initial focus on description. The adoption of a grafting approach also implies that the soft issues can be dealt with at the start and ignored thereafter. This fails to recognise that the organisational context is in a continual state of change and that the process of systems development is likley to accelerate this. The intervention of a team of analysts will give rise to soft issues throughout the life of the project but these may fall outside the scope of a hard methodology and thus fail to be accounted for.

CONCLUSION

Hard and soft systems approaches both have capabilities which are necessary for the effective development of computer based information systems. This paper has examined the grafting approach as a means of gaining the benefits of both hard and soft, and has found that some of the assumptions underpinning this endeavour are questionable. These inconsistencies are rooted in the differences between the views of organisational reality and in the modes of enquiry, and are likely to lead to very real problems on a methodological level.

The aim of this paper is not to urge that grafting should never be adopted under any circumstances but to point out inherent flaws in the process. If a car with poor brakes is the only available transport and the user is warned about the problem, a safe journey may be made if care is exercised. If the user is not warned, an accident is almost inevitable!

REFERENCES

Checkland, P.B.,1981, Systems Thinking, Systems Practice, John Wiley,Chichester.

Checkland, P.B. & Scholes, J.,1990, Soft Systems Methodology In Action, John Wiley, Chichester.

Downs, E., Clare, P., & Coe,I., 1988, Structured Systems Analysis and Design Methodology: Application and Context, Prentice Hall, Hemel Hempstead.

Miles, R.K., 1988, Combining 'Soft' and 'Hard' Systems
        Practice: Grafting or Embedding?, Journal of
        Applied Systems Analysis, Vol. 15, 55-60.

Vickers,G.,1965, The Art Of Judgement, Chapman & Hall,
        London.

Wilson, B.,1990, Systems: Concepts, Methodologies, and
        Applications (2nd.Edition), John Wiley,
        Chichester.

INTEGRATING HARD AND SOFT SYSTEMS THINKING:

THE USE OF THE 'CURRENT ACTIVITIES DESCRIPTION' IN OPIUM

Keith Sawyer and Patricia Trahern

Computer Studies Department
Bristol Polytechnic
Bristol, UK

INTRODUCTION

This paper begins by outlining the need to integrate hard and soft systems thinking, and then describes the nature and use of the Current Activities Description (CAD), which forms part of a hard/soft synthesising methodology called OPIUM (Organisational Performance Improvement and Understanding Methodology). The CAD model promotes a shift from subjective to objective depictions of a situation. Objectivity is used here to refer to socially (publicly) substantialised activities (see Jackson, 1982).

ADDRESSING THE HARD/SOFT DIVIDE

It can be generally said that systems methodologies fall into one of two categories, namely hard or soft. The social theory which underpins hard systems has been identified as functionalism and that of soft systems as interpretivism (Checkland, 1981). Each of these social theoretical approaches offers distinct and opposing frameworks to understand and change social arrangements. As a consequence, we find that in hard (functionalist) systems approaches, goals are predefinable and achieveable through a process of simplification using reductive techniques. Soft (interpretivist) systems approaches are based on enquiry and learning, enabling fresh insight to be gained about the problem situation.

Each approach offers useful contributions in the endeavour to understand and improve organisational performance, but each, taken separately, harbours deficiencies. For example, soft methodologies do not offer guidance for building and achieving goals, an essential component for business strategic planning; hard methodologies, on the other hand, lack guidance to tackle organisational social and political issues. We do not of course find that organisations fashion their activities in ways which render their complexity amenable to either a hard or soft methodology; yet, as described above, this is the way systems analysis is conducted because of the dichotomous nature of systems methodologies.

The development of OPIUM has resulted from the need to address the impoverishment of systems methodologies caused by this dichotomisation. There is no prescribed starting point or stipulated pathway through OPIUM; instead, practitioners are free to jump back and forth as they iterate between models. A manual on the use of OPIUM is in preparation (Sawyer, 1991). Efforts have been made elsewhere to relate hard and soft systems, such as Miles (1988) and Wood-Harper's Multiview (1990). Mathiassen and Stage (1990) also argue for a combined use of approaches, using the terms 'rational' and 'experimental' for 'hard' and 'soft'. Hard approaches increase uncertainty as a result of deeper abstractions of decomposition, and soft approaches increase complexity by extending the areas of exploration. Each approach, according to Mathiassen and Stage, requires the other as a counter-measure, as shown in figure 1.

## THE NATURE OF THE 'CURRENT ACTIVITIES DESCRIPTION' (CAD)

CADs enable public statements of activities to be derived from subjective depictions of organisational activities, such as an accounts or stores section. These apparently well-formed situations have a tendency to harbour social disharmony and uncertainty, and a soft, subjective approach is important to allow conflicting issues to become surfaced, understood, and alleviated. In the process of deriving an Objective Model from Subjective Models, learning and change takes place among the participants, bringing about an increased pronouncement of commonly shared meanings and understandings. The complexity and uncertainty of a situation may initially defy attempts to complete an Objective Model; on these occasions, other models in the OPIUM range, such as Personal Description Models (not covered in this paper) can be invoked to facilitate successful completion.

Participants receive instruction on Subjective Model construction, which entails a set of $7 \pm 2$ verb-fronted activities with logical flows; decision diamonds, with logical "yes" and "no" outflows can also be used. Figures 3 and 4 show examples of Subjective Models constructed by two participants. To ensure that the subjectivity of each participant is captured, it is important that participants do not communicate during the construction of their Subjective Models. The change facilitator assists in the establishment of the Objective Model, which should generally follow the $7 \pm 2$ ruling, to capture the same degree of detail.

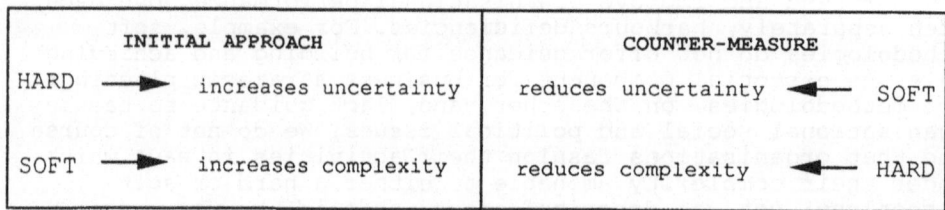

*Figure 1 - Hard and soft systems related to complexity and uncertainty*

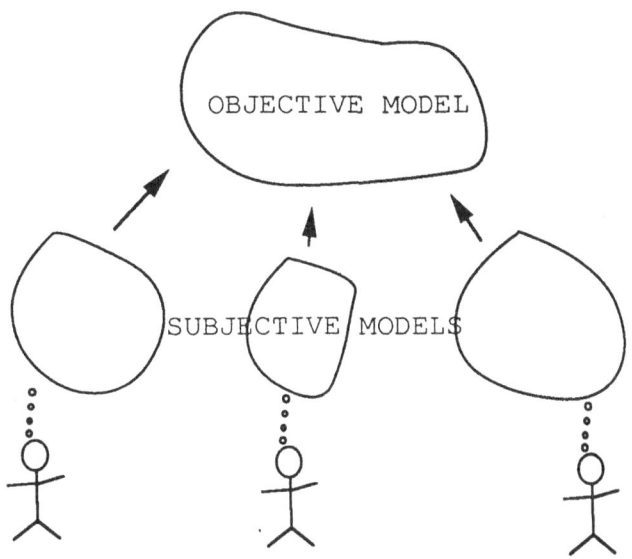

*Figure 2 - Producing an Objective Model from Subjective Models*

**THE CASE STUDY**

The case study involves a large brick making company which has two production areas, one for standards and the other for specials. The Specials Section, which is the focus of our attention, is responsible for hand-made and batch produced bricks. An office of five people controls the receipt of an enquiry or order, and follows the progress through the manufacturing process, keeping customers informed about progress and completion.

The major reason for the analysis was to identify the main reasons for poor communication between the factory and the Specials office, and to design an improved computer system which was able to faithfully track and report on the stages of manufacture.

All five people working in the Specials office constructed Subjective Models, but only two are included here because of space restrictions (figures 3 and 4). Experience in the office ranged from 8 years to 18 months, and a reasonably high level of cooperation was received. The following illustrates some of the points which arose during the exercise:

**1. Boundary Variances**

Different understandings arose concerning the operations specific to the Specials Section.
* The accounts section also received orders and then passed them on to Specials Section.
* Participants were divided on whether 'dispatch', and 'stock held' fell within their domain.

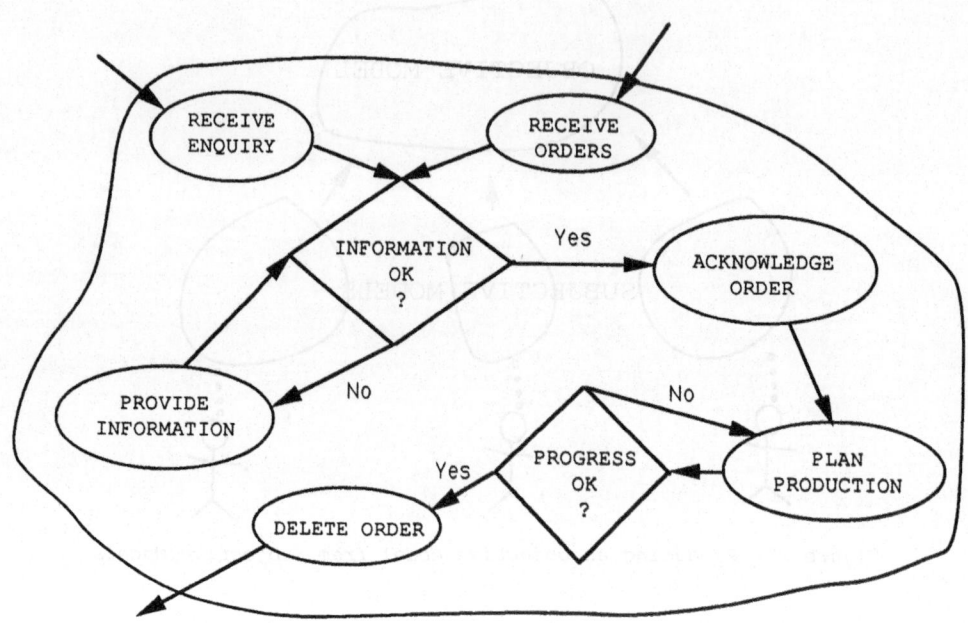

*Figure 3 - Subjective Model from Participant A*

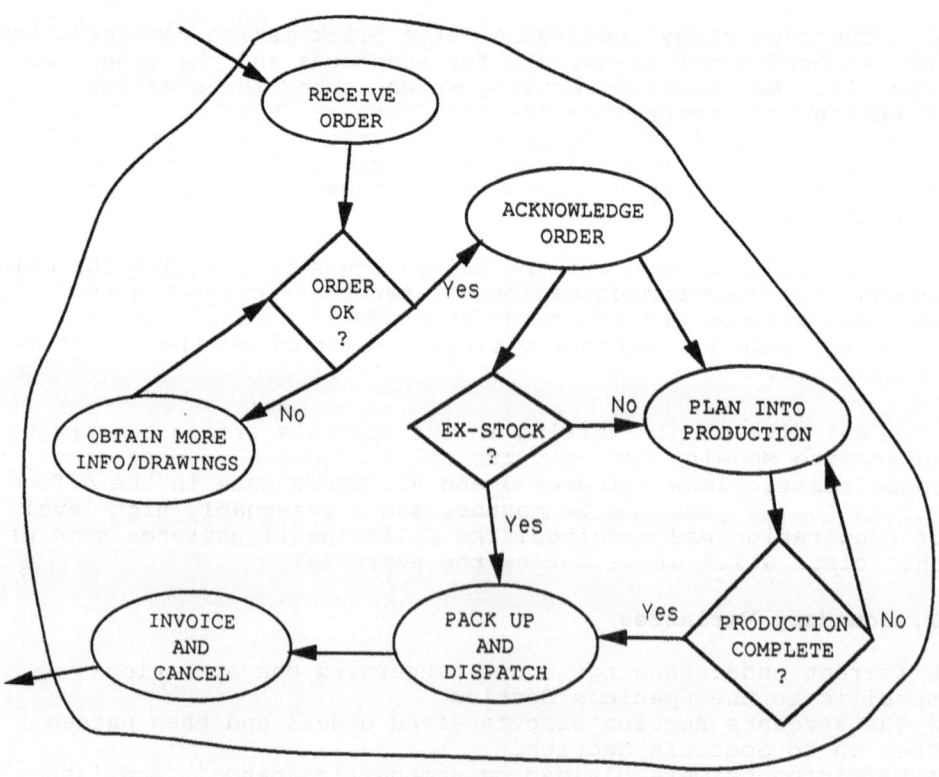

*Figure 4 - Subjective Model from partipant B*

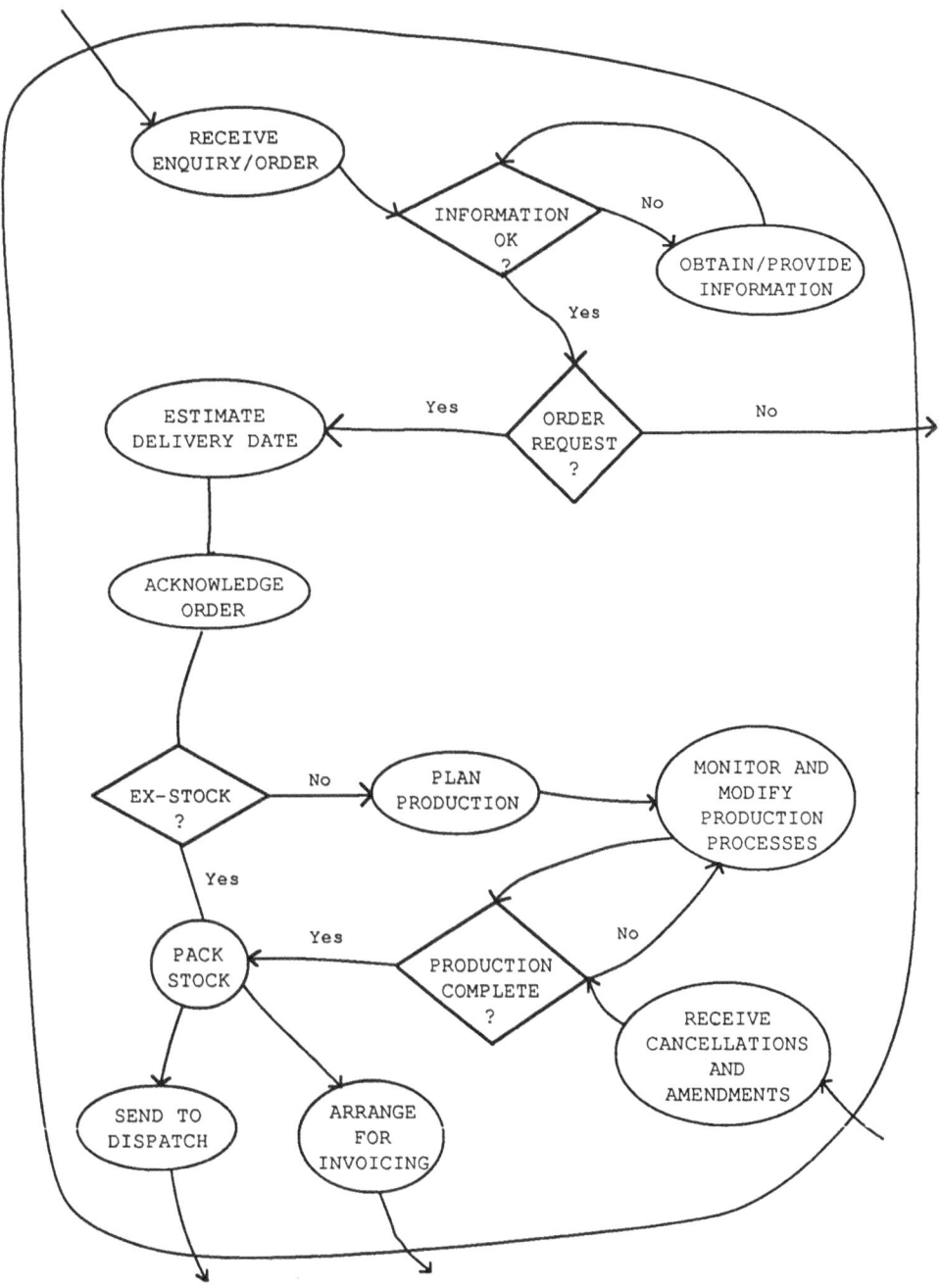

*Figure 5 - Objective Model of the Specials Section, constructed from
debate and learning derived from the Subjective Models*

## 2. Semantics

The meanings applied to some of the terminology differed:
* The term 'plan production' meant 'initial planning' to some, while to others it also meant 're-planning'.
* Two participants perceived 'ex-stock' to refer to the availability of the the entire order from stock, while for the others it meant that at least some stock was available.

## 3. Omissions

None of the participants included 'cancellations and amendments' in their Subjective Model, but during negotiation, these activities were accepted as very important.

## 4. Primacy Activities

There was an inclination for the particpants to highlight their activities stronger than others, despite urging impartiality.

## 5. Frequency

It was found that those activities which were done on a regular basis were more likely to be included (see Valusek and Fryback, 1985). For example, 'acknowledge order', is a frequent activity, and all included it in their Subjective Models.

## CONCLUDING NOTES

In this investigation, several of the Objective Model activities were unpacked, and this enabled iteration to take place between the levels, thereby facilitating the development of the Objective Model as a spinoff. All participants were involved in the negotiations and debate to construct the Objective Model. The learning and changes which followed resulted in a fresh cultural atmosphere. An important balance needs to be struck between freeing peoples potential for enhanced creativity and individualism, and shaping and developing shared perspectives and opinions.

## REFERENCES

Avison, D.E., and Wood-Harper, A.T., 1990, *Multiview, An Exploration in Information Systems Development*, Blackwell, Oxford.

Checkland, P.B., 1981, *Systems Thinking, Systems Practice*, John Wiley, Chichester.

Jackson, M.C., 1982, *The Nature of 'Soft' and 'Hard' Systems Thinking : The Work of Churchman, Ackoff and Checkland*, JASA, Lancashire.

Mathiassen, L. and Stage, J., 1990, *Complexity and Uncertainty in Systems Design*, Internal paper, University of Aalborg.

Miles, R. K., 1988, *Combining 'Soft' and 'Hard' Systems Practice: Grafting or Embedding?*, JASA.

Sawyer, K.,1990, *A Manual for Using OPIUM*, Technical Paper, Information Systems, Bristol Polytechnic.

Valusek,J.R., and Fryback, D.G., 1985, *Information Requirements Determination : Obstacles Within, Among and Between Participants*, ACM.

# INFORMATION SYSTEM DEVELOPMENT AND SOFT SYSTEMS THINKING:

# TOWARDS AN IMPROVED METHODOLOGY

Paul Keys and Mel Roberts

Department of Management Systems and Sciences
University of Hull, Hull HU6 7RX, UK

## INTRODUCTION

There has been considerable interest over the past five or six years in the use of Checkland's soft systems methodology as an aid to the development of information systems. Wilson (1984) and Wood-Harper et al. (1985) in particular have shown how soft systems thinking and its associated tools and techniques can be integrated with other approaches to information system development. In this paper these attempts are considered with a view to assessing how well the philosophy behind Checkland's original work has been continued in its application to the specific arena of information system analysis and design. On the basis of the conclusions drawn some steps are being taken towards the construction of an improved methodology and associated software to facilitate this way of tackling information systems development. The principles behind this extension of current practice are outlined.

The paper consists of three main sections. First, two attempts to integrate soft systems methodology and information systems analysis and design are briefly reviewed. Then a critical assessment is made in which a major focus is the way in which the fundamental notion of multiplicity of perspective, which is central to soft systems methodology, is addressed. Finally an outline is given of a methodology which is intended to improve the integration of soft systems methodology and structured systems analysis and design methods for the building of information systems.

## INTEGRATING SOFT SYSTEMS METHODOLOGY AND INFORMATION SYSTEMS ANALYSIS

Here two attempts to use soft systems thinking to aid the analysis and design of information systems are considered. These are chosen because they each highlight one aspect of the central issue of concern here, the way in which multiple perspectives are dealt with. Two approaches are looked at in some detail, those of Wilson (1984) and Wood-Harper et al. (1985).

Wilson (1984) develops a methodology for the specification of information requirements as part of his much wider exploration of how soft systems methodology might be used. The end result of his methodology is a statement of what information processing procedures are used by whom in order that information requirements are satisfied. The basis for the methodology is the belief that by understanding the system that uses the information to be a human activity system and to analyse this using the notions of soft systems thinking will lead to beneficial insights into the nature of the situation and of the information needs of the people within it. In a recent restatement of how soft systems thinking is seen to interact with information system creation, rather than design, Checkland and Scholes (1990) do not provide any major revision to these arguments.

The methodology, as stated in Wilson (1984, pp. 245-7) has eight stages. Of prime concern here are the first three of these. The starting point is to carry out an issue-based analysis of the situation in order that a relevant primary-task description can be determined. A primary-task model is then developed and validated. This model takes the form of a conceptual model and its associated CATWOE root definition. Each activity in the primary-task model is then considered to be related with an information transformation and from this a detailed analysis of information requirements and how to meet them can be made.

The key point here is the move from an issue-based analysis to a primary-task analysis of the situation. An issue-based analysis is concerned with exploring the implications of different Weltanschauungen so that a variety of different perspectives and attitudes towards a situation can be more deeply understood. A primary-task analysis focusses instead upon a "version of the situation which, it can be argued, is close to agreed perceptions of reality" (Wilson, 1984, p.86). A primary-task analysis is usually preceeded by an issue-based analysis as the question of what primary-task to use is one best dealt with through an issue-based analysis.

The second attempt to integrate soft systems thinking with information systems analysis and design which is to be considered here is that embodied in the Multiveiw methodology (Wood-Harper et al., 1985). Multiview brings together components of several other methodologies which, together, enables a multi-faceted view of an information systems project to be adopted. The methodology allows a system designer to produce a specification for a computer-based information system. Integral to the methodology is the need to consider how the changes to the technical system will affect and be affected by the social and organisational systems with which it interacts. It is necessary to adopt such a socio-technical approach so that the system produced is "complete in both technical and human terms" (Wood-Harper et al., 1985, p. 17).

The methodology has five stages, of which the first two are significant here. The first stage involves analysing the human activity system. This is done by constructing first of all a rich picture of the situation. This allows those involved to examine the various concerns surrounding the situation and to decide upon which area they will concentrate. The picture, in particular, "should yield both the primary tasks of the situation and the issues which surround those tasks" (p. 41). On the basis of what is discovered during the creation and analysis of the rich picture a conceptual model of the situation is constructed on the basis of an agreed root definition. In the second stage of the Multiview methodology the results of the analysis of the human activity system, the conceptual model, are used as the basis for a detailed analysis of information flows. This results in a description of the information system in terms of entities and flows which is then used as the basis for more detailed system specification and development.

As with Wilson's approach the move is made between a broad and 'messy' picture, literally, of the situation and a single well-defined statement about the situation of concern. Although not explicitly stated the process for making this transition seems to be similar to that adopted by Wilson. Themes are extracted from the rich picture and ways of understanding relevant to each of these established. This mirrors Wilson's use of an issue-based analysis and has the same aim, to enable a single root definition to be agreed upon. It is this agreed root definition which is the foundation for the conceptual model.

## CRITIQUE AND EXPLANATION

What is interesting about these three examples of soft systems methodology being applied to information systems analysis and design is the way in which the multiplicity of views about the nature of a situation is condensed into a single perspective. An important theoretical notion in the development and application of soft systems methodology generally is its acceptance that when faced with a situation different perceptions are not likely but guaranteed to exist once the subjectivist stance of Checkland is accepted. A feature of the above work is the focus upon the singular end result which is THE information system. This is, in one sense, only to be expected as an information system can be seen as a physical system carrying out identifiable and unambiguous tasks. However, an information system can also be seen to be serving different purposes for different users and playing a different role in the organisation for different observers and stakeholders. The production of a single statement of what an information system should be like seems to suggest that this issue may not be being adequately treated. For this reason it is worthwhile asking why it is that when soft systems methodology is used to aid in the design of information systems a single design for an information system is produced rather than several reflecting the different perspectives that may be brought to bear upon an information system?

Consideration of the work discussed above generates three possible explanations as to why multiple views are filtered out during the analysis of information systems. These will now be considered in turn.

The first reason can be said to be due to pragmatism. The difficulty in producing an information systems design from one conceptual model, allied with the practical need to put in place a working system, may be sufficient to persuade an analyst that to address more than one perspective to this level of detail is unwise. Thus it is the sheer amount of work which is required to yield a satisfactory design together with the resource constraints imposed on an analyst which makes working with more than one conceptual model impossible in practical terms.

This pragmatic argument is used to support the focus on a single conceptual model in Multiview. Wood-Harper et al. (1985, p. 49) refer to the difficulty of reconciling different perspectives at an early stage but note that "if one is not agreed, it will be even more difficult to agree on a final system". The cause of the increased difficulty is not stated and it could be due to either the increased complexity and specificity of the information system models that are being dealt with or the increased levels of committment to certain views caused by a prolonging of the focus on separate perspectives. Whatever the cause there is a clear case which might be able to be made for bringing the different views of a situation together as early as possible in an analysis.

A second reason may be labelled as contextual. There may be something in the context of the analysis which is specific to the nature of information systems and which allows a strong agreement of purpose and objective to be achieved. This characteristic may be less prominent in other contexts which must therefore involve a more detailed and lengthy exploration of the differing perspectives on the situation.

Wilson (1984, pp.85-6) argues that "there are instances ... particularly in relation to problems concerning ... information systems analysis" where a primary-task analysis is of value. This involves "choosing to model a version of the situation which, it can be argued, is close to agreed perceptions of reality." Thus there seems to be a quality associated with information systems which minimises the differences in perception and which can, after some negotiation aided by an issue-based analysis, enable them to be effectively disregarded.

The third reason is representational and Prior (1990) highlights this in his analysis of the use of soft systems methodology in information systems development. The way in which conceptual models and data flow models represent perceptions of reality encourages associations between them. As the focus is upon the structure and logic of the modelling tools the tendency to concentrate upon the ability of these tools to represent an actual system, the information system, rather than an abstract system, the human activity system it refers to, is difficult to conquer. As Prior states (p.67) the problem can be seen in terms of transferring "the activities of a conceptual model to the sphere of design where the object is to model the logic of what is intended to be an actual system".

These three possible explanations are not independent of each other and are to some extent mutually reinforcing. Accepting that situations can be seen from different viewpoints introduces a great deal of complexity into what is typically an already complicated set of data requirements and information flows. From the analyst's point of view the size of the task having to be dealt with can be reduced if legitimate reasons for doing so can be found. The need to complete a project within a timescale and resource constraints, the particular physical character of information systems and the possibilities offered by the representations used each provide such reasons.

The restrictions on information systems development can be eased in several ways. Currently available tools and techniques may be replaced by improved aids to information system development. These may allow several perspectives to be used to underpin several designs which may then be compared and used as the basis for an agreed final design. The ability to focus on an agreed view of the nature of an information system may take place at one level but different users may see the same information system in a variety of ways depending upon how it impinges upon their place and role in the organisation. The consequences of this may be reflected in the design of the information system that each individual would prefer. A more complex representational device may enable this complexity to be tackled effectively. An analyst may then be forced to recognise that the logic of the physical system is separate from rather than a consequence of the logic of the abstracted system of perceptions.

One key to making the switch from the current approach to one which emphasises the multiple perceptions to a greater extent is to develop tools to aid in the development of information systems. In the final section of this paper an outline of such a tool is presented.


TOWARDS AN IMPROVED METHODOLOGY

A programme of work being carried out at Hull University by the authors is directly concerned with the problem of how an improved methodology in line with the above arguments can be developed. One current piece of work has resulted in the following six stage methodology. This draws strongly on the previous attempts to combine soft systems thinking and information systems developement methods which are reviewed above. The six stages are presented below with the usual caveat that although they are ordered as a sequence of tasks in practice they have to be combined in a more complex fashion.

This approach places a considerable amount of work onto the analyst who has responsibility for producing a variety of models and analyses. As an aid to this process a Computer Aided Systems/Software Engineering (CASE) tool is being designed. This is based on CUSTOMIZER and EXCELERATOR/IS software which has been provided by EXCELERATOR UK Limited as part of a grant in support of this work. The software being developed, EXCELERATOR/SSM, will not only aid the above methodology but also offer a wide range of other information system design aids which EXCELERATOR/IS offers. For each of the six stages of the methodology the corresponding components of the CASE tool are also indicated below.

| Methodology | CASE Tool |
|---|---|
| 1. A rich picture is produced which describes the situation and serves to highlight issues of concern. | 1. The rich picture is specified in terms of a rich picture graph (RPG) containing rich piture objects (RPO) and rich picture connections (RPC). |
| 2. The issues are given formal meaning within a set of themes. | 2. The RPOs and RPCs are related to one or more themes (THM). |
| 3. From the set of themes a number of root definitions are constructed using the normal CATWOE conventions. | 3. Each THM is related to one or more root definitions (RDE). |

| 4. Each root definition is elaborated through a conceptual model. | 4. Each RDE is developed into a conceptual model graph (CMG) containing conceptual model objects (CMO) and conceptual model connections (CMC). |
| --- | --- |
| 5. From each conceptual model a set of user requirements is defined which state information needs relevant to the issues underpinning the root definition and conceptual model. | 5. Each activity in a CMG is related to a user requirement (URQ). |
| 6. The set of user requirements acts as a basis for designing the system to be implemented. | 6. The URQs are fed, as necessary, into an information system development methodology. |

This approach serves to place debate and the consequential narrowing of attention in a later phase of system developement than is the case in the methodologies discussed above. Debate and design decisions now are to be made about specific information needs and their provision rather than about how to see the situation which the information system is to serve. As yet this approach has not been tested and its costs and benefits remain to be established. However it offers a consistent and logically sound extension of the established approaches and will, if nothing else, provide further insights into the information system development process which these define.

REFERENCES

Checkland P.B. and Scholes J., 1990, "Soft Systems Methodology in Action", John Wiley, Chichester.
Prior R., 1990, Deriving data flow diagrams from a 'Soft Systems' conceptual model, Systemist, 12(2):65.
Wilson B., 1984, "Systems: Concepts, Methodologies and Applications", John Wiley, Chichester.
Wood-Harper A.T., Antill L. and Avison D.E., 1985, "Information Systems Definition: the Multiview Approach", Blackwell Scientific Publications, Oxford.

INFLUENCE DIAGRAMS IN INFORMATION SYSTEMS RESEARCH

Chee Sing YAP

Department of Information Systems and Computer Science
National University of Singapore
Singapore 0511

## INTRODUCTION

A major difficulty facing researchers in information systems (IS), management science and social science is coping with complexity. Systems such as business organizations are highly complex. Cause and effect are difficult to establish in system enquiry, and often the effect of a particular factor cannot be easily isolated from that of others. In IS research, for instance, it is difficult to determine whether higher computer usage leads to more profit or whether higher levels of profit lead to more computer usage, or whether such observed correlation is due to other factors. This problem is, however, not new or limited to IS research alone. In this paper, the author proposes the use of influence diagrams as a tool to cope with complexity and causality in IS research.

Influence diagrams, also known as causal maps, are diagrams showing what influence what. They are used extensively in system dynamics, cognitive mapping, social psychology, and several other fields (see, for example, Eden, 1985; Weick, 1979, Wolstenholme and Coyle, 1983). Influence diagrams, however, did not appear in the IS literature until recently (Montazemi and Conrath, 1986; Yap, 1986; Conrath et al, 1987; Swanson, 1987; Banville, 1990). This is surprising because influence diagrams appear to be well suited as a modelling tool in certain ill-structured IS research areas, including information requirements analysis, impact studies, and implementation research. This paper presents three case examples which demonstrate the relevance of influence diagrams to IS research. The influence diagrams of system dynamics are adopted in these studies.

## SYSTEM DYNAMICS

System dynamics is originally defined as 'the study of information feedback characterization of industrial enterprise to show how structure, amplification, and time delays interact to influence the success of the enterprise' (Forrester, 1961, 1969, 1971). More recently, this methodology has been refined and enhanced by Wolstenholme (1982) and a new definition is offered: 'A rigorous method of system description, which facilitates feedback analysis, usually via a continuous simulation model, of the effects of alternative system structure and control policies on system behaviour.'

This new perspective considers system dynamics as a very general two stage procedure for system enquiry applicable over a wide range of system types. The two stages are (1) the application of a stepwise method of system description for problem development and qualitative analysis; and (2) the application of continuous simulation techniques for quantified analysis to enhance the design of system structure and control rules. By considering the methodology in two stages, a researcher can model a system through influence diagrams showing causal relationships, and analyze the system qualitatively. This qualitative analysis of the diagrams can be seen as an end in itself. When more insight is required, the researcher can proceed to the second stage where the system is modelled quantitatively using a suitable simulation language (Wolstenholme, 1982). In IS research where the main interest is on understanding causal relationships, much insight can be gained through a qualitative analysis of influence diagrams alone.

A unique feature of the system dynamics methodology is the explicit modelling of feedback mechanisms. It is this explicit representation of feedback mechanisms, which exist in many social systems, and the insights that can be gained through this perspective, that the influence diagram approach in system dynamics should be exploited in IS research.

Wolstenholme and Coyle (1983) provide detailed guidelines and an algorithm for translating mental pictures into influence diagrams suitable for qualitative analysis. Table 1 shows a modified version of the steps involved in constructing influence diagrams (Wolstenholme and Coyle, 1983). In system dynamics, a distinction is made between a level (for example, a concern over the size of inventory), a rate (for example, the rate of consumption), and a composite variable (for example, productivity). For clarity, these distinctions can be ignored. The convention for influence diagrams as used in system dynamics is shown in Figure 1.

One strength of the present approach is that the analyst helps the client to express his 'theories' concerning a problem situation. This approach takes into consideration the fact that many IS research areas are ill-structured, such as in information requirements analysis, where it is likely to have differing views on a problem situation. The suggested approach also recognizes explicitly feedback mechanisms which exist in many social systems. Influence diagrams can be drawn quickly and revised easily and they are easy to explain to managers. Qualitative analysis of the model may reveal interesting mechanisms which may be counter-intuitive, as is often the case in system dynamics. Such analyses may also help the analyst to generate policy options which can then be used to achieve desirable results.

## INFLUENCE DIAGRAMS IN IS RESEARCH

Influence diagrams did not appear in the IS literature until the mid 1980s. The paper by Montazemi and Conrath (1986) is the first paper which the author came across where influence diagrams are explicitly used in an IS problem situation. The influence diagrams (or cognitive maps) of cognitive mapping were used in information requirements analysis. These cognitive maps, though differ from the influence diagrams of system dynamics, can be transformed into a system dynamics model (Eden et al, 1983). The authors concluded that this technique provide three benefits: it aids in the identification of irrelevant data, it can be used to evaluate factors that affect a given class of decisions, and it enhances the overall understanding of a decision maker's environment, particularly when it is ill-structured.

In a subsequent paper, Conrath et al (1987) examined the usefulness of cognitive mapping for evaluating information in an ill-structured

Table 1. Steps involved in constructing influence diagrams in IS research (adapted from Wolstenholme and Coyle, 1983)

---

1.  Identify the key variables giving rise to the observed symptoms of concern and to the need for enquiry.

2.  Identify some of the variables associated with the key variables.

3.  Devise an open-ended questionnaire based on an initial understanding of the problem.

4.  Conduct interviews and let the clients express their 'theories', using the questionnaire as a general guide.

5.  Draw influence diagrams based on materials gathered from interviews.

6.  Analyze influence diagrams and identify specific questions to be asked in second interviews.

7.  Conduct second interviews, using influence diagrams as a guide. Revise influence diagrams during interview.

8.  Combine influence diagrams.

9.  Carry out a qualitative analysis of the overall diagram to identify:

    (i)    further problems associated with the system;
    (ii)   specific relationships in the system which need further analysis;
    (iii)  controllable variables;
    (iv)   the general systemic impact of changes to controllable variables;
    (v)    the vulnerability of the system to changes in uncontrollable variables;
    (vi)   alternative groups of compromise changes which might lead to improvements in the system.

10. Organize a feedback session to present findings to the sponsor and interested parties.

---

environment. Although there is no definitive conclusion, the authors felt that the development of cognitive maps for decision makers is one way to identify the criteria for the design of computer-based information systems. More recently, Banville (1990) also attempted to use cognitive mapping to study legitimacy in the context of transferring information systems from one group to another.

In 1985, the author started using influence diagrams of system dynamics as a pictorial representation of the impact of specific information technologies and factors affecting their use in three case studies (Yap, 1986). Three examples taken from these case studies are given below.

Figure 2 shows an influence diagram developed by the author during the initial stage of a study concerning the use of electronic mail in a company. It was the result of a series of interviews with users (managers,

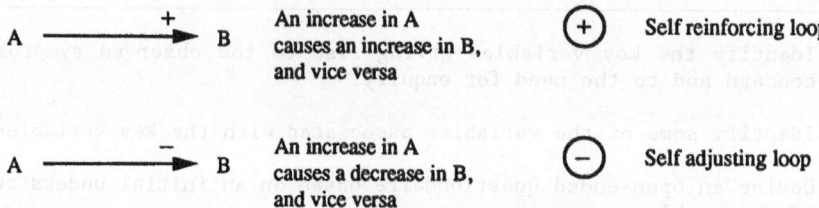

A ———— **+** ————▸ B    An increase in A causes an increase in B, and vice versa    ⊕ Self reinforcing loop

A ———— **−** ————▸ B    An increase in A causes a decrease in B, and vice versa    ⊖ Self adjusting loop

Figure 1. Conventions for influence diagrams

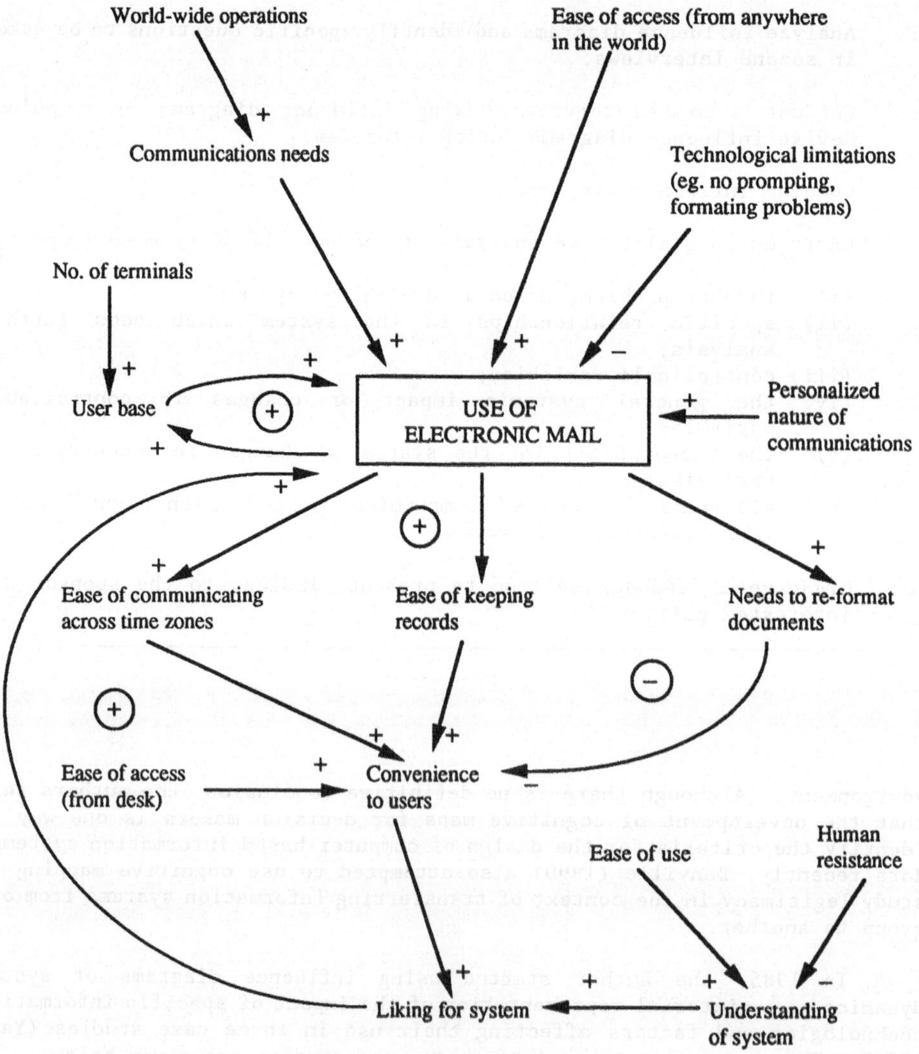

Figure 2. Use of electronic mail

consultants, and secretaries) in the company. Factors contributing and discouraging the use of electronic mail are shown, as are the feedback loops. During the presentation to the management, a further factor was identified, this being the personalized nature of electronic mail communications. The managers found the influence diagrams easy to understand. As shown in the diagram, an important factor affecting the level of use is the size of user base within the company, and this is related to the number of terminals available in the company. Thus, through a qualitative analysis of the diagram, it is possible to establish that to further encourage the use of electronic mail, more terminals need to be acquired. This can then form the basis for a policy option. Negative factors found from the interviews had also been shown in the diagrams, and it is easy to see that more widespread use of electronic mail was, to a certain extent, dependent on overcoming human resistance and the technological limitations.

Figure 3 shows the factors affecting the use of a computerized marketing database in another company. The company concerned had developed a marketing database on a microcomputer, but it was regarded by many within the company as a failure. Factors leading to this were identified and shown in the diagram. It is interesting to note that because the computerized database could not fulfil satisfactorily the continuing need for information, marketing managers in individual divisions began to set up their own database using index cards (shown as a loop at the bottom). Thus, there were two systems: one for the company, and one for the individual divisions, resulting in duplication of effort since both systems had to be maintained.

The failure of the system could be attributed to two factors: technological and administrative. The computer hardware and software chosen did not really have the capability of performing the designed tasks. Moreover, it was a single-user machine situated at the marketing division, even though it was meant to be used by managers from all divisions. Managers were discouraged from using it because they had to make a special 'trip' to the marketing division to use the machine. Their desktop terminals could not be used to access this database. The other main factor is administrative or managerial in nature. Although management regarded the marketing database as important, they did not wish to commit too many resources to it. Managers who had been contacted by an external organization were required to fill in a standard form which was then forwarded to the marketing division. There the data were added to the database by a secretary, but this task was given a low priority since the secretary had other work to do. As a result, there was always a backlog of forms to be processed. The information on the computer was thus not as up-to-date as it should have been. The database was also inadequate because some potentially useful information could not be included due to the limitation of the software. These factors formed a vicious circle (Masuch, 1985): as the database was thought to be not very useful, less importance was attached to it and backlog of work was allowed to build up, which reduced the usefulness of the database as a source for timely information.

Influence diagram was similarly used in a study on the use of computer aided drafting (CAD) facilities in a third company. The findings are shown in Figure 4. Here the factors leading to the use of CAD were the high volume of repetitive drawings and competitive pressure as some of the company's competitors had already used CAD facilities and there was a danger that it might lose business if it did not adopt this innovation. The company had held back investment because of the relatively high initial costs and long pay-back period. As shown in the diagram, there would be a direct impact on productivity due to faster production and easier

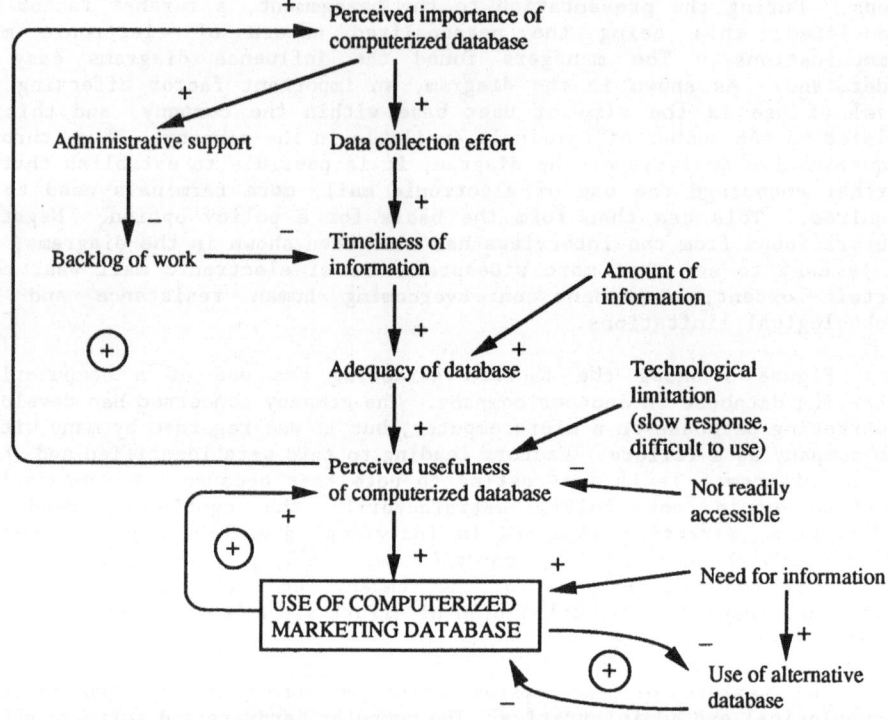

Figure 3. Factors affecting the use of computerized marketing database

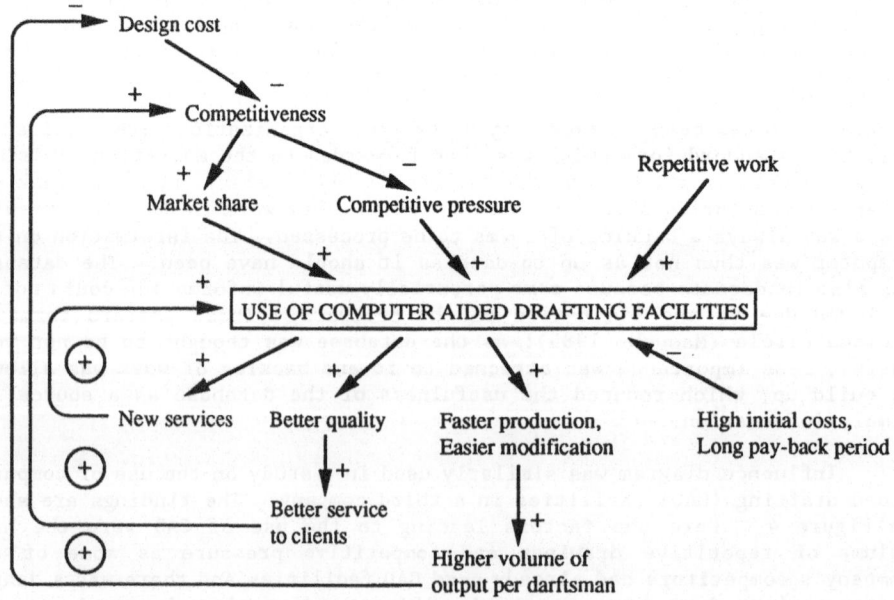

Figure 4. Impact of computer aided drafting (CAD) facilities and factors affecting their use

modification. This would increase the output per draftsman and reduce design cost, hence improve competitiveness and increase market share in the longer term. The quality of output and service to clients would also improve. Finally, the use of CAD opened up new markets for the company, as it could take on sub-contract work to utilize its excess capacity. Thus, three positive feedback loops were identified, and once the company had successfully exploited this technology, its use would continue to increase.

## POTENTIAL APPLICATION AREAS

The author believes that influence diagram is an effective tool for system description which should be exploited more fully in IS research. Four areas which can benefit from this are identified below:

1. As demonstrated by the author's work, influence diagrams can be used in impact studies. Through influence diagrams, one can show clearly how specific information technology affects an organization. Both positive and negative effects can be shown, as well as feedback loops which influence the level of usage. Swanson (1987) also suggests that such diagrams should be used more extensively in IS research concerning determinants and effects.

2. Influence diagrams can be used in implementation research to show the myriad of factors which cause the success or failure of a system. One example is the attempt by Banville (1990) to use cognitive mapping to build a representation of legitimacy in the context of organizational information systems transfer.

3. As shown in Montazemi and Conrath's (1986) work, influence diagrams have been found to be useful in information requirements analysis, particularly when the environment is ill-structured.

4. Influence diagrams can enhance critical success factor (CSF) analysis (Rockart, 1979) by showing explicitly the relationships between various factors and how each affects the success of an organization. Through qualitative analysis of influence diagrams, we can increase our understanding of why certain factors are critical to the success of an organization.

## CONCLUSION

Influence diagram as a system description tool is under-utilized in IS research. This paper has demonstrated the relevance of these diagrams in IS research through three examples. Influence diagrams were found to be highly valuable as a tool for qualitative analysis of causal relationships. Potential areas for application include impact studies, implementation research, information requirements analysis, and critical success factor analysis.

## ACKNOWLEDGEMENT

The author would like to thank Mr Geoff Walsham of Judge Institute of Management Studies, University of Cambridge and Dr Eric Wolstenholme of Univerity of Bradford for drawing my attention to the potential of influence diagrams in IS research.

# REFERENCES

Banville, C. (1990) Legitimacy and cognitive mapping: A study of the social dimension of information systems, in: "Proceedings of the IFIP TC8 WG 8.2 Working Conference on The Information Systems Research Arena of the 90's: Challenges, Perceptions, and Alternative Approaches," Copenhagen, Denmark, December 14-16, 1990.

Conrath, D.W., Montazemi, A.R. and Higgins, C.A. (1987) Evaluating information in ill-structured decision environments, Journal of the Operational Research Society, 38:375-385.

Eden, C. (1985) Perish the thought!, Journal of the Operational Research Society, 36:809-819.

Eden, C.; Jones, S. and Sims, D. (1983) "Messing About In Problems," Pergamon Press, Oxford.

Forrester, J.W. (1961) "Industrial Dynamics," MIT Press, Cambridge, Massachusetts.

Forrester, J.W. (1969) "Urban Dynamics," MIT Press, Cambridge, Massachusetts.

Forrester, J.W. (1971) "World Dynamics," Wright-Allen Press, Cambridge, Massachusetts.

Masuch, M. (1985) Vicious circles in organizations, Administrative Science Quarterly, 30:14-33.

Montazemi, A.R. and Conrath, D.W. (1986) The use of cognitive mapping for information requirements analysis, MIS Quarterly, 10(1):45-56.

Rockart, J.F. (1979) Chief executives define their own data needs, Harvard Business Review, March-April, 81-93.

Swanson, E.B. (1987) Information systems in organization theory: A review, in: "Critical Issues In Information Systems Research," R.J. Boland and R.A. Hirschheim, eds., John Wiley & Sons, Chichester.

Weick, K.E. (1979) "The social psychology of organizing," 2nd Edition, Addison-Wesley, Reading, Massachusetts.

Wolstenholme, E.F. (1982) System dynamics in perspective, Journal of the Operational Research Society, 33:547-556.

Wolstenholme, E.F. and Coyle, R.G. (1983) The development of system dynamics as a methodology for system description and qualitative analysis, Journal of the Operational Research Society, 34:569-581.

Yap, C.S. (1986) "Information technology in organizations in the service sector," Unpublished PhD Dissertation, Department of Engineering, University of Cambridge, UK.

# CONCLUSIONS FROM ACTION RESEARCH: THE MULTIVIEW EXPERIENCE

D. E. Avison* and A. T. Wood-Harper+

*Department of Accounting & Management Sciences
Southampton University
Southampton SO9 5NH, UK

+Department of Mathematics and Computer Science
University of Salford
Salford M60, UK

## INTRODUCTION

The Multiview framework is one attempt to bring together soft and hard approaches to information systems development. The analysis of human activity systems (Checkland, 1984, Checkland & Scholes 1990, and Wilson 1990) and socio-technical systems (Mumford 1981 and Land & Hirschheim 1983) has been wedded to the more conventional work on data analysis (Rock-Evans 1981 and Shave 1981) and structured analysis (Gane & Sarson 1979 and DeMarco 1979) so as to create a theoretical framework which attempts to take account of the different points of view of all the people involved in using a computer system and to reconcile issue-based with task-related aspects. The approach is more fully described elsewhere (Avison & Wood-Harper, 1990).

We have had various experiences of using the approach in a number of action research projects, six of which are discussed in the main text and in more detail in Wood-Harper (1989) and Avison (1990). Multiview is a contingency approach and these varied experiences and others have shaped and developed the framework and the techniques and tools used. The purpose of this paper is to discuss some of the lessons learnt and the emergent, generalisable conclusions made from these experiences.

## LESSONS FROM THE EXPERIENCE OF USING MULTIVIEW

### A blended, contingency approach can work in practice

Multiview is a blended approach to information systems development. It is unreasonable for an organisation to rely on any methodology, soft or hard, for all information systems development. One rigid approach can never be a full answer, because the tools and techniques appropriate for one set of circumstances may not be appropriate for others; the 'fuzziness' of some applications require an attack on a number of fronts; and as an information systems project develops, it takes on very different perspectives or 'views'. Multiview is an explorative structure, a loose framework with which to define and develop information systems, a contingent approach, in that alternative techniques and tools are available which may be chosen according to the dictates and requirements of each situation. The six cases describe Multiview being used in small, medium and large departments, for package and tailor-made solutions on micro, mini, and mainframe computers, representing large as well as comparatively small investments, and in situations where the roles of analysts and users varied.

## An information systems development methodology is complex and difficult to learn

The wide range and large number of techniques and tools that need to be described in an information systems development methodology make such a description long and complex and therefore difficult to learn and to master. Multiview is, perhaps, even more difficult because alternative techniques and tools are included. Further, it does not follow the usual rigid step-by-step description with deliverables, well defined, at each step, because of its contingent philosophy.

## The conventional descriptions of information systems methodologies are inappropriate

This methodology, as evidenced by the field work, does not, in practice, exhibit the step-by-step, top-down nature of conventional models and none of the applications have exactly followed the methodology as espoused in the main text. The users of the methodology will almost certainly find that they will carry out a series of iterations which are not shown in the model. Further, in the real-world cases undertaken, some phases of the methodology were omitted and others were carried out in a different sequence from that expected.

## The political dimension is important

The manipulation of power, that is, the political dimension, is important in real-world situations. This transcends the rationale of any methodology. Most of the cases showed decisions being made which were influenced by considerations beyond those that are implied by the Multiview methodology.

## Responsible participation is contingent

A high level of responsible participation, where appropriate, is a positive ingredient of successful information systems development. Nevertheless in one case managers refused to cooperate, arguing that it was the job of systems analysts to develop the applications and they did not want to get involved.

## The technical dimension is also important

Although the social and human aspects of systems development are stressed in Multiview, sometimes the use of Multiview has

592

exposed difficulties over technical decisions. Either the techniques were not well applied or there were major problems with the hardware and software used and interfacing them.

## Evaluation is difficult

In the first of these 'lessons', we argued that a blended contingency approach to information systems 'worked' in that the prototypes developed proved to be useful. This is a somewhat unsatisfactory way of justifying the approach, though evaluation is difficult in information systems work.

## CONCLUSIONS FROM THE EXPERIENCE OF USING MULTIVIEW

## The Multiview methodology is in a continuing state of development

This conclusion is not an attack on Multiview in isolation – all information systems development methodologies have limitations. Information Systems is a comparatively new discipline, the diversity of approaches is caused to some extent by the background and cultures of their authors and none is all-inclusive. The methodologies address a moving target in that the technology, along with techniques and tools supporting it, develop relentlessly. Multiview is part of a process of improving information systems practice.

## Developing an information system is contingent

Developing an information system is contingent on the methodology, the situation and the information systems development team. The team of users and analysts affect the perception of the situation and they interpret the methodology. The variety of possible interpretations reflect differences in the backgrounds and experiences of the analysts. In some of the applications not all of the stages of Multiview were used because of the situation. It is possible to envisage cases where it is deemed inappropriate to develop a computer-based information system. The framework was adapted in one case. In others, the different analysts interpreted the 'same situation' differently. In any situation where an information system might be appropriate, there are factors such as culture, language and education which have to be taken into consideration. Frequently particular techniques and tools are not appropriate to the problem situation. The systems analyst has to choose from a 'tool box' those techniques and tools appropriate for each situation, but within the framework of an approach such as Multiview (Avison, Fitzgerald & Wood-Harper, 1988). Without such a framework, the information systems are likely to be idiosyncratic and difficult to maintain, and therefore of variable value.

## The adoption of a contingency approach can lead to problems

Although developing an information system is contingent, there are a number of difficulties associated with adopting a contingency approach, in particular:
• Difficulties associated with the levels of expertise and the breadth of knowledge required from systems analysts. Conventional information systems development methodologies

are prescriptive, analysts are expected to follow a well-defined structure for all cases. The techniques and tools within each phase are also well-defined. The analyst following an approach such as Multiview needs to know which technique and tool from a range might be appropriate for a particular situation at a particular point of time and this requires wide experience.

Difficulties relating to the consistency of standards in organisations that adopt a contingency approach. Information systems will be developed using different techniques and tools and in different ways, dependent on the particular situation. This does mean a loss of common standards, something which is meant to be one of the practical benefits of using an information systems development methodology. Thus, in the latest version of Multiview, documentation and other standards are recommended and the techniques are now more in line with the best practice of information systems.

## There are problems as well as advantages of action research in information systems

The authors have developed Multiview in the tradition of action research which allows the researcher great potential to utilise the ideas of users and change concepts and methods as the work develops. Work takes place in its natural setting and this gives the researcher an insight into real-life practical areas. However there are disadvantages of action research. The lack of impartiality of the researcher has led to its rejection by a number of researchers and academic departments. The lack of scientific discipline in such research makes it difficult for the work to be assessed for the award of research degrees and for publication in academic journals. A particular difficulty lies in persuading research funding bodies that this type of research is as valid and as useful as conventional methods of scientific research. Further, although the researcher's intent is to conduct research while effecting change, the approach is sometimes branded with the description 'consultancy' and not research. The open-endedness of such research and the consequent flexibility necessary in writing a research proposal provide additional difficulties. Further, action research is context-bound as opposed to context-free. It is difficult to determine the cause of a particular effect, which could be due to environment (including its subjects), researcher, or methodology. This can mean that action research produces narrow learning in its context because each situation is unique and cannot be repeated. However, in our research there is an attempt to reconcile the narrow learning from action research with the need for generalisable research (Gummesson, 1988). It is hoped that the generalised findings from our work will result in other researchers and practitioners applying the Multiview theory in other problem situations and will take into consideration the learning that emerges in the complex process of developing an information system.

## Defining an information system can be considered as a social process

The process of going from 'thinking about the content of the information situation' by the information systems development team to the 'perceived content of the situation (including users)' can be examined within different social theories relevant to information systems definition. For example, the first dimension comprises assumptions about the nature of defining an information system ranging from objective to subjective. The second dimension relates to the assumptions about the degree of change to the information situation ranging from regulation to radical. These dimensions yield four quadrants or paradigms in which the following beliefs or assumptions can be located: ontological - about the nature of reality; information situation - about the behaviour of humans in the information situation; epistemological - about how knowledge is acquired; and methodological beliefs in the appropriate devices, acquiring knowledge about and intervening in the information situation.

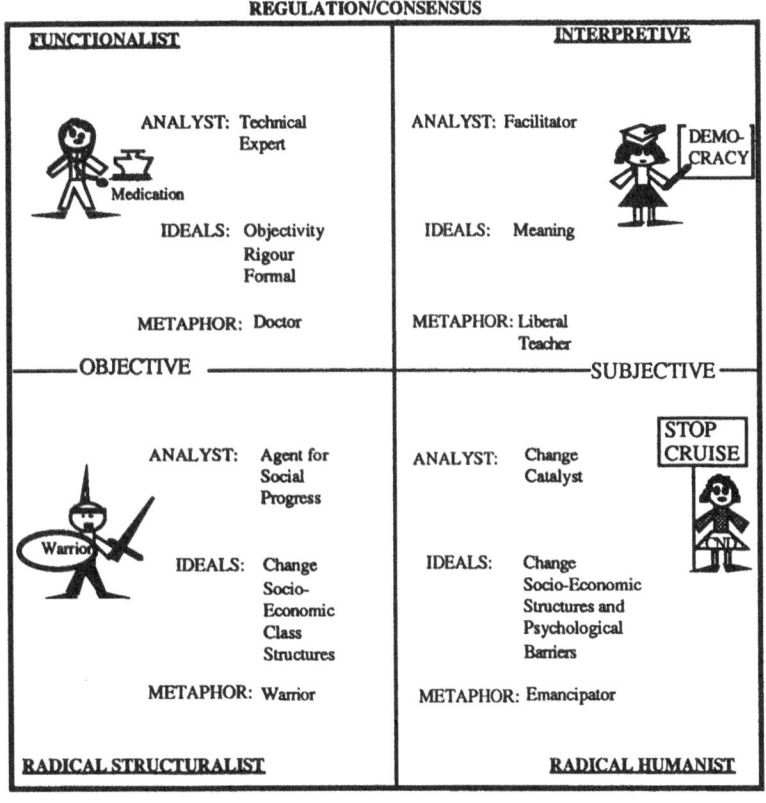

Figure 1. Roles, ideas and metaphors assumed when defining an information system (from Avison & Wood-Harper, 1990)

There are four paradigms (see figure 1). In the functionalist perspective, the information system consists of interactions which function independently of outside manipulation. The analyst assumes that he can readily understand the situation, indeed there is an assumption of rational behaviour by the actors which makes understanding easier. The systems are well controlled, can be well understood and can be formally defined.

In the interpretive perspective, it is assumed that the analyst is subjective and interprets the problem situation. He hopes that it will be possible to understand the intentions of the actors in the situation. Participation and involvement will be the best way to obtain detailed information about the problem situation, and later to be able to predict and control the situation.

In the radical structuralist view, the situation will appear to have a formal existence but require radical change due to, for example, contradictory and conflicting elements. The systems analyst regards himself as an agent for change and social progress, emancipating people from their socio-economic structures.

Finally, in the radical humanist view, the situation is seen as external and complex. There is an emphasis on participation to enable a rapport between the actors and lead to emancipation at all levels, for example, socio-economic and psychological.

The view taken of the role and the effect of the systems analyst on the problem situation will depend on the perspective: it might be as a 'technical expert' imposing good practices on the situation; or as 'facilitator' helping the users achieve their goals; or as an 'agent for social progress' imposing radical change on the situation; or finally as change analyst, encouraging the users to effect major change.

BIBLIOGRAPHY

Avison, D. E. & Wood-Harper, A. T. (1990) Multiview - An Exploration in Information Systems Development, Blackwell Scientific Publications, Oxford,

Avison, D. E., Fitzgerald, G. & Wood-Harper, A. T. (1988) Information Systems Development: A Tool-kit is not Enough, Computer Journal, 31, 4.

Avison, D. E. (1990) A Contingency Approach to Information Systems Development, PhD Thesis, Aston University, Birmingham.

Checkland, P. B. (1984) Rethinking a Systems Approach, In: Tomlinson & Kiss (1984).

Checkland, P. B. & Scholes, J. (1990) Soft Systems Methodology in Action, John Wiley, Chichester.

DeMarco, T. (1979) Structured Analysis and System Specification, Prentice-Hall, Englewood Cliffs.

Gane, C. P. & Sarson, T. (1979) Structured Systems Analysis: Tools and Techniques, Prentice-Hall, Englewood Cliffs.

Gummesson, E (1988) Qualitative Methods in Management Research, Studentlitteratur, Lund.

Land, F. F. & Hirschheim, R. (1983) Participative Systems Design: Rationale, Tools and Techniques, Journal of Applied Systems Analysis, 10.

Mumford, E. (1981) Participative Systems Design: Structure and Method, Systems, Objectives and Solutions, 1.

Rock-Evans, R. (1981) Data Analysis, IPC Press, London.

Shave, M. J. R. (1981) Entities, Functions and Binary Relations: Steps to a Conceptual Schema, Computer Journal, 24, 1.

Wilson, B. (1990) Systems: Concepts, Methodologies and Applications, 2nd ed., Wiley, Chichester.

Wood-Harper, A. T. (1989) Comparison of Information Systems Definition Methodologies: Action Research Multiview Perspective, PhD Thesis, University of East Anglia, Norwich.

# AMBITIOUS PROJECTS

R. H. Hombach

Department of Information Systems
The London School of Economics and Political Science
Houghton Street, London WC2A 2AE, UK

## INTRODUCTION

Some projects set out to create or modify enduring human-activity systems; others set out to do something else. Let us call the first sort "ambitious" and the others, "humble." "Ambition" and "humility" are technical terms in this context. In particular, there is no implication that ambitious projects are necessarily large, complex, or expensive; some are and some are not. The computerization of PAYE, Inland Revenue's programme of personal taxation, was a major project, in comparison with which the attempt to introduce draft-animal power into Mishamo, an African community with a tradition of hand ploughing, was tiny. Both, however, were ambitious projects. By the same token, humble projects are not necessarily small: the moon shot was a humble project.

Nonetheless, ambitious projects are fundamentally different from humble projects in ways that bear on how they ought to be designed, how they ought to be executed, and how they ought to be managed. If a project is attacked humbly when it ought to be attacked ambitiously, it will likely fail. Unfortunately, it can be very easy to lose sight of the fact that what one is after is to create or modify an enduring human-activity system. Thus, projects undertaken to improve information systems may be carried out as if all they had to deliver were new software; projects undertaken to improve agricultural systems may be carried out as if all they had to deliver were new technology. When they fail, it is by default.

In this paper, an analysis in systems terms is given of the concept, "ambitious project," the problem of failure by default is discussed, and exhortations are made for further research.

## ANALYSIS OF PROJECTS AS SYSTEMS

A system of any kind is a complex, controlled, structured entity whose properties derive from its components and their structure, but are not reducible to them. A human-activity system is a purposive, open system within whose components are undertaken human actions that together conduce to the purposes of the whole. Within the category of human-activity systems, let us distinguish between those that are intended to endure and those that are not:

**Definition 1.** An *ad hoc system* is a human-activity system with pre-fixed objectives and a limited lifespan; it is assembled only to achieve its objectives and then is disassembled.

Many sorts of things can be construed as *ad hoc* systems: meetings, jury trials, chess tournaments . . . and projects.

**Definition 2.** A *standing system* is a human-activity system with an indefinite lifespan; it has ongoing, sometimes evolving objectives and at least one constant objective: to endure.

Nations, families, ministries of agriculture, garden clubs, corporations, departments of corporations, and many other things are standing systems.

It is in the nature or their objectives that projects differ from other *ad hoc* systems.

A project is a transitory organisation of individuals dedicated to the attainment of a specific objective within a schedule, budget, and technical performance target. . . . [A]s an entity, a project is defined both in terms of the work that is being carried out within it (to achieve . . . its objective with technical quality) and in terms of what is being made available to it to achieve [its] objective (e.g., budget, timescale, quantity and quality of resources). (Berkeley *et al*, 1990, p.13)

In fact, a project has two kinds of objective. Besides its fixed or *defining* objectives, it also has an *ancillary* objective: to achieve its defining objectives according to a set of parameters governing its duration, costs, inputs, and *modus operandi*. It is with reference to its defining objectives that a project is judged *effective*; it is with reference to its ancillary objective that it is judged *efficient*. But the ancillary objective, unlike the defining objectives, is revisable. We stipulate that *the setting and any revision of its ancillary objective are parts of the project*. This is to say that projects are self-regulatory.

**Definition 3.** A *project* is an *ad hoc* system with at least one defining objective and with an ancillary objective whose setting and revision are actions belonging to the project.

Some projects--a family project to organize the garage, say--are carried out very informally. In other cases--a government project to add lanes to a motorway--certain project actions must be formally agreed by some specified subset of the agents of the project; they must be *authorized*. Setting or revising the ancillary objective are actions of this sort. We shall stipulate that *authorization, itself, is part of the project*.

**Definition 4.** A *formal project* is a project, some of whose actions must be preceded by authorizations, which are also actions belonging to the project.

Self-regulation, including authorization-dependent self-regulation, is a form of system control. Control bespeaks an information system. All human-activity systems contain *information-related activity* by which information is selectively acquired, stored, processed, formatted, and distributed. How this activity is carried out is an essential determinant of the system's success. A systemic description of how it occurs will be said to be a formulation of the *associated information-system*. On this account, a human-activity system and its associated information-system are not separable entities, nor is the latter an element or subsystem of the former. The associated information-system is a limited description of the human-activity system, itself; it is the human-activity system described "information-wise."

**Definition 5.** An *associated information-system* is a systemic description (a model) of the information-related activity of a human-activity system.

Information-system study, in so far as it pertains to associated information-systems, has been directed at standing systems, especially organizations. But *ad hoc* systems, also being human-activity systems, have associated information-systems too. They are, perforce, *ad hoc*. In particular, projects have associated information systems.

**Definition 6.** A *project information-system* is a model of the information-related activity of a project.

## HUMBLE AND AMBITIOUS PROJECTS

**Definition 7.** A *humble project* is a formal project, none of whose objectives is to produce a standing system.

**Definition 8.** A *preparatory project* is a humble project one of whose defining objectives is to produce an *ad hoc* system.

**The Moon Shot.** The publicly announced objective of the moon shot was to put American astronauts on the moon and bring them back safely (defining objective) by the end of the 1960s (part of the ancillary objective). If this is a fair statement of the true defining objective, the moon shot was *not* intended to produce a standing system as it would have been if, for example, it had aimed to establish a moon colony. Thus, it is an example of a humble project. The moon shot was a research project; it aimed for knowledge. It was also an R&D project; it aimed for technologies and physical products. It may be that it was, or was part of, a preparatory project for some subsequent part of the American space-programme. And certainly it was an adventure (political and otherwise); it aimed for a feat. The concatenation of all these aims, however, make it no other than humble.

In general, formal projects of any sort are responses to perceived problems or opportunities. What sort of project, if any, *ought* to arise depends on what it would take to solve the one or to capitalize on the other. If what it would take is new knowledge, new technology, new physical products, or a feat, then as a rule, if a project is called for, it will be a humble project.

**Definition 9.** An ambitious project is a formal project at least one of whose defining objectives is to produce a standing system, either by creation *de novo* or by modification of an existing standing system.[1]

**The Computerization of PAYE.** The defining objectives of the computerization of PAYE (Pay As You Earn) have been given as the following:

--to increase the efficiency of PAYE, in particular through a reduction in staff costs;

---

[1]"Produce" should be understood in this disjunctive sense throughout the paper.

--to improve the service to the public through greater accuracy, reliability and
   speed of response to communications;
--to provide up-to-date facilities for staff and offer greater job satisfaction;
--to create a system offering greater flexibility for the implementation of future
   change, either within the present tax structure or in more far-reaching reforms
   of personal taxation. (Morris and Hough, 1987, p.161)

In short, this project aimed to make PAYE, a standing system, more cost-efficient, more
effective, better operating, and more flexible. It is an example of an ambitious project.

Notice that the term "computerization" does not appear in the above statement of
defining objectives. It is suggested that this is appropriate. Computerization is part of the
*modus operandi*, and hence part of the *ancillary objective*; it would have been quite possible
to computerize PAYE without making the desired changes that were the real defining
objectives of the project. Now, it follows from Definition 3 that a project retains self-identity
through changes in its ancillary objective. This implies that all ideas of computerization
might have been abandoned and yet the PAYE project continue--which is a significant
violation of standard notions! However, standard notions are not always the best guide.

To computerize any standing system is, in effect, to alter its associated information
system. Any information-technology (IT) project aimed, in effect, at modifying an associated
information-system is ambitious; this follows from Definitions 5 and 9. But to recognize such
projects can be difficult if one goes by names. Sometimes what are called "IT projects,"
"computer projects," "software-development projects," etc., are ambitious, but sometimes they
are not; these names are also used for humble projects. For example, ambitious projects may
be called "software-development projects" because they produce bespoke software. But so
also may be humble R&D projects that produce software packages.

This lack of distinguishing terminology is not, itself, pernicious, but it is symptomatic
of something else that is: a widespread insensitivity to the differences between the successful
production of hardware or software and the production of an effective standing system. This
can be generalized to *all* technologies. Any project whose defining objective is to produce
a technology is humble. An ambitious project may contain such a project as a subproject;
this is perfectly compatible with Definition 9, which allows for complex projects. But an
ambitious project always aims for something more--to produce a standing system--and no
standing system is ever produced from technology alone.

## KEEPING THE DEFINING OBJECTIVE FOREMOST

In theory, commitment to ambition or humility seems logically to precede even commitment to there being a project. It comes of understanding the nature of the problem or
opportunity that might eventually suggest a project. Presumably, it falls within the realm of
what is called "prefeasibility." This suggests two ways in which humble projects might be
carried out when ambitious projects are called for, the scenario for failure by default. A
commitment to humility might mistakenly be made during prefeasibility; or a commitment
to ambition might appropriately be made during prefeasibility, but subsequently subverted.

That the first sort of problem might arise suggests that the concept of ambition is a
candidate for treatment by soft systems-analysis. How often it *does* arise is not clear. Nor
is it clear that it is always possible, in practice, to make a commitment to ambition *ex ante*.
These are questions for further research.

The second sort of problem, on the other hand, is ubiquitous: an occasion calls for production of a standing system; a project intended to produce a standing system is authorized; a project that will *not* produce a standing system is designed; this project is carried out competently as designed; and it is subsequently accounted a failure.

**Environment Sanitation at La Quebradita.** La Quebradita was a high-rise housing complex in Caracas, administered and served by several public agencies and community organizations. No particular agency was responsible for maintaining the grounds, which were littered and unsightly. "Agreement was reached on a proposal to hold an "environmental sanitation campaign" in the apartment community as a joint effort between the module agencies and the various community organizations in the complex . . ." (Gomez and Myers, 1983, p.100). But what was planned and carried out was a highly publicized and very successful clean-up project. The "voluntary" committee subsequently appointed to continue effort did not function, and the environment reverted to slum condition soon after.

Variations on this story are legion, and they occur in every sphere in which projects are launched. Only part of what they are about is bad choice of *modus operandi*, for projects should have available to them a fail-safe with which to correct this. An ambitious project is like a gamble where one is allowed to cheat. What one bets on is that to meet the ancillary objective will produce the desired standing system. If, during the course of the project, it begins to look as if the ancillary objective will not do, one can change it. How could anyone lose? By taking seriously either of two standard bits of project mythology:

**The Myth of Life Cycle.** According to this,

[e]very project, no matter of what kind or for what duration, essentially follows the activity sequence of prefeasibility/feasibility, design and contract negotiation, implementation, handover and in-service support. . . . [T]his life-cycle of projects is relatively straightforward--a single sequence from prefeasibility to handover . . . . (Morris and Hough, 1987, p.4)

This implies that the ancillary objective cannot be revised, for then the design, and possibly the feasibility, stage would have to be revisited. To accept the standard account of life cycle is to abandon any notion of control that would keep a project's ancillary objective congruent with its defining objective.

**The Myth of Management Virtue.** According to this, project management is, in effect, a challenge to preserve the originally agreed budget, schedule, and *modus operandi* against all contingencies. Managers are deemed praiseworthy (or worthy of hire) to the extent that they succeed. What is usually meant by "project management" in this context pertains to the implementation of some project design, from which it follows that what is meant by "the project" is some particular design. Strangely, this view is incompatible with the standard notion of life cycle, but current project mythology manages to incorporate both.

The ancillary objectives of ambitious projects *must* be construed as revisable. In fact, Definitions 3 and 4 go beyond this. They stipulate that modifications of the ancillary objective and authorizations of such modifications are actually project activities. As such, they are candidates for project management and their information-related activities are candidates for inclusion in the project information-system. It must be part of an ambitious project that information relevant to possible incongruence between ancillary and defining objectives be regularly sought--a management function--and that such information be made available on a timely basis--a function of the project-information system.

**Project Information-Systems.** Given Definitions 5 and 6, the better the design of project information-systems, the better the design of the projects with which they are associated. Good procedures are available for controlling project activity related to ancillary objectives, that is, related to budget, schedule, and *modus operandi*; there is even some software support for this. But except for incorporating these procedures, project information-systems are usually not designed at all. To this extent, the projects themselves are undesigned. If they succeed, they succeed fortuitously. But we do not yet know how to control project activity so that it is not subverted from the production of a desired standing system. A theory of ambitious-project information-systems does not exist.

**The Structure and Management of Ambitious Projects.** It follows from our analysis of ambitious projects that their life cycle cannot follow the standard account. From this it follows that no theory of project management based on the standard life-cycle can be true. So far we know that any correct account of the life cycle of ambitious projects must have feed-back loops, and we know that authorization is a project activity. What theory of the structure and management of ambitious projects does this entail?

**The Design of Ambitious Projects.** Even if one keeps one's eyes squarely on the right goal, ambitious projects are hard to carry out successfully: how to engineer enduring human-activity systems, or enduring changes is such systems, is simply not well understood. This is what especially bedevils projects like the introduction of draft-animal power into Mishamo, mentioned in the introduction to this paper.[2] These are the so-called "development projects," where "development is an increase in potential . . . It has more to do with how much one can do with whatever one has than with how much one has" (Ackoff, 1989, p.6). All development projects aim at producing standing systems; they are all ambitious.

Projects, whether ambitious or humble, are successful only if they are effective, that is, only if they deliver the product for which they were assembled. But an ambitious project delivers a standing system. It is successful only if that standing system is effective. How to design projects so as to achieve this "once-removed" effectiveness is not attacked by systems study, management study, or information-systems study as explicitly as seems warranted.

The attempt to introduce draft-animal power into Mishamo failed. If we can put a man on the moon, why can't we . . . ?

## REFERENCES

Ackoff, R. L., 1989, "From data to wisdom," *Journal of Applied Systems Analysis*, 16:5.
Berkeley, D., de Hoog, R., and Humphreys, P., 1990, *Software Development Project Management: Process and Support*, Ellis Horwood, London.
Gomez H. and Myers, R. A., 1983, "The service module as a social development technology," in *Bureaucracy and the Poor*, Korten, D. C. and Alfonso F. B., ed., Kumarian Press, West Hartford.
Morris, P. W. G., and Hough, G. H., 1987, *The Anatomy of Major Projects*, John Wiley & Sons, Chichester.

---

[2]The author was formerly an officer of the implementing agency for the project that established Mishamo, a permanent settlement of Burundian refugees in Tanzania.

# IMPLEMENTATION OF PROJECT MANAGEMENT SYSTEMS THROUGH PROTOTYPING

Wita Wojtkowski, Margaret A. Barrett, Judith M. Barton, and
W. Gregory Wojtkowski

Lancashire Polytechnic
Business Information Management
Preston, U.K.

Boise State University
Computer Information Systems and Production
Management
Boise, Idaho, U.S.A.

## INTRODUCTION

This paper considers implementation of project management systems through the use of prototyping.

Project management systems (PMS) have been implemented with various degrees of success in organizations. The implementation of a PMS is typically realized with a "top-down", or "bottom-up" approach [Robey, 1987, Miller and Freisen, 1982], though in many organizations there is no methodical approach at all [Robey 1987]. The reasons for failing to implement PMS are well documented [Murphy et al., 1983; Avots, 1969; Robey and Farrow, 1982]

We propose that elemental causes for PMS implementation failures recommend the use of *prototyping* methodology [Jasany, 1990; Wojtkowski and Wojtkowski, 1989] as a reasonable strategy for constructive and efficient implementation of PMS.

Specifically, the goal of this analysis is to understand 1) the common properties of general PMSs and 2) the possibility of implementation of PMSs through prototyping process.

The paper is organized as follows. In the first section we describe, in brief, the concept of prototype, the prototyping processes and the basic notions of project management systems. In the second part we discuss project management systems implementation success factors that are enhanced by the use of prototyping.

## PROTOTYPES AND PROTOTYPING OF INFORMATION SYSTEMS

The prototyping approach is not new. In manufacture engineering, prototyping refers to a well defined phase in the production process. Its most prominent characteristic is an extensive use of working models of the system under development, that is, prototypes. Until recently, the construction of prototype information system was burdensome, because of the lack of powerful development tools. However, the advancing automation of the information system development process is bringing changes to this area. For, example, fourth generation languages (4GLs) and CASE tools are having radical consequences for the way in which information systems are developed [Baker, 1990]. Currently, in software engineering, prototyping is used for rapid systems development [Willis et al., 1988]. For the purpose of this paper we define prototyping as:

A working model of (or parts of) an information system, which emphasizes specific aspects of the system

Central to this definition is the 'model' concept. In a prototype there is no attempt at completeness. The function of the prototype is to throw light on the certain aspects of the information system. The particular emphasis is on these aspects of the system for which there is most uncertainty.

There are variety of information systems prototypes. For example, mock-up of the information system prototype models only the external appearance of a system (screens, reports, dialogues); no action is taken with any data input. Functional prototypes allow to store data and to perform operations on them. A prototype may model either only part of the system, or the entire system.

The term 'prototyping' has singular meaning when applied to the development of information systems. We define it here as:

An approach for establishing a systems requirements definition that is characterized by a high degree of iteration, by a very high degree of user participation in the development process and by an extensive use of prototypes.

The chief premises of prototyping are that prototypes constitute a better means of communication with intended users of the system, and that iteration is necessary to channel the inevitable learning process in the right direction. This is because users cannot indicate what requirements and wishes they have for a system if they have no first-hand experience of it. In essence, the use of the prototype deepens both intended system's users' and builders' insight into the problem and supports the learning process all must eventually undergo.

As there are many forms of prototypes, there are many forms of prototyping [Budde et al., 1984]. For example, the kind of prototyping that models the user interface of the system is generally referred to us user-interface prototyping. If some real functionality is added to the prototype, the term functional prototyping is used. Prototyping methodology has been tested and used extensively in the analysis, development and implementation of varied types of systems [Rowen, 1990; Owen, 1989; Boar, 1986; Cerveny, 1986, Miller, 1982]. Peters and Waterman [1982] list it as number one of eight basic management principles employed by "excellent" companies. "A bias for action: a preference for doing something-anything-rather than sending a question through cycles and cycles of analyses and committee reports". A basic schema of the prototyping process is summarized in Figure 1.


PROJECT MANAGEMENT SYSTEMS

To clarify the meaning of project management (PM) two definitions are offered. As defined by the American Association of Cost Engineers PM represents:

The utilization of skills and knowledge in coordinating the organizing, planning, scheduling, directing, controlling, monitoring, and evaluating of prescribed activities to ensure that the stated objectives of a project, manufactured product, or service are achieved.

As defined by the Project Management Institute, PM is:

The art of directing and coordinating human and material resources throughout the

604

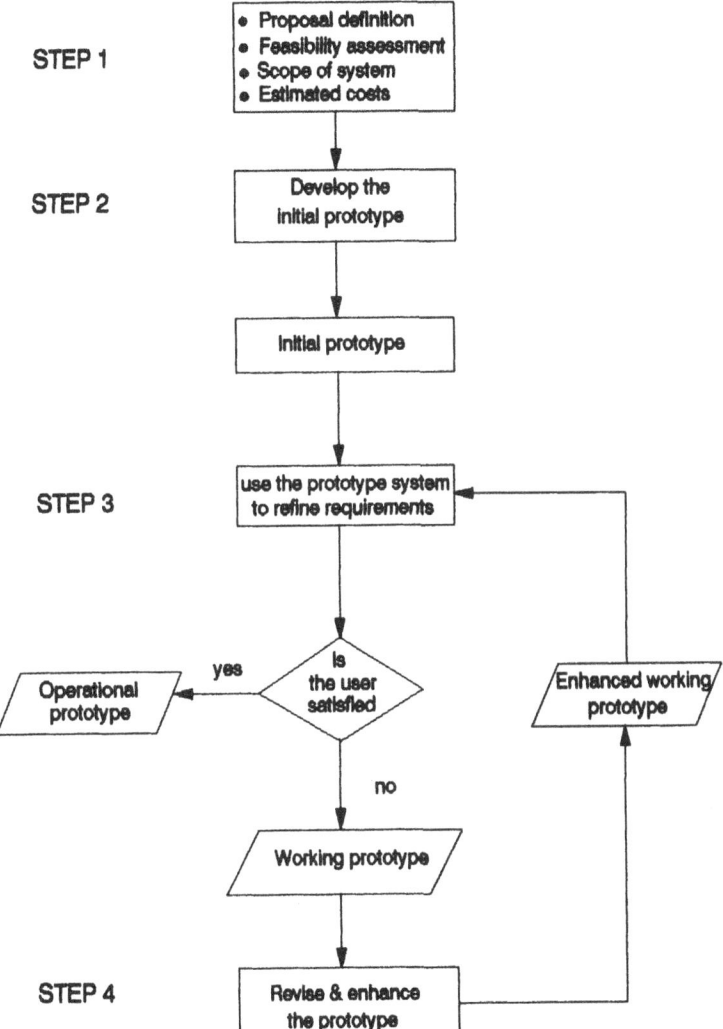

Figure 1. Basic schema for the Prototyping Process

life of a project by using modern management techniques to achieve predetermined objectives of scope, cost, time, quality, and participant satisfaction.

Although specific components of a given PM have to be defined to meet distinctive needs of a particular organization, there exists a generic PM system (PMS) model defined in terms of its subsystems, that many have been using as a common reference [Cleland, 1990].

This generic PMS includes:

*Planning Subsystem* (Objectives, Goals, Strategies)
*Information Subsystem* (Time, Cost, Performance)
*Control Subsystem* (Standards, Comparison, Corrective Action)
*Technique & Methodologies Subsystem* (Scheduling, Costing, Modeling, Programming)
*Facilitative Organizational Subsystem* (General Managers, Resource Managers, Project Managers: Authority, Responsibility, Accountability)
*Cultural Ambience Subsystem* (Values, Attitudes, Traditions, Beliefs, Behaviour)
*Human Subsystem* (Motivation, Communication, Negotiation)

PMS generic archetypes such as this serve two main purposes:- to give project management means of controlling the process, and to make the process surveyable and thus teachable.

PMS IMPLEMENTATION SUCCESS FACTORS THAT ARE ENHANCED BY THE USE OF PROTOTYPING

The reasons for failing to implement PMS are well documented [Murphy et al, 1983; Avots, 1969]. Two of the most cited ones are [Robey and Farrow, 1982]:

An attempt at achieving too much at one time
Lack of purposeful involvement of the potential project users in the formation of the PMS

We propose that elemental causes for PMS implementation failures recommend the use of prototyping methodology [Jasany, 1990] as a reasonable strategy for constructive and efficient implementation of PMS. Figure 2 depicts an example of possible place of prototyping for project requirements definition.

PMS by its very nature requires continual modification, change, and updating [Brousseau, 1988]. The implementation of PMS has to be understood as a process not an static assignment. In addition, according to Hunter and Stickney [1983], most projects require team efforts of different specializations. This holds true for profit and non-profit sectors of the economy: in research and development, construction, aerospace, finance, manufacturing and service sectors. The prototyping methodology shows itself to be very useful when team effort is required [Boehm et al., 1987]. For that reason prototyping can serve as a useful implementation vehicle. We suggest that implementation success factors enhanced by the use of prototyping will include:-

- better reporting
- client involvement in all project phases
- greater team participation, commitment and influence
- easier project budgeting and cost control
- more realistic estimating
- easier change control procedures
- minimization of organizational bureaucracy

Four factors have been identified as common to all projects regardless of characterization [Stewart, 1972]. These are:-

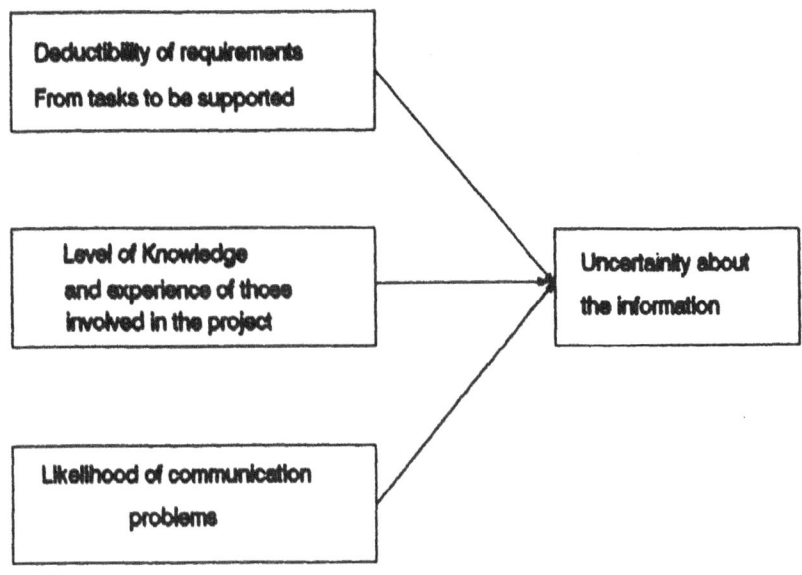

Figure 2. Possible uncertainty determining factors

1.	Scope
2.	Degree of Familiarity
3.	Complexity
4.	Stake

Here too, we propose that for specific projects these factors can be corroborated with efficiency through the use of prototyping.


CONCLUSION

The requirements definition phase is the most important, and at the same time the most difficult, part of any project. The correctness, unambiguity, completeness and consistency of requirements definition strongly determines the long-term success of the project. Establishing good specifications is not a trivial matter. Prototyping combining traditional methods with the use of working models [Jasany, 1990; Swartout, W. and Blazer, 1982] enables 'clients' of the project to obtain the clearer view with respect to planned project, and to communicate any potential objectives more clearly to future project executioners. The most important novel feature that the prototyping adds to implementation of PMS is the use of working models as means of communication and learning. It is important to recognize, though, that each prototype can represent a significant investment, warranting serious cost considerations.

REFERENCES

Avots, I., 1969, "Why Does Project Management Fail,"
	California Management Review, Vol. 12, No. 1, pp. 77-82.
Baker, R., 1990, "CASE Method: Tasks and Deliverables",
	Addison-Wesley, Reading.
Boar, B.D., 1986, "Application Prototyping: A Life Cycle Perspective,"
	Journal of Systems Management, Vol. 37, No. 2, pp. 25-31.
Boehm, B. W., Gray, T., E., and Seewaldt T., 1987, "Prototyping Versus Specifying:
	A Multiproject Experiment," IEEE Transaction on Software Engineering,
	Vol. SE-10, No. 3, pp.8-11.

Bourke, M., 1986, "Actual Experiences in Prototyping," Prototyping: State of the Art Report, Pergamon Infotech Ltd., Maidenhead Berkshire, England, pp. 15-26.

Brousseau, J., 1988, "Project Management: Look Before You Leap," Computing Canada, Vol. 14, No. 9, p. 25.

Budde, R., Kuhlenkamp, K., Mathiassen, L., and Zullinghoven, H., 1984, "Approaches to Prototyping", Springer-Verlag, Berlin.

Cerveny, P.R., Garity, E. J., and Sanders, G. L., 1986, "The Application of Prototyping to System Development: A Rationale and Model," Journal of Management Information Systems, Vol. 3., No. 2., pp. 52-62.

Hunter, M., B., and Stickney F. A., 1983, "Overview of Project Management Applications," in Project Management Handbook, Eds. Cleland, D.I., and King, W.R., Van Nostrand Reinhold, New York, pp. 644-688.

Jasany, L. C., 1990, "Develop Application Programs in 120 Days," Automation, Vol. 37, No. 2, pp.34-35.

Miller, D., and Freisen, P.H., 1982, "Structural Change and Performance: Quantum vs Piecemeal-Incremental Approaches," Academy of Management Journal, Vol. 25, No. 4 pp. 867-892.

Murphy, D.C., Baker B.N., and Fisher D., 1974, "Determinants of Project Success," National Technical Information Services, Springfield, VA22151, #N-74-30392.

Owen, D. E., 1989, "Prototyping: Essence of Pragmatic IS Development," Information Strategy: The Executive's Journal, Vol. 5, No. 2, pp. 21-25.

Peters, T.J. and Waterman, R.H., 1982, "In Search of Excellence" - Lessons from America's Best Run Companies, Warner Books, New York.

Robey, D., 1987, "Implementation and Organizational Impacts of Information Systems," Interfaces, Vol. 17, No. 3, pp.72-84.

Robey, D., and Farrow, D.,1982, "User Involvement in Information System Development: A Conflict Model and Empirical Test", Management Science, Vol. 28., No. 1, pp. 73-84.

Rowen, R. B., 1990, "Software Project Management Under Incomplete and Ambiguous Specifications," IEEE Transactions on Engineering Management, Vol. 37. No.1, pp.10-21.

Stewart, John W., 1972, "Making Project Management Work," Business Horizons. Swartout,W., Blazer, R., 1982, "On the Inevitable Intertwining of Specification and Implementation", Communications of the ACM - Vol. 25, No. 1.

Willis, T. H., Huston, R., and d'Ouville, E.L., 1988, "Project Manager's Responsibilities in Prototyping System Analysis and Design Environment," Project Management Journal, Vol. 19, No.1, pp.56-60.

Wojtkowski W., G., and Wojtkowski, W.,1990 "Application Programming with Fourth Generation Languages, Boyd and Fraser, Boston.

The manufacturer's authorised representative in the EU is Springer
Nature Customer Service Centre GmbH, Europaplatz 3, 69115 Heidelberg,
Germany. If you have any concerns regarding our products, please
contact ProductSafety@springernature.com

Printed and bound by CPI Group (UK) Ltd, Croydon, CR0 4YY
23/04/2026
02095625-0016